W9-CDX-742

g-in-Publication Data

onic equipment. 2. Micropro-

629.2′549 86-17016

al/production supervision and
erior design: Linda Zuk
r design: Photo Plus Art
nufacturing buyer: Rhett Conklin

A division of Simon & Schuster
Englewood Cliffs, New Jersey 07632

Printed in the United States of America

10 9 8 7 6 5 4 3 2 1

ISBN 0-8359-9353-1 025

Prentice-Hall International (UK) Limited, *London*
Prentice-Hall of Australia Pty. Limited, *Sydney*
Prentice-Hall Canada Inc., *Toronto*
Prentice-Hall Hispanoamericana, S. A., *Mexico*
Prentice-Hall of India Private Limited, *New Delhi*
Prentice-Hall of Japan, Inc., *Tokyo*
Prentice-Hall of Southeast Asia Pte. Ltd., *Singapore*
Editora Prentice-Hall do Brasil, Ltda., *Rio de Janeiro*

The Automo

Don Knowles

Library of Congress Catalogin
Knowles, Don. (date)
The automotive computer
Includes index.
1. Automobiles—Electr
cessors. I. Title.
TL272.5.K57 1987
ISBN 0-8359-9353-1

Editor
int
Cov
Ma

A Reston Book
Prentice-Hall, Inc.
Englewood Cliffs, New Jersey 07632

Contents

Preface

In 1946 the world's first electronic digital computer was dedicated. This computer weighed 30 tons and filled a very large room. Vacuum tubes were used in the computer and generated excessive heat, which often caused premature failure.

An electronics revolution began in 1947 when the transistor was invented at Bell Telephone Laboratories. The transistor was much smaller and generated less heat than the vacuum tube it replaced.

Silicon chip development in 1959 accelerated the electronics revolution. Electronics engineers discovered that the transistor could act as its own circuit board, and so the integrated circuit, or silicon chip, was born. Each year since 1959 engineers have approximately doubled the number of electronic components on a single chip. For example, 5000 transistors may be designed into the chip in a digital watch, and the chip in a small computer could contain 100,000 transistors. A computer designed with silicon chips would be 30,000 times cheaper than the world's first vacuum-tube computer. The silicon chip computer uses only a fraction of the electrical energy required by the original vacuum-tube computer while performing 200 times as many decisions.

The electronics revolution has affected every segment of business and industry. Electronics engineers in the automotive industry have developed computer technology to meet emission standards, improve fuel economy and performance, and provide a wide range of convenience functions. In the 1980s, computer technology has expanded rapidly in the automotive industry. Statistics from Sun Electric Corporation indicate that in 1983 18 percent of the cars in the United States had on-board computers. This source also estimates that 94 percent of the U.S. cars will have on-board computers in 1990. The electronics revolution in the automotive industry has created an unprecedented demand for information in this area. This book was written to meet that demand.

Electrical fundamentals and computer basics are provided in this book. Many current automotive computer systems are explained in detail. Service and diagnostic procedures on these systems are also described. Computer control functions, such as air conditioning and transmission shift, are included. Many other computer applications such as turbocharger control, voice alert systems, trip computers, electronic instrumentation, antilock brakes, and suspension systems are clearly described and illustrated.

Computer-controlled diesel injection and emission systems are being used to meet increasingly stringent California diesel particulate emission standards. These computer-controlled diesel systems are described here.

One of the latest developments in automotive electrical systems is the use of multiplex data transmission. Various types of these data transmission systems are extensively described. An explanation of an automotive test computer completes the book.

I would like to express my appreciation to all the companies who granted us permission to use their diagrams in the book. I would also like to thank the publishing staff for their excellent cooperation.

Don Knowles

1

Electricity and Electric Circuits

Practical Completion Objectives

1. Test for an open circuit with a 12V test light.
2. Test for a grounded circuit with 12V test light.
3. Use a compass to test for a grounded or shorted circuit.
4. Test for a shorted condition in an electromagnet.
5. Use Ohm's Law formula to calculate volts, ohms, or amperes in a series circuit.
6. Use Ohm's Law formula to calculate volts, ohms, or amperes in a parallel circuit.
7. Calculate the total resistance in a series or parallel circuit.

Atomic Structure

Elements

An element is defined as a liquid, a solid, or a gas that contains only one type of atom. Copper is an element that is widely used in automotive wiring and electrical components.

Atoms

The smallest particle into which an element can be divided and still retain its characteristics is an atom. An atom cannot be seen with the most powerful electron microscope that magnifies millions of times.

Compounds

A compound is a liquid, a solid, or a gas that contains two or more types of atoms. Water is a compound that consists of hydrogen and oxygen atoms.

Molecules

A molecule is the smallest particle of a compound in which the characteristics of the compound are maintained. A water molecule contains two hydrogen atoms and one oxygen atom, as illustrated in Figure 1-1, and the chemical symbol for water is H_2O.

Parts of the Atom

At the center, or nucleus, of most atoms there are two extremely small particles known as ''protons'' and ''neutrons.'' Arranged in different orbits, or rings, around the nucleus there are other small particles called ''electrons.'' The electrons orbit the nucleus in much the same way as the planets orbit the sun. Seven rings located various distances from the nucleus explain the configuration of most atoms. The outer ring is called a ''valence ring,'' and any electrons on this ring are referred to as ''valence electrons.'' A copper atom, as shown in Figure 1-2, contains 29 protons and 35 neutrons in the nucleus, and 29 electrons orbiting the nucleus.

The maximum number of electrons on the first ring is 2, while second and third rings may have a

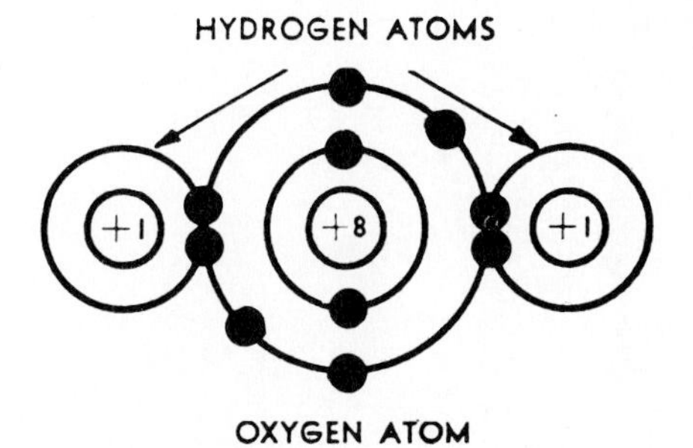

Figure 1-1 Water Molecule. (Courtesy of General Motors Corporation)

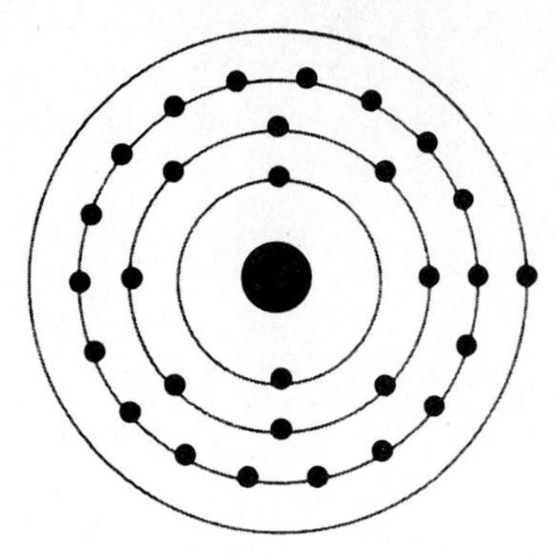

Figure 1–2 Copper Atom. (Courtesy of General Motors Corporation)

maximum of 8 and 18 electrons respectively. Since a copper atom has 29 electrons, there is 1 electron on the valence ring of the atom. The number of valence electrons is important because this determines the electrical characteristics of the element. Electrons have a negative electrical charge, and protons are positively charged. Neutrons do not have an electrical charge, but they add weight to the atom. If an atom is in balance, it will have the same number of protons and electrons. Centrifugal force tends to move the electrons away from the atom. However, the negatively charged electrons are held in their orbits by the attraction of the positively charged protons. Some atoms have the same number of protons and neutrons, but many of the heavier atoms have more neutrons.

A proton is 1840 times heavier than an electron. Therefore, electrons are very light and can be moved easily from one atom to another in some elements.

Periodic Table

All known elements are listed on the periodic table according to their number of protons. For example, hydrogen has 1 proton, and it is number 1 in the periodic table. The hydrogen atom contains 1 proton

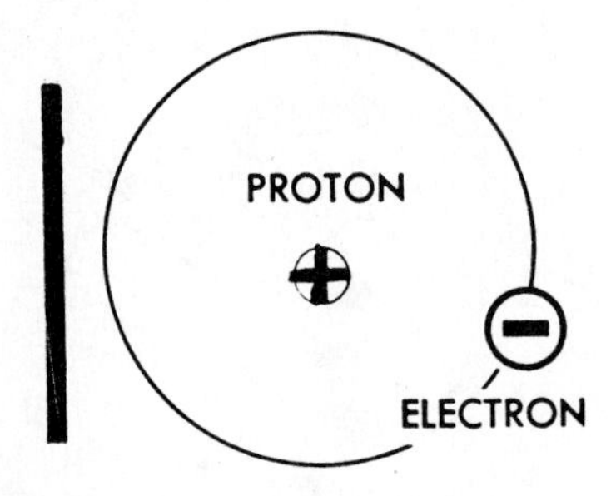

Figure 1–3 Hydrogen Atom. (Courtesy of General Motors Corporation)

and 1 electron, but it does not have any neutrons (Figure 1–3).

The atomic number of each element is determined by the location of the element in the periodic table. For example, oxygen has an atomic number of 8, and it is number 8 in the periodic table, which indicates that each oxygen atom has 8 protons and 8 electrons. Some of the other well-known elements in the periodic table are silver, number 47; gold, number 79; and uranium, number 92. Many periodic tables list 103 known elements.

Electricity

Electrical Conductors

If an element has one, two, or three valence electrons, these electrons will move from one atom to another. Copper, silver, or gold atoms have one valence electron. Aluminum atoms have three valence electrons, and this element is a reasonably good conductor.

Electrical Insulators

Some elements will not conduct electricity easily, and these are classified as insulators. These elements have five or more valence electrons on each atom. When an atom has five or more valence electrons, they do not move easily from one atom to another.

Semiconductors

When an element has four valence electrons, it is classified as a semiconductor. Carbon, silicon, and germanium are in the semiconductor category. Some of these semiconductors, such as silicon, have unusual properties when they are combined with other elements. Silicon is used in the manufacture of diodes and transistors.

Electron Movement

There are three requirements for electron movement in an electric circuit.

1. A massing of electrons (high voltage) at one point in the circuit.
2. A lack of electrons (low voltage) at another point in the circuit.
3. A complete circuit.

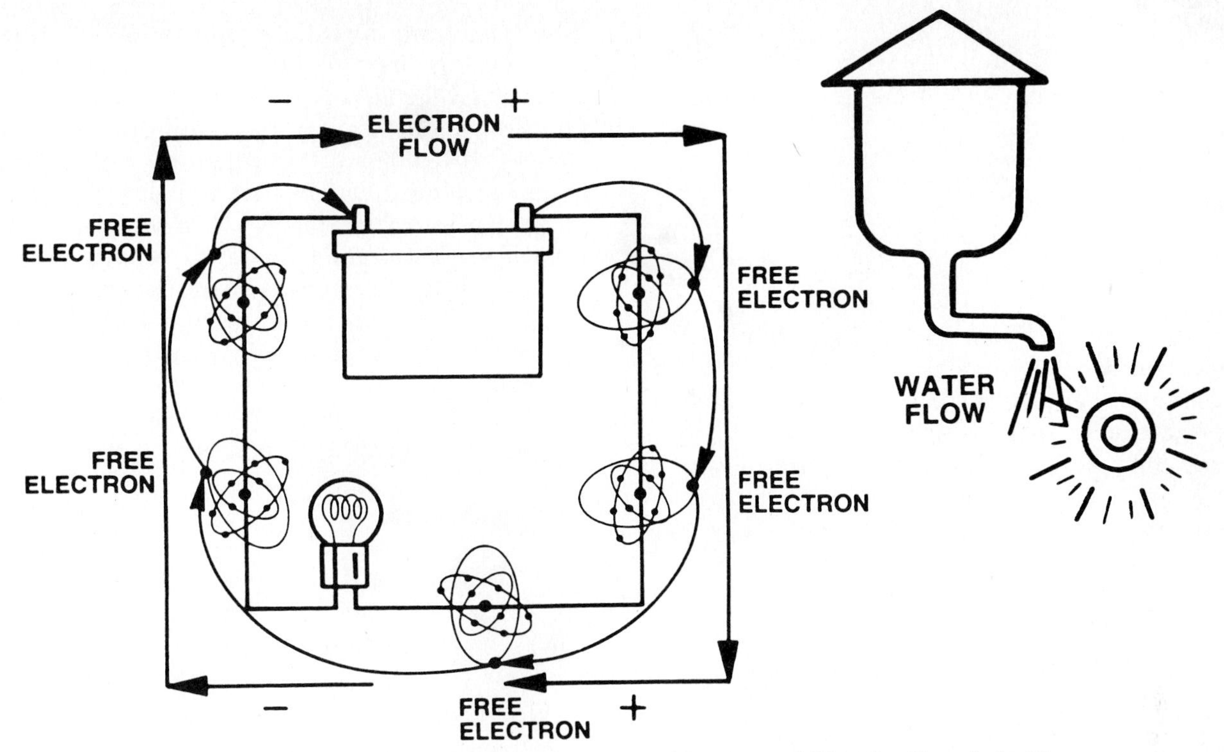

Figure 1–4 Electron Movement. [Courtesy of Chrysler Canada Ltd.]

A voltage, or electrical-pressure difference, is created by a massing of electrons at one point in the circuit and a lack of electrons at another point in the circuit. Chemical action in a lead acid battery causes a massing of electrons on one set of plates and a lack of electrons on the other set of plates. When the battery is connected to a complete circuit, electrons will begin to move through the circuit from one set of battery plates to the other. The electrons are actually massed on the negative battery plates, and electron movement is from the negative battery terminal through the external circuit to the positive battery terminal. However, in the automotive industry, the conventional current-flow theory is often used, in which current flow is from the positive battery terminal to the negative terminal. A basic electric circuit is illustrated in Figure 1–4, in which electrons are moving from the positive battery terminal to the negative terminal.

Only the valence electrons move from atom to atom through the conductor. When a copper atom loses a valence electron, it has an excess of one positively charged proton. This positive attraction will immediately move a negatively charged valence electron from another atom. If a copper atom receives a free valence electron, thus having two valence electrons, one of the valence electrons will be repelled immediately to another atom. The flow of electric current can be defined as ''the controlled mass movement of valence electrons from atom to atom in a conductor.''

Electrical Measurements

Volt

A volt is a measurement for electrical-pressure difference in a circuit. If a voltmeter is connected across the terminals of a lead acid battery, the meter may read 12.6V. This indicates that there is 12.6V electrical pressure at one battery terminal in relation to 0V pressure at the other terminal. The electrical pressure, which is measured in volts, forces the electrons or current to flow through the circuit, as shown in Figure 1–5.

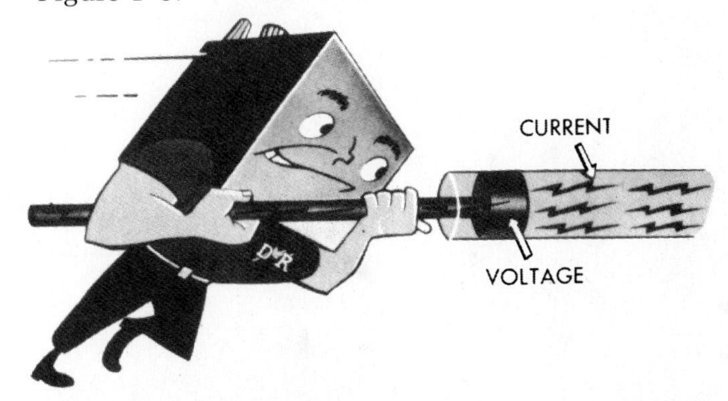

Figure 1–5 Volts in an Electric Circuit. [Courtesy of General Motors Corporation]

Ampere

The electron movement, or current flow, in a circuit is measured in amperes.

Ohms

Electrical resistance, or opposition to current flow, is measured in ohms (Ω). A water pipe offers a resistance to the flow of water. In much the same way, electrical components, such as a light filament, offer a resistance to the flow of current. A small filament containing fine wire has a higher resistance to current flow than a large filament made from heavier wire. The resistance in a light filament is compared to the resistance in a water pipe in Figure 1–6.

Ohm's Law

Ohm's Law states that the current flow in a circuit is directly proportional to the voltage and inversely proportional to the resistance. This means that an increase in voltage will create a corresponding increase in amperes, while a voltage decrease will reduce the amperes. When the resistance in a circuit is increased, the amperes will be reduced; if the resistance is lowered, the amperes will increase.

Ohm's Law Formula

In the Ohm's Law formula, voltage is expressed as E, which is an abbreviation for electromotive force (EMF). Amperes are indicated by induction (I) in this formula, and ohms are represented by resistance (R). Ohm's Law formula is often expressed as $E/I \times R$. If two values in a circuit are known, this formula may be used to calculate the unknown value. The amperes in a circuit may be determined by dividing the resistance into the volts. Therefore, $I = E/R$. When the voltage is unknown, it may be calculated by multiplying the amperes times the resistance, and thus $E = I \times R$. The ohms in a circuit may be determined by dividing the amperes into the volts, and therefore $R = E/I$.

Types of Electric Circuits

Series Circuit

In a series circuit, the resistances are connected so that the same current will flow through all the resistances, as indicated in Figure 1-7.

Four facts about series circuits can be summarized as follows:

1. The total resistance is the sum of all resistances in the circuit.
2. The same amperes must flow through each resistance.
3. A specific amount of voltage is dropped across each resistance.
4. The sum of the voltage-drops across each resistance must equal the source voltage.

In the circuit shown in Figure 1–7 the total resistance would be 2 + 5 + 4 + 1 = 12Ω.

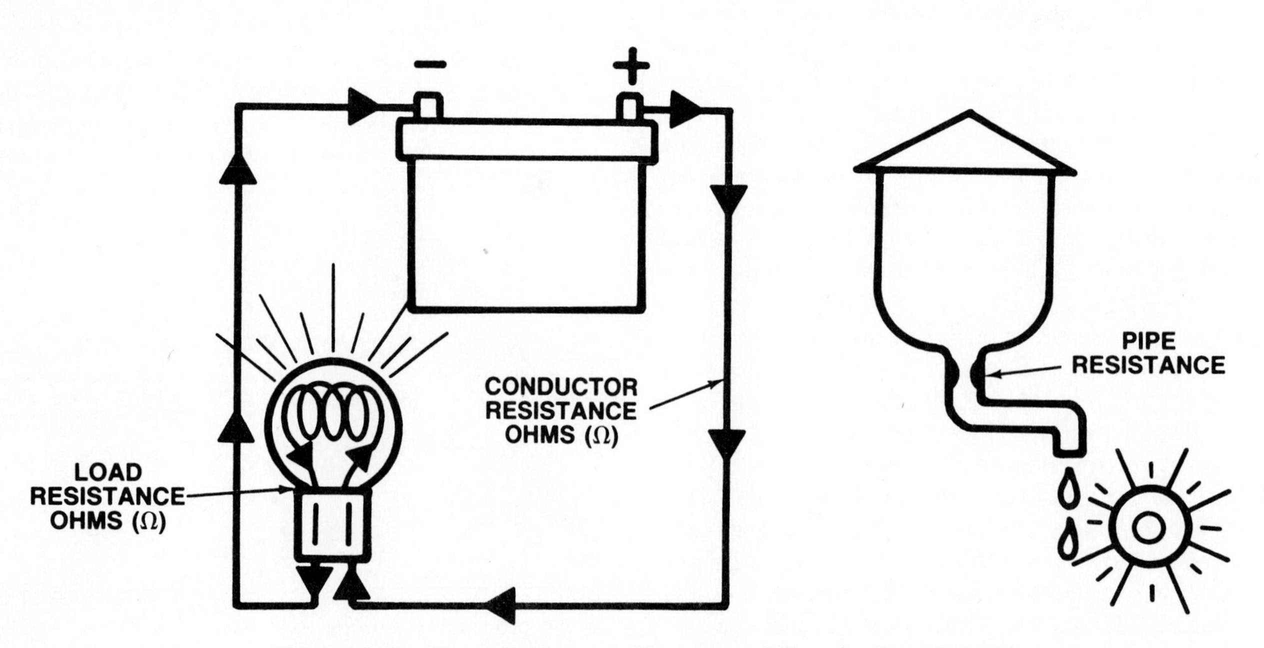

Figure 1–6 Ohms Resistance. (Courtesy of Chrysler Canada Ltd.)

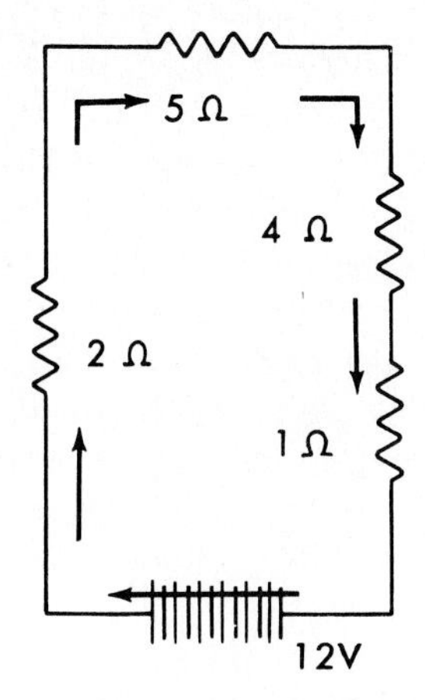

Figure 1–7 Series Circuit. [Courtesy of General Motors Corporation]

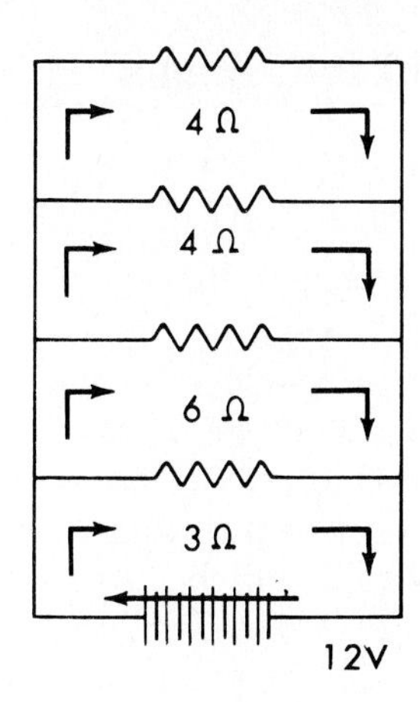

Figure 1–8 Parallel Circuit. [Courtesy of General Motors Corporation]

The current flow would be 12V ÷ 12Ω = 1A. This current would flow through each resistance.

The voltage drop across the resistors would be 1A × 2Ω = 2V and 1A × 5Ω = 5V, 1A × 4Ω = 4V, 1A × 1Ω = 1V.

The sum of the voltage drops across each resistance would be 2V + 5V + 4V + 1V = 12V.

Parallel Circuits

In a parallel circuit, each resistance forms a separate path for current flow. The automotive electrical system has many resistances connected parallel to the battery, as illustrated in Figure 1–8.

Four important facts that must be understood about a parallel circuit are the following:

1. The total resistance is always less than the lowest value resistor in the circuit.
2. The current flow in each parallel branch of the circuit will be determined by the resistance in that branch, and the sum of the branch current flows is equal to the current flow leaving the source and returning to the source.
3. Equal full-source voltage is applied to each resistance.
4. Equal full-source voltage is dropped across each resistance.

The easiest way to calculate the total resistance in a parallel circuit is to determine the current flow through each parallel branch and then add these current flows to obtain the total current.

In the parallel circuit shown in Figure 1–8, the current flow through each resistor would be: 12V ÷ 3Ω = 4A, 12V ÷ 6Ω = 2A, 12V ÷ 4Ω = 3A, 12V ÷ 4Ω = 3A. The total current flow would be 4A + 2A + 3A + 3A = 12A. Therefore, the total resistance would be 12V ÷ 12A = 1Ω.

Series–Parallel Circuits

Some automotive electrical circuits are series–parallel circuits. This type of circuit contains resistances that are connected in series with the parallel resistances in the circuit. The instrument-panel light circuit on many cars is an example of a series–parallel circuit. Each instrument-panel light is connected parallel to the battery, but a variable resistor in the headlight switch is connected in series with the panel lights. When the headlight switch control knob is rotated, the variable resistor reduces the voltage applied to the instrument panel lights, which reduces their brilliance. A series–parallel circuit is shown in Figure 1–9.

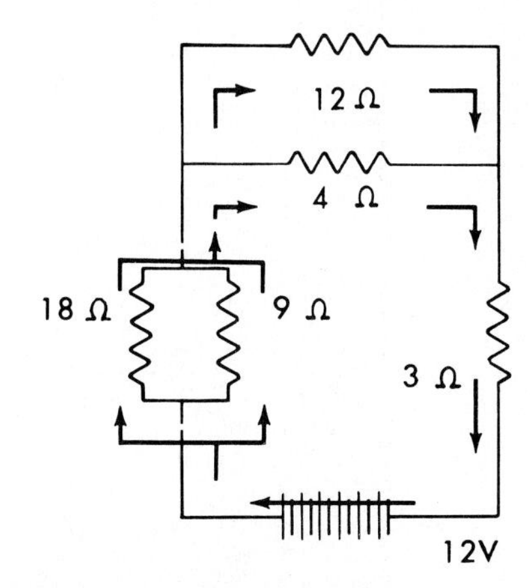

Figure 1–9 Series–Parallel Circuit. [Courtesy of General Motors Corporation]

Electric Circuit Defects

Open Circuit

An open circuit may be defined as an unwanted break in an electric circuit. This defect could occur in a wire or in an electric component such as a light filament. If an open circuit occurs in a series circuit, the current flow will be stopped in the entire circuit as shown in Figure 1–10.

An open circuit in one parallel branch of a parallel circuit will only stop the current flow in the parallel branch when the defect is located as outlined in Figure 1–11.

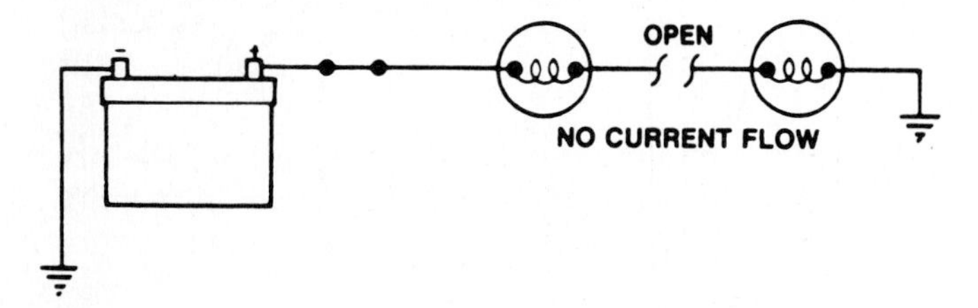

Figure 1–10 Open Circuit in a Series Circuit. (Courtesy of Chrysler Canada Ltd.)

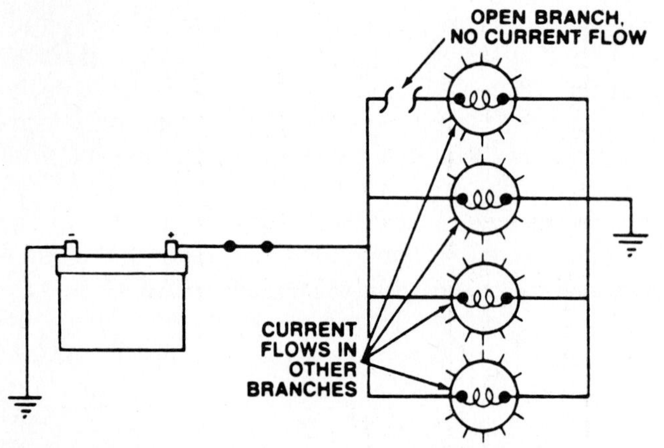

Figure 1–11 Open Circuit in a Parallel Circuit. (Courtesy of Chrysler Canada Ltd.)

Grounded Circuit

A grounded circuit may be defined as an unwanted copper-to-metal connection. Insulation is located on the connecting wires in an electric circuit. If the insulation is worn, the copper wire may touch the metal of the vehicle chassis and create a grounded condition. When this occurs, the current will flow directly through the defective ground connection without flowing through the electrical components in the circuit. Under this condition, excess current flow will probably burn out the fuse in the circuit.

Shorted Circuit

A shorted circuit may be defined as an unwanted copper-to-copper connection in an electrical circuit. When the insulation is worn on two adjacent wires and the wires contact each other, a short circuit is created, as illustrated in Figure 1–12.

A short circuit may cause the current to flow from one circuit into a second circuit when the switch in the second circuit is off. (Electrical defects in electromagnets are discussed later in this chapter.)

Circuit Diagnosis

Test Meters

An ammeter is used to measure current flow in a circuit. The ammeter has a very low internal resistance, and it must be connected in series in the circuit. If an ammeter is connected in parallel across the terminals of a battery, the meter will be damaged by excessive current flow. The ammeter in Figure 1–13 is connected in series.

A voltmeter has very high internal resistance, and it must be connected in parallel as shown in Figure 1–14.

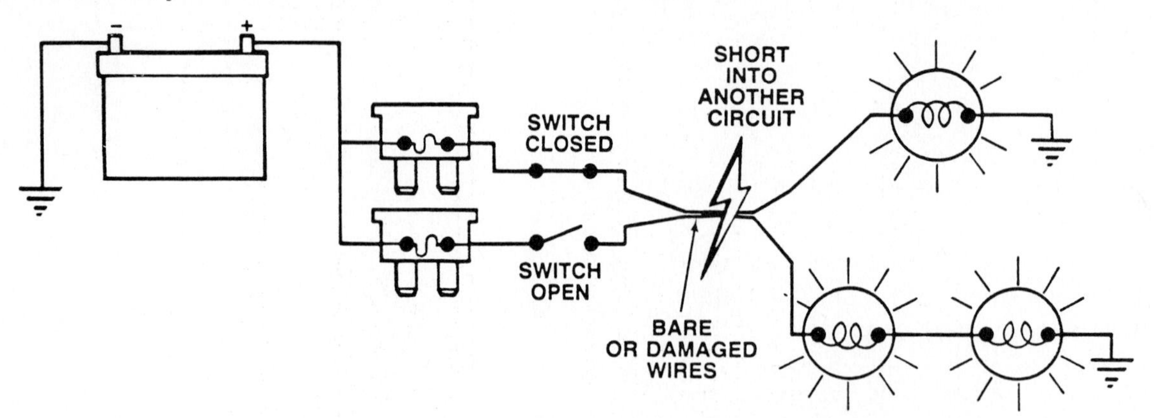

Figure 1–12 Shorted Circuit. (Courtesy of Chrysler Canada Ltd.)

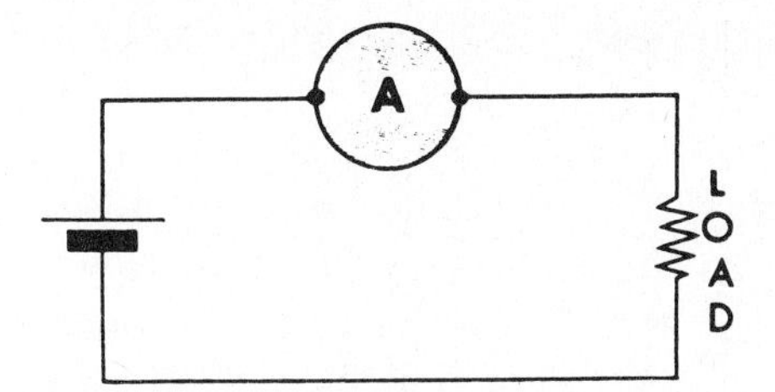

Figure 1–13 Ammeter Connected in Series. [Courtesy of General Motors Corporation]

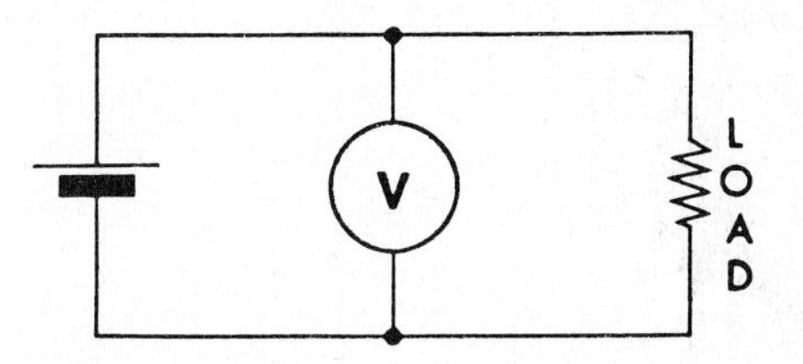

Figure 1–14 Voltmeter Connected in Parallel. [Courtesy of General Motors Corporation]

The ohmmeter is self-powered by an internal battery. This type of meter must never be connected to a circuit in which current is flowing or meter damage will result. When the resistance of an electrical component is being measured, disconnect the component from the circuit and connect the ohmmeter leads to the terminals on the component as pictured in Figure 1–15.

Most ohmmeters have several different switch positions such as X1, X10, and X1,000. The X1 switch position provides a meter scale reading of 0 to 100Ω. If the X10 position is selected, the scale will indicate 0 to 1,000Ω, and the X1,000 position will provide a meter reading of 0 to 100,000Ω. The correct scale must be used for the resistance of the component being tested. If the resistor in Figure 1–15 has a specified value of 1.25Ω, the X1 meter scale should be selected. Many ohmmeters require calibration on each scale by connecting the leads and rotating the calibration control until the pointer is in the 0 position.

Test for Open Circuits

A 12V test light may be used to test for an open circuit in conventional electric circuits. Do not use a 12V test light to test computer-controlled circuits, because these circuits operate on a very low current flow. The test-light current may damage the computer or other components in the circuit. (Refer to Chapters 3, 4 and 6 for computer system diagnosis.) With the switch in the circuit turned on, test for power at different locations in the circuit with the test lamp. Power will not be available immediately after the open circuit. The open circuit test procedure is illustrated in Figure 1–16.

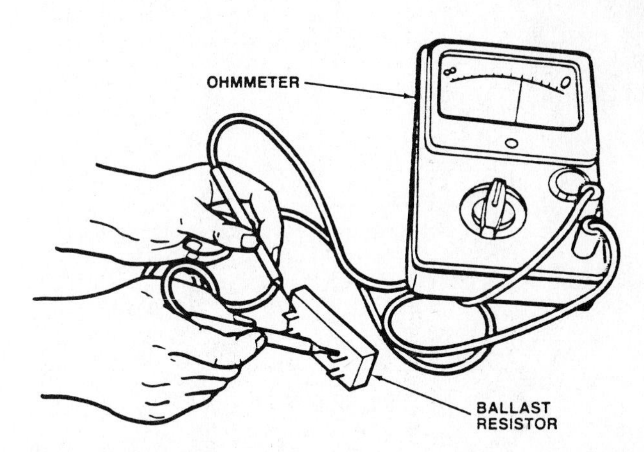

Figure 1–15 Resistance Test with Ohmmeter. [Courtesy of Chrysler Canada Ltd.]

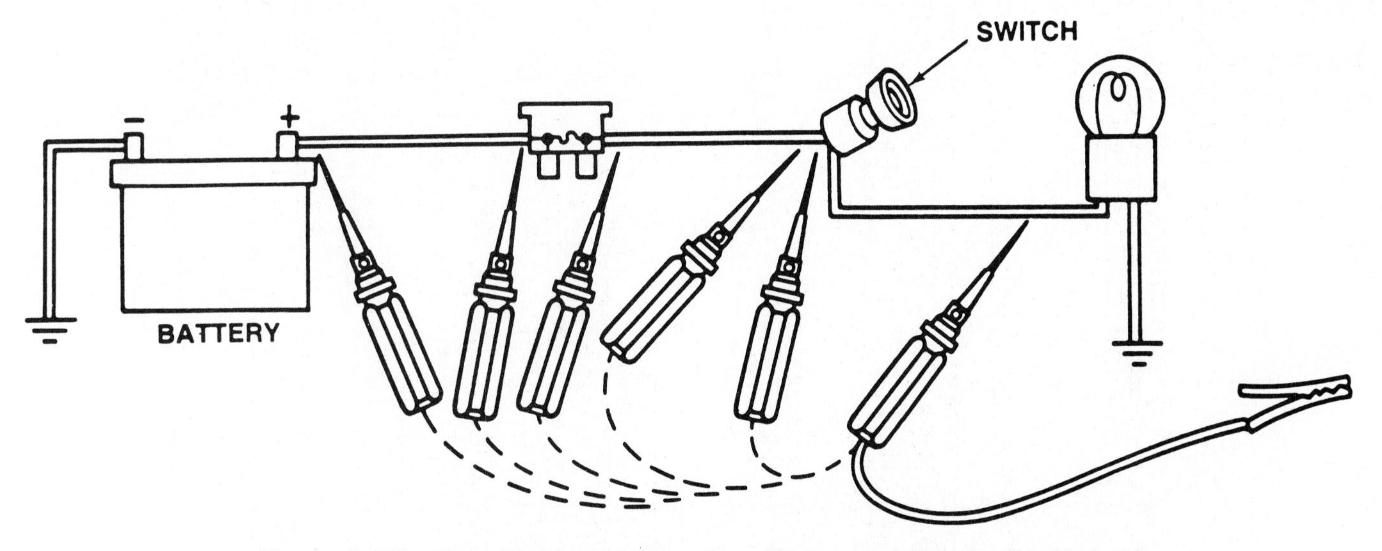

Figure 1–16 Open Circuit Test Procedure. [Courtesy of Chrysler Canada Ltd.]

Test for Grounded Circuits

When a grounded circuit test is being performed, the 12V test lamp should be installed in place of the circuit fuse, and the load should be disconnected from the circuit as shown in Figure 1–17.

If the test light is on, the circuit is grounded. The next step is to disconnect the circuit connectors starting at the load. When the test light goes out, the grounded circuit is after the connector that was disconnected last.

Compass Test for Shorted or Grounded Circuits

When this test is performed, a circuit breaker should be installed in place of the circuit fuse. Then a compass should be moved along the wiring harness. Each time the circuit breaker closes, the compass needle will deflect. When the compass is moved past the grounded location, the needle will no longer deflect. The same test procedure may be used to test for a shorted condition between the wires in a harness.

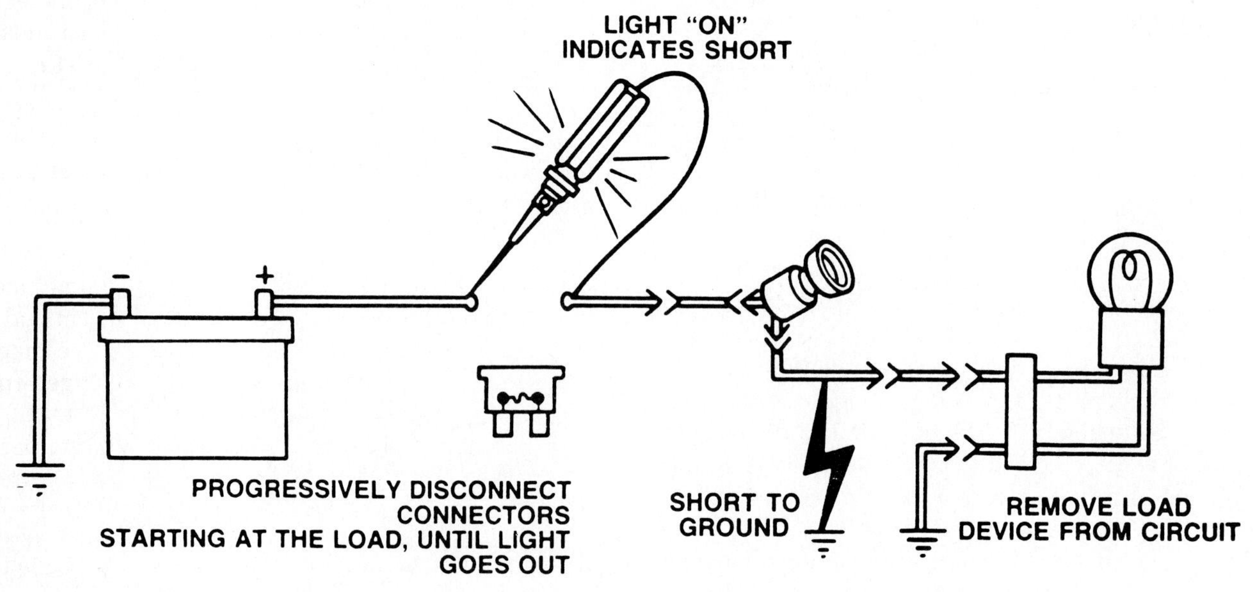

Figure 1–17 Grounded Circuit Test Procedure. [Courtesy of Chrysler Canada Ltd.]

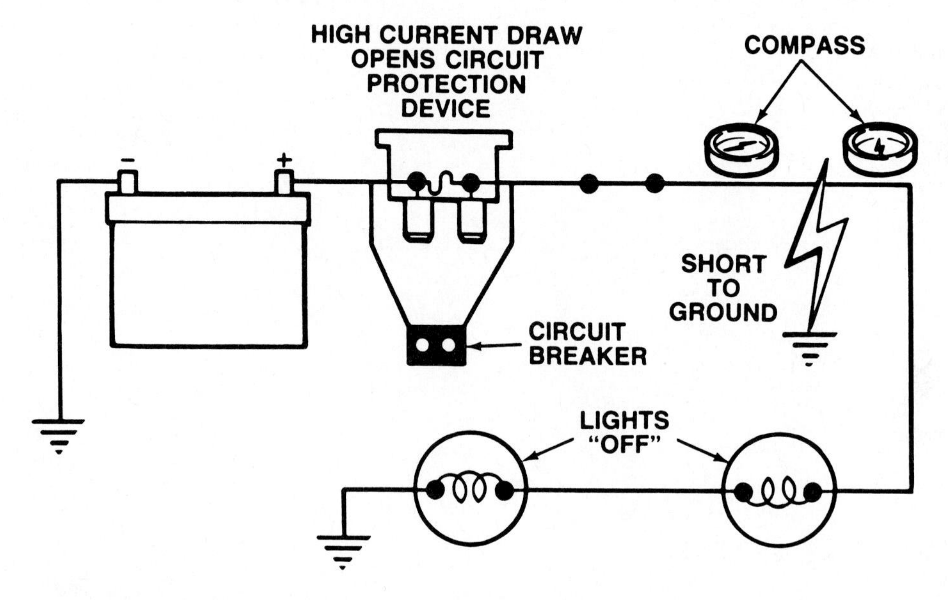

Figure 1–18 Compass Test Procedure for a Grounded or Shorted Circuit. [Courtesy of Chrysler Canada Ltd.]

When the shorted location in a wire is passed with the compass, the needle will no longer deflect because the current will be flowing from the shorted location to other wires that are connected into a different harness. (The test procedure for a shorted condition in an electromagnet is discussed later in this chapter.) The compass test procedure is pictured in Figure 1–18.

Voltage Drop

When current flows through a resistance, the amount of voltage that is dropped across the resistance will depend on the amount of resistance and the current flow. If the resistance increases, the voltage drop will also increase. An increase in current flow will also result in more voltage drop. The connecting wires in an electric circuit will have a very low voltage drop across them. Excessive resistance in the wires will result in a higher voltage drop across the wires. Voltage drop is only present when current is flowing. For example, in Figure 1–10, 12V would be available up to the open circuit, because there is no current flow and therefore no voltage drop across the first lamp. The voltmeter may be used to measure voltage drop. With current flowing in the circuit, the voltmeter may be connected across a resistor, or wire, to measure the voltage drop, as indicated in Figure 1–19.

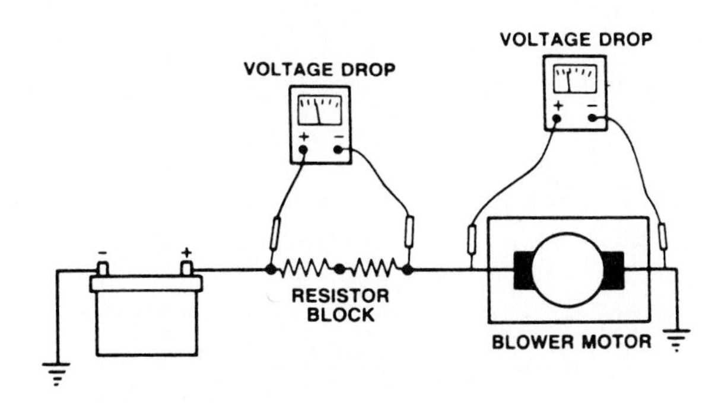

Figure 1–19 Voltmeter Connected to Measure Voltage Drop. (Courtesy of Chrysler Canada Ltd.)

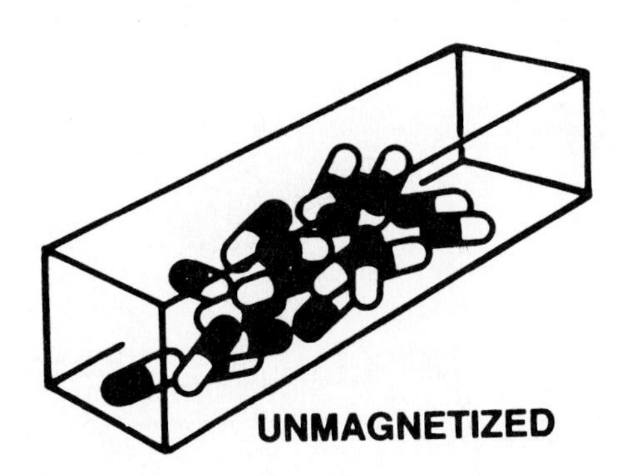

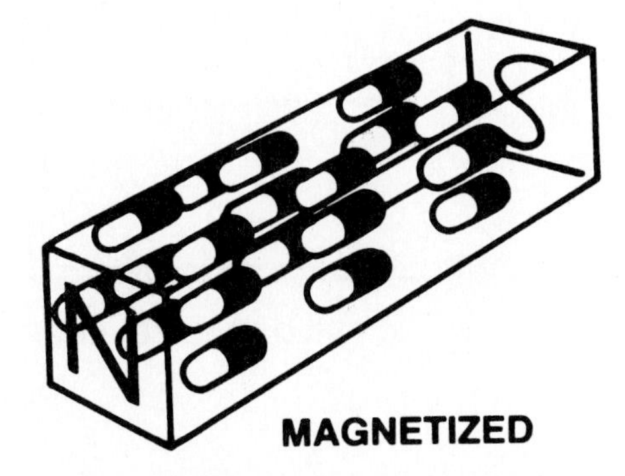

Figure 1–20 Magnetized and Unmagnetized Metal. (Courtesy of Chrysler Canada Ltd.)

Electromagnetism

Permanent Magnets

Certain metals such as iron, nickel, and cobalt can be magnetized. When a metal becomes magnetized, the molecules are aligned with the poles of the magnet, whereas the molecules are randomly arranged in unmagnetized metal as illustrated in Figure 1–20.

Hard steel will retain magnetism for long periods of time, and this type of magnet may be referred to as a *permanent magnet*. Soft iron will lose its magnetism immediately. An invisible field of force surrounds each magnet, with the lines of force moving from the north pole to the south pole as pictured in Figure 1–21.

If the permanent magnet is bent into a "U" shape, the lines of force will be more concentrated because the poles are closer together. When the unlike poles of two magnets are brought near each other, the magnets will be attracted because the lines of force around both magnets are moving in the same direction. Like magnetic poles repel because the lines of force surrounding the magnets are moving against each other, as shown in Figure 1–22.

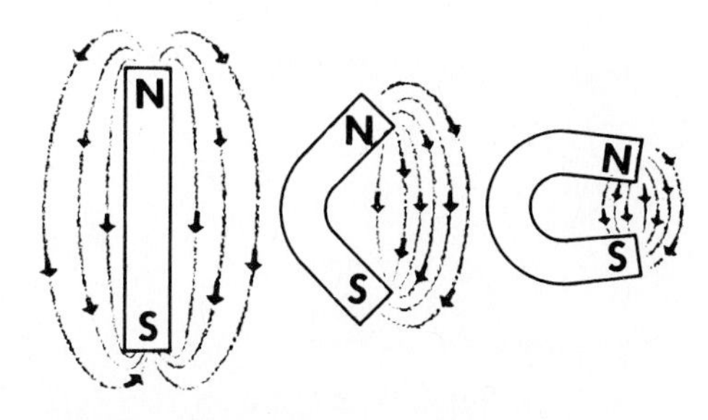

Figure 1–21. Permanent Magnets. (Courtesy of General Motors Corporation)

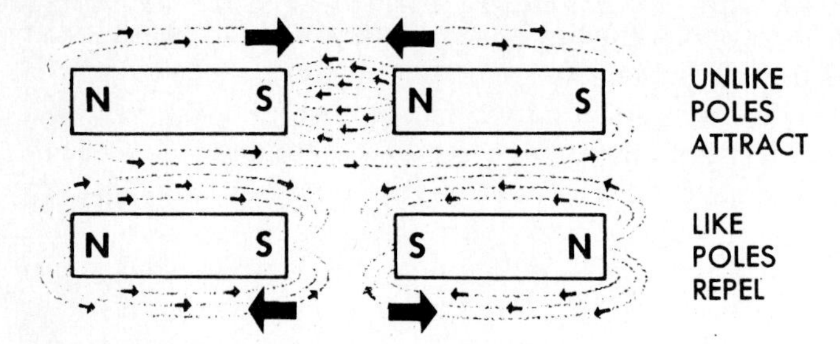

Figure 1–22 Magnetic Attraction and Repelling Action. [Courtesy of General Motors Corporation]

Electromagnets

When current flows through a wire, an invisible magnetic field surrounds the wire. This field is concentric to the conductor, and an increase in current flow will create a corresponding increase in magnetic strength. The right-hand rule can be used to determine the direction of current flow. If the fingers of the right hand are placed in the direction of the lines of force surrounding the conductor, the thumb will point in the direction of current flow, as indicated in Figure 1–23.

When the current flow through the conductor is reversed, the magnetic field will change directions. If two adjacent conductors have current flow through them in opposite directions, the magnetic field will be in opposite directions. Under this condition the magnetic lines of force directly between the conductors will be moving in the same direction, which results in a concentration of magnetic lines of force in this area, and the conductors will be pushed apart as illustrated in Figure 1–24.

If current flows through two adjacent conductors in the same direction, the magnetic fields around both conductors will be moving in the same direction. However, in the area directly between the conductors, the magnetic lines of force will be moving in opposite directions. Magnetic lines of force moving in opposite directions in the same space cannot exist. Therefore, the magnetic field will distort and surround both conductors, which tends to pull the conductors together, as shown in Figure 1–25.

When current flows through a coil of wire, it becomes an electromagnet with a north pole and a south pole at the ends of the coil. An iron core placed in the center of the coil concentrates the lines of force. The strength of an electromagnet is determined by the current flow through the coil and the number of turns in the coil. This magnetic strength is expressed in ampere turns, which is calculated by multiplying the amperes times the turns. The direction of the magnetic field around an electromagnet is determined by the direction of current flow, as illustrated in Figure 1–26.

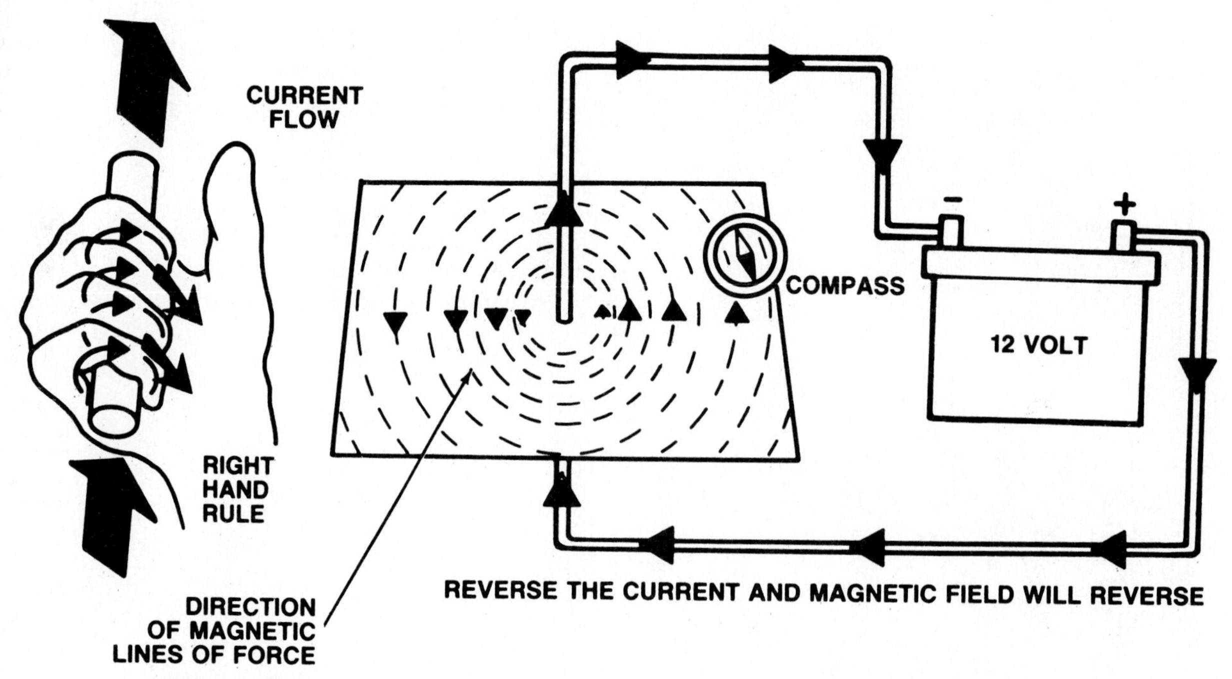

Figure 1–23 Magnetic Field Around a Current-Carrying Conductor. [Courtesy of Chrysler Canada Ltd.]

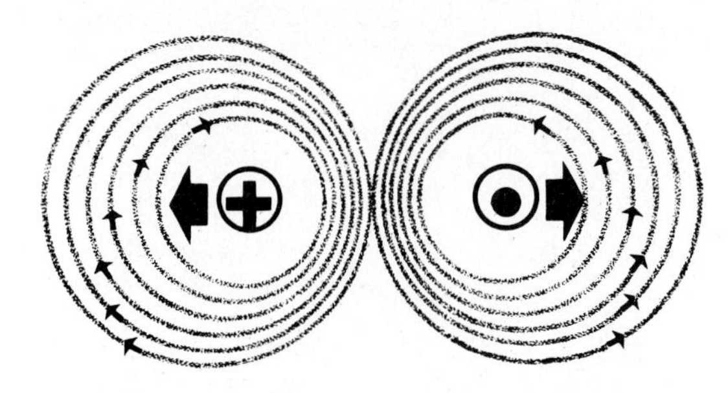

Figure 1–24 Adjacent Conductors with Current Flow in Opposite Directions. (Courtesy of General Motors Corporation)

Defects in Electromagnets

The defects in electromagnets are described as follows:

1. Open circuit—unwanted break.
2. Grounded circuit—unwanted copper-to-metal connection.
3. Shorted circuit—unwanted copper-to-copper connection

The windings in an electromagnet are coated with an insulating material that prevents the coils of wire from touching. Insulating paper may be located between each layer of turns. A shorted circuit occurs when the insulating material is melted on the coils and some of the coils touch. A shorted circuit results in the following:

1. Less effective coils, or turns.
2. Reduced resistance.
3. Increased current flow.
4. The same magnetic strength.

The reduction in the number of turns will tend to decrease the magnetic strength, but the increase in current flow will offset this effect. Therefore, the magnetic strength of an electromagnet with a shorted condition will remain the same if the electromagnet is the only resistance in the circuit. When a shorted circuit occurs in an electromagnet, the increased current flow will usually damage some of the other components in the circuit. For example, a shorted alternator field winding may result in a damaged voltage regulator. A shorted condition may be tested by connecting an ohmmeter to the electromagnet. If the coil is shorted, the resistance will be less than specified by the manufacturer.

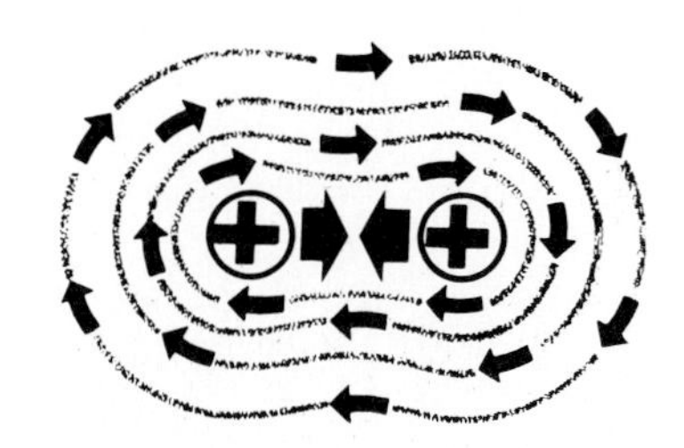

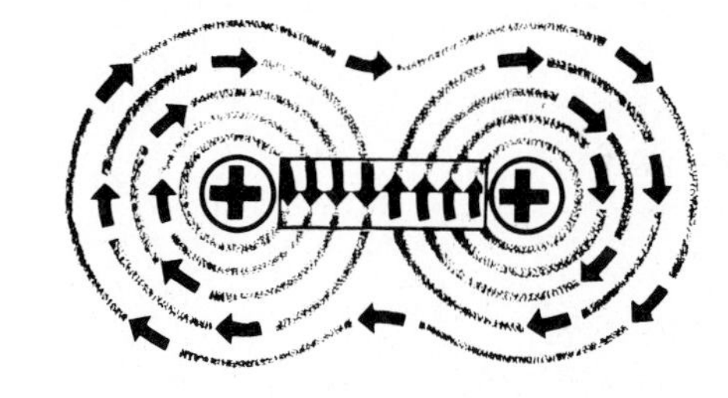

Figure 1–25 Adjacent Conductors with Current Flow in the Same Direction. (Courtesy of General Motors Corporation)

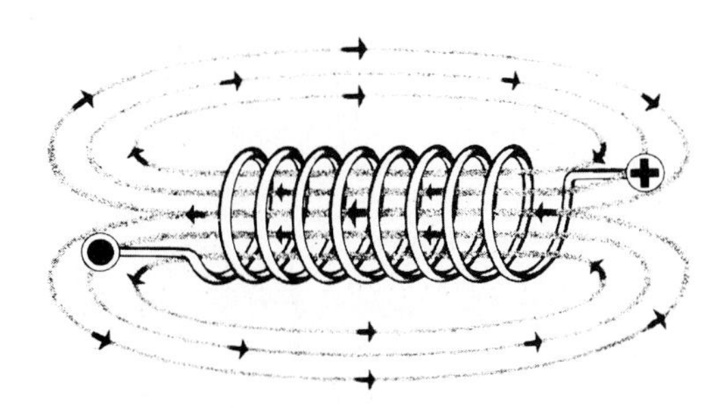

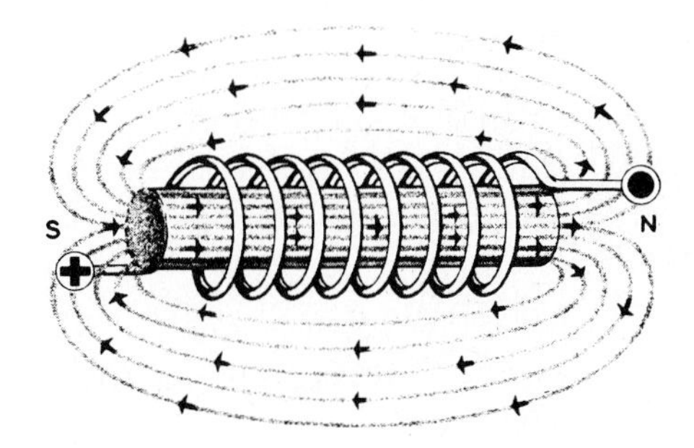

Figure 1–26 Magnetic Fields Surrounding an Electromagnet. (Courtesy of General Motors Corporation)

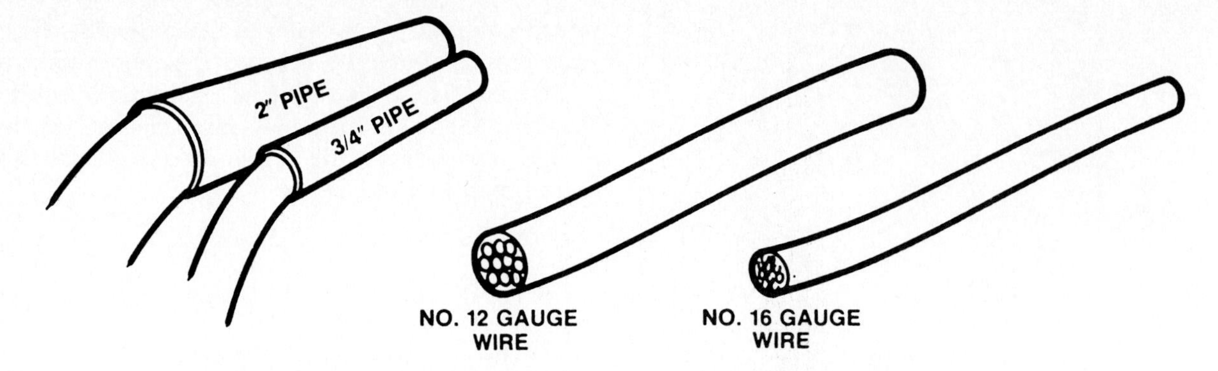

Figure 1-27 Wire Gauge Sizes. (Courtesy of Chrysler Canada Ltd.)

Basic Electric Components

Wires

Standard copper wires are used in most automotive electric circuits. The gauge size of the wire indicates the diameter of the wire. A 16-gauge wire has a smaller diameter than a 12-gauge wire, as indicated in Figure 1-27.

The resistance of a wire is determined by its size, length, temperature, and the type of conductor. A large-diameter copper wire has less resistance than a smaller wire, and therefore the larger wire has a greater current-carrying capacity. The resistance of a wire increases as its length is increased. When the temperature of copper wires is increased, they will have a higher resistance.

Relays

A relay contains a set of contacts mounted above the relay core. The magnetism developed by the fine winding on the relay core controls the action of the contacts. A relay is often connected in the horn circuit. When the horn ring is depressed, a switch in the steering wheel completes the circuit from the relay winding to ground. Under this condition the relay coil magnetism will close the contacts, and current will be supplied through the contacts to the horn. The lower current through the relay winding is used to control the higher current in the horn circuit as outlined in Figure 1–28.

Circuit Breakers, Fuses, and Fuse Panels

Circuit breakers or fuses are used as safety devices to protect electric wiring and circuit components. A circuit breaker contains a set of contacts and a bimetal strip. This strip is made from two metals that expand at different rates when they are heated or cooled. Therefore, the bimetal strip will bend as it is subjected to temperature changes. If a defect occurs, such as a grounded circuit or a shorted circuit, excessive current flow will heat the bimetal strip and cause the contacts to open. If this action did not take place, the wires or circuit components would be overheated and damaged. When the bimetal strip cools, the contacts will close again, as shown in Figure 1–29.

Older vehicles use round fuses in the fuse panel. The amp rating of each fuse and the circuit identification is stamped on the fuse panel. A fuse panel with round fuses is illustrated in Figure 1–30.

Most cars use plug-in fuses at the present time. The color code on these fuses indicates the amp rating. This code and the test procedure are explained in Figure 1–31.

Other components, such as signal-light flashers, horn relay, buzzers, and circuit breakers, may be plugged into the fuse panel as indicated in Figure 1–32.

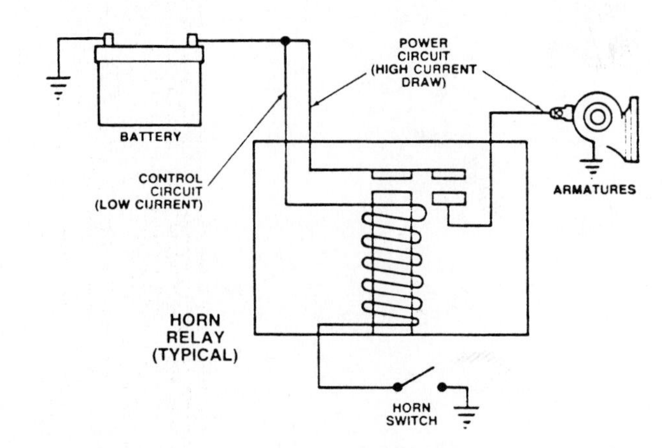

Figure 1–28 Horn Relay Circuit. (Courtesy of Chrysler Canada Ltd.)

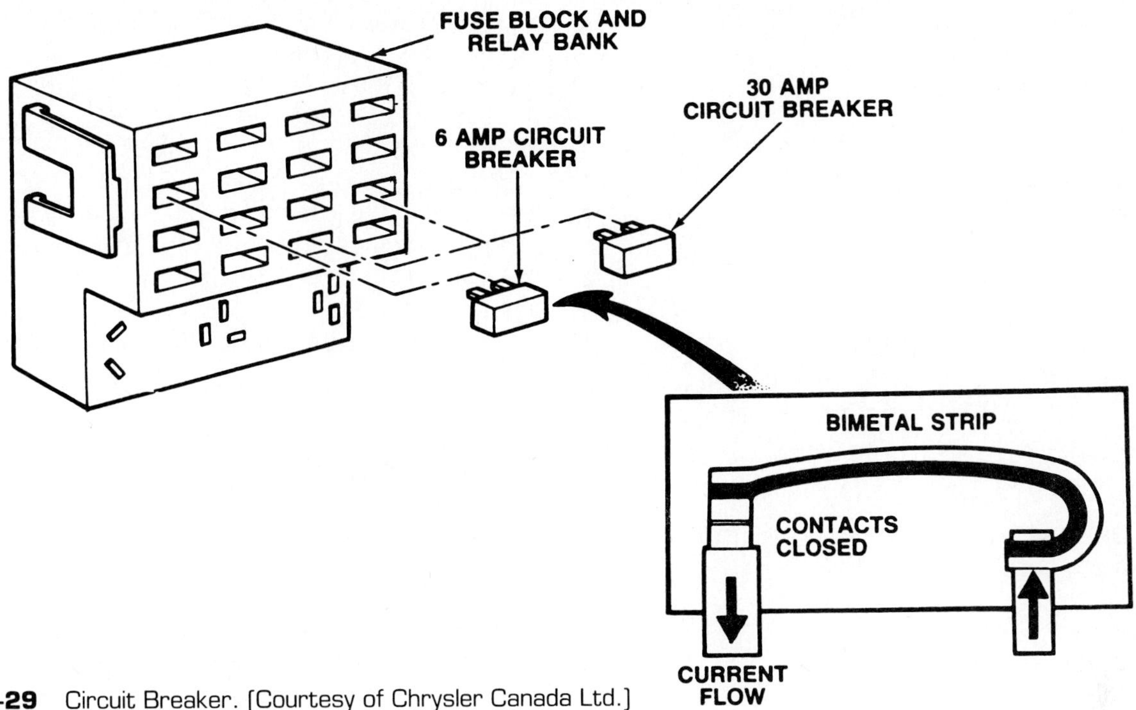

Figure 1–29 Circuit Breaker. [Courtesy of Chrysler Canada Ltd.]

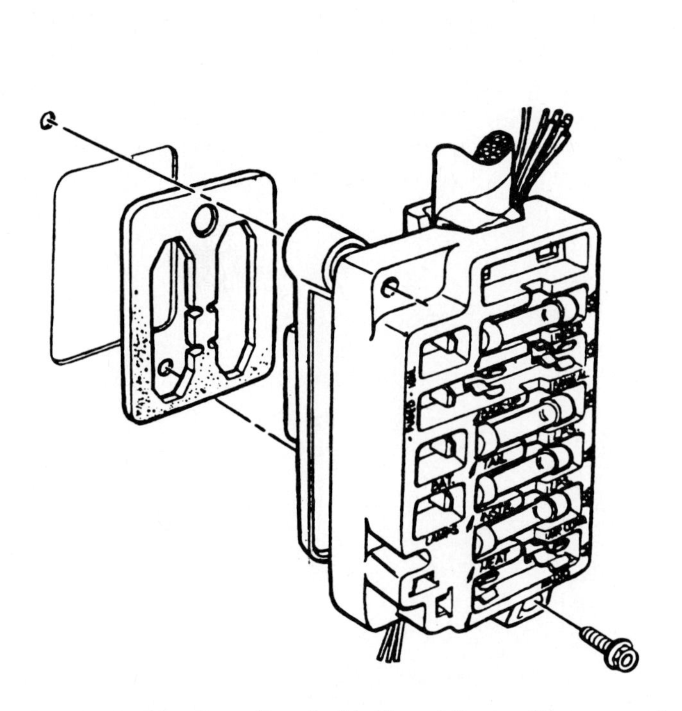

Figure 1–30 Fuse Panel with Round Fuses. [Courtesy of General Motors Corporation]

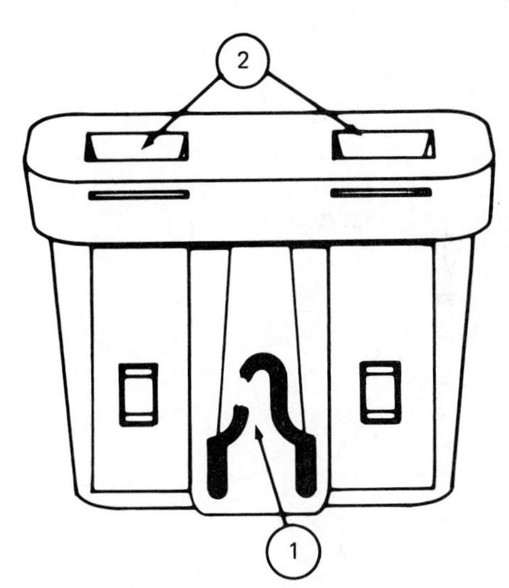

To test for blown mini-fuse

(1) Pull fuse out and check visually

(2) With the circuit activated, use a test light across the points shown

Mini-fuse color codes

Rating	Color
5 amp	Tan
10 amp	Red
20 amp	Yellow
25 amp	White

Figure 1–31 Plug-in Fuse. [Courtesy of General Motors Corporation]

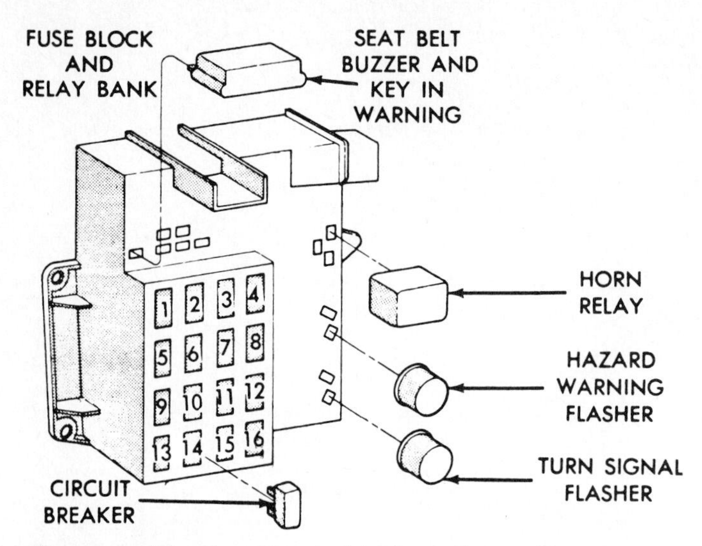

Figure 1–32 Fuse Panel with Plug-in Fuses. [Courtesy of Chrysler Canada Ltd.]

Resistors

Various types of resistors are used in automotive electric circuits. Some resistors, such as ignition resistors, or blower motor resistors, are made from a coil of wire. Variable resistors are used in headlight switches to reduce the brilliance of the instrument-panel lights. This type of resistor is called a "rheostat." Carbon resistors are used in some computers and modules. A variety of resistors is shown in Figure 1–33.

Capacitors

Most automotive capacitors contain two aluminum-foil plates that are separated by insulating paper. One plate is connected to the capacitor lead while the other plate is grounded to the case. The capacity of a capacitor is measured in farads or microfarads (MFD). In many circuits the capacitor is used to block direct current (DC) voltage. A capacitor is used in most alternators to protect the diodes from high induced voltages. Capacitors can also be used to suppress radio interference. The primary ignition circuit on a point-type ignition system uses a capacitor to prevent point arcing and provide faster magnetic collapse in the ignition coil. The design and basic operation of a capacitor is outlined in Figure 1–34.

Diodes and Transistors

Diode Operation

Most diodes and transistors are made from silicon. This element is classified as a semiconductor because it has four valence electrons. If two pieces of silicon are melted together, the valence rings on the atoms will join, and eight electrons will be located on the valence rings. When this occurs, a silicon crystal has been created, as pictured in Figure 1–35.

When a diode or transistor is manufactured, a silicon crystal is combined with other elements. Phosphorus is an element with five valence electrons, and boron atoms contain three valence electrons. In the diode manufacturing process a small silicon crystal wafer is coated with boron on one side and phosphorus on the other side. The wafer is then heated, and these two elements melt into the silicon. One excess, or free, electron will be provided for each atom of phosphorus that melts into the silicon because the

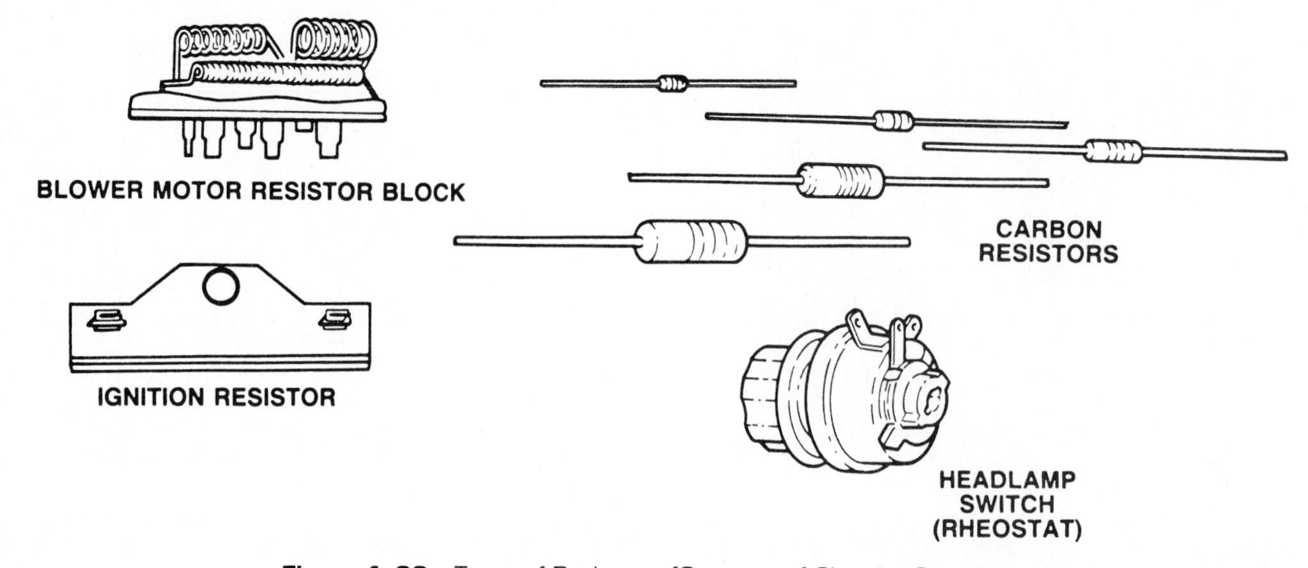

Figure 1–33 Types of Resistors. [Courtesy of Chrysler Canada Ltd.]

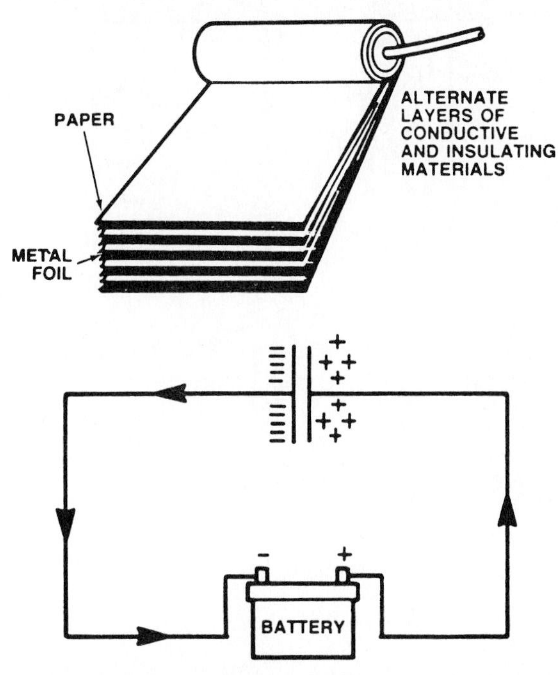

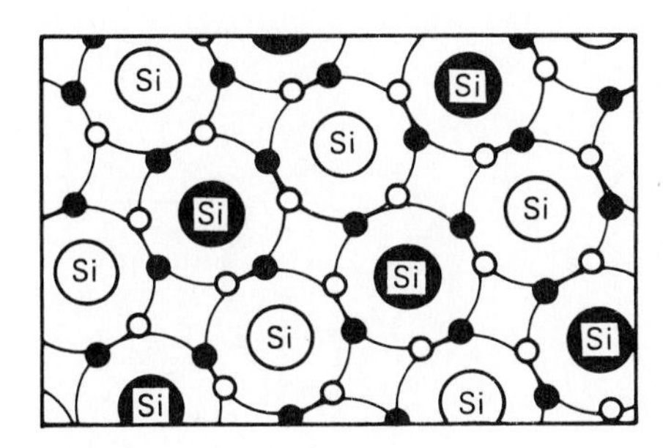

Figure 1-34 Capacitor and Capacitor Operation. [Courtesy of Chrysler Canada Ltd.]

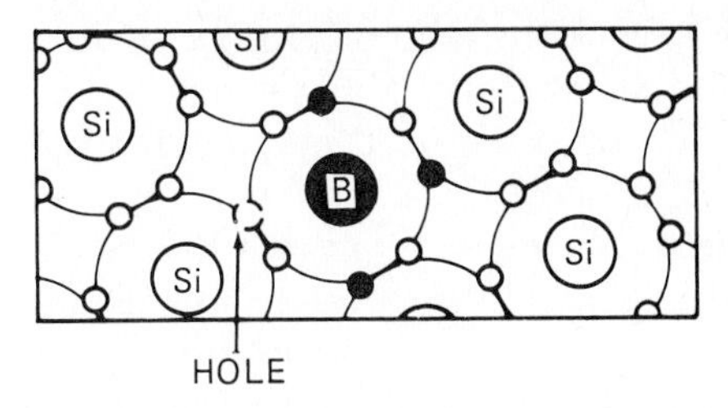

Figure 1–37 Positive Material. [Courtesy of General Motors Corporation]

outer rings of the atoms will join, and there are nine valence electrons in the two elements. The maximum number of valence electrons is eight. Therefore, a negative (N) type of material that has an excess of electrons is made when silicon is combined with phosphorus, as illustrated in Figure 1–36.

When boron is combined with silicon, a positive material with a lack of electrons is created, because the two elements have a total of seven valence electrons, and eight electrons are required to complete the valence ring. A positive material is shown in Figure 1–37.

Electrons will flow through a diode when a positive terminal of a voltage source is connected to the positive side of the diode and the negative source terminal is connected to the negative side of the diode. This type of connection is referred to as a ''forward bias,'' as indicated in Figure 1–38.

A diode blocks electron movement when the negative source terminal is connected to the positive

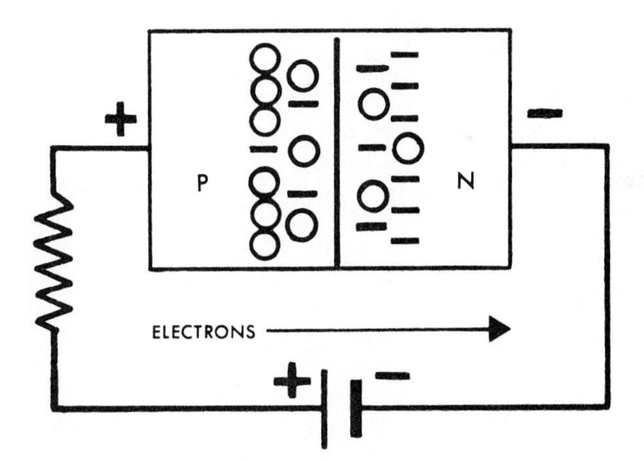

Figure 1–35 Silicon Crystal. [Courtesy of General Motors Corporation]

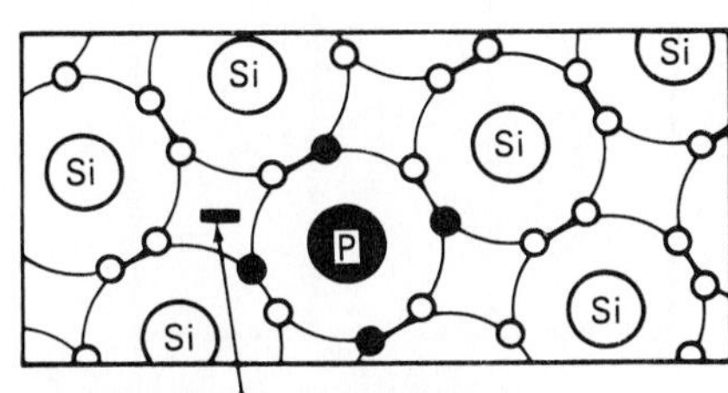

Figure 1–36 Negative Material. [Courtesy of General Motors Corporation]

Figure 1–38 Forward Bias Connection. [Courtesy of General Motors Corporation]

side of the diode and the positive source terminal is attached to the negative side of the diode. ''Reverse bias'' is the term used for this type of connection, as indicated in Figure 1–39.

A diode is sometimes referred to as a P/N junction. In the forward direction a diode will only conduct a certain amount of current flow. If the current flow is above the diode rating, the diode will be damaged. The peak inverse voltage (PIV) is the highest voltage that a diode will block in the reverse direction. If the PIV is exceeded, the diode may break down and conduct current, which will likely cause diode damage.

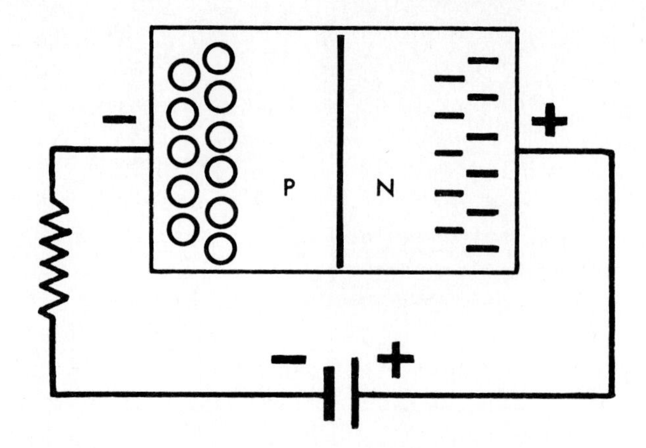

Figure 1–39 Reverse Bias Connection. [Courtesy of General Motors Corporation]

Zener Diode

A zener diode will break down at a specific voltage in the reverse direction. This type of diode is used in electronic voltage regulators and ignition modules. Zener diodes are not damaged by reverse current flow. A zener diode is shown in Figure 1–40.

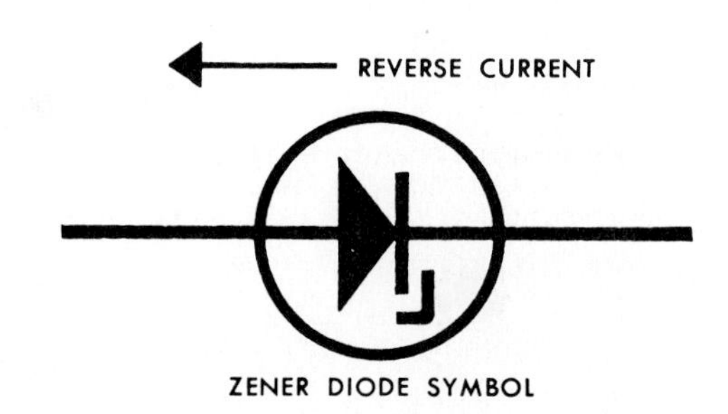

Figure 1–40 Zener Diode. [Courtesy of General Motors Corporation]

Transistors

A transistor contains three semiconductor materials: emitter, base, and collector. In some transistors the emitter is an N material, the base is a P material, and the collector is an N material. This type of transistor

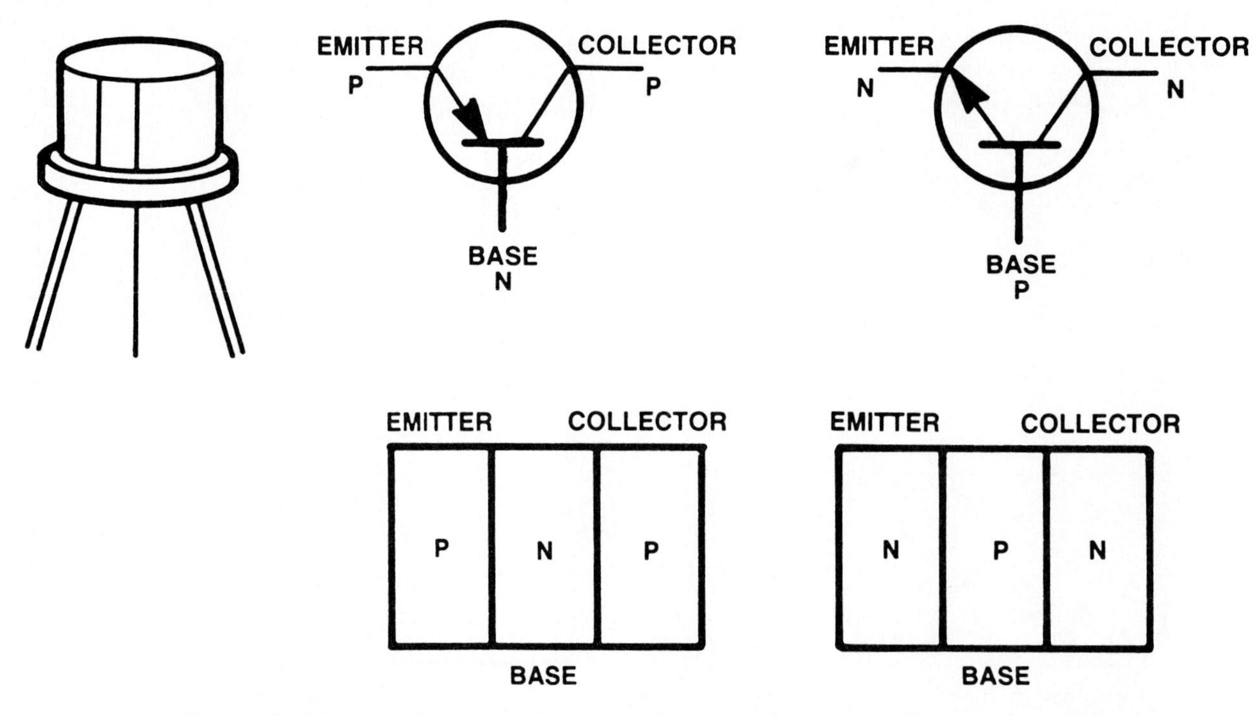

Figure 1–41 Transistors and Transistor Symbols. [Courtesy of Chrysler Canada Ltd.]

is called an ''NPN transistor.'' A PNP transistor may also be used in many circuits. Both types of transistors and the transistor symbols are illustrated in Figure 1–41.

In order to start current flowing through a transistor, a forward bias connection must be made from a source of voltage to the emitter and base terminals. When current begins to flow through the emitter-base circuit, a much higher current can flow through the emitter-collector circuit, as outlined in Figure 1–42.

If the base circuit is opened, the free electrons in the N collector material will be attracted to the outer edge of the collector, and a high resistance will develop immediately at the base-collector junction. This will stop the current flow in the emitter-collector circuit, as shown in Figure 1–43.

Therefore, the low current in the base circuit can be used to control the higher current in the collector circuit. The current flow in the emitter-base circuit may be stopped by opening a switch, or a set of contacts, in the base circuit. Another method of switching off the current flow in the emitter-base circuit is to create a reverse bias voltage at the emitter-base terminals, as illustrated in Figure 1–44.

A transistor can be defined as an electronic switch. The transistor can switch a circuit on and off much faster than a set of contacts or a mechanical switch.

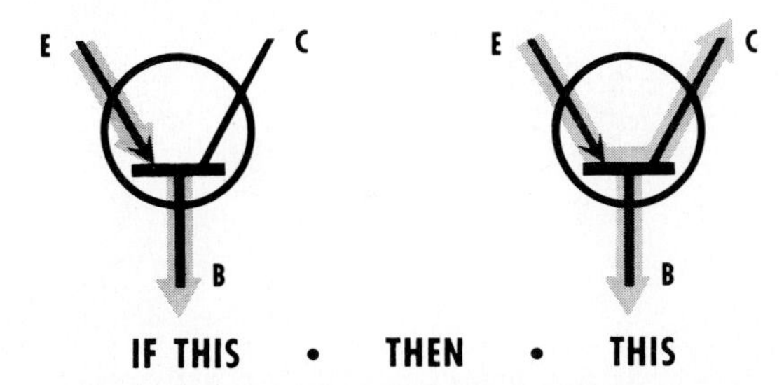
E
C
B
E
C
B
IF THIS • THEN • THIS
WITH VERY LITTLE CURRENT FLOW IN THE EMITTER-BASE CIRCUIT, A GREATER CURRENT CAN BE ESTABLISHED IN THE EMITTER-COLLECTOR CIRCUIT.

Figure 1–42 Transistor Operation, Emitter-Base Circuit Forward-Biased. (Courtesy of General Motors Corporation)

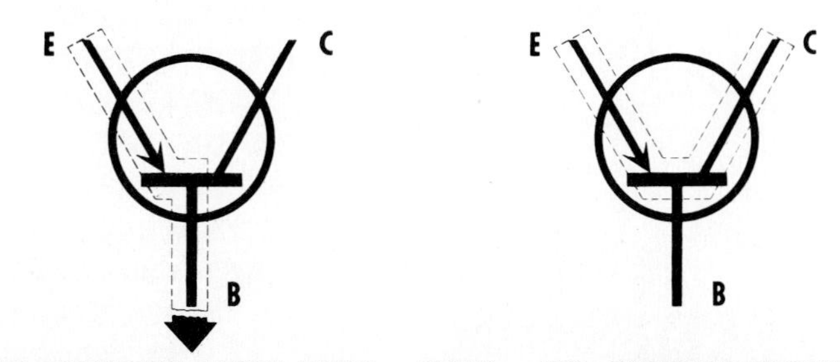
E
C
B
E
C
B
IF NO CURRENT FLOW HERE • THEN • NO CURRENT FLOW HERE
WITH NO CURRENT FLOW IN THE EMITTER-BASE CIRCUIT, THERE IS NO CURRENT FLOW IN THE EMITTER-COLLECTOR CIRCUIT.

Figure 1–43 Transistor Operation with Base Circuit Open. (Courtesy of General Motors Corporation)

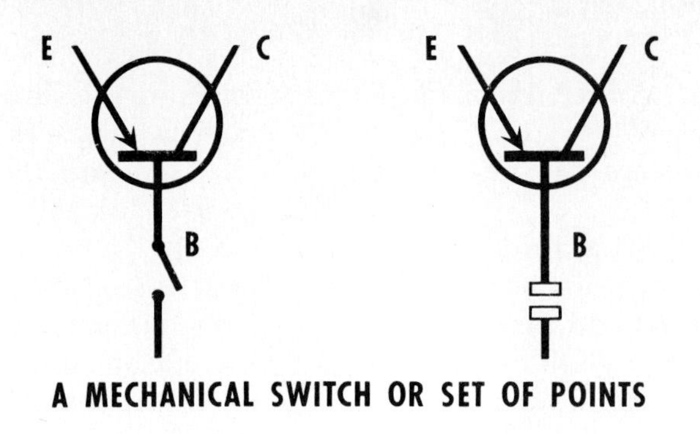

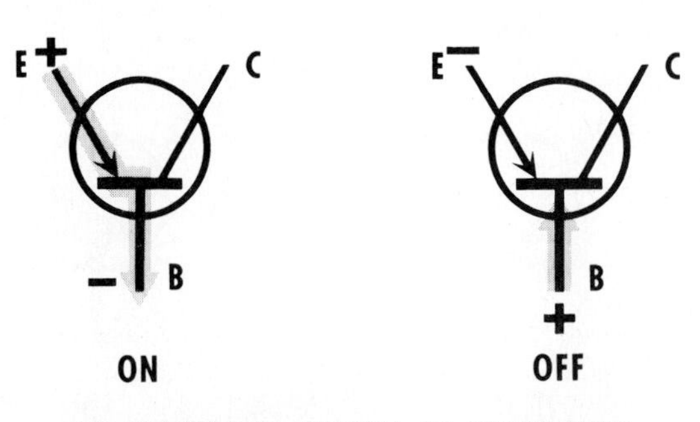

Figure 1–44 Methods of Switching a Transistor On and Off. (Courtesy of General Motors Corporation)

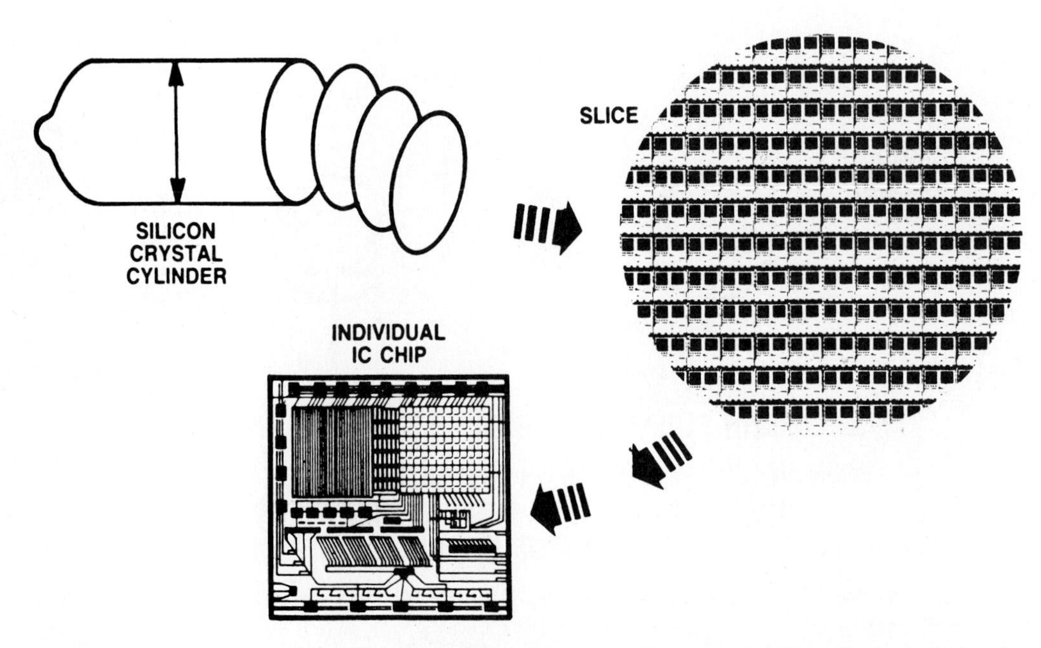

Figure 1–45 Integrated Circuit Chip Manufacturing. (Courtesy of Chrysler Canada Ltd.)

Integrated Circuits

Integrated circuits (ICs) are silicon chips that contain hundreds, or thousands, of solid-state components. Silicon crystal slices are used to manufacture ICs. A photographic printing and diffusion process is used to construct the circuits on individual chips in the slice, then the slice is cut into individual chips. One chip may contain hundreds of resistors, diodes, transistors, capacitors, and interconnecting circuits. The IC chip is a basic building block for many electronic voltage regulators, ignition modules, and computers. The IC chip manufacturing process is shown in Figure 1–45.

A list of electric and electronic symbols is provided in Figure 1–46.

+	POSITIVE
−	NEGATIVE
	GROUND
	FUSE
	CIRCUIT BREAKER
	CAPACITOR
Ω	OHMS
	RESISTOR
	VARIABLE RESISTOR
	SERIES RESISTOR
	COIL
	STEP UP COIL

	CONNECTOR
	MALE CONNECTOR
	FEMALE CONNECTOR
YYY	MULTIPLE CONNECTOR
	DENOTES WIRE CONTINUES ELSEWHERE
	SPLICE
J2 2	SPLICE IDENTIFICATION
	OPTIONAL WIRING WITH / WIRING WITHOUT
	THERMAL ELEMENT (BI-METAL STRIP)
	"Y" WINDINGS
88:88	DIGITAL READOUT
	SINGLE FILAMENT LAMP

	OPEN CONTACT
	CLOSED CONTACT
	CLOSED SWITCH
	OPEN SWITCH
	CLOSED GANGED SWITCH
	OPEN GANGED SWITCH
	TWO POLE SINGLE THROW SWITCH
	PRESSURE SWITCH
	SOLENOID SWITCH
	MERCURY SWITCH
	DIODE OR RECTIFIER
	BY-DIRECTIONAL ZENER DIODE

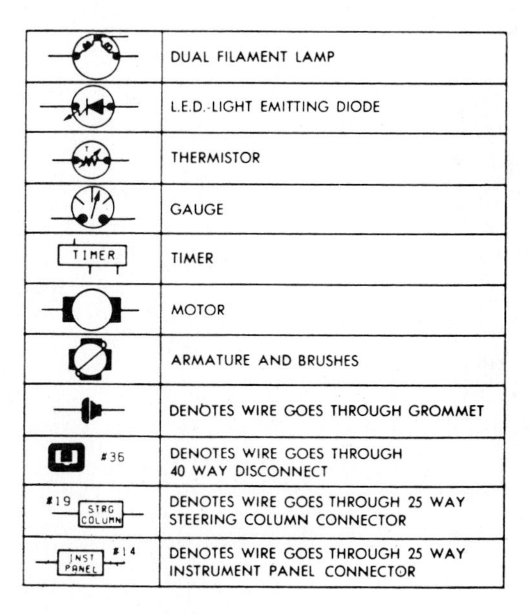

	DUAL FILAMENT LAMP
	L.E.D.-LIGHT EMITTING DIODE
	THERMISTOR
	GAUGE
TIMER	TIMER
	MOTOR
	ARMATURE AND BRUSHES
	DENOTES WIRE GOES THROUGH GROMMET
#36	DENOTES WIRE GOES THROUGH 40 WAY DISCONNECT
#19 STRG COLUMN	DENOTES WIRE GOES THROUGH 25 WAY STEERING COLUMN CONNECTOR
INST PANEL #14	DENOTES WIRE GOES THROUGH 25 WAY INSTRUMENT PANEL CONNECTOR

Figure 1–46 Electric and Electronic Symbols (Courtesy of Chrysler Canada Ltd.)

Test Questions

1. An atom is the smallest particle of an ________ .
2. An element with five valence electrons on each atom will be a good conductor. T F
3. The three requirements for electron movement are:
   (a) ________
   (b) ________
   (c) ________
4. Electrical pressure is measured in ________ .
5. When the voltage in a circuit is decreased, the current flow will ________ .
6. If the resistance in a circuit is increased, the current flow will ________ .
7. A shorted circuit may be defined as an unwanted copper-to- ________ connection.
8. When phosphorus is combined with a silicon crystal, a ________ type of material is formed.
9. The current flow in the collector circuit of a transistor is ________ than the current flow in the base circuit.
10. When the emitter-base circuit of a transistor is connected to a voltage source with a reverse bias connection, there will be current flow in the emitter-collector circuit. T F
11. Calculate the following values from the series circuit in Figure 1–47.
    (a) total resistance
    (b) current flow
    (c) voltage drop across R1
    (d) voltage drop across R2
    (e) voltage drop across R3

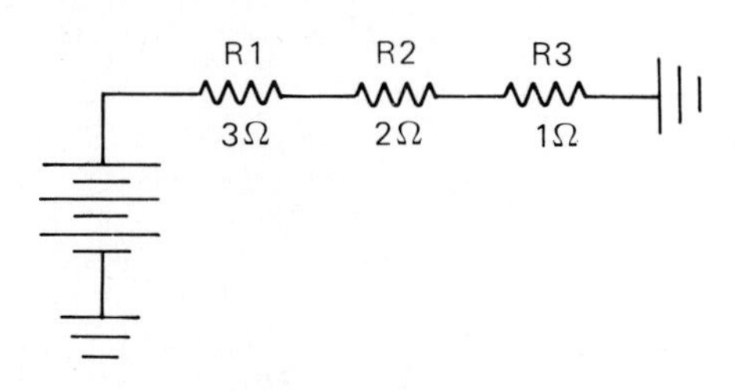

Figure 1–47

12. Calculate the following values from the parallel circuit in Figure 1–48.
    (a) current flow in R1
    (b) current flow in R2
    (c) current flow in R3
    (d) current flow in R4
    (e) total current
    (f) total resistance

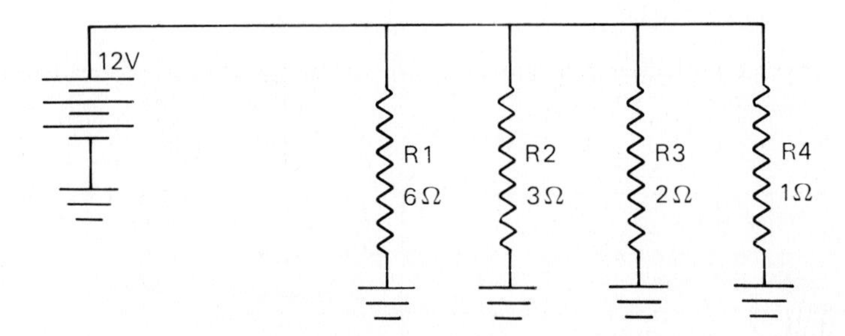

Figure 1–48

2

Automotive Computers

Practical Completion Objectives

1. Demonstrate the ability to explain basic computer signals.
2. Demonstrate an understanding of basic computer operation.
3. Demonstrate a knowledge of various data-transmission methods.

Computers

Voltage Signals

An analog signal is continuously variable. This type of signal can be any voltage within a certain range. For example, a throttle-position sensor (TPS) provides a voltage signal that varies continuously between 0V and 5V in relation to throttle opening. (This sensor is explained in Chapter 3.)

Digital signals can only be represented by specific voltages within a specific range. For example, 1V, 2V, or 3V would be allowed in a digital signal, but 1.73V or 2.66V would not be allowed. Digital signals are used when information refers to two conditions such as on and off or high and low. This type of signal is referred to as a "digital binary signal," which is limited to two voltage levels. One of these voltage levels is positive voltage, and the other level is a zero level. The digital binary signal is a square-wave signal. An analog signal is compared to a digital binary signal in Figure 2–1.

The computer uses digital signals in a code that contains ones and zeros. The number one represents the high voltage of the digital signal, and zero represents zero volts. Each zero and each one is referred to as a "bit" of information, and eight bits are called a "word." A combination of eight binary digital signals are contained in each word, as illustrated in Figure 2–2.

Digital binary codes are used inside a computer and also between the computer and other electronic components. Thousands of bits are used by a computer to store an infinite variety of information and

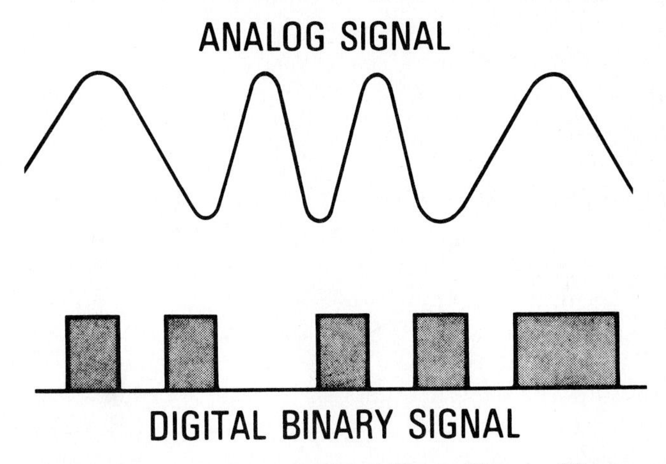

Figure 2–1 Analog and Digital Binary Signals. (Courtesy of General Motors Corporation)

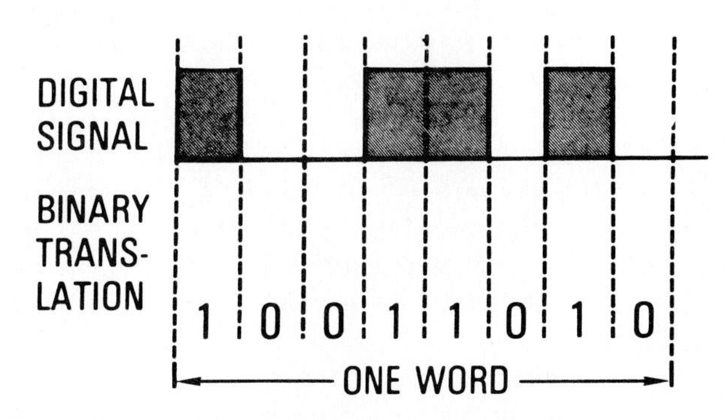

Figure 2–2 Eight Digital Binary Codes in a Word. (Courtesy of General Motors Corporation)

to communicate effectively with other system components. For example, if the air conditioner is turned on, a 10100110 may be received by the automotive computer. In response to this signal, the computer will provide a slight increase in idle speed.

Computer Programs

A computer receives information, makes decisions regarding this information, and takes some action as a result of the decisions. When it performs these functions, the computer follows a detailed list of instructions called a "program." Programs break down each computer task into its most basic parts. A program may contain hundreds of steps depending on the complexity of the computer functions.

Microprocessor

The microprocessor is the brain of the small computer. This brain does calculations and makes decisions. The other components in the computer support the microprocessor. A microprocessor cannot store information; therefore, a computer requires storage devices called "memories."

Memories

Information that is stored permanently in a computer is placed in a read-only memory (ROM). The programs that control the microprocessor are stored in the ROM. Instructions can be read from the ROM by the microprocessor, but it cannot write any new information into the ROM.

Random-access memories (RAMs) are often used for temporary storage. The microprocessor can write information into the RAM and also read information out of the RAM. If the microprocessor receives information it requires to make several different decisions, the measurement will be written into the RAM and then read out each time it is needed. A RAM can have either a volatile or nonvolatile memory. A volatile memory is erased when power is switched off, whereas a nonvolatile memory holds its information when power is removed. A RAM with a volatile memory is used in many automotive computers. This type of memory is connected to the battery at all times. However, if the battery becomes completely discharged, or if the cables are removed, the RAM will lose its information.

Many automotive computers contain a programmable read-only memory (PROM). This type of

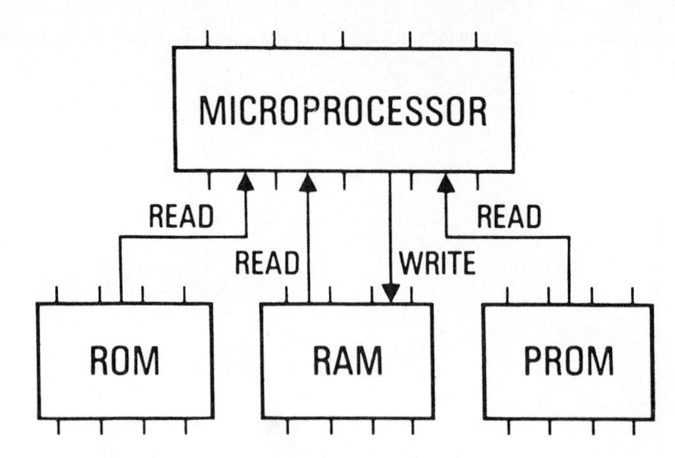

Figure 2-3 Microprocessor with Memories. (Courtesy of General Motors Corporation)

memory is removable from the computer, whereas ROMs and RAMs are installed permanently. The microprocessor can read information from the PROM, but it cannot write information into this memory. The PROM contains information that varies in each model of car. New information can be added to the PROM without changing the entire original program. (Refer to Chapter 3 for an explanation of computers with a PROM.) A microprocessor with different memories is shown in Figure 2-3.

Computer Clocks

Thus far we have seen that a computer contains a microprocessor that communicates with several memories. The language used for communication is a digital binary code, which contains long strings of ones and zeros. A computer must use some method to determine when one message ends and the next information begins. For example, the computer must be able to distinguish a 01 signal from a 0011 input, as indicated in Figure 2-4.

Computers contain clock generators that provide constant steady pulses that are one bit in length. The memories watch these clock pulses when they are sending or receiving information. Therefore, the memories and the microprocessor know how long each bit is supposed to last. In this way the computer can distinguish between a 01 and a 0011 signal. A microprocessor with a clock and various memories is shown in Figure 2-5.

Interfaces

A computer requires interfaces to handle the incoming and outgoing information. These interfaces pro-

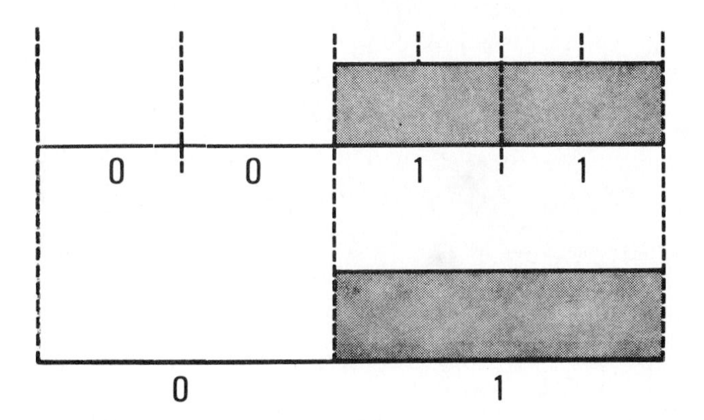

Figure 2-4 Different Digital Binary Code Messages. (Courtesy of General Motors Corporation)

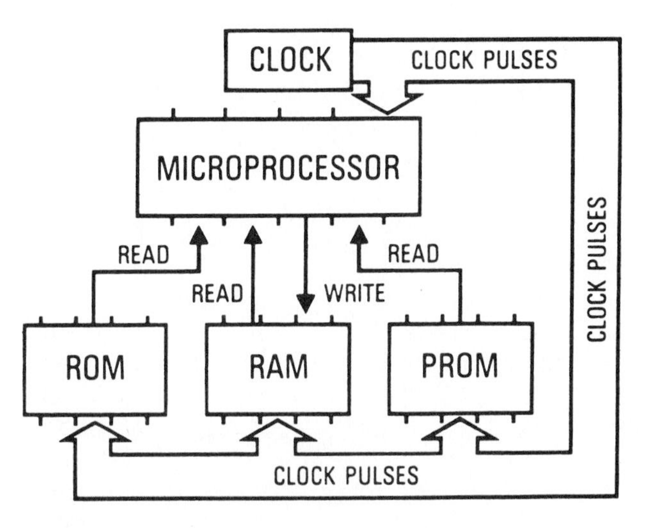

Figure 2-5 Microprocessor with Memories and Clock. (Courtesy of General Motors Corporation)

tect the delicate electronics in the computer from high induced voltages in electrical circuits. The interfaces also translate input and output signals. When the computer receives an analog signal, the interface converts this signal to a digital binary code for the microprocessor. If the computer controls output devices with a digital signal, the output interface translates the computer language into digital output signals. The computer must know the operating conditions of the system that it is controlling. Various sensors usually contain thermistors or potentiometers. A thermistor is a temperature-sensitive resistor, and a potentiometer is a variable resistor. The throttle-position sensor contains a potentiometer that is connected to the carburetor throttle-shaft. This potentiometer sends an analog signal to the computer in relation to throttle opening. Many computer systems have a thermistor in the coolant-temperature sensor that sends an analog signal to the computer in relation to engine coolant temperature. The computer sends a reference voltage to many of the sensors, and the sensors transmit an analog voltage signal back to the computer. (Refer to Chapters 3 and 4 for a complete explanation of automotive computer systems.) A complete computer with interfaces is illustrated in Figure 2–6.

Types of Computer Data

Serial Data

Many automotive computer systems now have more than one computer. These computers use various methods to transmit data to each other. Wires connected between computers for data transmission are referred to as ''data links.'' When the serial-data-transmission method is used between two computers, all the data words are sent one after the other. The computers are synchronized to the data being sent and clocked together. Serial data transmission is shown in Figure 2–7.

Duplex Serial Data

If duplex serial data transmission is used between two computers, a single two-way data link is used. Serial data can be sent or received by either computer. An external clock line between the two computers controls the sending and receiving functions of each computer. When the clock pulse is high, computer 1 is programmed to send data while computer 2 listens to, or receives, the data. During low clock pulses, computer 2 transmits data and computer 1 receives the information. Duplex serial data lines are used on some cars between the body computer module (BCM) and the electronic climate control head. (Refer to Chapter 6 for an explanation of the BCM system.) Duplex serial data transmission is shown in Figure 2–8.

Peripheral Serial Bus Data

In a peripheral serial bus data-link system, dual data links are used between two computers. One of these data links is always used to send data from computer

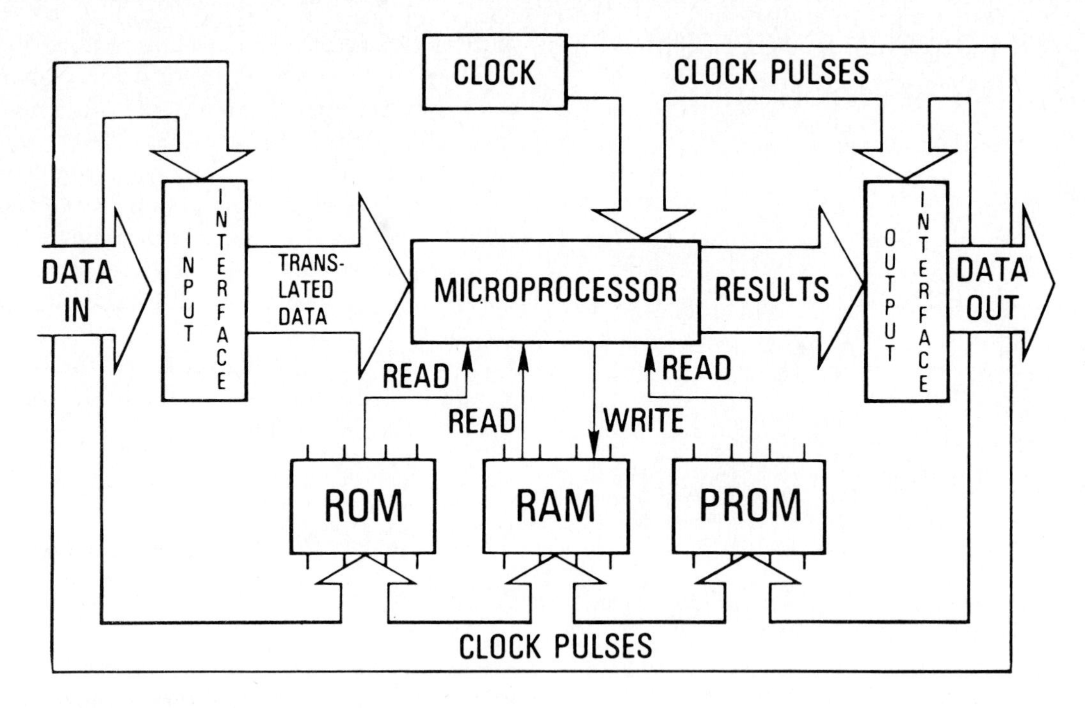

Figure 2-6 Computer with Memories, Clock, and Interfaces. (Courtesy of General Motors Corporation)

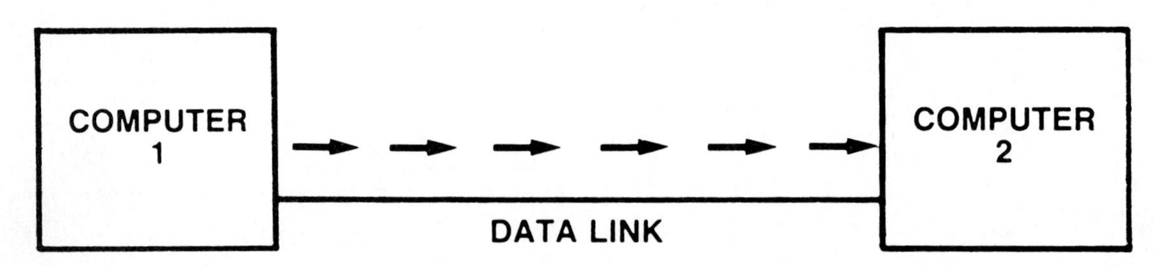

Figure 2-7 Serial Data Transmission. (Courtesy of General Motors Corporation)

1 to computer 2, whereas the other data link transmits data from computer 2 to computer 1 as indicated in Figure 2–9.

Peripheral serial bus data links are used between the body control module (BCM) and the electronic control module (ECM) on some cars. (Refer to Chapter 14 for an explanation of multiplex wiring systems with data links.)

Synchronous Data

Synchronous data systems have a constant data flow with regular clock pulses. In these systems, ''sync'' pulses occur at regular clock intervals. These sync pulses signal the computer that a new piece of data is starting. Synchronous data transmission is pictured in Figure 2–10.

Asynchronous Data

In some computer systems, data is transmitted only when it is required rather than being sent continuously. This type of data transmission is referred to as ''asynchronous data.'' A start pulse informs the computer prior to data transmission. A stop pulse informs the computer that data transmission is completed. The computer may use a parity check to ensure that the total word, or words, have been sent correctly. When the stop pulse is received, data transmission stops until the next start pulse signal. The asynchronous data transmission method is shown in Figure 2-11.

Parallel Data

When the parallel data-transmission method is used, separate data lines are connected from several switches to the computer. These switches may be turned on and off simultaneously to send specific data

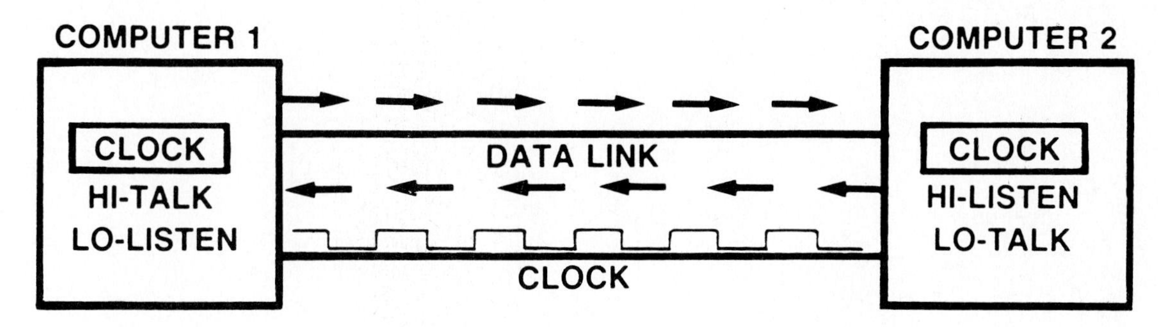

Figure 2-8 Duplex Serial Data Transmission. [Courtesy of General Motors Corporation]

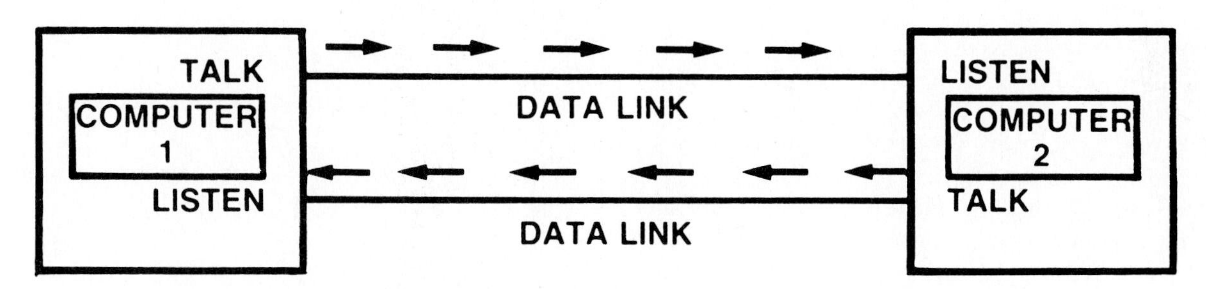

Figure 2-9 Peripheral Serial Bus Data Link. [Courtesy of General Motors Corporation]

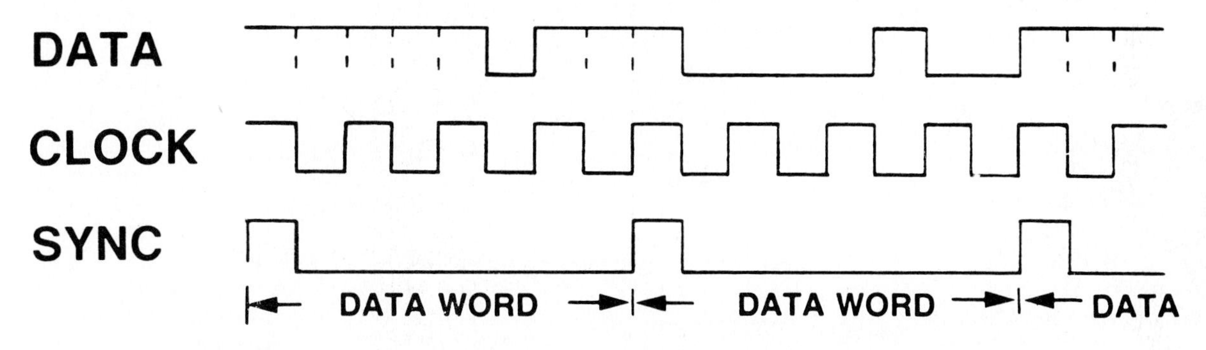

Figure 2-10 Synchronous Data. [Courtesy of General Motors Corporation]

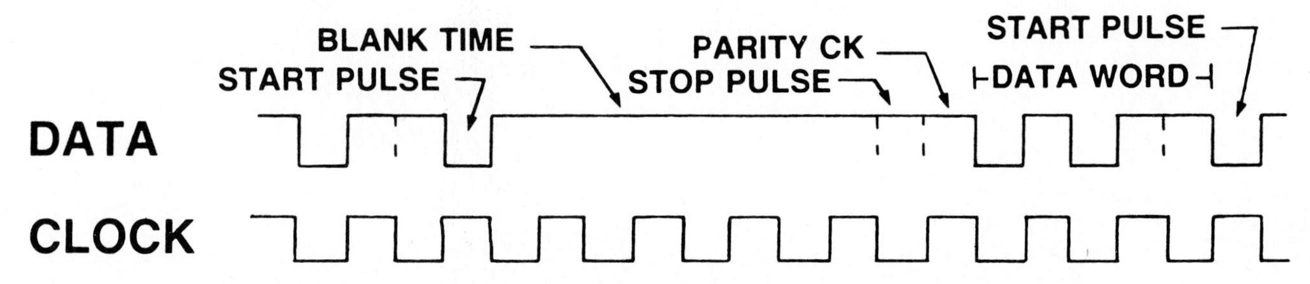

Figure 2-11 Asynchronous Data. [Courtesy of General Motors Corporation]

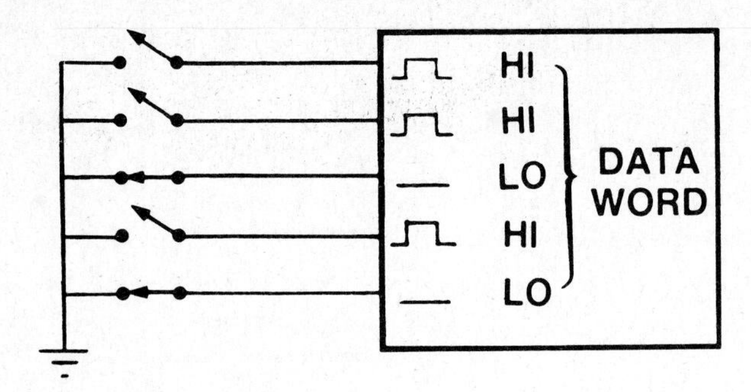

Figure 2-12 Parallel Data. (Courtesy of General Motors Corporation)

signals to the computer. Parallel data transmission is pictured in Figure 2–12.

Voice Synthesizers

Operation

Voice synthesizers are used in voice-alert systems that audibly warn the vehicle operator regarding specific abnormal conditions. The voice synthesizer is a microprocessor, or computer, that stores electronic voice signals in its memory. Input signals to the computer indicate abnormal conditions in each monitored system. When one of these signals is received, the computer selects the correct response from its memory. In some systems, the electronic voice signals from the computer are converted into an audible message through the radio speaker nearest the driver. (Voice-alert systems are discussed in Chapter 9.) The operation of a voice synthesizer is shown in Figure 2–13.

Test Questions

1. An analog signal is continuously variable.
   T F
2. A microprocessor can write information into a random-access memory. T F
3. In computer language a word represents ______ ______ bits of information.
4. Computers use a clock generator to recognize different input signals. T F
5. An analog signal is changed to a digital signal in the ______________ of a computer.
6. Asynchronous data is transmitted continuously.
   T F
7. In a peripheral serial bus data-link system, data may be sent through the data links in both directions. T F

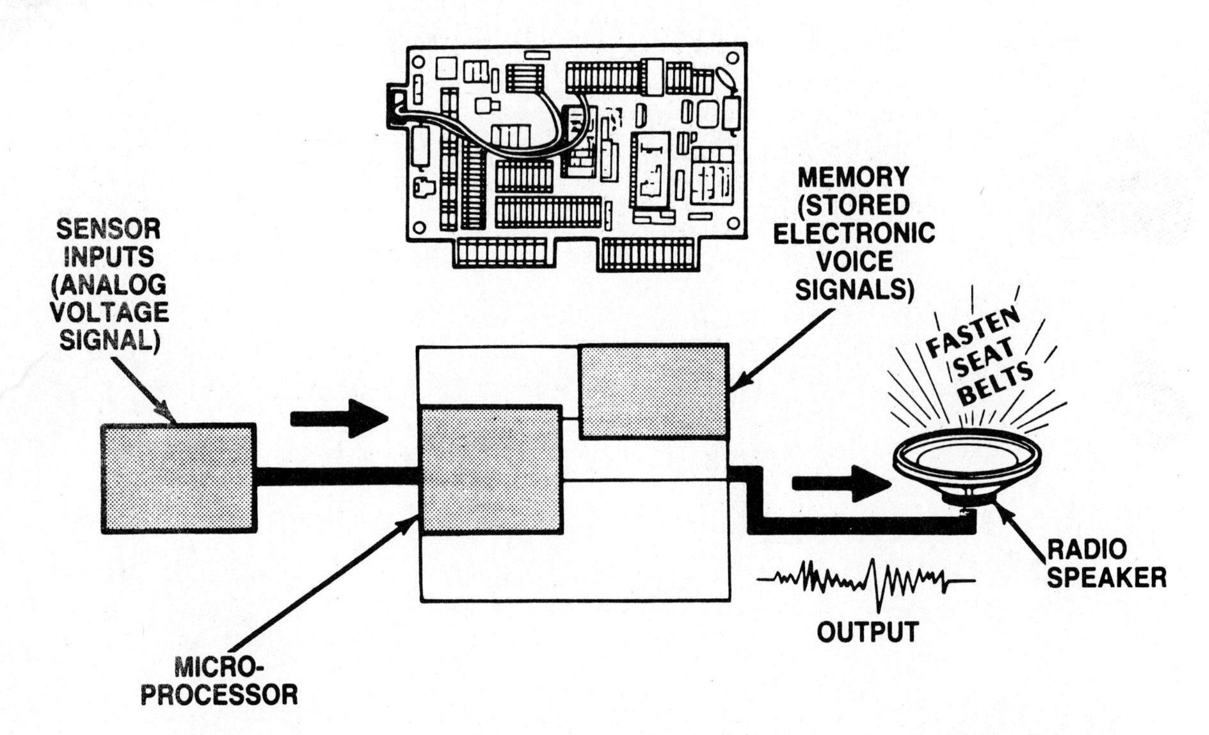

Figure 2–13 Voice Synthesizer Operation. (Courtesy of Chrysler Canada Ltd.)

3

Computer-Controlled Carburetor Systems

Practical Completion Objectives

1. Diagnose computer command control (3C) systems using the self-diagnostics in the system.
2. Perform a system performance test on a 3C system.
3. Diagnose Chrysler oxygen-feedback systems.

General Motors Computer Command Control (3C) System

Electronic Control Module (ECM)

The computer used with the 3C system is referred to as an electronic control module (ECM). A removable programmable read-only memory (PROM) is located in the ECM, as illustrated in Figure 3–1.

The ECM is the control center of the 3C system. It constantly monitors information it receives from the various sensors and provides commands in order to perform the correct functions. Specific information about each vehicle is programmed into the PROM so the ECM can be tailored to meet the requirements of the specific engine, transmission, and differential combination of a particular vehicle. The ECM receives signals from the following devices:

1. Oxygen (O_2) sensor.
2. Throttle-position sensor (TPS).
3. Coolant temperature sensor.
4. Manifold absolute pressure (MAP) and barometric pressure (BP) sensors.
5. Vehicle-speed sensor (VSS).
6. Distributor pickup coil, crankshaft position, revolutions per minute (rpm).

The ECM, in turn, controls the following:

1. Mixture-control (MC) solenoid.
2. Electronic spark-timing (EST) system.
3. Exhaust-gas recirculation (EGR) valve.
4. Air injection reactor (AIR) pump.
5. Torque-converter clutch (TCC) lockup.
6. Early fuel evaporation (EFE) system.
7. Idle speed control (ISC) motor.

Sources of ECM Input

Oxygen (O_2) Sensor An exhaust gas oxygen (O_2) sensor is located in one of the exhaust manifolds. Rich air–fuel ratios supply fuel to mix with all the oxygen entering the engine. Rich air–fuel ratios create low oxygen levels in the exhaust because excess fuel mixes with all the oxygen entering the engine. When lean mixtures are used, oxygen levels in the

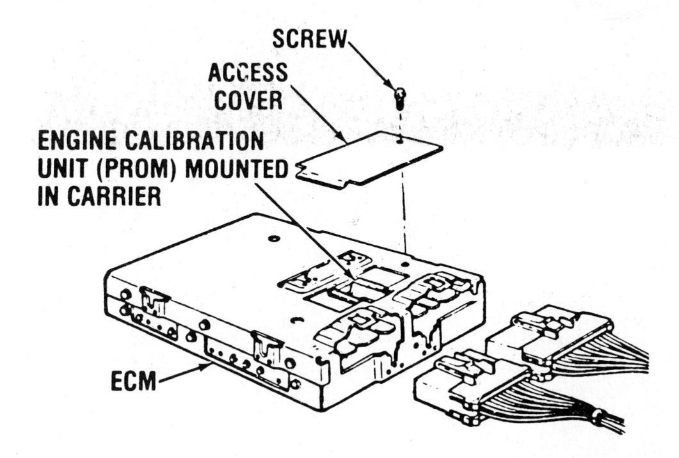

Figure 3–1 Electronic Control Module (ECM). (Courtesy of General Motors Corporation)

exhaust are high because of the lack of fuel entering the cylinders. Exhaust gas is applied to the outside of the oxygen-sensing element, and atmospheric pressure is supplied to the inside of the sensing element, as shown in Figure 3–2.

Rich mixtures and low levels of oxygen in the exhaust cause the sensing element to generate higher voltage, as shown in Figure 3–3.

High levels of oxygen in the exhaust and lean mixtures result in low oxygen sensor voltage because high oxygen levels are present on both sides of the sensing element. The oxygen sensor signal to the ECM is used to control the air–fuel ratio in the carburetor. Unleaded gasoline must be used to obtain satisfactory oxygen sensor operation.

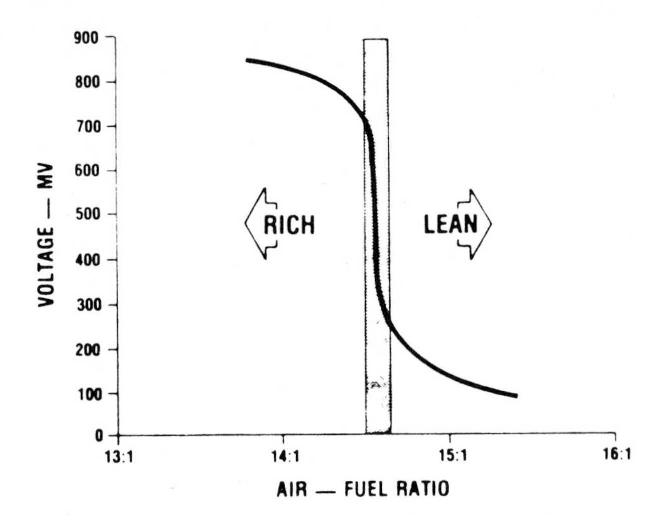

Figure 3–3 Oxygen Sensor Voltage Output. [Courtesy of Sun Electric Corporation]

Throttle-Position Sensor (TPS) The throttle-position sensor is a variable resistor operated by the accelerator pump linkage, as illustrated in Figure 3–4.

A reference voltage is supplied to the TPS from the ECM. The TPS output varies in relation to throttle opening. At wide-open throttle and sudden acceleration, the TPS signal calls for mixture enrichment.

Coolant Temperature Sensor The resistance of the coolant sensor varies in relation to coolant temperature. The ECM operating mode is selected from the coolant sensor signal. Open-loop operation occurs during engine warmup, when a richer mixture is necessary. In the open-loop mode, the ECM maintains the carburetor mixture. Closed-loop operation occurs when the coolant sensor reaches a predetermined temperature. The ECM controls the air–fuel ratio in response to an oxygen sensor signal in the closed-loop mode. The ECM may also use the coolant sensor signal to assist in regulating ignition, air injection, and exhaust-gas recirculation. Figure 3–5 illustrates a coolant temperature sensor.

Manifold Absolute Pressure Sensor A reference voltage is supplied to the manifold absolute pressure (MAP) sensor from the ECM, as illustrated in Figure 3–6.

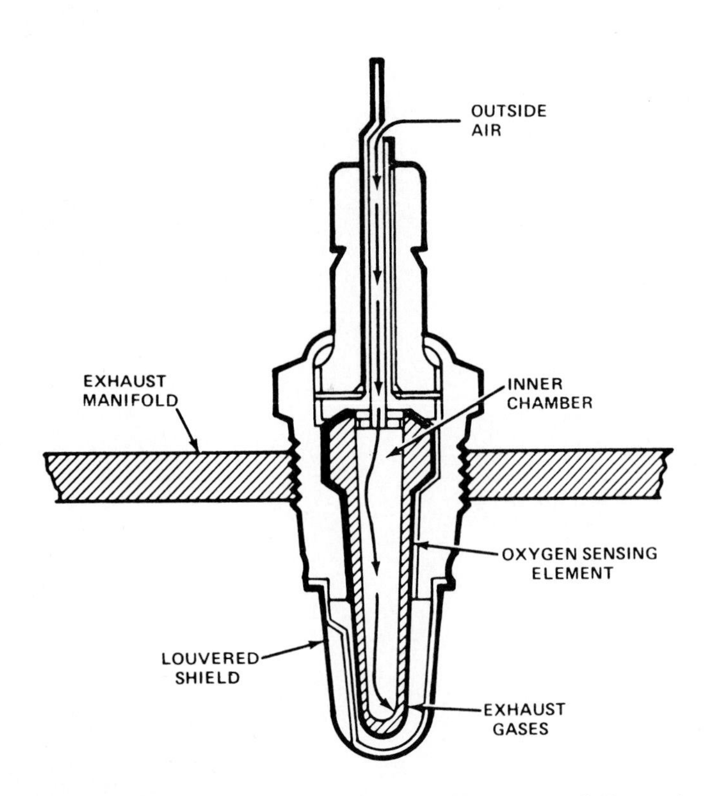

Figure 3–2 Oxygen (O_2) Sensor. [Courtesy of General Motors Corporation]

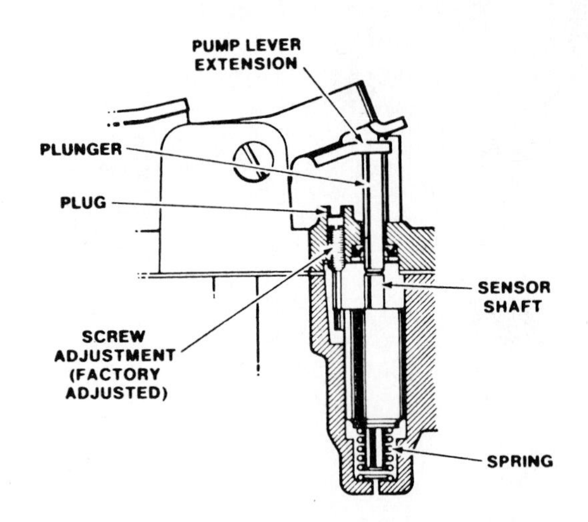

Figure 3–4 Throttle-Position Sensor [TPS]. [Courtesy of General Motors Corporation]

The voltage output signal from the MAP sensor to the ECM varies in relation to manifold vacuum. Spark control is managed by the ECM in response to the MAP sensor signal. Other ECM functions may be affected by the MAP sensor signal. A barometric pressure (BP) sensor may be used with the MAP sensor on some applications.

Vehicle-Speed Sensor (VSS) The speedometer cable rotates a reflector blade past a light-emitting diode and photo cell in the VSS, as shown in Figure 3–7.

Reflector-blade speed and VSS output signal are directly proportional to vehicle speed. The ECM controls torque-converter clutch lockup from the VSS signal.

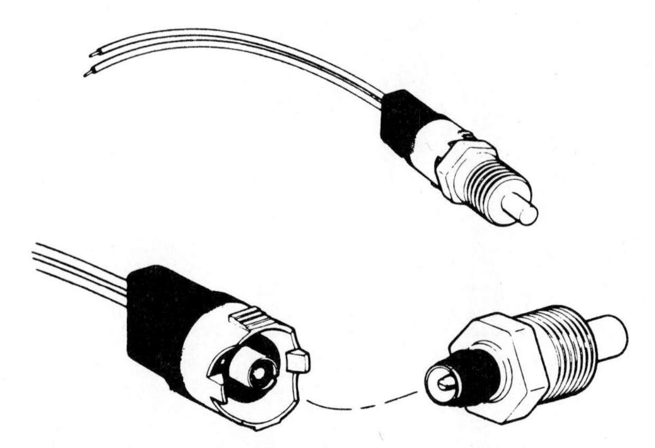

Figure 3–5 Coolant Temperature Sensor. (Courtesy of General Motors Corporation)

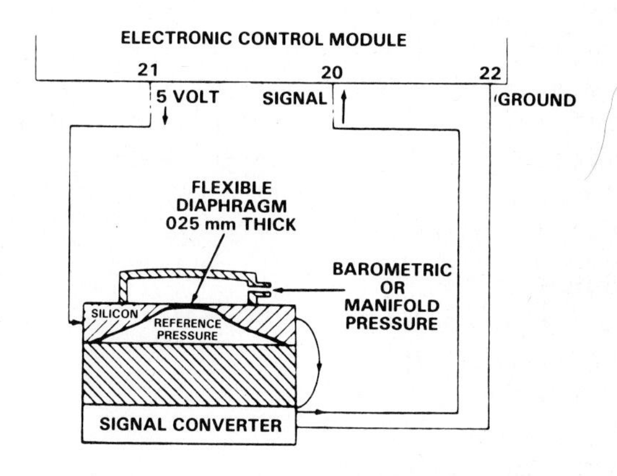

Figure 3–6 Manifold Absolute Pressure (MAP) Sensor. (Courtesy of General Motors Corporation)

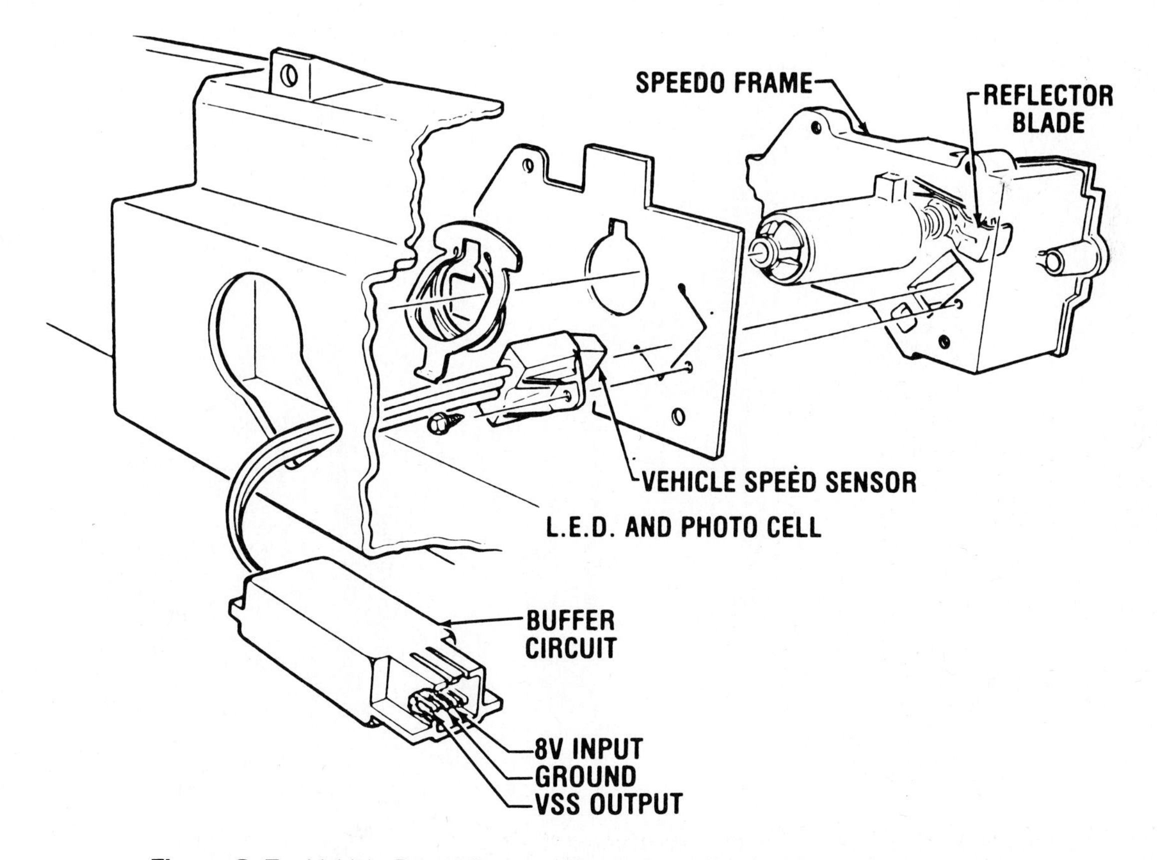

Figure 3–7 Vehicle-Speed Sensor. (Courtesy of General Motors Corporation)

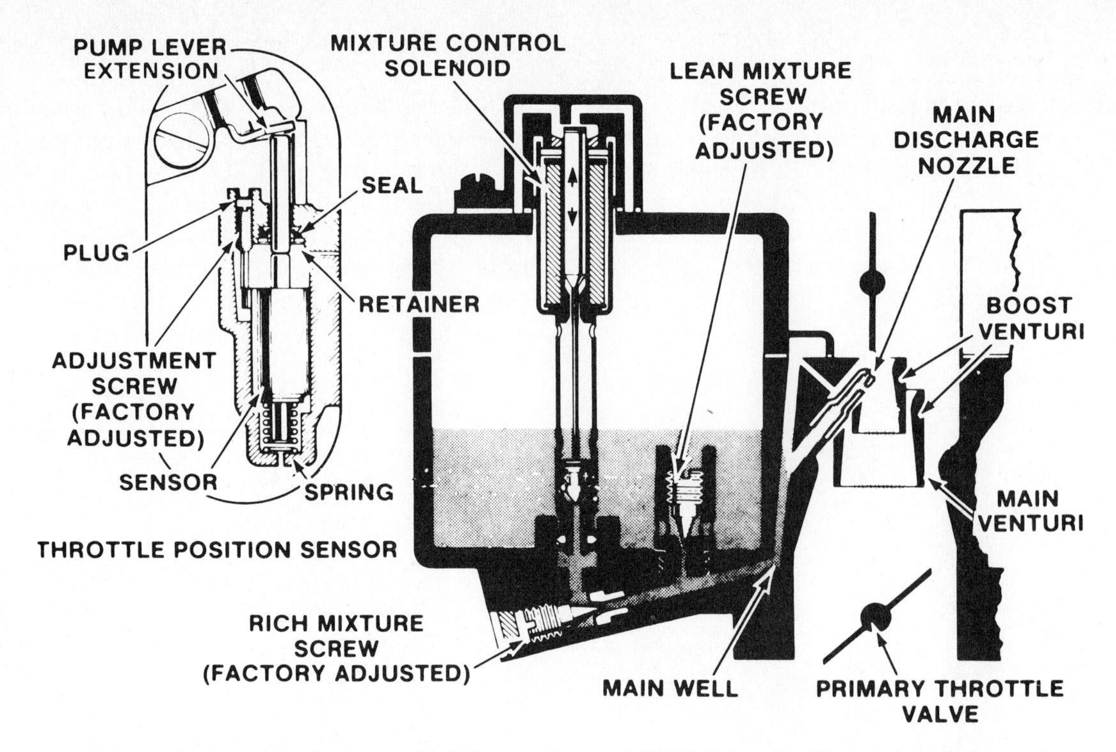

Figure 3-8 Main System with Mixture Control (MC) Solenoid. (Courtesy of General Motors Corporation)

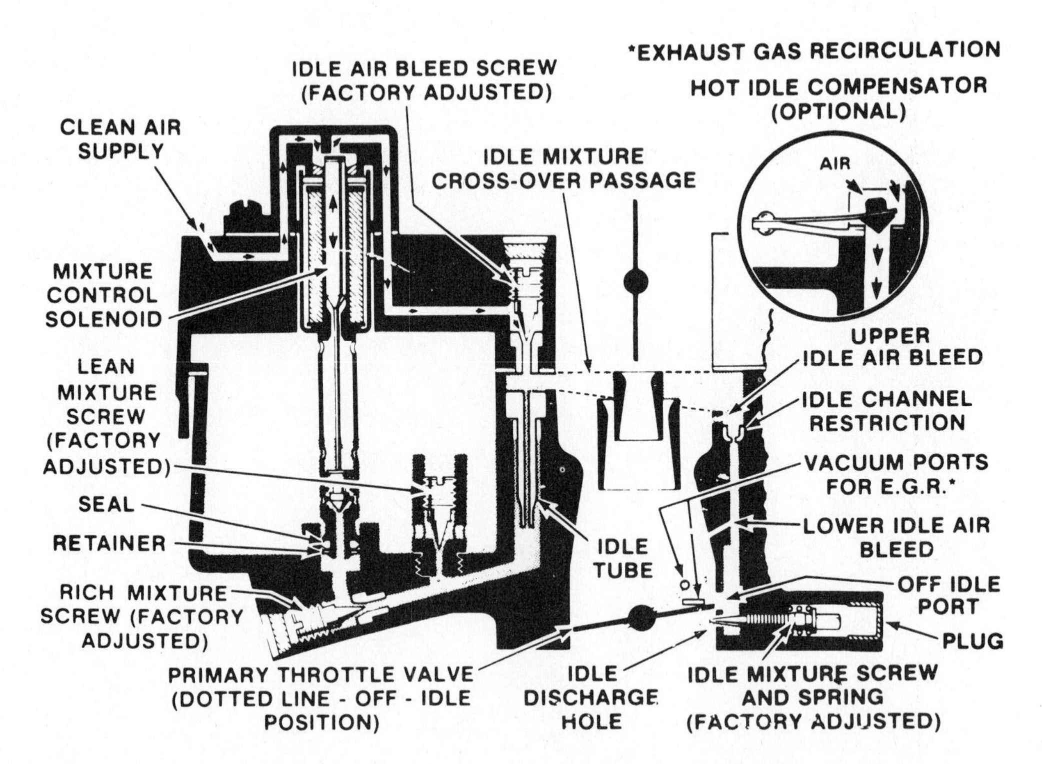

Figure 3-9 Idle System with Mixture Control (MC) Solenoid. (Courtesy of General Motors Corporation)

ECM Control Functions

Mixture Control Solenoid Management The 3C carburetor contains a mixture control (MC) solenoid that is controlled by the ECM. The MC solenoid is spring-loaded in the upward position. When the ECM energizes the solenoid winding, the plunger moves downward. This plunger movement controls a tapered valve in the main system and an air bleed in the idle system. Fuel for the main system is supplied through the lean mixture screw. Some fuel for the main system is also supplied through the tapered valve on the MC solenoid plunger and the rich mixture screw as shown in Figure 3–8.

When the ignition switch is turned on, voltage is supplied to one end of the MC solenoid winding, and the other end of the winding is connected to the ECM. If the oxygen sensor provides a rich signal to the ECM, the ECM will ground the MC solenoid winding. When this occurs, the MC solenoid plunger moves downward and the tapered valve closes, which shuts off the fuel flow through the rich-mixture screw. This action results in a leaner air–fuel mixture. A lean oxygen sensor signal to the ECM causes the ECM to deenergize the MC solenoid winding. This causes upward plunger movement and additional fuel flow through the tapered valve and rich-mixture screw into the main system, and thus a richer air–fuel mixture is provided. The ECM energizes the MC solenoid winding 10 times per second, and the "on-time" of the winding can be varied by the ECM to maintain the air–fuel ratio at 14.7:1 under most operating conditions.

Air is bled into the idle system past the top of the MC solenoid plunger. Upward plunger movement results in a reduced air flow into the idle system and a richer air–fuel mixture. If the solenoid plunger is moved downward by the ECM, additional air will flow into the idle system and a leaner air–fuel mixture results. Therefore, the ECM and the MC solenoid control the air–fuel mixture in the idle system or main system. The MC solenoid and the idle system are illustrated in Figure 3–9.

Most defects in the MC solenoid, oxygen sensor, or the ECM result in upward movement of the MC solenoid plunger, which causes a rich air–fuel mixture and excessive fuel consumption. The MC solenoid shown in Figures 3–8 and 3–9 is from a Varajet carburetor. This type of MC solenoid plunger is not adjustable.

Ignition Management Conventional distributor advances are discontinued in the 3C system, and the ignition spark advance is controlled by the ECM. (Refer to Chapter 5 for a description of 3C ignition systems.)

Torque-Converter Clutch Lockup A lockup type of converter contains a pressure plate with a narrow strip of friction material attached to the front of the plate. This plate is splined to the turbine assembly, as indicated in Figure 3–10.

Converter lockup takes place when oil enters the converter hub and forces the friction disc against the front of the converter. In the lockup mode, the flywheel is connected through the front of the converter to the friction disc, turbine, and transmission input shaft. The converter becomes unlocked when oil is directed through the hollow input shaft and the friction disc is forced away from the front of the converter. Oil flow to the converter is controlled by an apply valve in the transmission. The apply valve is controlled by the TCC solenoid.

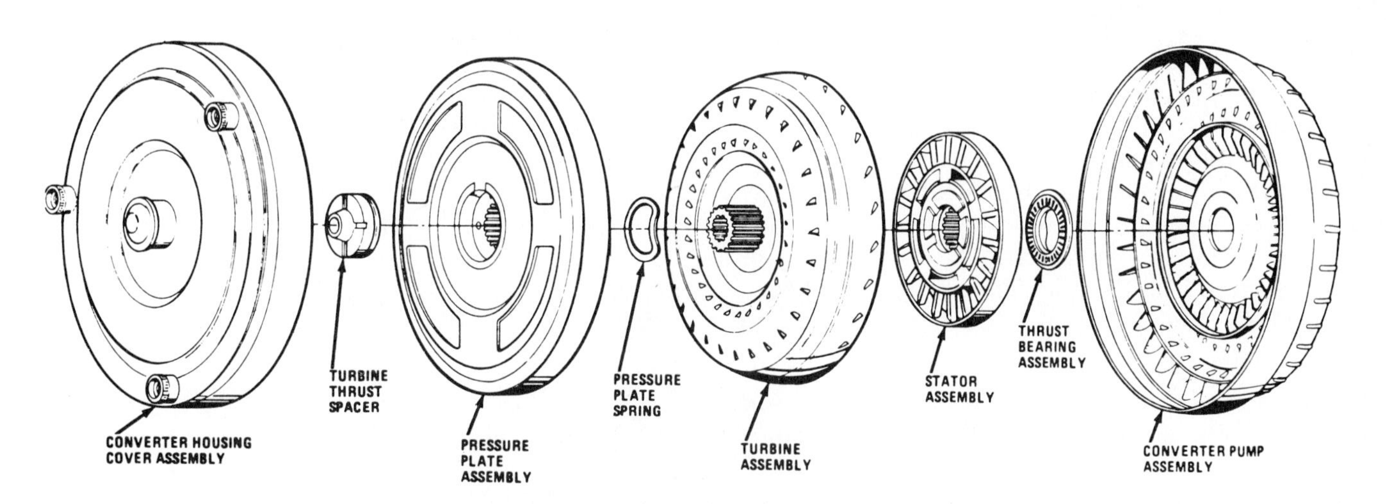

Figure 3–10 Lockup Torque Converter. (Courtesy of General Motors Corporation)

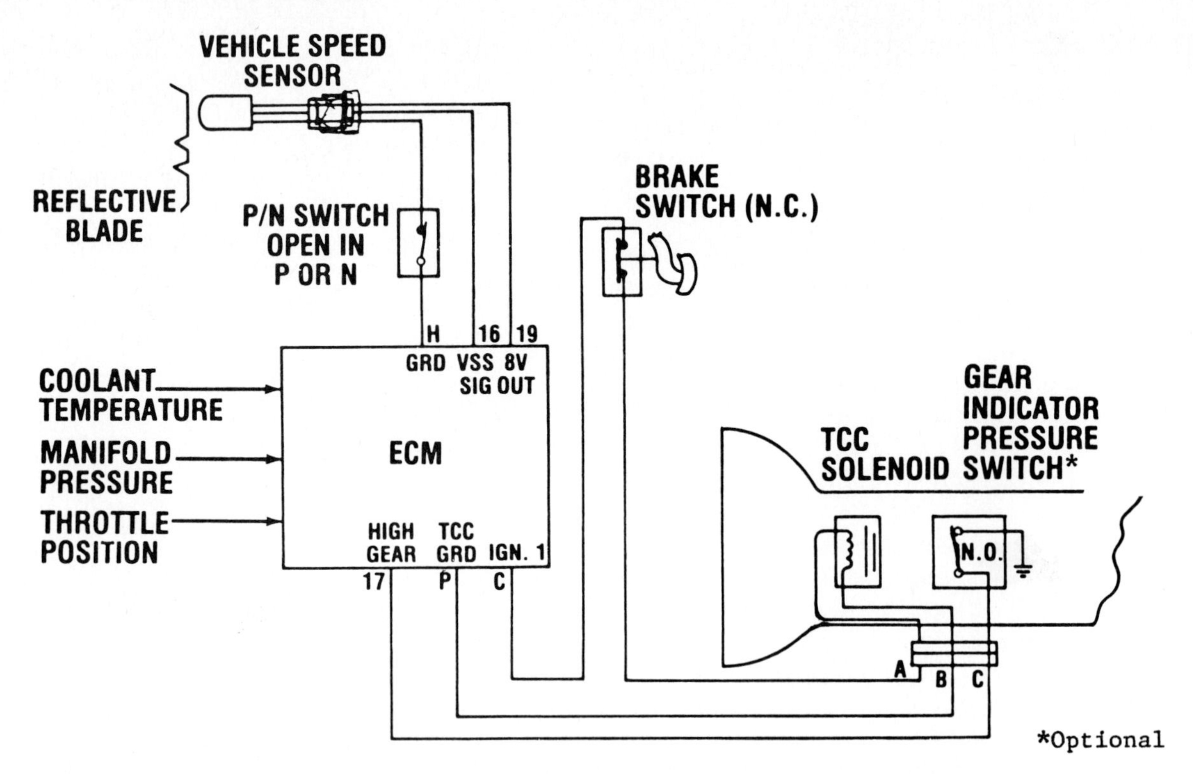

Figure 3–11 Torque-Converter Clutch Circuit. (Courtesy of General Motors Corporation)

When the vehicle reaches a predetermined speed in high gear, the computer grounds the TCC solenoid winding. Once the TCC solenoid is energized, a hydraulic bleed port in the direct-clutch circuit closes, and the direct-clutch oil pressure rises. The increase in oil pressure moves the converter apply valve and directs oil in the converter hub to lock up the converter. Other input signals to the computer may affect TCC lockup, as indicated in Figure 3–11.

Early Fuel Evaporation (EFE) Management The heat-riser valve in the EFE system is operated by a vacuum diaphragm, as shown in Figure 3–12.

A cold coolant sensor signal causes the ECM to energize the EFE solenoid and apply vacuum to the power actuator. If the heat-riser valve is closed, the intake manifold will become heated as exhaust gas is forced through the intake manifold crossover passage. The ECM will deactivate the EFE solenoid at a specific coolant temperature and allow the heat-riser valve to open. The EFE vacuum is also supplied to the air-injection system, as illustrated in Figure 3–13.

Air Injection Reactor (AIR) Management The AIR system supplies air from the air pump to the exhaust ports or to the catalytic converter, as illustrated in Figure 3–14.

Air flow to the exhaust ports lowers emission levels during warmup and reduces oxygen sensor and converter warmup time. Oxygen is necessary for converter operation once the converters have reached operating temperature.

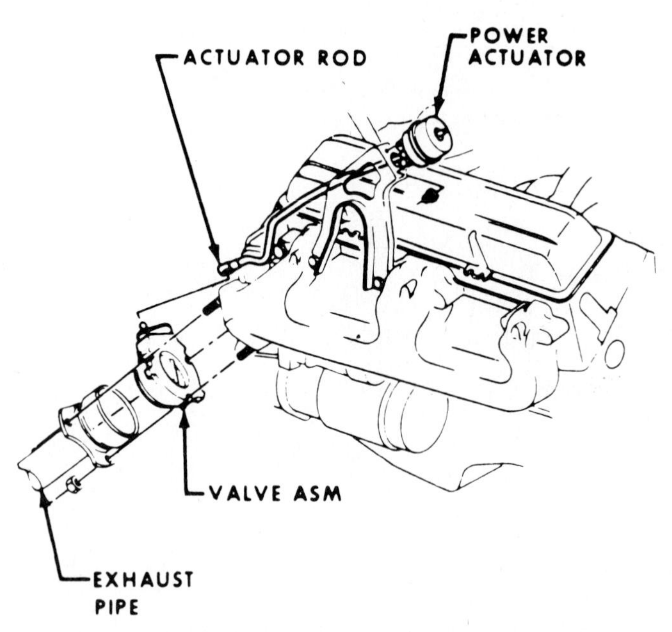

Figure 3–12 Early Fuel Evaporation (EFE) System. (Courtesy of General Motors Corporation)

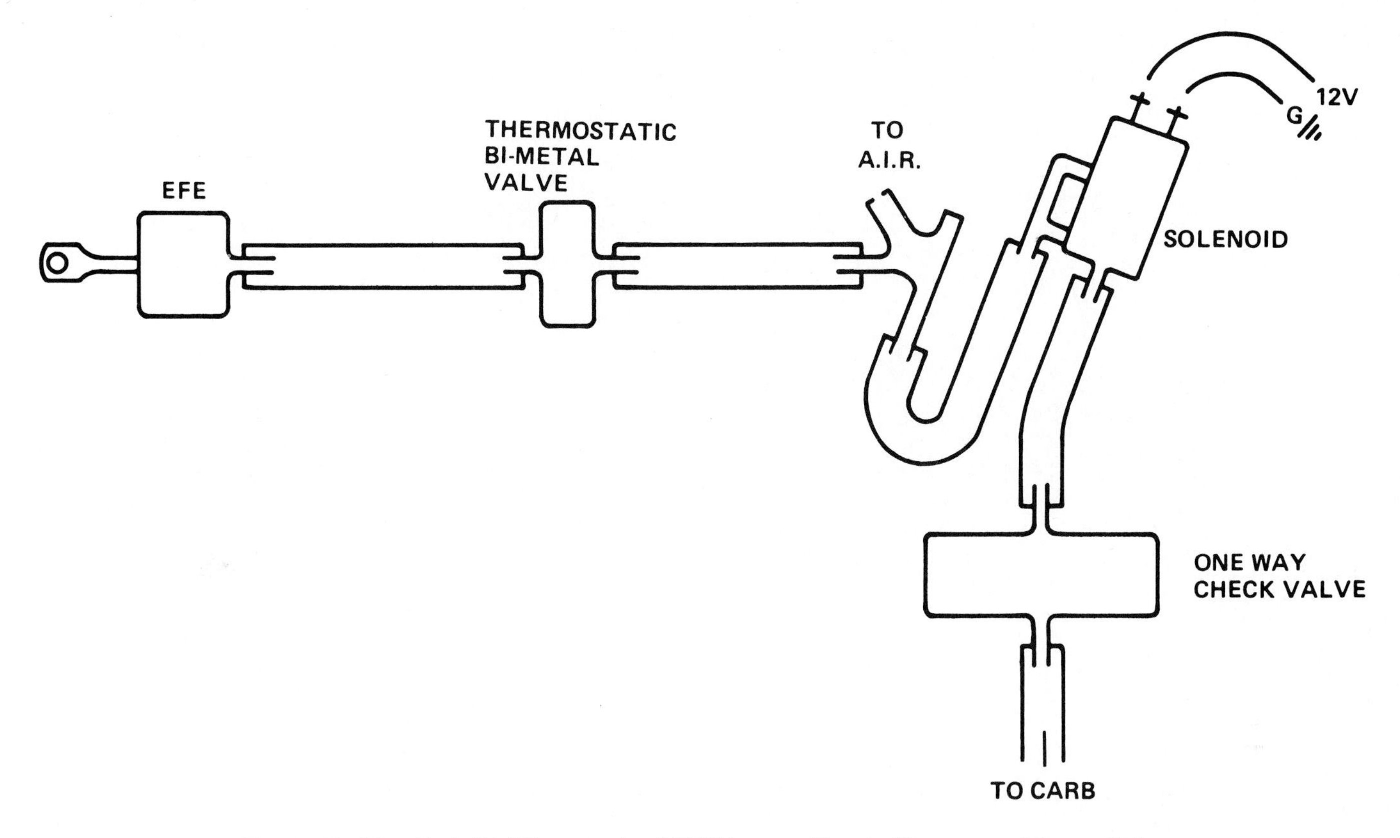

Figure 3–13 Early Fuel Evaporation [EFE] Vacuum Circuit. [Courtesy of General Motors Corporation]

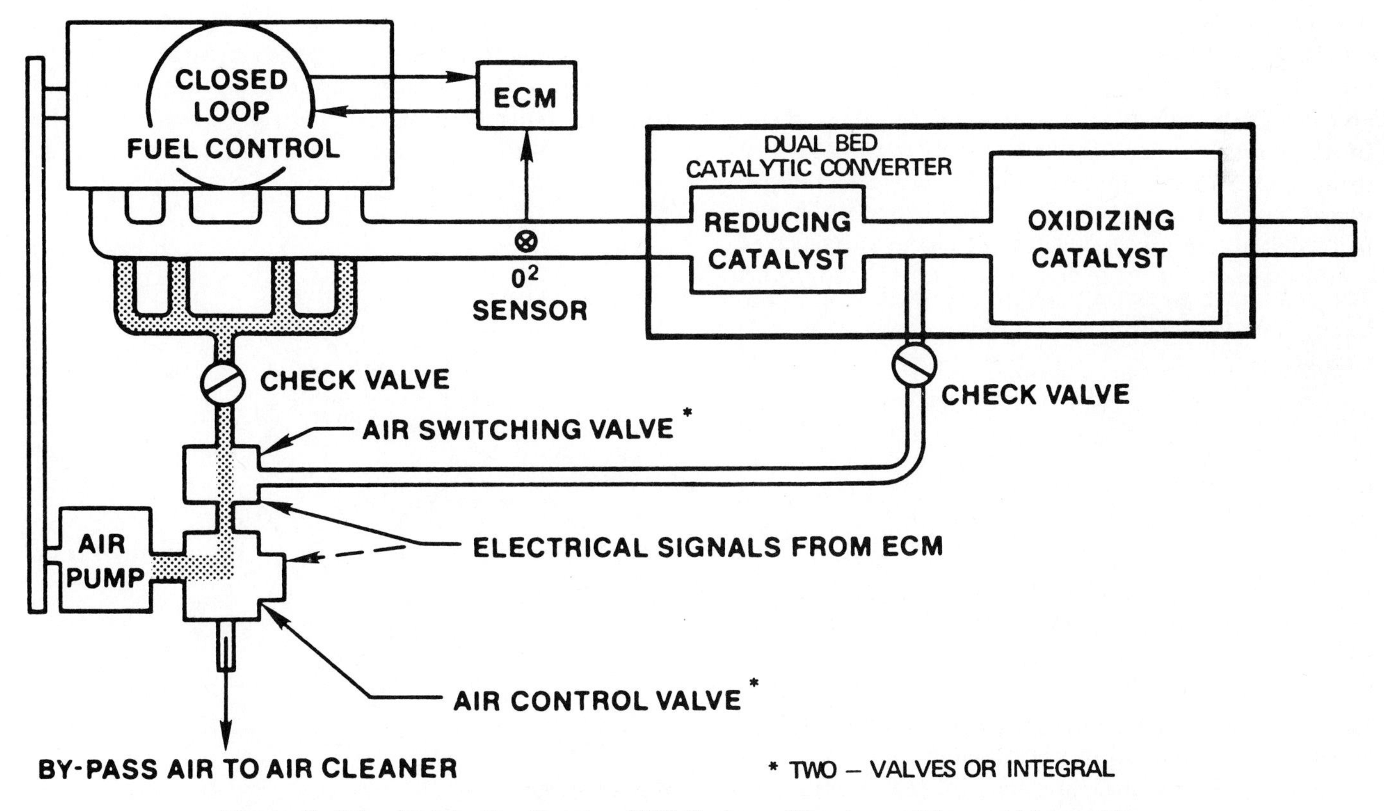

Figure 3–14 Air Injection Reactor [AIR] System. [Courtesy of General Motors Corporation]

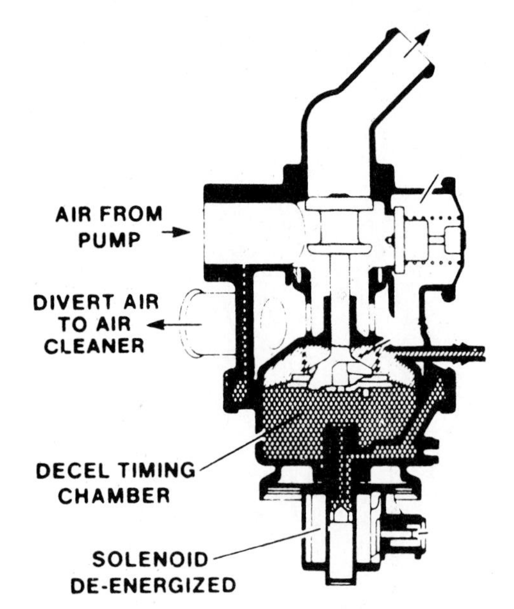

Figure 3–15 Air Injection System Reactor (AIR) Divert Valve. (Courtesy of General Motors Corporation)

A divert valve and a switching valve are used to control air flow from the AIR pump. During engine warmup, the ECM deenergizes the divert valve solenoid winding and enables AIR pump pressure to move the divert valve diaphragm and spool valve upward. Air from the pump is diverted to the air cleaner when the spool valve is in the upward position, as indicated in Figure 3–15.

When the ECM energizes the divert solenoid winding, the solenoid valve shuts off AIR pump pressure applied to the divert valve diaphragm, and the diaphragm and spool valve are able to move to the downward position. As a result, air is directed from the AIR pump to the air-switching valve. High manifold vacuum on deceleration can lift the divert valve diaphragm and cause air from the pump to exhaust to the air cleaner.

When the engine coolant is cold, vacuum from the EFE system is applied to the lower air-switching valve vacuum port and shuts off the flow of air from the AIR pump to the air-switching valve diaphragm. As a result, the diaphragm and spool valve remain in the downward position, and air from the AIR pump is directed to the exhaust ports. Vacuum to the lower air-switching valve port is shut off when the coolant is warm and AIR pump pressure can force the diaphragm and spool valve upward. The flow of air through the air-switching valve will be directed to the converter when the spool valve is in the upward position, as indicated in Figure 3–16.

Manifold vacuum applied to the upper air-switching valve port could lift the diaphragm and spool valve on deceleration with a cold engine. Air will be directed momentarily to the converters to prevent manifold backfiring.

Exhaust Gas Recirculation (EGR) Management The EGR valve directs exhaust from the exhaust system to the intake manifold and lowers nitrous oxide (NOx) emissions. Vacuum to the EGR valve is controlled by a solenoid operated by the ECM. The ECM shuts off EGR vacuum when the coolant is cold by energizing the solenoid. Warm engine coolant conditions signal the ECM to deactivate the solenoid and thus allow vacuum to the EGR valve. The vacuum port connected to the EGR system is always above the throttles. Vacuum should never be applied to the EGR valve until the throttles are opened. An EGR valve that is open at idle speed will cause rough idling. Figure 3–17 illustrates the EGR valve and the vacuum-control circuit.

Idle Speed Control (ISC) Management The idle speed control (ISC) motor is a small reversible motor that is operated by the ECM to control idle speed. When the engine coolant is cold, the fast idle cam will determine the fast idle speed. The ISC motor will extend slightly more than normal when a cold signal is sent from the coolant temperature sensor to the ECM. The air conditioner clutch-switch signals the ECM to maintain engine idle speed when the air conditioning compressor clutch is engaged. If the transmission selector is moved from drive to park, the park/neutral (P/N) switch signals the ECM to reduce idle speed to specifications. The reference signal from the distributor will send a revolutions-per-minute (rpm) signal to the ECM as long as the engine is run-

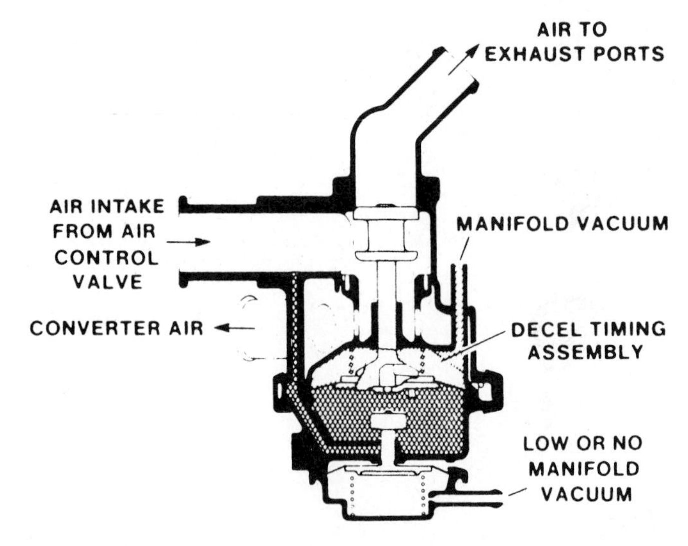

Figure 3–16 Air-Switching Valve. (Courtesy of General Motors Corporation)

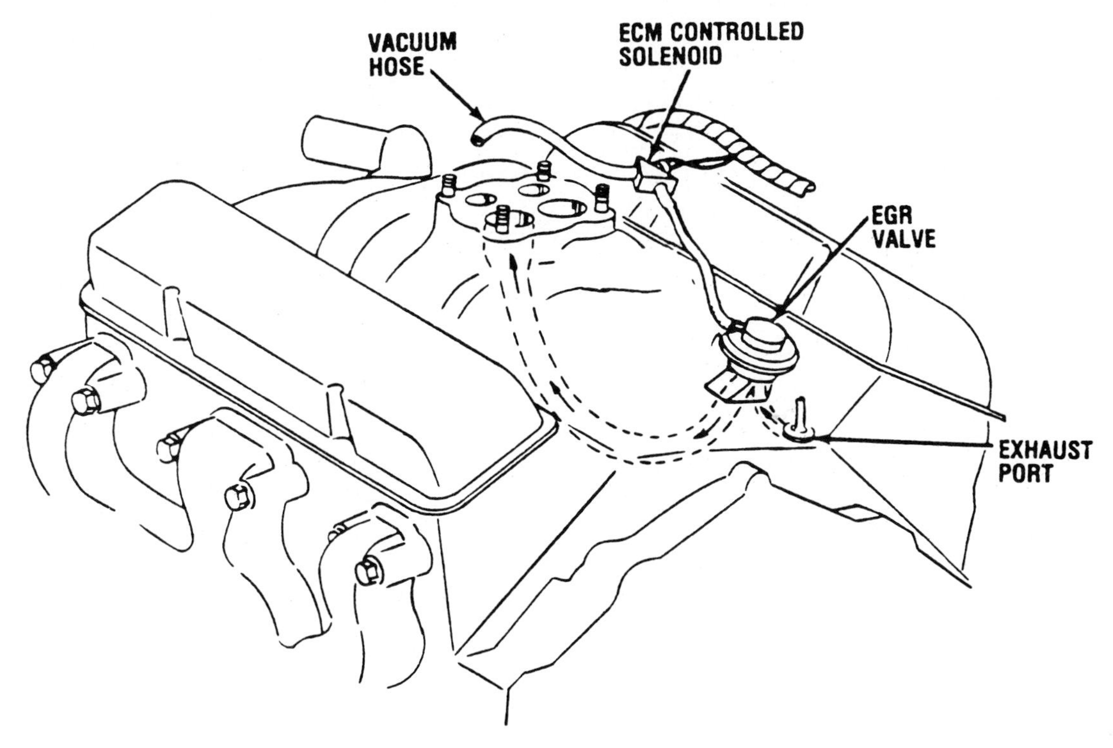

Figure 3–17 Exhaust Gas Recirculation (EGR) Valve. (Courtesy of General Motors Corporation)

ning. A low battery-voltage signal to terminal R on the ECM will cause the ECM to increase idle speed to assist in recharging the battery. When a hot signal is received from the coolant temperature sensor, the ECM will extend the ISC motor plunger and increase the idle speed so the engine temperature will decrease.

The ISC motor does not require adjusting because the ECM determines the correct idle position. A tang on the throttle level contacts the ISC motor plunger and closes a throttle switch in the ISC motor. The throttle switch is connected to the ECM, and the ECM will not control the ISC motor unless the throttle switch is closed. The system must be in the closed-loop mode before the ECM will control the ISC motor. During a cylinder output test, the coolant sensor must be disconnected to put the system in open-loop operation; otherwise the ISC motor will try to correct the drop in speed as each cylinder is shorted out. See Figure 3–18 for the ISC motor and related ECM circuit.

Diagnosis of Computer Command Control (3C) Systems

Self-Diagnosis A check-engine light on the instrument panel is energized by the ECM whenever certain system defects occur. Service personnel can ground a diagnostic test terminal in an assembly line communications link (ALCL) connector under the dashboard, as indicated in Figure 3–19.

Various diagnostic connectors can be used, depending on the year and make of the vehicle. The check-engine light should be on for a few seconds each time the engine is started. Part-throttle operation for 5 minutes is sometimes required before the check-engine light will indicate a system defect.

When the diagnostic test terminal is grounded with the ignition switch on, the check-engine light will begin flashing a code 12 and all other trouble codes that are stored in the computer memory. The codes will be given in numerical order and repeated three times. One flash followed by a pause and two more flashes in quick succession indicates code 12, as shown in Figure 3–20.

The initial code 12 that is provided when the diagnostic test lead is grounded with the ignition switch on indicates that the ECM is capable of diagnosing the system. The ECM of most 3C systems manufactured before 1982 will perform the following functions when the diagnostic test lead is grounded with the ignition switch on:

1. Flash a code 12 and any other codes stored in the computer memory.
2. Send a 30° dwell signal to the mixture control (MC) solenoid.
3. Energize all solenoids controlled by the ECM.

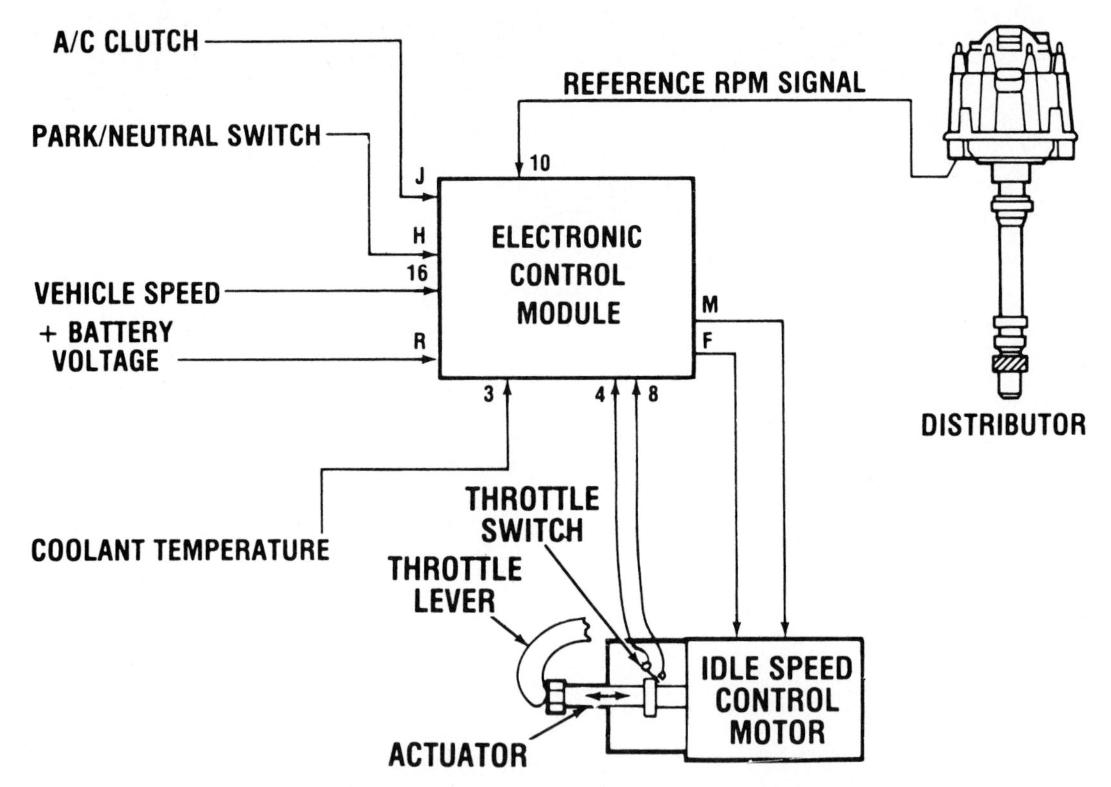

Figure 3–18 Idle Speed Control [ISC] Motor. [Courtesy of General Motors Corporation]

The 1979 and early 80 vehicles (Gen 1) use a single pigtail hanging near the ECM connector. This lead usually has a green connector on it.

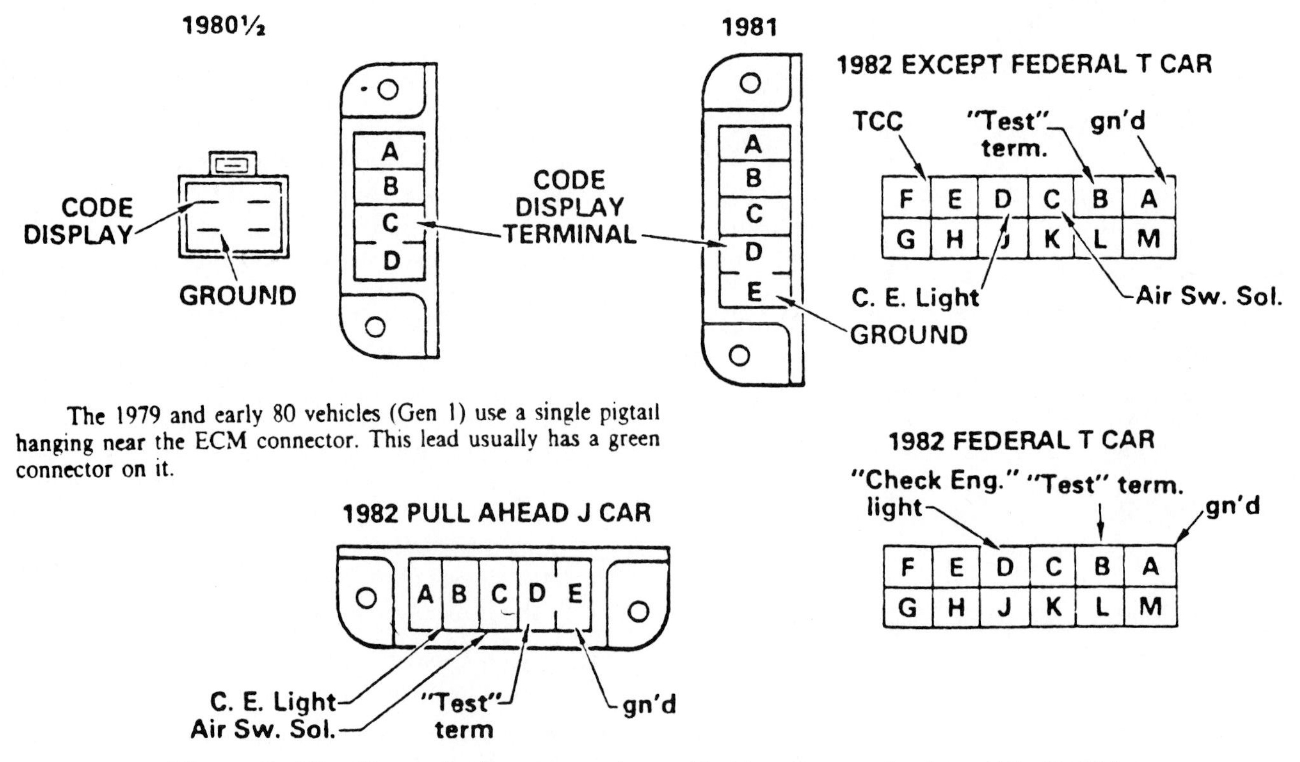

Figure 3–19 Diagnostic Connector, Assembly Line Communications Link [ALCL]. [Courtesy of General Motors Corporation]

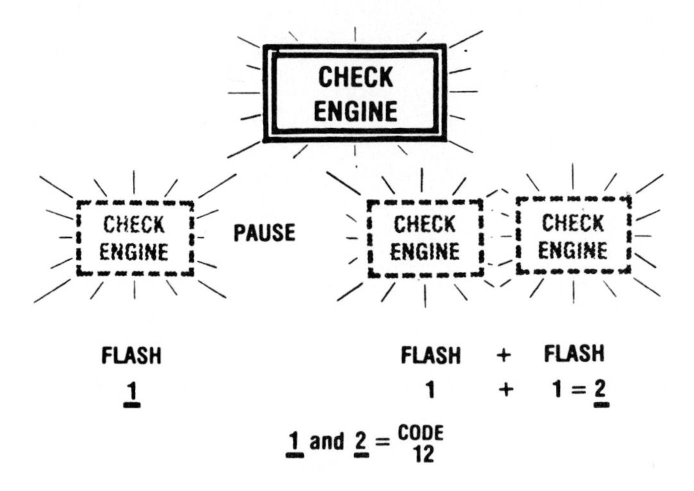

Figure 3-20 Check-Engine Light. (Courtesy of General Motors Corporation)

4. Remove the open-loop mode time when the engine coolant is cold.
5. Remove start-up enrichment and blended enrichment.

If the diagnostic test terminal is grounded with the engine running, the ECM will function normally except that the system will go into closed-loop operation immediately when a warm engine is started and the oxygen sensor is hot.

When the diagnostic test lead of most 1982 and later 3C systems is grounded with the ignition switch on, the ECM will perform the following functions:

1. Flash code 12 and any other trouble codes stored in the ECM memory that would indicate system defects.
2. Send a 30° dwell signal to the mixture control (MC) solenoid.
3. Energize all solenoids controlled by the ECM.
4. Pulse the idle speed control motor in and out.

When the diagnostic test lead is grounded with the engine running, the ECM will follow another routine:

1. No trouble codes will be set.
2. The open-loop timer, start-up enrichment, and blended-enrichment functions of the ECM will be taken out.
3. The electronic spark timing (EST) circuit in the ECM will provide a fixed spark advance that may be used to set initial timing.

The check-engine light will be on as long as a system defect exists. When an intermittent defect occurs, the check-engine light will go out once the defect disappears. However, the trouble code caused by the defect will be stored in the computer memory. Figure 3-21 indicates the possible trouble codes.

Code 44 would be caused by an excessively lean exhaust gas oxygen (O_2) sensor signal. The actual defect may be in the O_2 sensor or the wires from the sensor to the ECM. A code 44 might also be displayed, because the ECM is defective and therefore unable to receive the O_2 sensor signal. If the carburetor delivers an extremely lean mixture, it could cause

The trouble codes indicate problems as follows:

12 - No reference pulses to the ECM. This code is not stored in memory and will only flash while the fault is present. Normal code with ignition "ON," engine not running.

13 - Oxygen Sensor Circuit - the engine must run up to five (5) minutes at part throttle, under road load, before this code will set.

14 - Shorted Coolant Sensor Circuit - The engine must run up to two (2) minutes before this code will set.

15 - Open Coolant Sensor Circuit - The engine must run up to five (5) minutes before this code will set.

21 - Throttle Position Sensor Circuit - The engine must run up to 25 seconds, at specified curb idle speed, before this code will set.

23 - Open or grounded M/C Solenoid Circuit.

24 - Vehicle Speed Sensor (VSS) Circuit - The car must operate up to five (5) minutes at road speed before this code will set.

32 - Barometric Pressure Sensor (BARO) Circuit low.

34 - Manifold Absolute Pressure (MAP) or Vacuum Sensor Circuit. The engine must run up to five (5) minutes, at specified curb idle speed, before this code will set.

35 - Idle Speed Control (ISC) Switch Circuit shorted. (Over 1/2 throttle for over two (2) sec.)

41 - No distributor reference pulses at specified engine vacuum. This code will store.

42 - Electronic Spark Timing (EST) Bypass Circuit grounded or open.

43 - ESC retard signal for too long; causes a retard in EST signal.

44 - Lean Oxygen Sensor indication - The engine must run up to five (5) minutes, in Closed Loop, at part throttle before this code will set.

44 & 55 - (At same time) - Faulty Oxygen Sensor Circuit.

45 - Rich System indication - The engine must run up to five (5) minutes, in closed loop, at part throttle before this code will set.

51 - Faulty calibration unit (PROM) or installation. It takes up to 30 seconds before this code will set.

54 - Shorted M/C Solenoid Circuit and/or faulty ECM.

55 - Grounded V ref (term. 21), faulty Oxygen Sensor or ECM.

Figure 3-21 3C Trouble Codes. (Courtesy of General Motors Corporation)

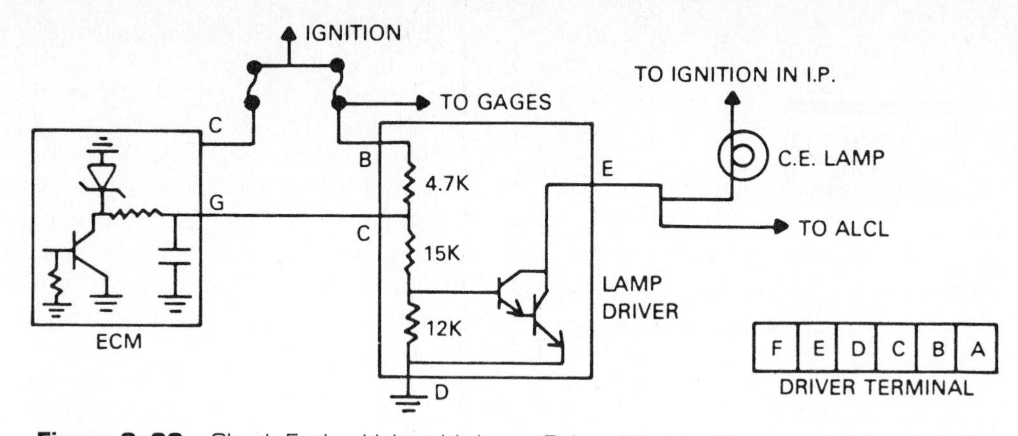

Figure 3-22 Check-Engine Light with Lamp-Driver Module. [Courtesy of General Motors Corporation]

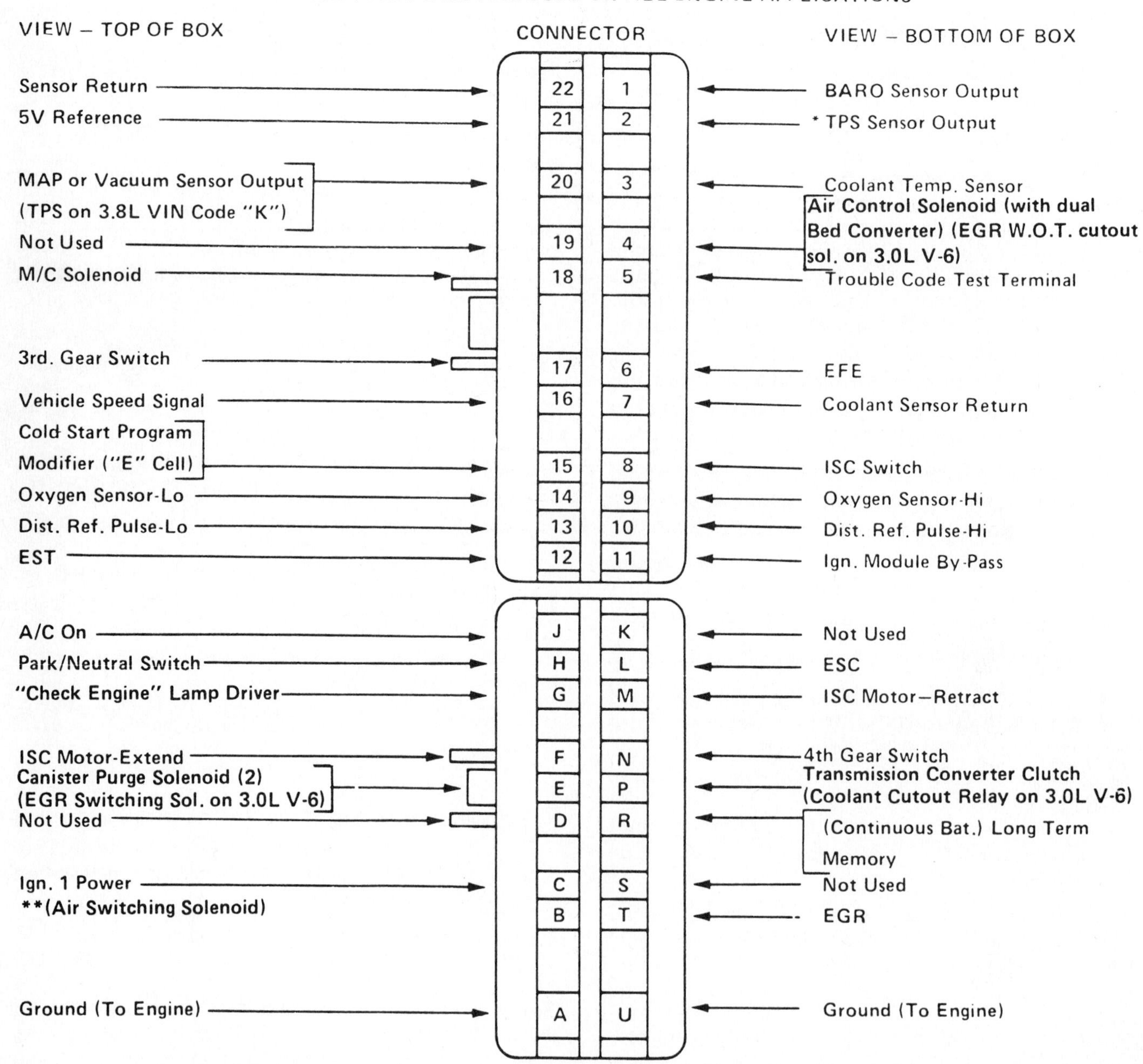

Figure 3-23 Electronic Control Module [ECM] Terminals. [Courtesy of General Motors Corporation]

a code 44 to be set in the computer memory. Most trouble codes indicate a defect in a specific circuit and not necessarily in the sensor that is affected.

In some systems, the check-engine light is connected to terminal G on the ECM, and the light is operated by the ECM. In other models, a lamp-driver module is connected between the check-engine light and the ECM terminal G, as shown in Figure 3–22.

The lamp-driver module is taped to the wiring harness. The check-engine light may be tested by grounding terminal D in the diagnostic connector, which may be referred to as an assembly line communications link (ALCL). On some "J" cars, terminal B must be grounded to test the check-engine light. If the check-engine light does not come on when terminal D or B is grounded, the gauge fuse, check-engine light bulb, or connecting wires are defective.

The trouble codes may be cleared from the computer memory by disconnecting the ECM fuse in the fuse panel for 10 seconds. This fuse supplies power to terminal R on the ECM, as illustrated in Figure 3–23.

Dwellmeter A dwellmeter may be connected from the mixture control (MC) solenoid connector to ground, as pictured in Figure 3–24.

The dwellmeter control switch should be located in the 6-cylinder position regardless of the engine being tested. One end of the MC solenoid winding is supplied with 12V when the ignition switch is on. The ECM controls the air–fuel mixture by completing the circuit from the MC solenoid winding to ground 10 times per second. A dwellmeter is normally used to measure the number of degrees that the distributor cam rotates while the ignition points are closed. Specifically, the dwellmeter is measuring the length of time that current flows through the primary ignition circuit. During 3C system diagnosis the dwellmeter is connected from the ECM side of the MC solenoid winding to ground. The dwellmeter will measure the on-time of the MC solenoid winding, as illustrated in Figure 3–25.

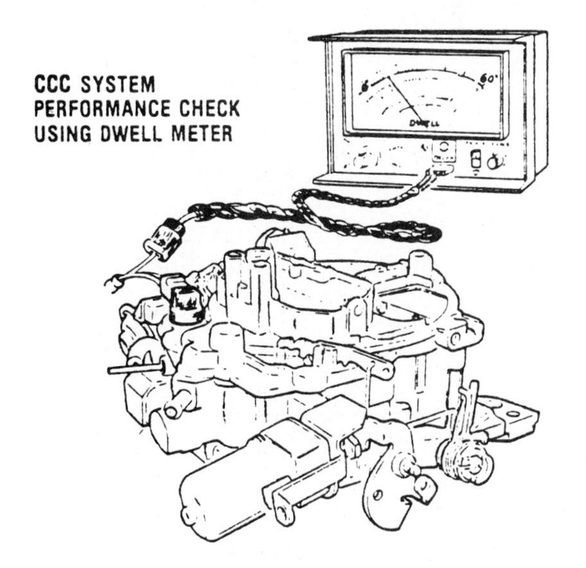

Figure 3–24 Dwellmeter Test Connections to the Mixture Control (MC) Solenoid. (Courtesy of General Motors Corporation)

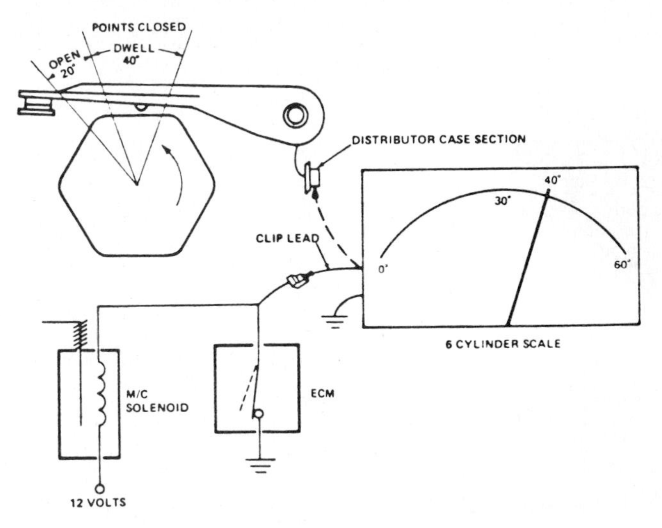

Figure 3–25 Dwellmeter Diagnosis of 3C System. (Courtesy of General Motors Corporation)

The ECM varies the on-time of the MC solenoid to control the air–fuel mixture in the carburetor. Upon receiving a "rich" signal from the oxygen sensor, the ECM will energize the MC solenoid for a longer period of time. If the ECM calls for a leaner air–fuel ratio, it may energize the MC solenoid for 90 percent of the time on each cycle. The MC solenoid plunger will then stay down longer, and a leaner air–fuel ratio will be provided. The 6-cylinder scale on the dwellmeter reads from 0° to 60°. When the ECM energizes the MC solenoid winding for 90 percent of the time in each cycle, the dwellmeter will register 54° (90 percent of 60 is equal to 54), as indicated in Figure 3–26.

A lean oxygen sensor signal to the ECM will cause the ECM to decrease the on-time of the MC solenoid winding. If the ECM provides an on-time of 10 percent on each cycle of the MC solenoid, the solenoid plunger will remain in the upward position for 90 percent of the time on each cycle. The upward plunger position provides a richer air–fuel ratio. When the engine is at normal operating temperature, and the 3C system is in closed loop, the dwellmeter reading should vary continuously between 6° and 54°. The variation indicates that the ECM and the sensors are controlling the air–fuel mixture in the normal manner.

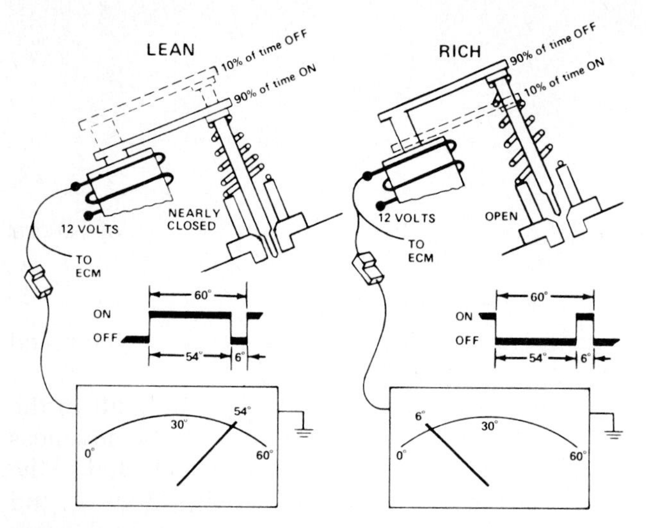

Figure 3–26 Dwellmeter Readings with Lean and Rich Air–Fuel Mixtures. (Courtesy of General Motors Corporation)

Under certain operating conditions such as a cold engine, the system will be in the open-loop mode, and a fixed reading will be provided on the dwellmeter because the ECM is maintaining a slightly richer air–fuel ratio. In open-loop operation, the ECM is no longer controlling the MC solenoid from the oxygen sensor signal. Wide-open throttle (WOT) operation will also cause a fixed reading on the dwellmeter. The oxygen sensor may cool down during prolonged idle operation and cause the system to go into the open-loop mode.

A high reading on the dwellmeter always indicates that the system is trying to provide a leaner air–fuel ratio; therefore the oxygen sensor signal to the ECM will indicate a rich mixture. A lean mixture signal from the oxygen sensor to the ECM will be indicated by a low dwellmeter reading as the system tries to provide a richer air–fuel ratio. The relationship between the dwellmeter readings and MC solenoid cycling is illustrated in Figure 3–27.

If the carburetor is partly choked when the engine is idling, the dwellmeter pointer should move upscale. This indicates that the system is in closed loop and that it will react immediately to correct the rich air–fuel mixture.

System-Performance Test Before proceeding with the diagnosis of the computer command control (3C) system, check all other sources of performance or economy complaints such as ignition components, engine compression, and the exhaust-gas recirculation (EGR) valve. The absence of trouble codes in the ECM does not necessarily mean that the system is working normally. Certain defects in the system will not be indicated by trouble codes. The following system-performance test may be conducted to determine if the system is working normally:

1. Run the engine until it reaches normal operating temperature.
2. Ground the diagnostic test terminal with the engine running.

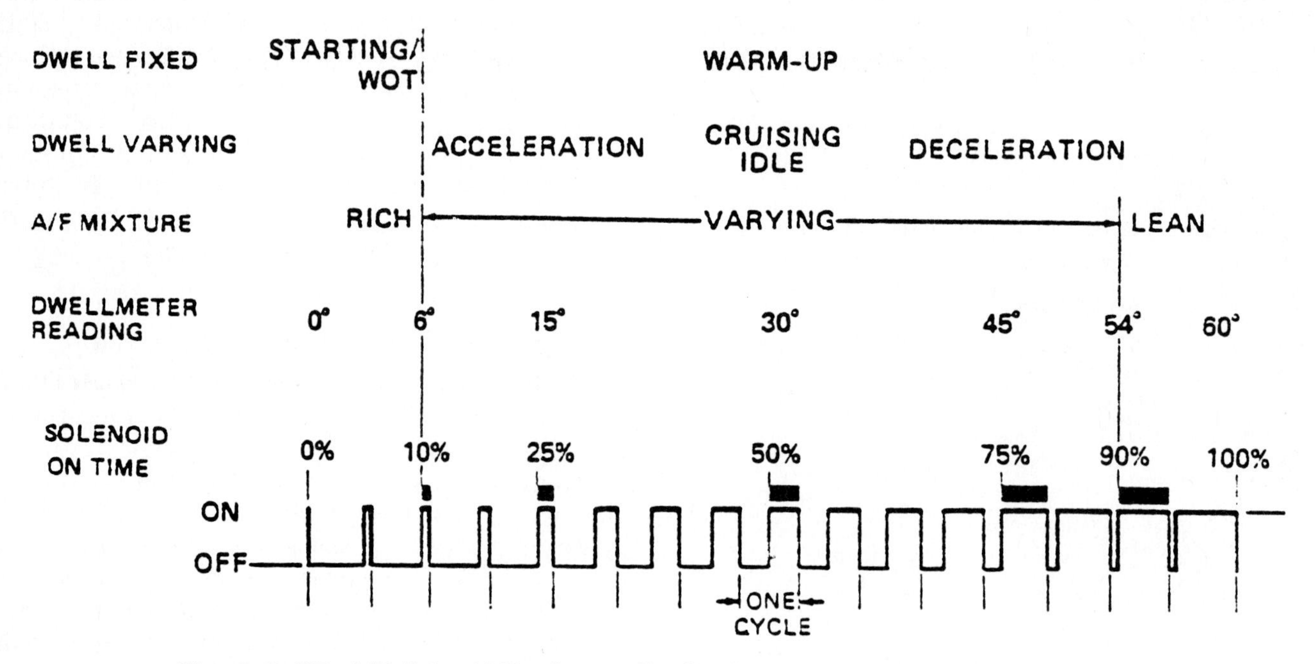

Figure 3–27 MC Solenoid Dwellmeter Reading Interpretation. (Courtesy of General Motors Corporation)

3. Connect a tachometer from the distributor "tach" terminal to ground.
4. Disconnect the mixture control (MC) solenoid and ground the solenoid dwell terminal. Once the solenoid is disconnected, the plunger will move upward and provide a very rich air–fuel ratio.
5. Operate the engine at 3000 rpm and reconnect the MC solenoid. The grounded MC solenoid dwell lead will energize the solenoid winding continuously. The plunger should then move downward, and the air–fuel mixture should become leaner. If the engine slows down 100 rpm or more, the MC solenoid is working normally. A decrease of less than 100 rpm in engine speed indicates that the MC solenoid requires servicing.
6. Disconnect the grounded MC solenoid dwell lead before slowing the engine to idle speed. Reconnect the MC solenoid and observe the dwellmeter. If the ECM and the sensors are working normally, the dwellmeter will vary continuously between 6° and 54°. If the MC solenoid dwell reading is in a fixed position or out of the specified range, the system is defective.

Ignition Timing Before basic ignition timing is checked, the four-wire distributor connector must be disconnected, as pictured in Figure 3–28.

On 3C systems manufactured in 1982 or later, the diagnostic test terminal may be grounded with the engine running when basic ignition timing is being checked. With the distributor disconnected or the diagnostic test terminal grounded, the ignition module supplies a fixed timing signal. The ignition module

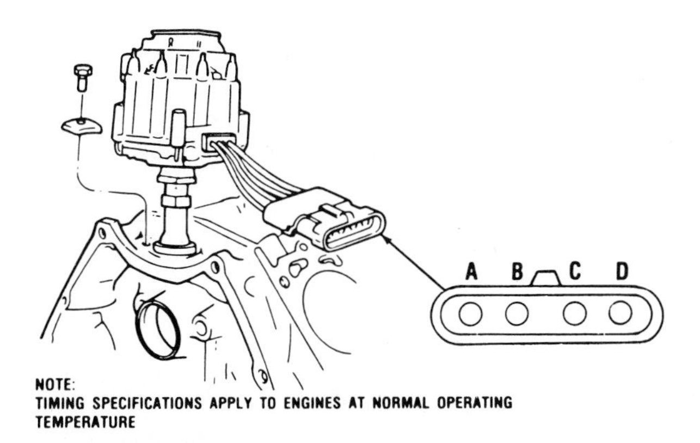

A
B
C
D
NOTE:
TIMING SPECIFICATIONS APPLY TO ENGINES AT NORMAL OPERATING TEMPERATURE
• EST TIMING TO BE SET WITH FOUR WIRE DISTRIBUTOR HARNESS DISCONNECTED FROM CONTROLLER
INITIAL IGNITION TIMING SETTING TO BE WITHIN +1/2° OF SPECIFIED TIMING. RECHECKS WITH A KNOWN GOOD TACHOMETER AND TIMING LIGHT SHOULD BE WITHIN +2° OF SPECIFICATION OR THE TIMING SHOULD BE RESET.
(CENTERLINE OF SLOT IS T.D.C.,

Figure 3–28 Distributor Four-Wire Connector. [Courtesy of General Motors Corporation]

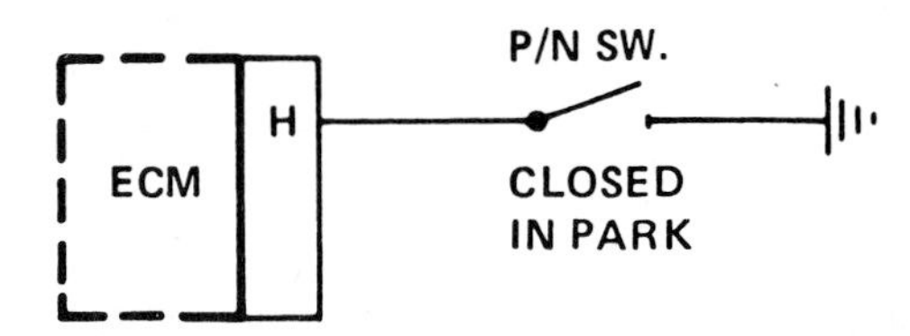

Figure 3–29 Park/Neutral Switch. [Courtesy of General Motors Corporation]

will supply some advance in relation to engine speed under this condition.

The electronic spark timing (EST) circuit in the ECM is working normally if the timing advances when the distributor connector is reconnected or the diagnostic test terminal is disconnected from ground with the engine idling in drive. A defective park/neutral switch, or a defective wire connected from the ECM to the park/neutral switch, could prevent the ECM from providing spark advance. Figure 3–29 shows the park/neutral switch and the related ECM circuit. Other defects that could result in a loss of spark advance would be indicated by trouble codes in the ECM.

If there is an open circuit in the compensated ignition spark timing signal, the ignition system will fail about 5 seconds after the engine is started. The start mode will operate to start the engine. Engine stalling can be expected when the 5V bypass disable signal is received but the compensated ignition spark timing signal is not available to keep the engine running. The four-wire distributor connector should be disconnected when diagnosing a complaint of stalling 5 seconds after starting the engine. If the stalling problem is corrected by disconnecting the four-wire connector, then the ECM, or compensated ignition spark timing wire, has an open circuit. The compensated ignition spark timing wire is connected to terminal 12 on the ECM. (A complete description of 3C ignition systems is provided in Chapter 5.)

Ford Electronic Engine Control IV Systems Used with 2.3L High Swirl Combustion Engines

Electronic Control Assembly

The computer in the electronic engine control IV (EEC IV) system is referred to as an "electronic control assembly" (ECA). Input signals from various sensors are sent to the ECA, which performs specific output functions. When the ignition switch is turned on, current flows through the power relay winding

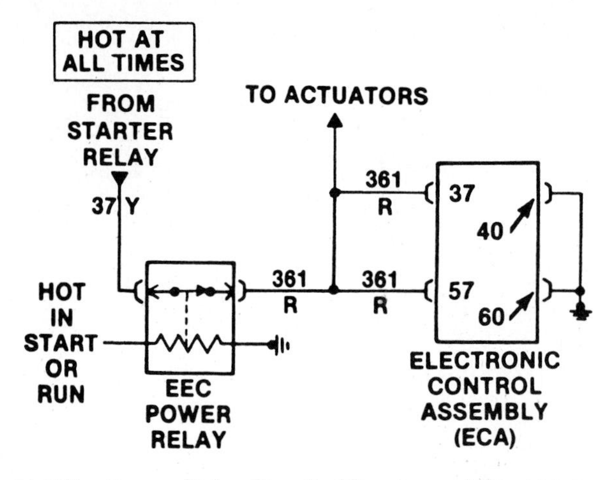

Figure 3–30 Power Relay Circuit. (Courtesy of Ford Motor Co.)

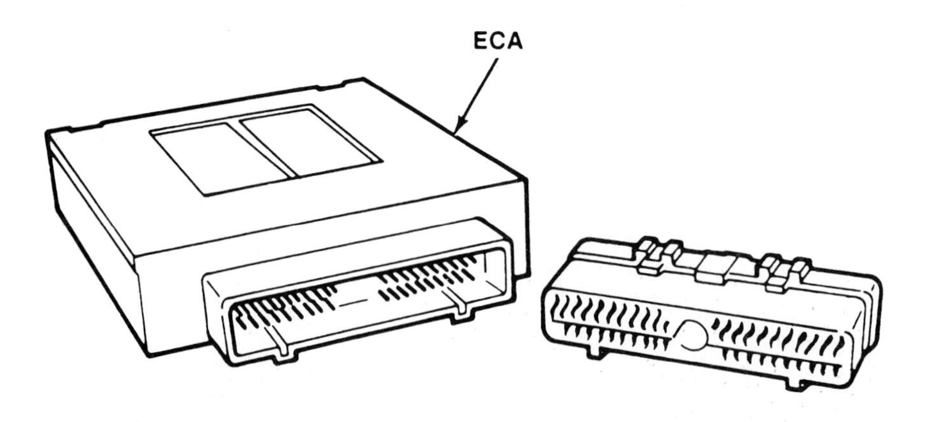

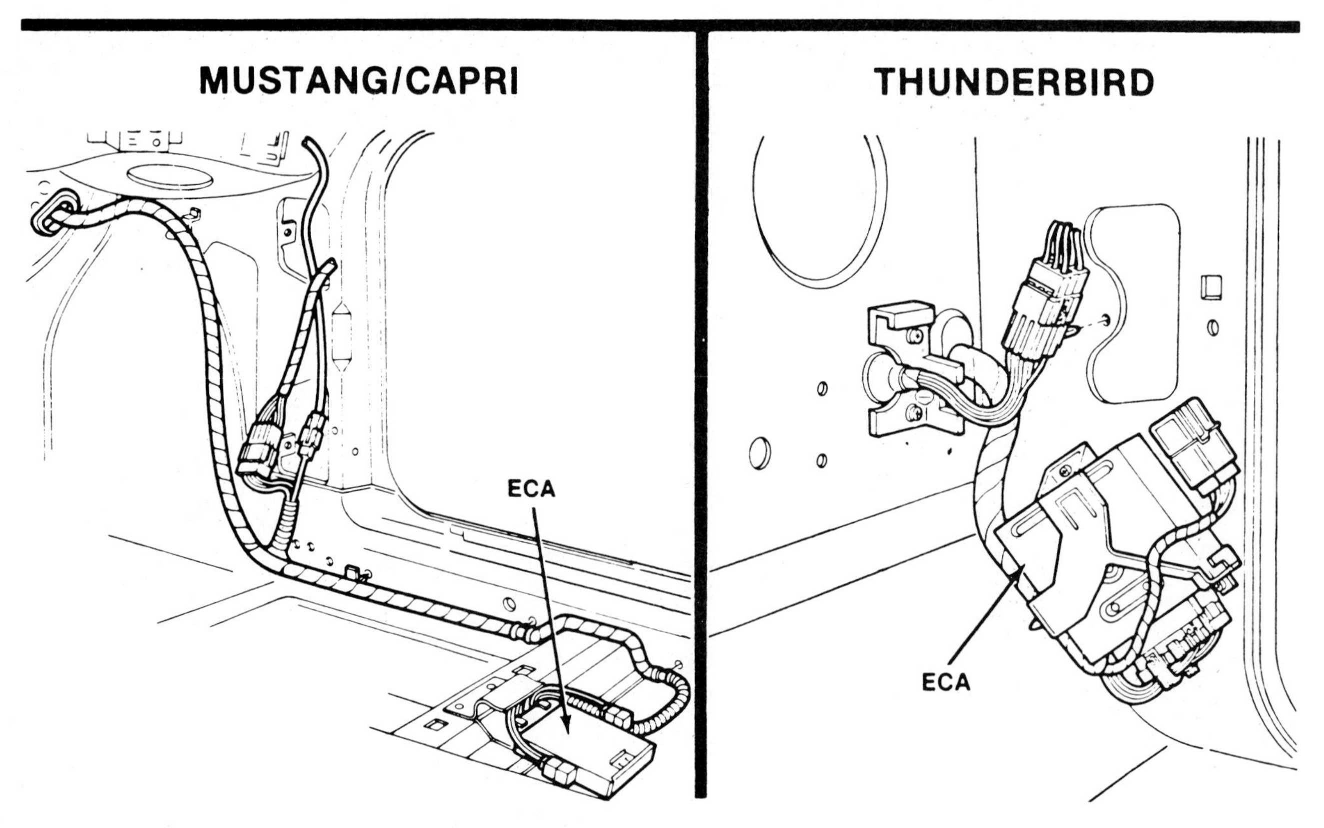

Figure 3–31 Electronic Control Assembly. (Courtesy of Ford Motor Co.)

to ground. The coil magnetism closes the relay contacts, and voltage is supplied through these contacts to the ECA, as indicated in Figure 3–30. An ECA is illustrated in Figure 3–31.

Sources of ECA Input

Many of the same input sensors such as the exhaust-gas oxygen (EGO) sensor, engine-coolant temperature (ECT) sensor, throttle-position sensor (TPS), manifold absolute pressure (MAP) sensor, and the air conditioner (A/C) clutch compressor signal are similar to the sensors in other systems. The exhaust-gas recirculation (EGR) position sensor is a variable resistor that sends a signal to the ECA in relation to EGR valve position. A Hall Effect switch in the distributor is referred to as a "profile ignition pickup" (PIP). (The PIP with the EEC IV ignition system is explained in Chapter 5.) All of the inputs and outputs in the EEC IV system are displayed in Figure 3–32.

Output Control Functions

Air–Fuel Ratio Control

The feedback control solenoid (FCS) contains an electrically operated vacuum solenoid with a movable plunger. The FCS controls the amount of vacuum that is supplied to a carburetor diaphragm, and it also controls the amount of air that is bled into the idle circuit, as shown in Figure 3–33.

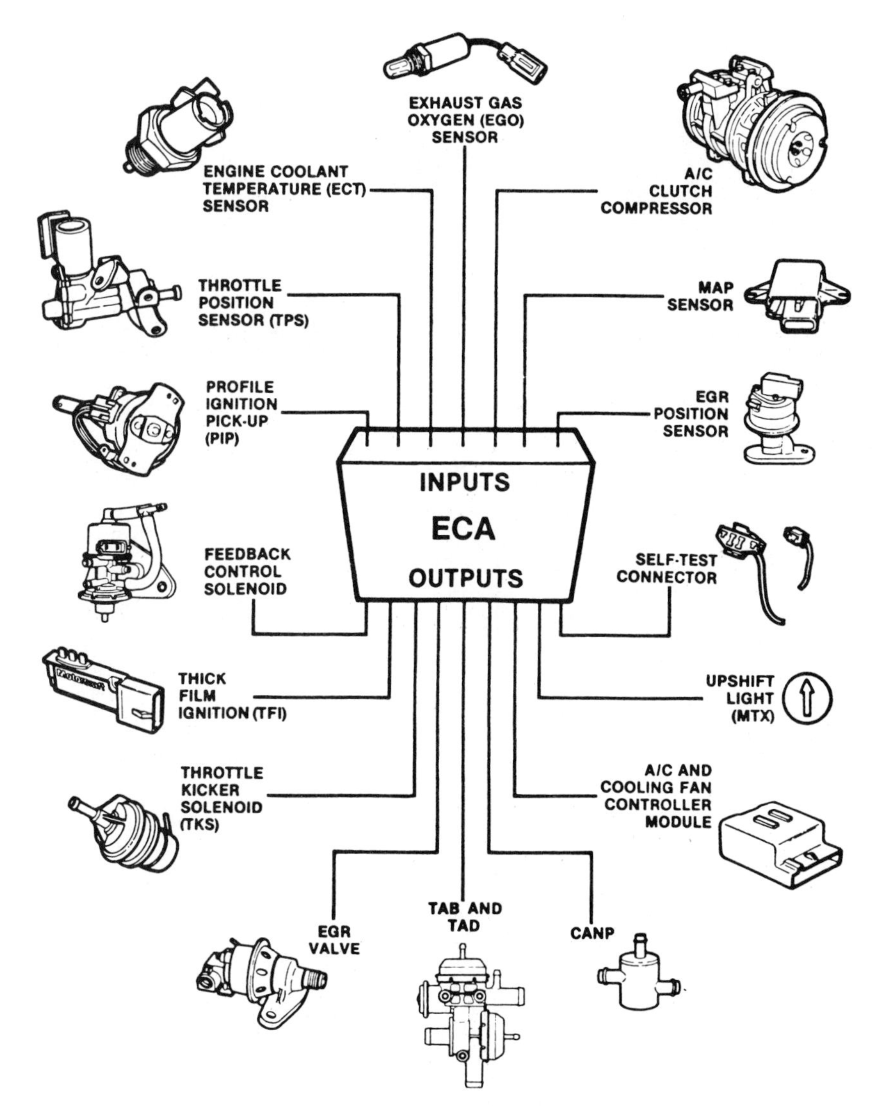

Figure 3–32 EEC IV Inputs and Outputs 2.3L High Swirl Combustion (HSC) Engine. (Courtesy of Ford Motor Co.)

The ECA energizes the FCS winding 10 times per second. Output vacuum from the FCS to the carburetor diaphragm varies from 2 to 5 in hg. When the winding is not energized, this vacuum output will be very low, and energizing the winding will cause the output vacuum to increase. The ECA varies the on-time of the FCS winding to supply the correct output vacuum and an air–fuel ratio of 14.7 : 1. If the oxygen sensor provides a rich mixture signal, the ECA will energize the FCS winding, which increases the output vacuum from the FCS to the carburetor diaphragm. This vacuum increase will lift the carburetor diaphragm and actuator, which allows the feedback metering valve to reduce the fuel flow to the main system. A lean oxygen sensor signal results in reduced FCS winding on-time and output vacuum that allows the carburetor diaphragm and actuator to move downward. This action opens the feedback metering valve and supplies more fuel to the main system as indicated in Figure 3–34.

When the FCS supplies more vacuum to the carburetor diaphragm, it also supplies more air to the idle system, which leans the air–fuel mixture if the engine is idling. The idle feedback air bleed is shown in Figure 3–35.

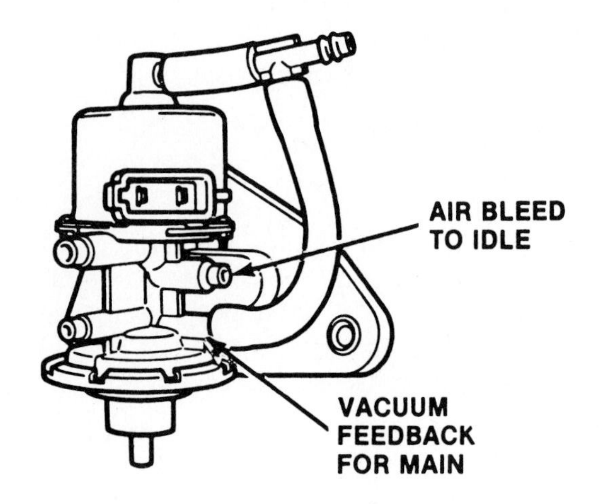

Figure 3–33 Feedback Control Solenoid. (Courtesy of Ford Motor Co.)

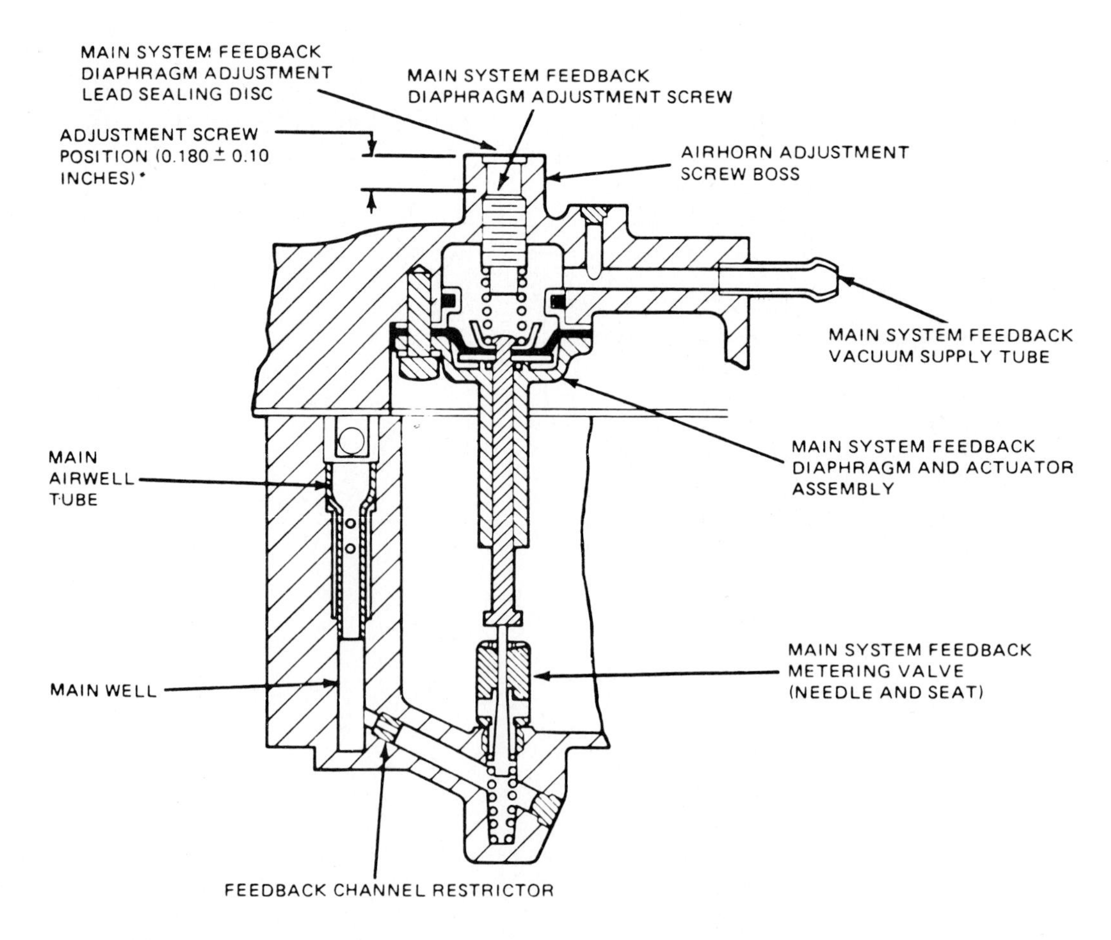

Figure 3–34 Feedback Diaphragm and Actuator. (Courtesy of Ford Motor Co.)

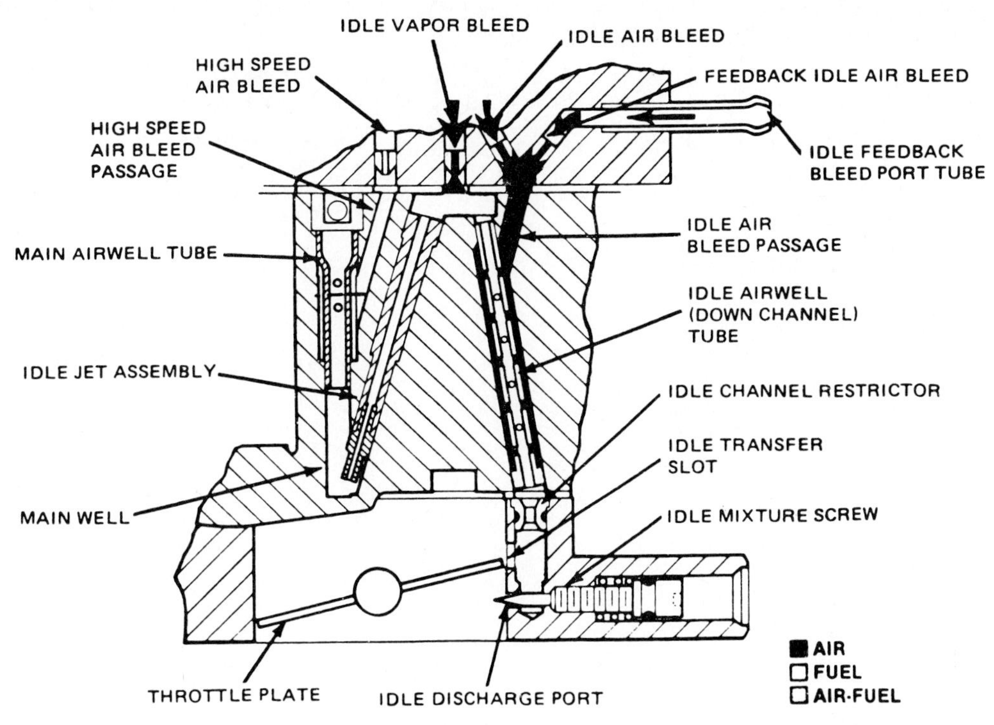

Figure 3–35 Idle Feedback Air Bleed. (Courtesy of Ford Motor Co.)

Thick Film Ignition Module

The thick film ignition (TFI) module and spark advance control are discussed in Chapter 5.

Throttle-Kicker Solenoid

When the air conditioner is turned on, the ECA will energize a throttle-kicker solenoid (TKS), and vacuum will be supplied through the solenoid to the throttle-kicker actuator on the carburetor. This action will increase the idle speed. The ECA may energize the TKS under other operating conditions such as cold engine coolant. However, the fast idle cam still provides the correct fast idle speed during engine warmup.

Exhaust-Gas Recirculation

The ECA operates two solenoids in the vacuum system connected to the exhaust-gas recirculation (EGR) valve. If more EGR flow is required, the ECA will supply power to both solenoids. When this occurs, the vent solenoid will close and the vacuum solenoid will open, which supplies more vacuum to the EGR valve, as illustrated in Figure 3–36.

When the input signals inform the ECA that the EGR flow should be reduced, the ECA will open the circuit to both solenoids. Under this condition the EGR control solenoid will close and the EGR vent solenoid will open, which vents the vacuum in the system and allows the EGR valve to close.

Thermactor Pump Air Control

The ECA operates the dual air control solenoids to control the thermactor pump air flow. Vacuum applied to the thermactor air bypass (TAB) valve and thermactor air divert (TAD) valve is turned on and off by the dual air control solenoids. The thermactor pump air flow has three possible routings:

1. Downstream air, which is injected into the catalytic converter.
2. Upstream air, which is injected into the exhaust ports.
3. Bypass air, which is bypassed to the atmosphere.

In the bypass air mode, the ECA deenergizes the TAB and TAD solenoids. Under this condition, vacuum is not applied to the TAB or TAD valves, and air flow from the pump is bypassed to the atmosphere as shown in Figure 3–37.

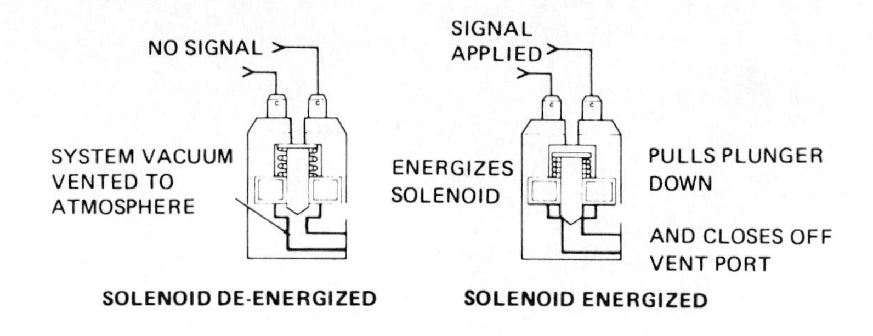

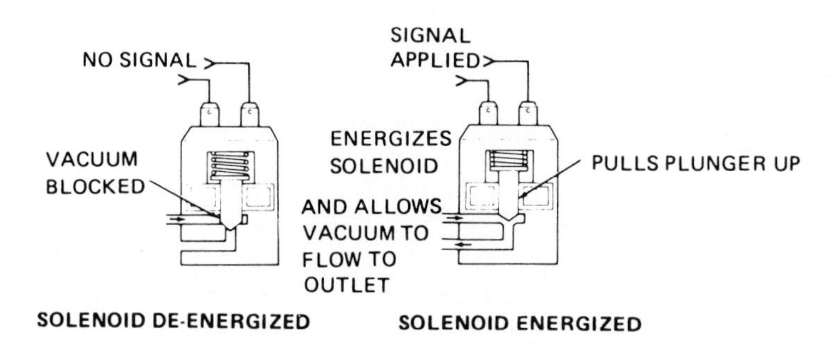

Figure 3–36 EGR Control Solenoid and Vent Solenoid. [Courtesy of Ford Motor Co.]

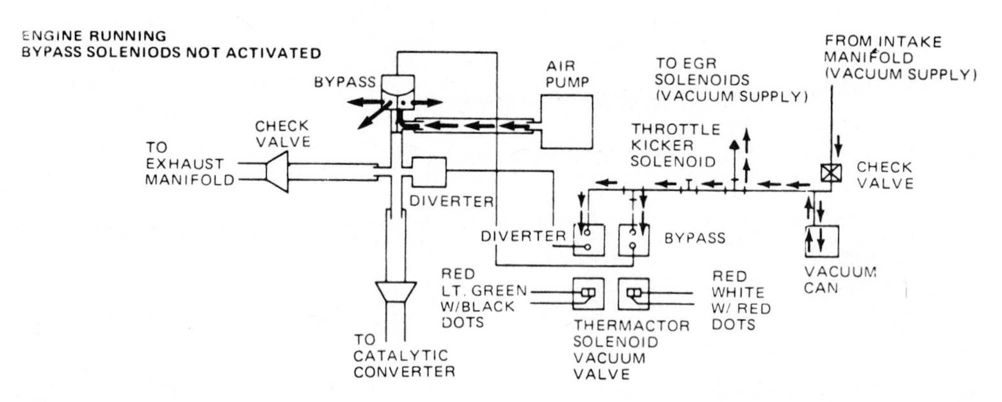

Figure 3–37 Thermactor Pump Air Flow Bypassed to the Atmosphere. [Courtesy of Ford Motor Co.]

The bypass mode will occur during periods of prolonged idle, under wide-open throttle conditions, or when the oxygen sensor signal is extremely low.

When the engine coolant is cold, the ECA will energize the TAB and TAD solenoids, and manifold vacuum will be applied through the solenoids to the TAB and TAD valves. Under this condition thermactor pump air will be diverted to the exhaust ports as indicated in Figure 3–38.

Air from the thermactor pump will be diverted to the exhaust ports to lower hydrocarbon (HC) emissions during the engine warmup period.

When the input sensors signal the ECA to provide downstream injection, the ECA will energize the TAB solenoid and deenergize the TAD solenoid. Manifold vacuum will be maintained through the TAB solenoid to the TAB valve, but the manifold vacuum to the TAD will be shut off. When this occurs, the thermactor pump air flow will be diverted to the catalytic converter as illustrated in Figure 3–39.

Canister Purging

The ECA energizes the canister purge (CANP) solenoid, which controls the flow of fuel vapors from the canister to the intake manifold. Activation of the

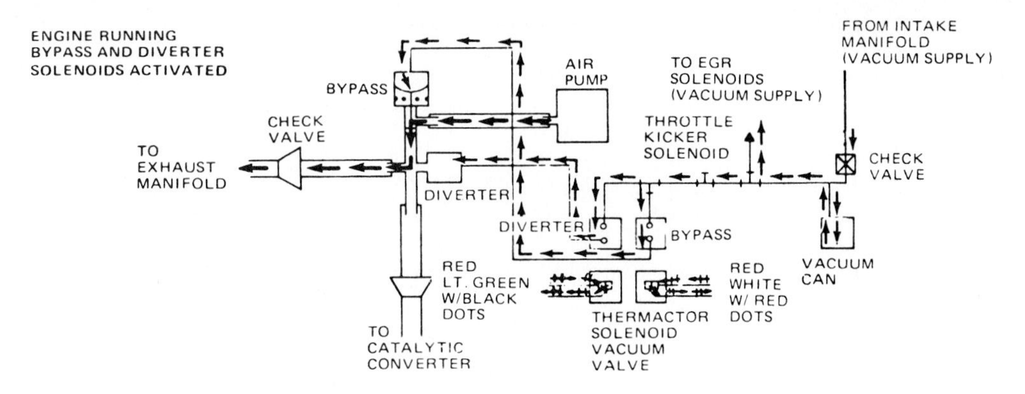

Figure 3–38 Thermactor Pump Air Flow Diverted to the Exhaust Ports. (Courtesy of Ford Motor Co.)

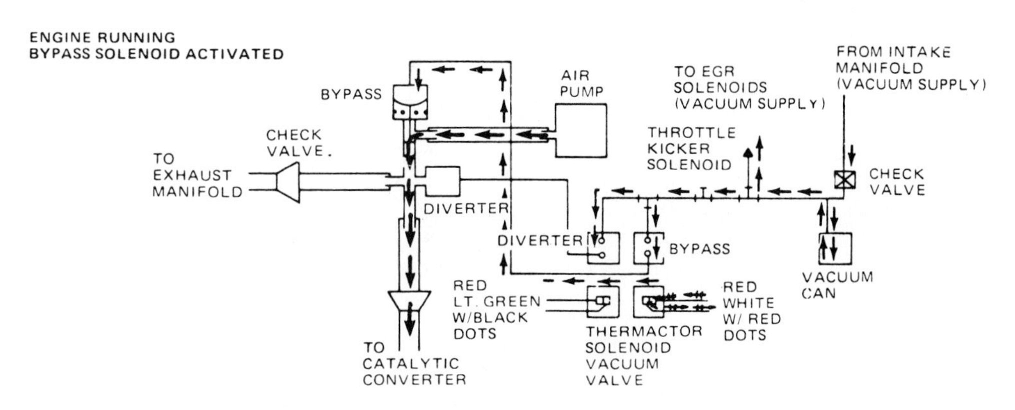

Figure 3–39 Thermactor Pump Air Flow Diverted to the Catalytic Converter. (Courtesy of Ford Motor Co.)

CANP solenoid by the ECA will occur when the following conditions are present.

1. Engine coolant temperature must be in the normal operating range.
2. Engine revolutions per minute (rpm) must be above a calibration value.
3. The time since the engine was started must be above calibration value.
4. The engine is not in the closed throttle mode.

Upshift Light

On vehicles equipped with manual transaxles the upshift light will be illuminated by the ECA when an upshift to the next highest gear will provide optimum fuel economy.

Self-Test Connector

Diagnostic procedures are performed when a voltmeter or digital tester is connected to the self-test connector. (Diagnosis of Ford EEC IV systems with computer-controlled carburetors or fuel injection is explained in Chapter 4.)

Air Conditioning (A/C) and Cooling Fan Controller Module

A cooling fan and A/C controller module is used with the EEC IV system on the 2.3L engine to control the operation of the cooling fan and the A/C compressor clutch. The module is mounted under the right side of the instrument panel. Input signals are sent to the module from the coolant temperature switch, the ECA, and the stop-lamp switch. The stop-lamp switch signal is only used on vehicles that are equipped with an automatic transaxle (ATX). Output signals from the module control the operation of the engine cooling fan and the A/C compressor clutch, as illustrated in Figure 3–40.

When the module receives a signal from the clutch cycling pressure switch, the module will send

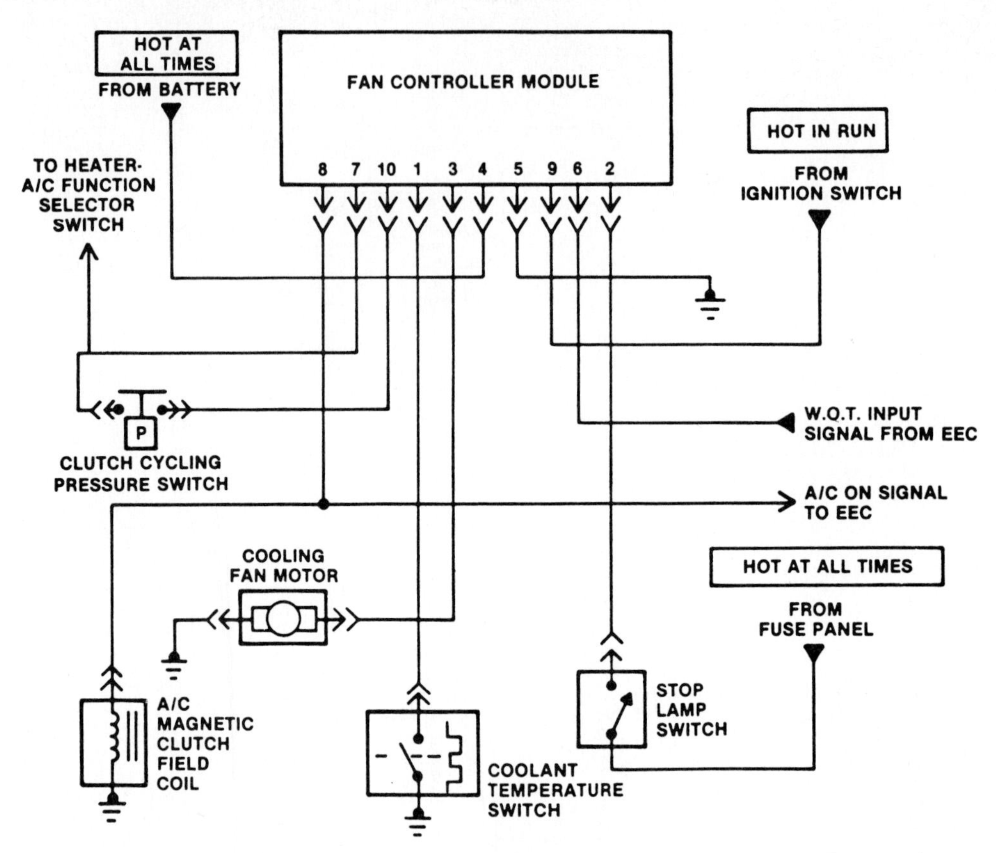

Figure 3–40 Cooling Fan and A/C Compressor Controller Module. (Courtesy of Ford Motor Co.)

an output signal to apply the A/C compressor clutch. During periods of wide-open throttle (WOT) operation, the ECA applies an input signal to the module that disables the signal to the compressor clutch. This WOT signal from the ECA also deenergizes the cooling fan motor if the engine coolant temperature is below 210°F (97°C). The coolant temperature switch applies a ground signal to the module if the coolant temperature exceeds 210°F (97°C). This ground signal overrides the WOT signal from the ECA and prevents the cooling fan from shutting off.

The ECA also contains a time-out feature for the WOT signal. After approximately 30 seconds, the WOT signal is stopped even if the WOT condition is still present. However, slow recycling from WOT to part throttle and back to WOT will again shut off the compressor clutch for 30 seconds.

When the brake pedal is applied on vehicles that are equipped with an automatic transaxle (ATX), an input signal is sent from the brake lamp switch to the module. This will result in a disable signal being sent from the module to the cooling fan and the A/C compressor clutch for 3 to 5 seconds to prevent engine stalling on deceleration. Disabling the compressor clutch and the cooling fan at WOT provides more engine power.

Control Modes

The crank mode is used by the ECA while the engine is being cranked or if the engine stalls and the ignition switch is left in the start position. When the engine is being cranked, fuel control is in open loop, and the ECA sets ignition timing at 10° to 15° before top dead center (BTDC). The EGR solenoids are not energized, and therefore the EGR valve is closed. Thermactor air is upstream to the exhaust ports, and the canister purge system is off. The crank mode is a special program that is used to aid engine starting.

The underspeed mode is used to provide additional fuel enrichment and increased air flow to the engine to help it recover from a stumble. If the engine speed drops below 600 rpm, the ECA will enter the underspeed mode. Operation in this mode is similar to the crank mode.

In the closed-throttle mode, the various input sensor signals are evaluated by the ECA, and the ECA

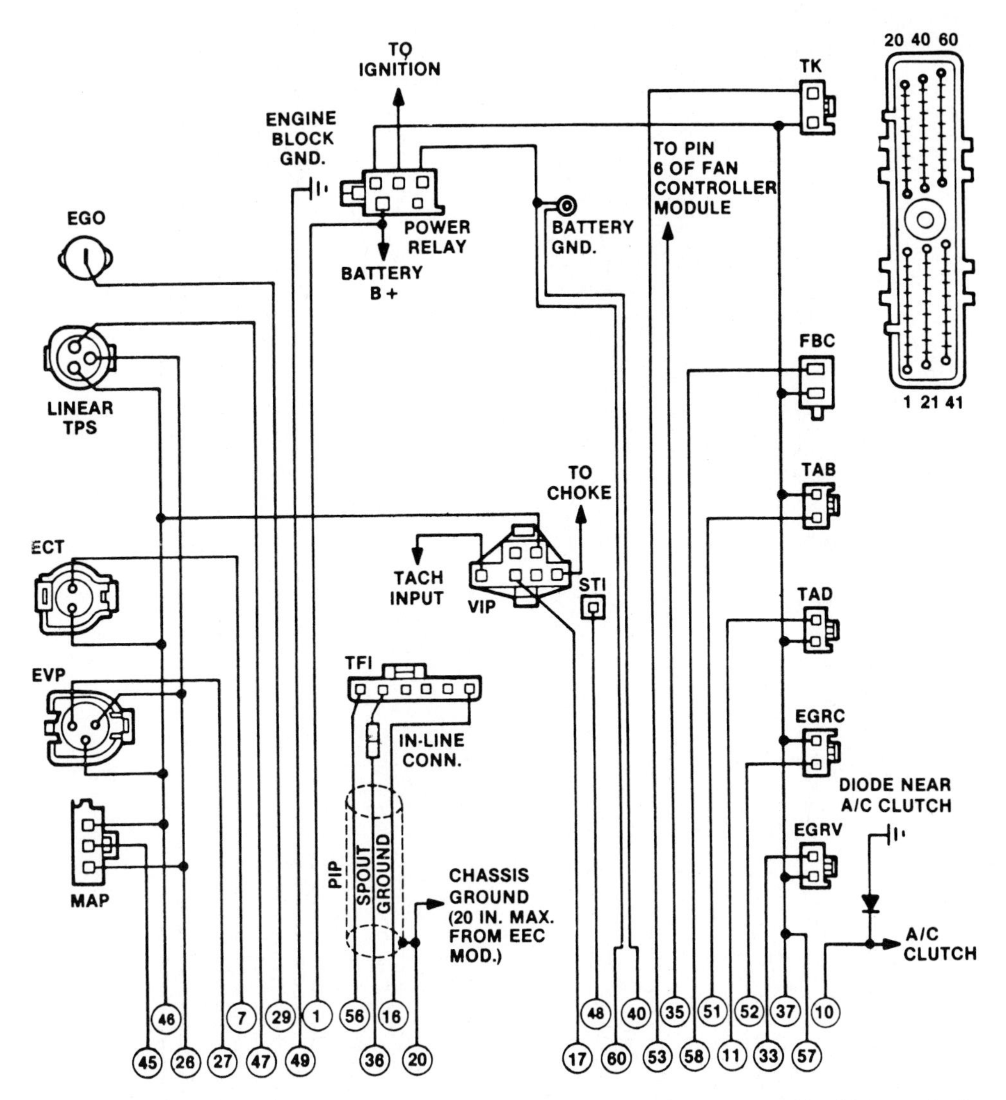

Figure 3–41 EEC IV 2.3L Engine Wiring Diagram. (Courtesy of Ford Motor Co.)

determines the correct air–fuel ratio and ignition timing. As long as the exhaust-gas oxygen (EGO) sensor signal is maintained, the system will remain in closed loop. If the EGO sensor cools off and its signal is no longer available, the system will go into open-loop operation. The ECA will activate the throttle-kicker system if the engine coolant is above a predetermined temperature, or if the air conditioning is on. The throttle kicker will also be activated for a few seconds after the engine is started. During the closed-throttle mode, the EGR valve and the canister purge system are turned off. The closed-throttle mode is used during idle operation or deceleration.

The part-throttle mode is entered during moderate cruising speed conditions. In this mode the system will operate in closed loop as long as the EGO sensor signal is available. The throttle kicker will be activated to provide a dashpot function when the engine is decelerated. The ECA will activate the EGR solenoids and the EGR valve will be open.

When the wide-open throttle (WOT) mode is entered, the system switches to open-loop operation, and the ECA operates the feedback solenoid to provide a slightly richer air–fuel ratio. The WOT signal from the ECA to the fan controller module will cause the module to shut off the cooling fan and the A/C clutch to improve engine performance.

When an electrical defect occurs that prevents the ECA from performing its normal modes of operation, the system enters the limited operation strategy (LOS) mode.

A complete wiring diagram of the EEC IV system used on the 2.3L HSC engine is provided in Figure 3–41.

Ford Electronic Engine Control IV (EEC IV) Systems Used with 2.8L V6 Engines

Input Sensors

Many of the same sensors are used on the EEC IV system on the 2.8L V6 engine and the 2.3L engine. The knock sensor, idle tracking switch (ITS), air-charge temperature (ACT) sensor, and the neutral/start switch are additional input signals on the 2.8L V6 engine. The knock sensor sends a signal to the ECA if the engine begins to detonate. When the ECA receives this signal it will retard the ignition spark advance. The amount of retard depends on the severity and duration of the detonation, and the maximum amount of retard is 8°.

The idle tracking switch (ITS) is an integral part of the idle speed control motor. This switch signals the ECA when the throttle is closed. When the throttle is closed, the ITS switch will be open. Under this condition the ECA will operate the idle speed control motor to control the idle speed. When the throttle is opened, the ITS switch will close and the ECA will no longer operate the idle speed control motor.

The air-charge temperature (ACT) sensor is mounted in the air cleaner. This sensor sends a signal to the ECA in relation to the air intake temperature. The ACT sensor signal is used by the ECA to control the choke opening when the air intake temperature is cold. Once the engine is warmed up, the ACT signal is used by the ECA to control idle speed. When the air intake temperature is below 55°F (13°C), the ACT signal will cause the ECA to direct thermactor air to the atmosphere.

The neutral/start switch signal is used on vehicles that are equipped with an automatic transaxle. This switch is closed when the transaxle is in neutral or park, and it is open in all other transaxle selector positions. The ECA increases the idle speed 50 rpm to improve idle quality when it receives a signal from the neutral/start switch that indicates the transaxle selector is in neutral or park. All of the input signals and output controls in the EEC IV system on the 2.8L V6 engine are illustrated in Figure 3–42.

Output Controls

The EEC IV system on the 2.8L engine uses many of the same output controls that are used on the EEC IV

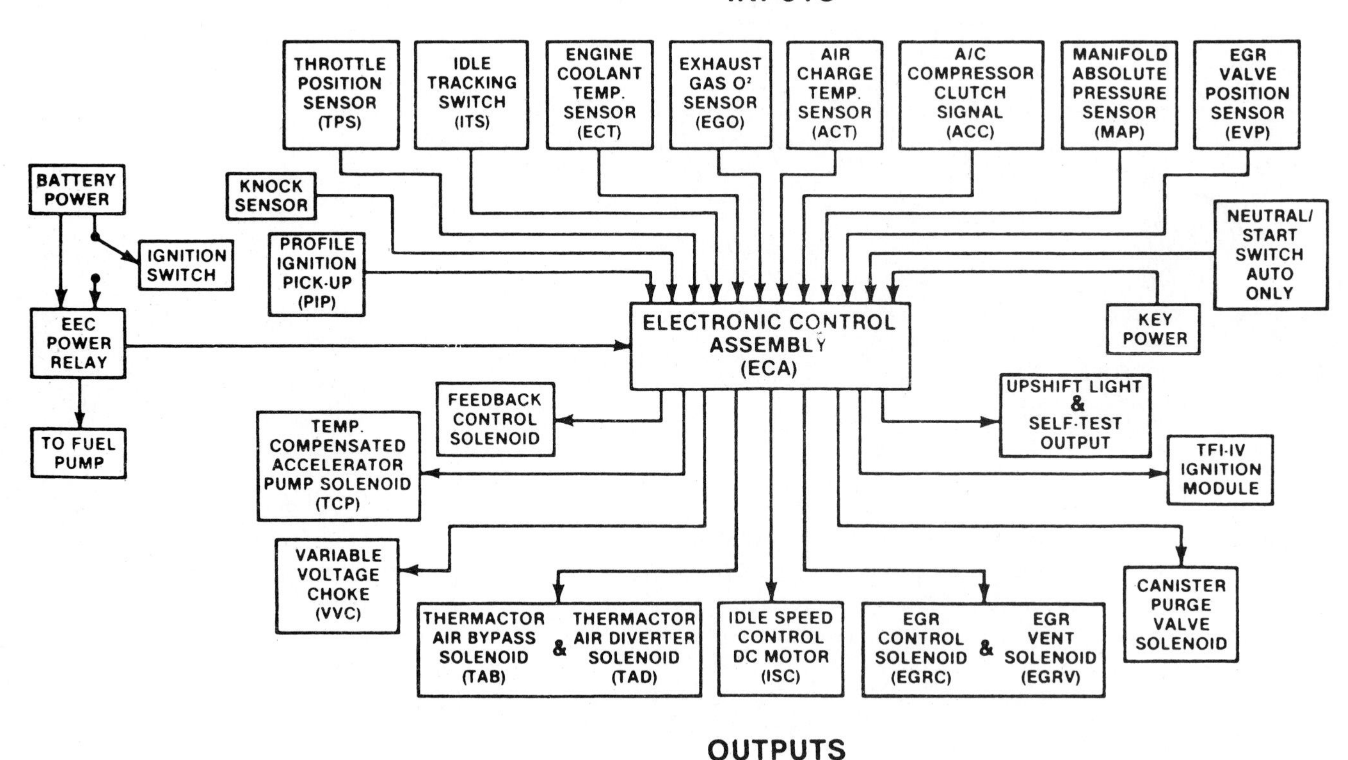

Figure 3–42 EEC IV Input Signals and Output Controls, 2.8L V6 Engine. (Courtesy of Ford Motor Co.)

system on the 2.3L engine. Additional output controls on the 2.8L engine would include the temperature-compensated accelerator pump (TCP) solenoid, variable-voltage choke (VVC), and the idle speed control (ISC) motor. A different type of feedback control (FBC) solenoid is used with the 2150A two-barrel carburetor on the 2.8L V6 engine. The ECA operates the FBC plunger, which controls the amount of air flow into the idle and main circuit air bleed passages as pictured in Figure 3–43.

When the exhaust-gas oxygen (EGO) sensor signals the ECA that the air–fuel ratio is too rich, the ECA will operate the FBC solenoid plunger to allow more air into the idle and main circuits to provide a leaner air–fuel ratio. The ECA operates the FBC solenoid to maintain an air–fuel ratio of 14.7:1 when the engine is at normal operating temperature and the system is operating in closed loop.

The temperature-compensated accelerator pump (TCP) allows delivery of a large pump capacity to improve cold engine performance and a smaller pump capacity during warm engine operation. The TCP contains a bypass bleed valve that is operated by a vacuum diaphragm. A vacuum solenoid, controlled by the ECA, applies vacuum to the vacuum diaphragm in the TCP. When the engine coolant is cold, the vacuum solenoid shuts off the vacuum applied to the TCP and thus allows the bypass valve to remain closed. Under this condition, full accelerator pump capacity is delivered through the accelerator pump system. The TCP system is shown in Figure 3–44.

At normal engine coolant temperature the ECA energizes the TCP solenoid and applies vacuum to the TCP diaphragm. When this occurs, the bypass valve will be open, and some fuel will be returned past the bypass valve to the float bowl each time the engine is accelerated.

The variable-voltage choke (VVC) is an electric choke that is controlled by the ECA and a solid-state power relay. A variable duty cycle is used by the ECA and power relay to control the opening of the choke. The VVC and the power relay are pictured in Figure 3–45.

As the air intake temperature decreases, the ECA and the power relay reduce the on-time of the voltage that is supplied to the choke. This will allow the choke to remain on longer. When the choke cover temperature is above 80°F (27°C), the ECA and the power relay supply voltage to the choke continuously to assure that the choke remains open.

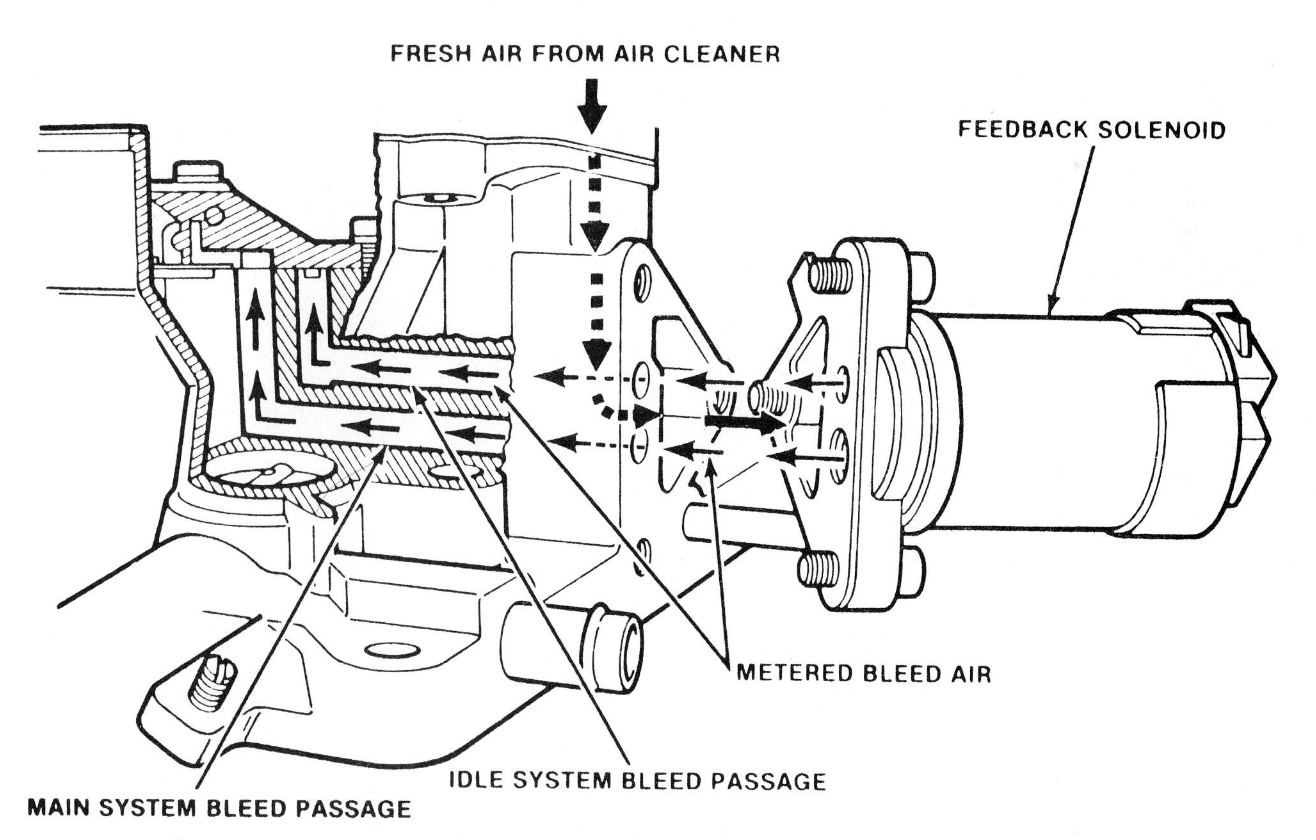

Figure 3–43 Feedback Control (FBC) Solenoid, 2.8L V6 Engine. (Courtesy of Ford Motor Co.)

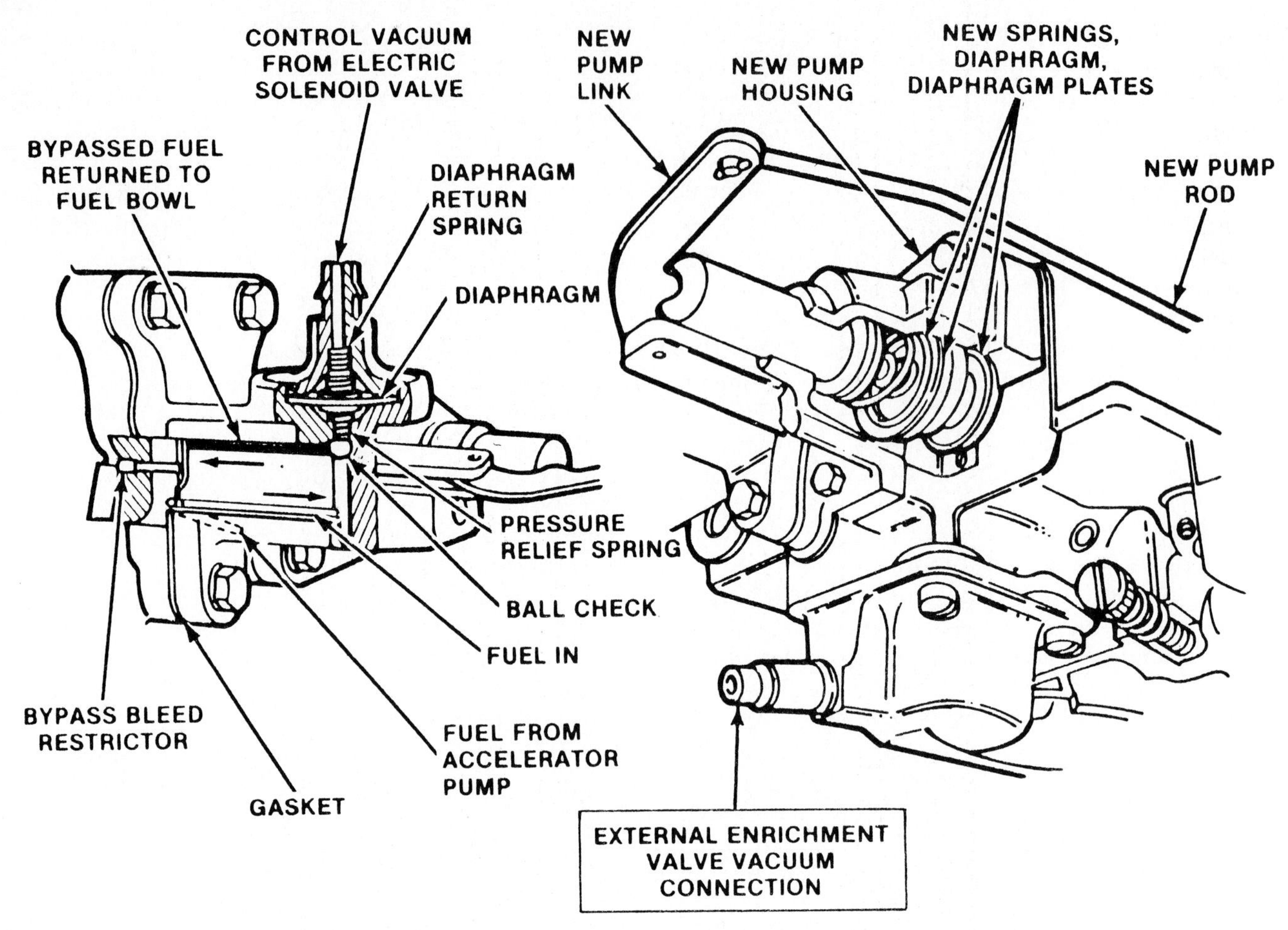

Figure 3–44 Temperature-Compensated Accelerator Pump (TCP). (Courtesy of Ford Motor Co.)

The idle speed control (ISC) motor is a direct current (DC) electric motor that is used to control idle speed after the choke is open. An ISC motor is shown in Figure 3–46.

A conventional fast idle cam that is operated by the choke spring provides the fast idle speed when the engine coolant is cold. The ISC motor will not move past preset limits, and it will not move if the idle tracking switch (ITS) signal to the ECA indicates that the throttle is not in the idle position. When the engine is shut off, the ISC motor will retract and close the throttle to prevent the engine from dieseling. After the engine stops running, the ISC motor will extend to its maximum travel for restart purposes. A time-delay circuit in the power relay supplies voltage to the ECA for 10 seconds after the ignition switch is turned off. This allows the ECA to extend the ISC motor after the ignition switch is turned off. When the engine is started, the ISC motor will be extended for a brief time to provide a faster idle speed. The time period when the ISC motor is extended depends on coolant temperature. Under cruising speed condi-

Figure 3–45 Variable-Voltage Choke (VVC) and Power Relay. (Courtesy of Ford Motor Co.)

tions the ISC motor will be extended to provide a dashpot action when the engine is decelerated. If the engine overheats, or if the battery voltage is low, the ECA will operate the ISC motor to increase the idle speed. The idle speed will also be increased by the ISC motor if the air conditioner is turned on or if the automatic transaxle is placed in neutral or park.

Most of the solenoids and relays in the EEC IV system are located behind a plastic cover on the firewall or in the fender shield. The ECA is located in the passenger compartment under the right kick pad. A complete wiring diagram of the EEC IV system used on the 2.8L V6 engine is illustrated in Figure 3–47.

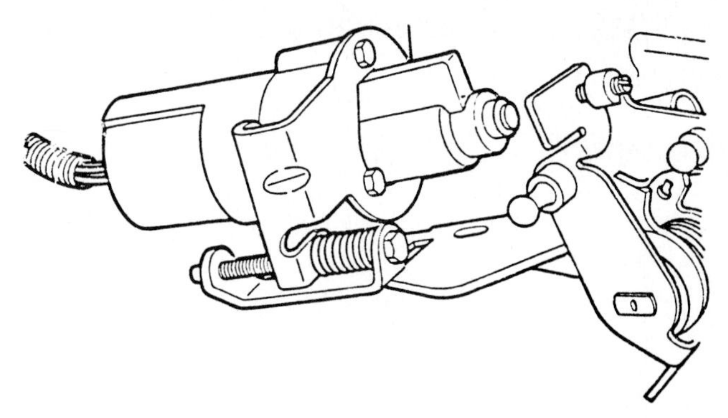

Figure 3–46 Idle Speed Control (ISC) Motor. (Courtesy of Ford Motor Co.)

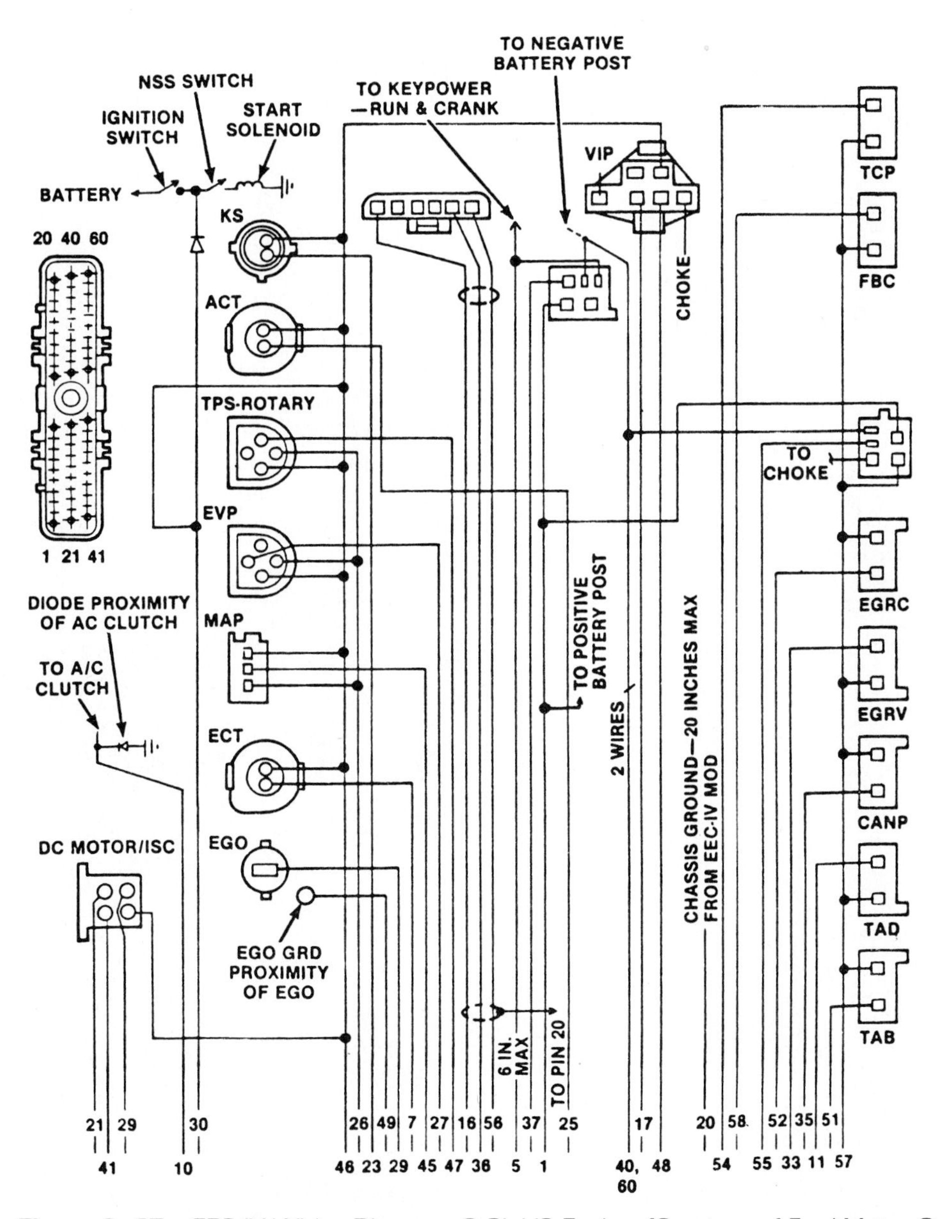

Figure 3–47 EEC IV Wiring Diagram, 2.8L V6 Engine. (Courtesy of Ford Motor Co.)

Chrysler Oxygen Feedback Carburetor Systems

Operation

In the oxygen (O_2) feedback carburetor system, the electronic fuel-control computer controls the on-time of the carburetor feedback solenoid winding. When the oxygen sensor sends a signal to the electronic fuel-control computer that indicates an excessively rich air–fuel ratio, the computer increases the on-time of the carburetor feedback solenoid and provides a leaner mixture in the main carburetor system, as illustrated in Figure 3–48.

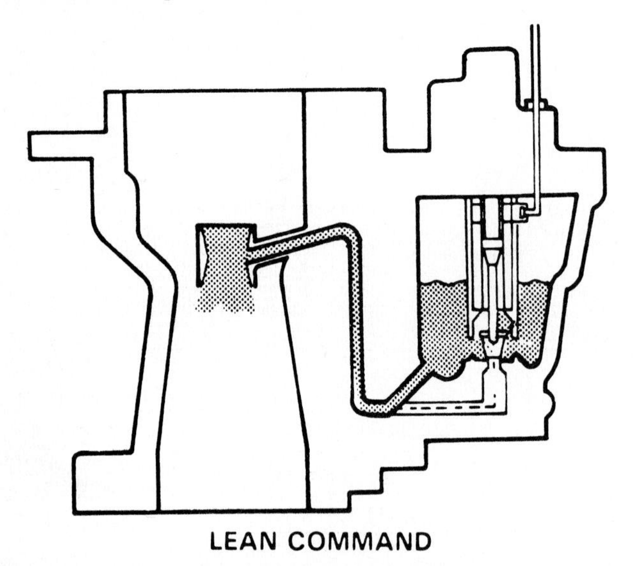

Figure 3–48 Carburetor Feedback Solenoid Lean Command. [Courtesy of Chrysler Canada Ltd.]

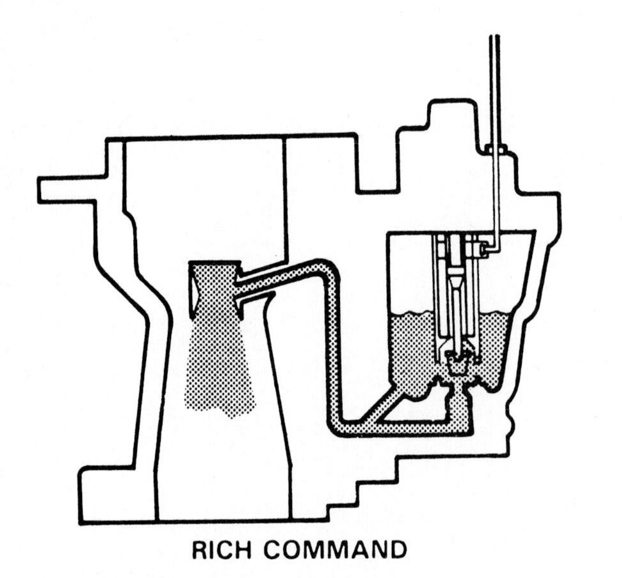

Figure 3–49 Carburetor Feedback Solenoid Rich Command. [Courtesy of Chrysler Canada Ltd.]

The return spring on the carburetor feedback solenoid holds the solenoid plunger in the upward position. When the electronic fuel-control computer energizes the carburetor feedback solenoid winding, the solenoid plunger is held in the downward position as indicated in Figure 3–48. If the oxygen sensor sends a lean air–fuel ratio signal to the electronic fuel-control computer, the computer will decrease the on-time of the solenoid winding. This will allow the solenoid plunger to remain in the upward position for a longer period of time and thus provide a richer air–fuel ratio, as shown in Figure 3–49.

The carburetor feedback solenoid also controls the air flow into the idle system. Downward movement of the solenoid plunger results in an increase in air flow into the idle system, and a leaner air–fuel ratio will be provided at idle speed. The air flow into the idle system is reduced when the solenoid plunger is in the upward position; therefore a richer air–fuel ratio will occur at idle speed. The air bleed from the top of the solenoid plunger into the idle circuit is illustrated in Figure 3–50.

In the closed-loop mode the electronic fuel-control computer operates the carburetor feedback solenoid in response to the oxygen sensor signal to provide an air–fuel ratio of 14.7:1. The electronic fuel-control computer and the electronic spark-advance computer are combined in one unit. The input sensor signals that are used by the combined computers are pictured in Figure 3–51.

The air pump system uses an air-switching valve to direct the air flow from the pump to the exhaust

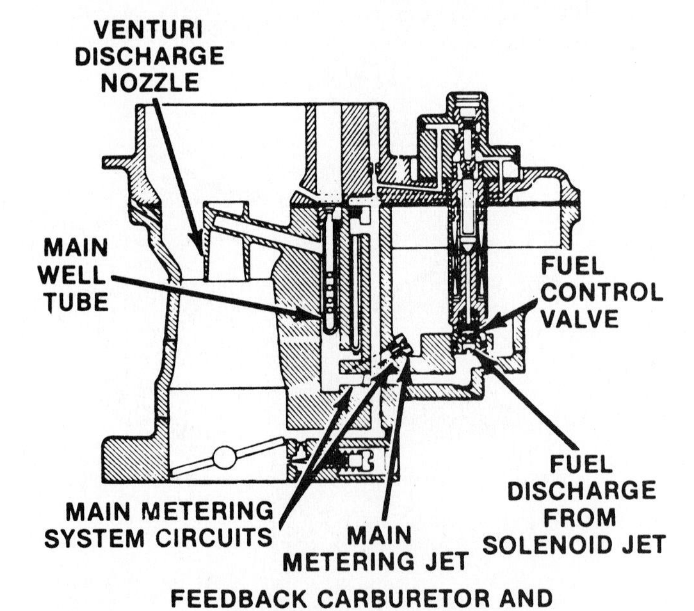

Figure 3–50 Feedback Solenoid with Air Bleed to Idle Circuit. [Courtesy of Chrysler Canada Ltd.]

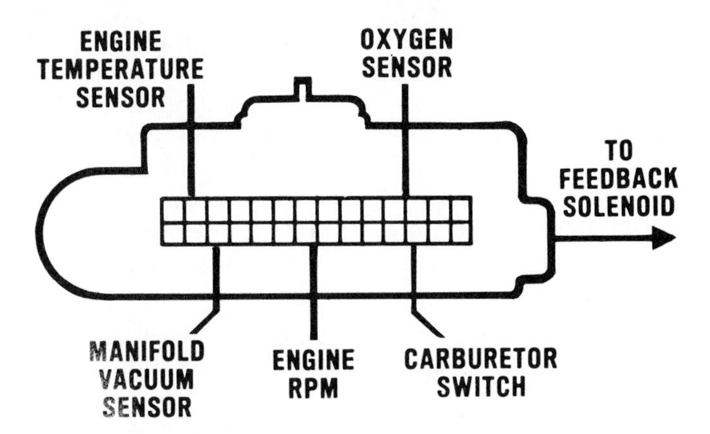

Figure 3–51 Computer Input Sensor Signals. (Courtesy of Chrysler Canada Ltd.)

ports when the engine coolant is cold. The air-switching valve directs air flow from the air pump to the catalytic converter when the engine is at normal operating temperature. A coolant-controlled vacuum switch applies manifold vacuum to the air-switching valve when the engine coolant is cold. On some engines that are equipped with an oxygen feedback system, the electronic fuel-control computer operates a vacuum solenoid winding to control the manifold vacuum that is applied to the air-switching valve. When the engine coolant is cold, the electronic fuel-control computer energizes the vacuum solenoid winding and thus allows vacuum through the solenoid to the air-switching valve. When manifold vacuum is applied to the air-switching valve, the air flow from the pump will be directed to the exhaust manifold, as shown in Figure 3–52.

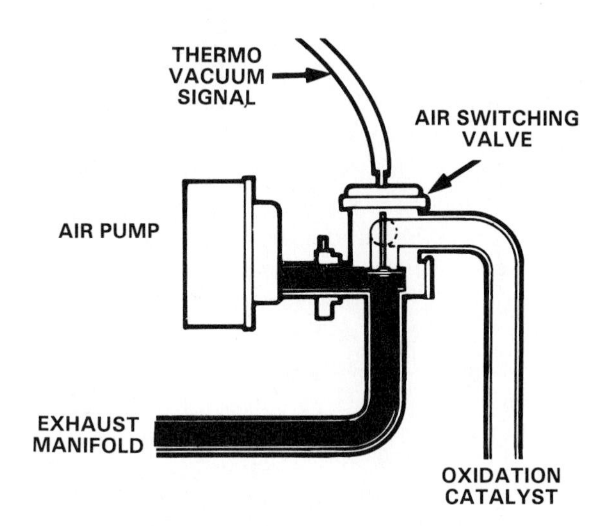

Figure 3–52 Air-Switching Valve Operation. (Courtesy of Chrysler Canada Ltd.)

The electronic fuel-control computer deenergizes the vacuum solenoid when the engine coolant is at normal operating temperature. Under this condition the vacuum solenoid will shut off the vacuum applied to the air-switching valve, and the air flow from the pump will be directed through the air-switching valve to the catalytic converter.

(An explanation of the ignition system and complete wiring diagrams for the Chrysler O_2 feedback system are provided in Chapter 5.)

Diagnosis of Oxygen Feedback Carburetor Systems

Diagnosis of Temperature Sensors When an O_2 feedback system is being diagnosed, all other sources of performance and economy complaints should be checked first, such as the electronic spark-advance system and engine compression.

Oxygen feedback carburetor systems may use a dual-terminal thermistor type of temperature sensor, or a single-terminal sensor to send a coolant temperature signal to the electronic fuel-control computer. A charge temperature switch may be mounted in the intake manifold, and therefore it sends a signal to the computer in relation to the air–fuel mixture temperature. When an ohmmeter is connected across the terminals on the dual-terminal temperature sensor, the ohmmeter should read 500 to 1000Ω if the engine coolant is cold, or above 1300Ω if the coolant is at normal temperature, as indicated in Figure 3–53.

If a single-terminal temperature switch is used, connect an ohmmeter from the switch terminal to ground. The ohmmeter should read zero if the engine coolant is cold, as illustrated in Figure 3–54.

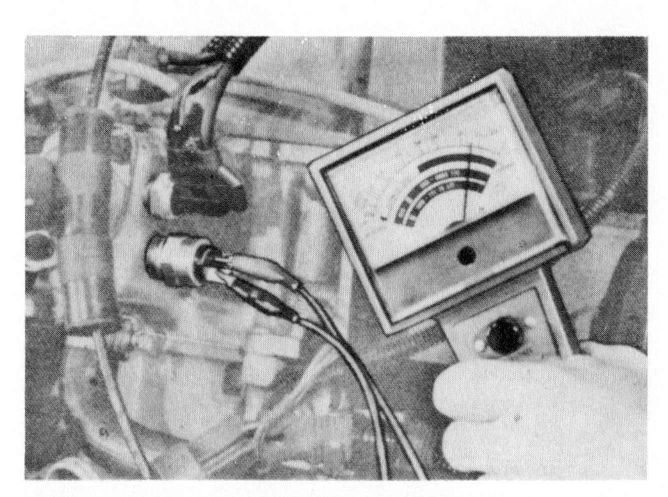

Figure 3–53 Dual-Terminal Temperature Sensor Test. (Courtesy of Chrysler Canada Ltd.)

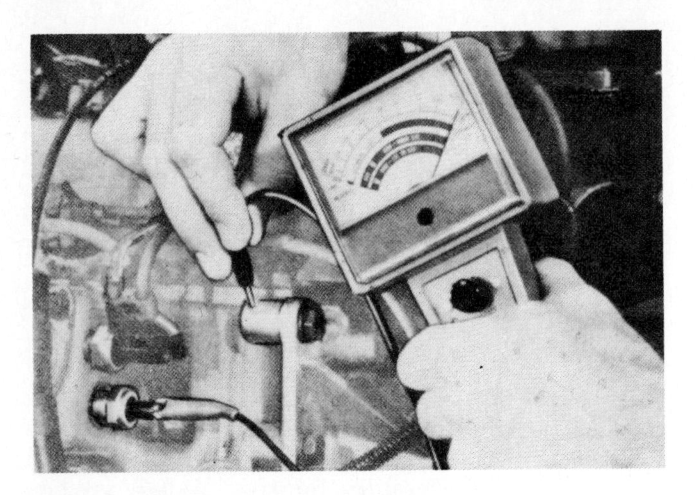

Figure 3–54 Single-Terminal Temperature Sensor Test. [Courtesy of Chrysler Canada Ltd.]

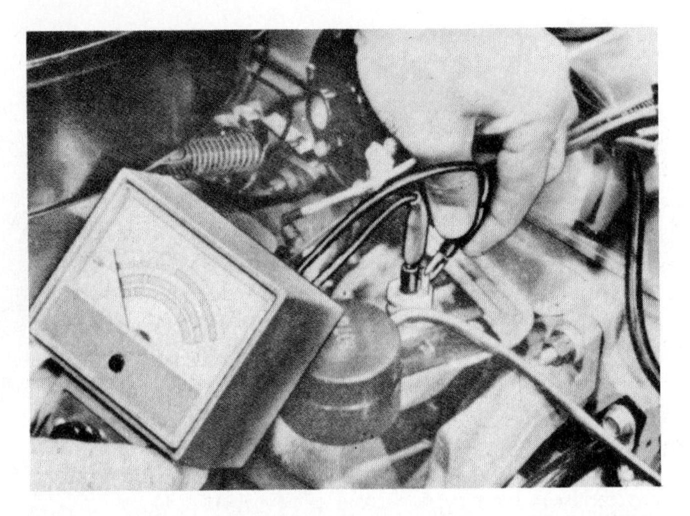

Figure 3–55 Charge Temperature Sensor Test. [Courtesy of Chrysler Canada Ltd.]

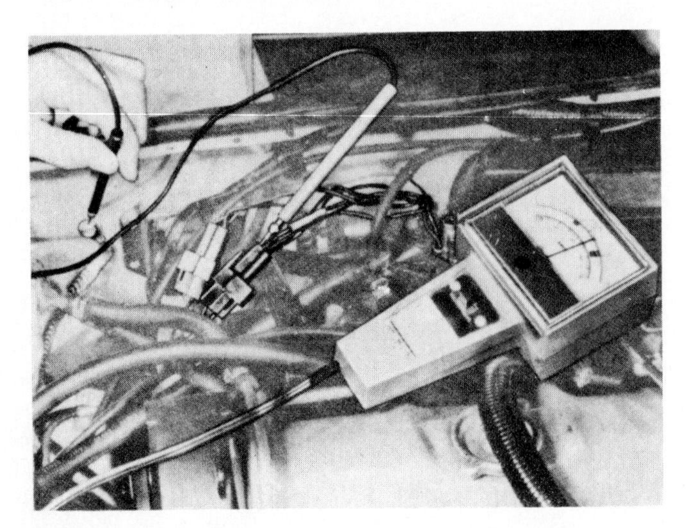

Figure 3–56 Carburetor Feedback Solenoid Test. [Courtesy of Chrysler Canada Ltd.]

When the engine coolant is at normal operating temperature, the ohmmeter reading should be infinite. A charge temperature sensor may be tested in the same way as the single-terminal coolant temperature sensor, as shown in Figure 3–55.

If the engine is cold, the ohmmeter reading should indicate zero ohms, and an infinite ohmmeter reading should occur when the engine is at normal temperature.

Diagnosis of Air Pump System When the engine is at normal operating temperature, the air flow from the air pump should be directed to the catalytic converter. During engine warmup this air flow should be injected to the exhaust manifold. The air-switching valve may be tested by disconnecting the vacuum hose and the air outlet hoses from the switching valve. When a hand-operated vacuum pump is used to apply 16 in hg (54 kPa) to the air-switching valve with the engine running, air flow should be directed from the outlet port that is connected to the exhaust manifold. If the vacuum pump is disconnected, air flow should be directed from the outlet port connected to the catalytic converter. An air-switching valve that does not operate as specified should be replaced.

Diagnosis of Carburetor Feedback Solenoid The following test procedure should be used to diagnose the carburetor feedback solenoid.

1. Connect a voltmeter from the blue wire that is connected to the carburetor feedback solenoid and ground as illustrated in Figure 3–56. The voltmeter should read battery voltage with the engine running. The blue wire is connected from the ignition switch to the feedback solenoid. On some applications, a set of oil-pressure switch contacts is connected in series with the feedback solenoid. If the voltmeter reading is less than specified, the oil-pressure switch, or the blue wire, is defective.
2. Apply 16 in hg (54 kPa) to the vacuum transducer on the computer. Operate the engine until it reaches normal temperature and install an insulator between the carburetor switch and the throttle linkage. The insulator must be thick enough to provide 2000 rpm on four-cylinder engines, or 1500 rpm on six- and eight-cylinder engines, as illustrated in Figure 3–57.
3. Disconnect the connector from the carburetor feedback solenoid. The engine speed should increase 50 rpm (as pictured in Figure 3–58) because the feedback solenoid will move upward and provide a richer air–fuel ratio.

4. With the blue wire connected to the feedback solenoid, connect a jumper wire from the green wire terminal on the solenoid to ground. The engine speed should decrease 100 rpm compared to the engine speed with the connector disconnected, because the feedback solenoid will move to the downward position and provide a leaner air–fuel ratio. If the speed-changes in steps 3 and 4 are not within specifications, the carburetor should be repaired or replaced.

Diagnosis of Computer Wiring On computers with one double and one single connector, disconnect the 6-terminal single connector and connect terminal 15 to ground with the engine idling. If the computer is equipped with two double connectors, disconnect the 12-terminal double connector and connect terminal 11 to ground, as shown in Figure 3–59.

Terminal 15, or terminal 11, is connected through the green wire to the carburetor feedback solenoid. When either of these terminals is connected to ground, the feedback solenoid should move downward and the engine speed should decrease 50 rpm. If there is no change in engine speed, and tests at the feedback solenoid are satisfactory, repair the open circuit or the grounded condition in the green wire that is connected from terminal 11 (or 15) to the feedback solenoid.

Diagnosis of Fuel-Control Computer When testing the fuel-control computer, operate the engine at normal temperature and 1500 rpm on six- and eight-cylinder engines or 2000 rpm if the vehicle has a four-cylinder engine. Use a hand vacuum to apply 16 in hg (54 kPa) to the vacuum transducer in the computer. Use a probe to connect the positive voltmeter lead to the green wire at the feedback solenoid, and connect the negative voltmeter lead to ground. Disconnect the oxygen sensor connector and connect one hand to the sensor wire from the computer and touch the other hand to the negative battery terminal.

When this connection is completed, the voltmeter should read over 9V, and the engine speed should increase 50 rpm. If there is no change in engine speed and the voltmeter reads less than 9V, connect one hand to the oxygen sensor wire connected to the computer and touch the other hand to the battery positive post. The engine speed should decrease at least 50 rpm, and the voltmeter should read less than 3V. If there is no change in voltage and engine speed, replace the computer.

Diagnosis of the Oxygen Sensor The engine speed and test conditions must be the same when testing the oxygen sensor or the fuel-control computer. Re-

Figure 3–57 Carburetor Feedback Solenoid Test Preparation. (Courtesy of Chrysler Canada Ltd.)

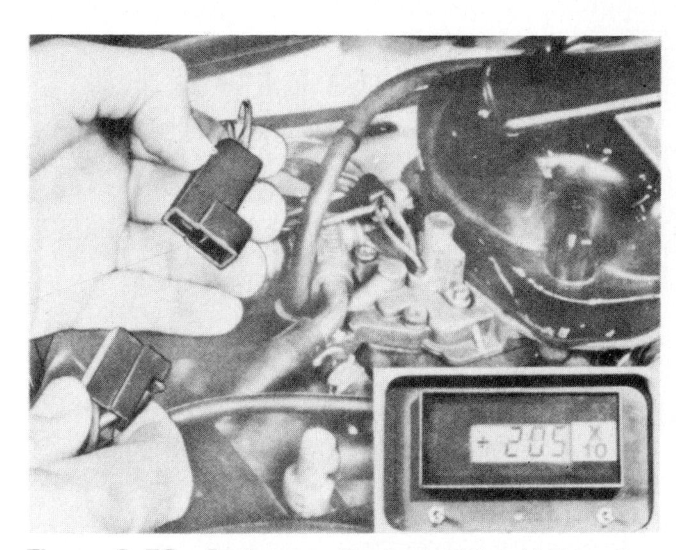

Figure 3–58 Carburetor Feedback Solenoid Test Connector Disconnected. (Courtesy of Chrysler Canada Ltd.)

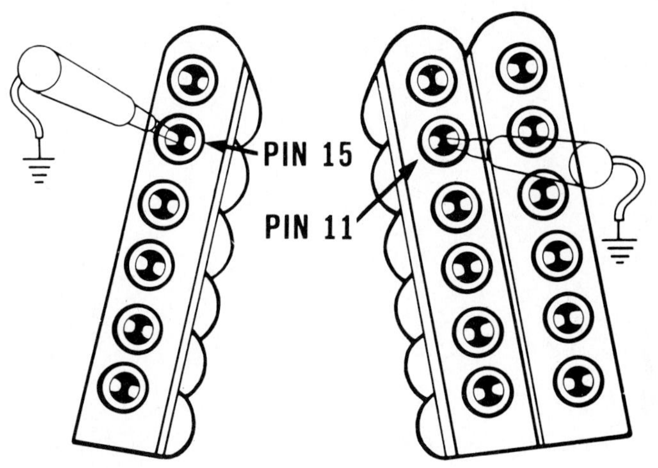

Figure 3–59 Computer Wiring Test. (Courtesy of Chrysler Canada Ltd.)

connect the oxygen sensor wire and move the choke to the 75 percent closed position with the air cleaner removed. The computer should react within 10 seconds to provide a voltmeter reading of 3V or less, as shown in Figure 3–60.

Reinstall the air cleaner and remove the positive crankcase ventilation (PCV) hose from the PCV valve. Cover the PCV hose with your thumb and slowly uncover the hose to create a large manifold leak. Within 10 seconds the computer should react to provide a voltmeter reading of 9V or over. Each oxygen sensor test should be limited to 90 seconds. If the voltmeter readings are not correct, the oxygen sensor should be replaced.

Chrysler O_2 Feedback Systems with Self-Diagnostic Capabilities

Diagnosis On later model O_2 feedback systems, the computer has self-diagnostic capabilities. If certain defects occur in the system, fault codes will be stored in the computer memory. When the system is being diagnosed, a digital readout tester must be connected to a diagnostic connector near the right front strut tower. This tester is also used to diagnose electronic fuel-injection (EFI) systems. (Diagnosis of EFI systems is explained in Chapter 4.) A complete wiring diagram of the O_2 feedback system with self-diagnostic capabilities is illustrated in Figure 3–61, and the readout tester connection is shown in Figure 3–62.

Fault-Code Diagnosis When a driveability complaint is being diagnosed, the engine compression

Figure 3–60 Oxygen Sensor Test. (Courtesy of Chrysler Canada Ltd.)

and the ignition system should be checked prior to a diagnosis of the O_2 feedback system. The wiring harness connections and vacuum hoses should always be checked before the fault-code diagnosis. Proceed with the fault-code diagnosis as follows:

1. With the engine at normal operating temperature, and the ignition switch in the off position, connect the digital readout tester to the diagnostic connector.
2. Place the fast idle screw on the highest step of the fast idle cam to open the carburetor switch.
3. Move the read/hold tester switch to the read position.
4. Turn on the ignition switch and do not cycle the switch on and off.
5. When 00 appears on the tester display, set the read/hold switch to the hold position.
6. Record all the fault codes indicated on the tester display.

Actuation Test Mode

A fault code indicates a defect in a certain circuit. For example, code 17 proves that a problem exists in the electronic throttle-control solenoid circuit. When this code is provided, it indicates a defect in this solenoid, or in the connecting wires between the solenoid and the computer. This code may also indicate that the computer is not capable of operating the electronic throttle-control solenoid. The actuation test mode may be used to diagnose a certain component in the system when a fault code indicates a defect in this component. Proceed with the actuation test mode as follows:

1. When the fault-code diagnosis is completed, the tester will display 55. At this time press and hold the actuation test mode button on the tester, which causes different actuation test mode (ATM) codes to be displayed on the tester.
2. When the desired ATM code appears on the tester, release the ATM button. This action will cause the computer to operate the component indicated by the ATM code. For example, if a fault code 17 indicates a problem in the electronic throttle-control solenoid, ATM code 97 may be used to activate this solenoid. In the ATM mode the technician may listen to the component or perform specific voltage tests to locate the exact defect.

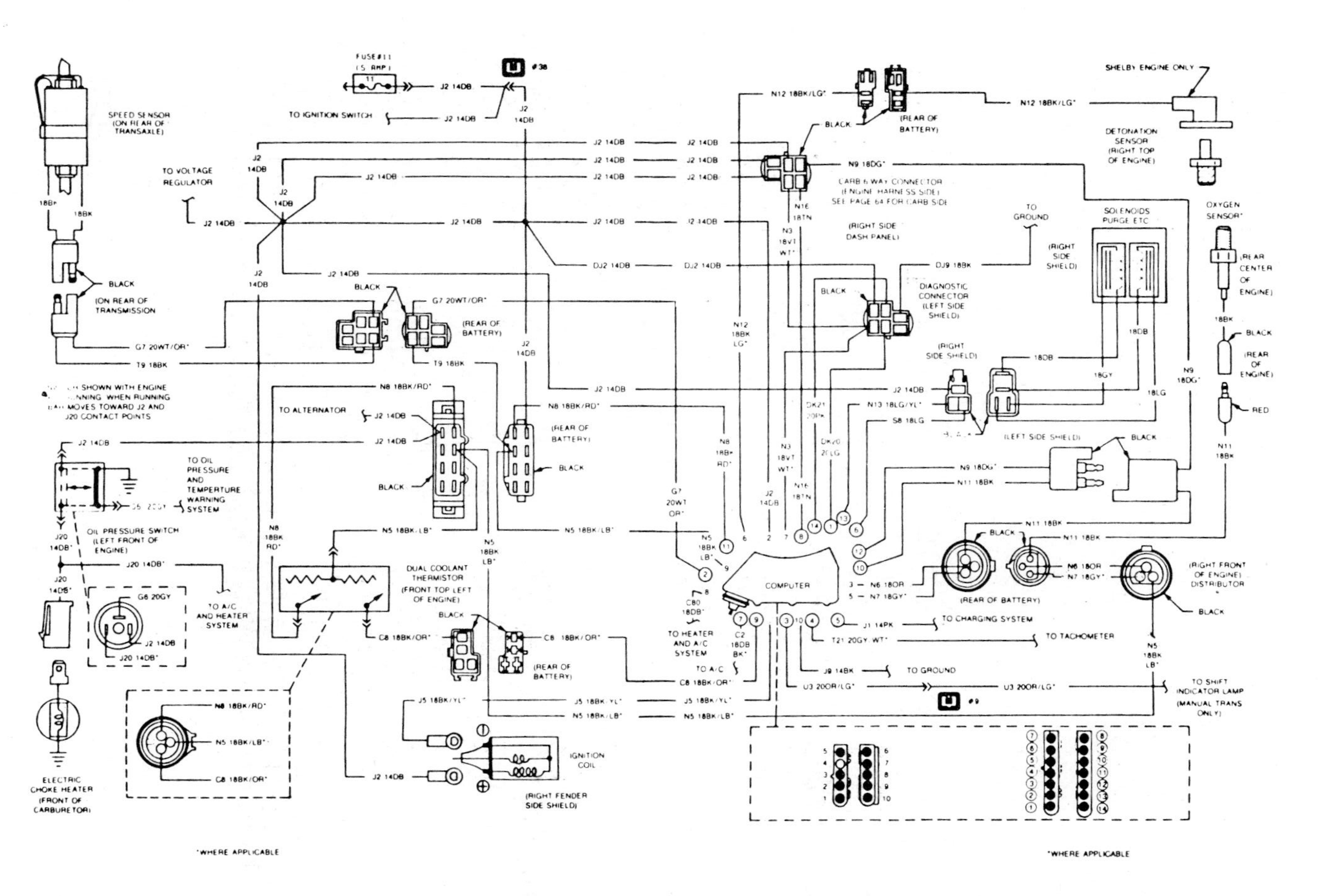

Figure 3-61 Chrysler O_2 Feedback System with Self-Diagnostic Capabilities. [Courtesy of Chrysler Canada Ltd.]

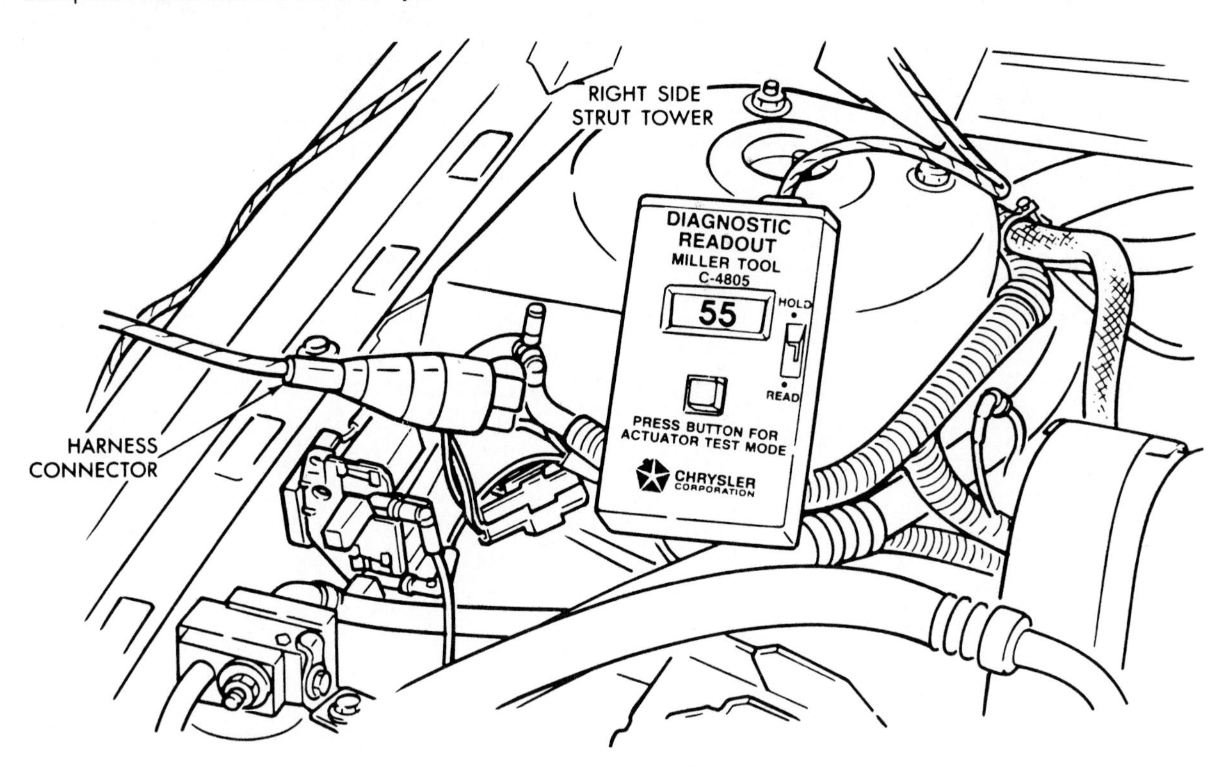

Figure 3–62 Digital Readout Tester Connection. [Courtesy of Chrysler Canada Ltd.]

3. The computer will operate each component indicated by an ATM code for 5 minutes. Each ATM code test will be stopped if the ignition switch is turned off.

Switch Test Mode

Certain switches in the O_2 feedback system may be tested in the switch test mode. This test mode must be entered after the fault-code test mode. The procedure for the switch test mode is the following:

1. When 55 is displayed on the tester at the end of the fault-code diagnosis, turn off a switch input to the computer such as the air conditioning switch.
2. Press the ATM button and immediately set the read/hold switch to the read position.
3. When 00 appears on the tester display, turn on the air conditioning switch.
4. If this switch input is received by the computer, 88 will appear on the display. The display reading will return to 00 when the air conditioning switch is turned off. Each computer input switch may be tested in the same way.

Fault codes may be erased from the computer memory by disconnecting the 14 terminal for 10 seconds with the ignition switch off. The fault codes, monitored conditions, and ATM codes are shown in Table 3–1.

Test Questions

Questions on 3C Systems

1. When the air–fuel ratio becomes richer, the oxygen sensor voltage ______________.
2. Leaded gasoline may be used in a vehicle that has an oxygen sensor in the exhaust system. T F
3. If the electronic control module (ECM) increases the on-time of the mixture-control solenoid winding, the air–fuel ratio will become ______________.
4. Air flow from the air pump should be directed to the catalytic converter when the engine is operating at normal temperature and light load cruising speed conditions. T F
5. If the check-engine light comes on when the engine is running, it indicates a defect in the 3C system. T F
6. When checking basic timing, the ______________ should be disconnected.

TABLE 3–1 **Diagnostic Codes, O_2 Feedback System.**

Code	Type	Engine	Circuit	When Monitored By The Computer	When Put Into Memory	ATM Test Code
11	Fault	1.6 - 2.2	Carburetor O_2 Solenoid	All the time when the engine is running.	If the O_2 solenoid circuit does not respond to computer commands.	91
12	Not Used				Disregard this code.	
13	Fault	2.2	Canister Purge Solenoid	All the time when the engine is running.	If the canister purge solenoid circuit does not turn on and off at the correct time.	93
14	Indication	1.6 - 2.2	Battery Feed for Computer Memory	All the time when the ignition switch is on.	If the battery was disconnected within the last 20-40 engine starts.	None
16	Not Used				Disregard this code.	
17	Fault	2.2	Electronic Throttle Control Solenoid	All the time when the engine is running.	If the throttle circuit control solenoid does not turn on and off at the correct time.	97
18	Fault	2.2	Vacuum Operated Secondary Control Solenoid (V.O.S.)	All the time when the engine is running.	If the V.O.S. control solenoid circuit does not turn on and off at the correct time.	98
21	Fault	1.6 - 2.2	Distributor Pickup Coil	During engine cranking.	If there is no distributor signal input at the computer.	None
22	Fault	1.6 - 2.2	Oxygen Feedback System	All the time when the engine is running in closed loop operation.	If the oxygen feedback system stays rich or lean longer than 5 minutes.	None
24	Fault	1.6 - 2.2	Computer	All the time when the engine is running.	If the vacuum transducer fails.	None
25	Fault	2.2	Radiator Fan Temperature Sensor	All the time when the engine is running.	If the engine temperature sensor circuit indicates 100°F or less and the radiator fan temperature does not agree or changes too fast to be real.	None
26	Fault	1.6 - 2.2	Engine Temperature Sensor	All the time when the engine is running.	If the engine temperature sensor circuit does not read 100°F after 30 minutes from when the engine started. Also, if the circuit is shorted or changes too fast to be real.	None
28	Fault Manual Trans Only	1.6 - 2.2	Speed Sensor	During vehicle deceleration from 2000 to 1800 rpm for 3 seconds, engine temperature above 150°F and vacuum above 21.5 inches.	If speed sensor circuit does not indicate between 2 and 150 mph.	None
31	Fault	1.6 - 2.2	Battery Feed for Computer Memory	All the time when the ignition switch is on.	If the engine has not been cranked since the battery was disconnected.	None
32 33	Fault	1.6 - 2.2	Computer	Upon entry into the diagnostic mode.	If computer fails.	None
55	Indication	1.6 - 2.2			Indicates end of diagnostic mode.	None
88	Indication	1.6 - 2.2			Indicates start of diagnostic mode. **NOTE:** This code must appear first in the diagnostic mode or fault codes will be inaccurate. Indicates switch is on in switch test mode.	None
00	Indication	1.6 - 2.2			Indicates that the diagnostic readout box is powered up. Indicates switch is off in switch test mode.	None

(Courtesy of Chrysler Canada Ltd.)

7. If the system is working normally, the mixture-control solenoid dwell reading will be varying continuously between:
   (a) 10° and 30°.
   (b) 30° and 50°.
   (c) 6° and 54°.
8. A mixture-control solenoid dwell reading that is fixed at 55° indicates that the system is trying to provide a ________ air–fuel mixture.

Questions on Electronic Engine Control Systems

1. On an EEC IV system used on a 2.3L engine, the cooling fan and air conditioning compressor clutch are operated by the ________.
2. The upshift light is turned on when a shift to the next highest gear will provide optimum ________.
3. The temperature-compensated accelerator pump supplies more fuel when the engine is:
   (a) operating at normal temperature.
   (b) overheated.
   (c) operating during warmup.
4. When the temperature decreases, the duty cycle of the variable voltage choke:
   (a) decreases.
   (b) remains the same.
   (c) increases.

Questions on Chrysler Feedback Systems

1. If the electronic fuel-control computer receives a rich mixture signal from the oxygen sensor it will ________ the on-time of the carburetor feedback solenoid.
2. The manifold vacuum that is applied to the air-switching valve is controlled by a coolant-controlled vacuum switch or a ________ vacuum solenoid.
3. If the air flow is directed from the air pump to the exhaust manifold with the engine at normal temperature, the oxygen sensor signal will indicate a ________ air–fuel ratio.
4. When manifold vacuum is applied to the air–switching valve, the air flow from the air pump will be diverted to the ________.
5. When the engine is operating at normal temperature and 1500 rpm, if the carburetor feedback solenoid is disconnected, the engine speed should ________.
6. If the choke is moved to the 75 percent closed position with the engine running with a voltmeter connected from the green wire on the feedback solenoid to ground, the voltmeter reading should ________ because the computer ________ the on-time of the feedback solenoid.

4

Computer-Controlled Fuel-Injection Systems

Practical Completion Objectives

1. Diagnose General Motors throttle body injection systems using self-diagnostics in the system.
2. Diagnose General Motors port-type injection systems using the self-diagnostics in the system.
3. Test fuel pump pressure on throttle body or port-type injection systems.
4. Use the self-diagnostics in Ford EEC IV systems to diagnose system complaints.
5. Perform a diagnosis of Chrysler single-point or multipoint injection systems using the fault codes.
6. Observe safety precautions when work is performed on fuel-injection systems.

General Motors Throttle Body Injection Systems

Throttle Body Injection Assemblies

In a throttle body injection system, the carburetor is replaced with a throttle body injection assembly. This assembly may contain a single injector, as illustrated in Figure 4–1.

When a single injector is used, the injector system is referred to as "electronic fuel injection" (EFI) or "throttle body injection" (TBI).

Some larger displacement engines have dual injectors in the throttle body assembly, as shown in Figure 4–2.

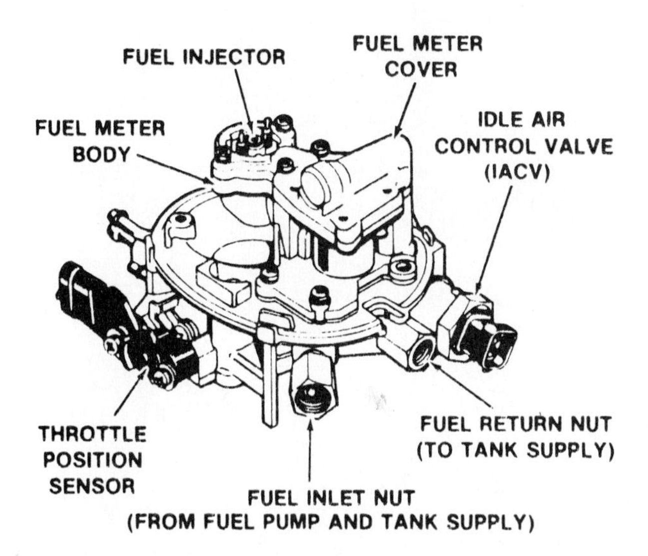

Figure 4–1 Throttle Body Injector (TBI) Assembly. (Courtesy of General Motors Corporation)

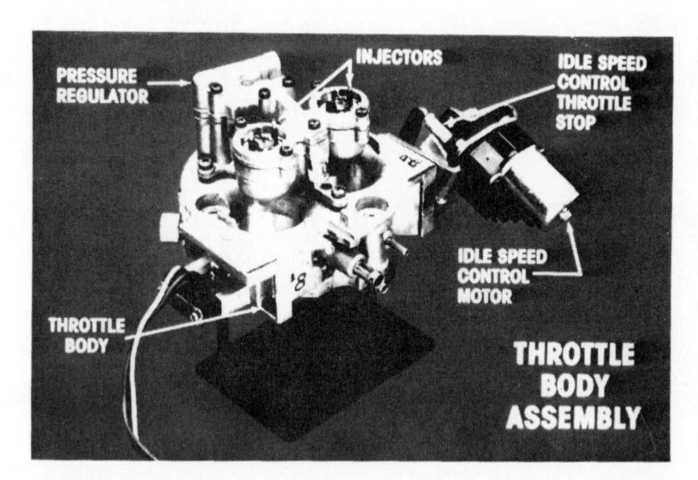

Figure 4–2 Digital Fuel Injection (DFI), Throttle Body Injector Assembly. (Reprinted with Permission Society of Automotive Engineers, © 1980.)

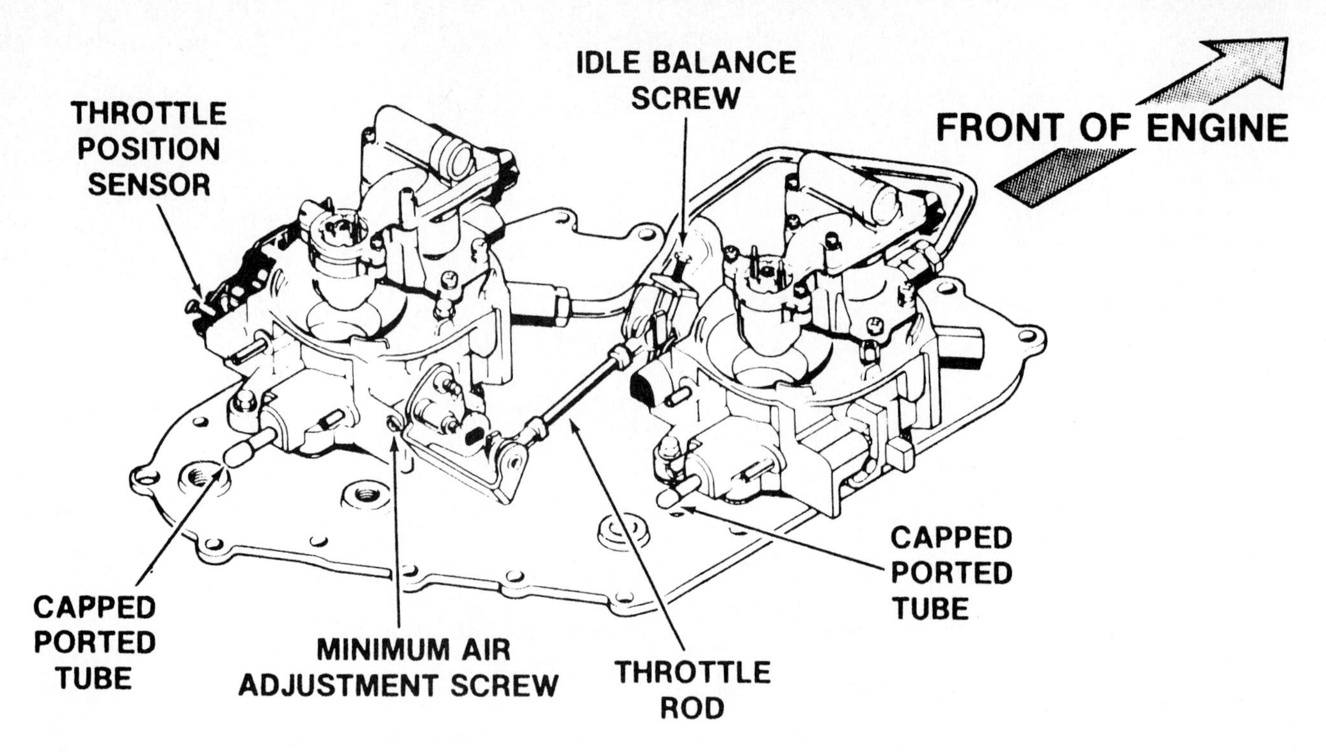

Figure 4–3 Crossfire Injection (CFI) Throttle Body Injector Assembly. (Courtesy of General Motors Corporation)

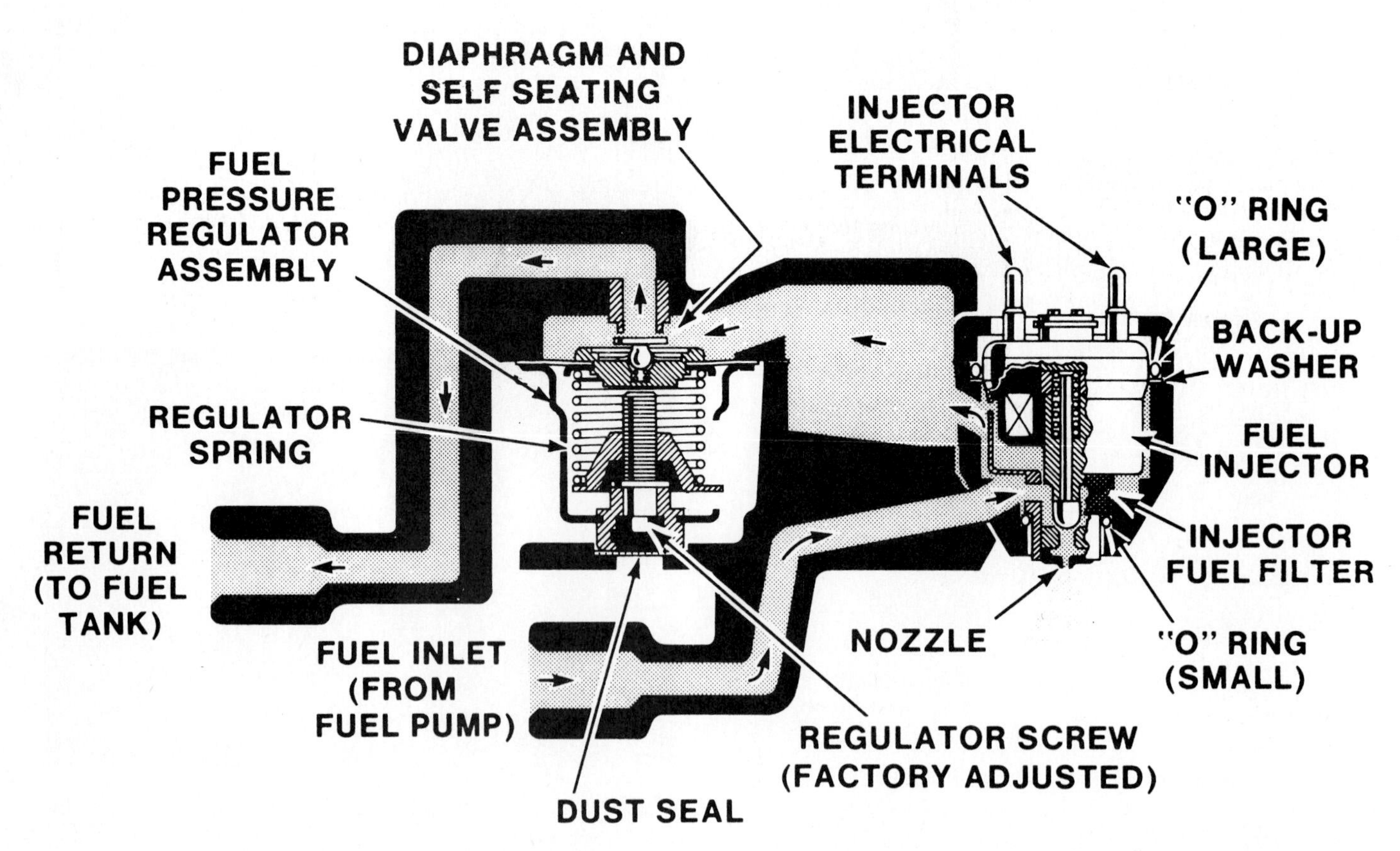

Figure 4–4 Internal Design of Throttle Body Injector Assembly. (Courtesy of General Motors Corporation)

Such systems are called "digital fuel injection" (DFI) or "digital electronic fuel injection" (DEFI). Other throttle body injection systems use a pair of single injectors that are spaced in an offset position on a special intake manifold, as pictured in Figure 4–3.

When offset single injectors are used, the injection system is called crossfire injection (CFI).

Regardless of the type of throttle body injector assembly used, the internal structure is similar: gasoline is forced in the fuel inlet from the fuel pump, and it then flows past the injector(s) to the pressure regulator; when fuel pressure reaches 10 psi (69 kPa), the pressure-regulator spring will be compressed, the pressure-regulator valve will be able to open, and any excess fuel will be returned to the fuel tank. Pressurized fuel surrounds the injectors at all times, as illustrated in Figure 4–4.

A return spring on the injector plunger holds the injector closed. The injector will be opened when the solenoid coil surrounding the plunger is energized by the electronic control module (ECM). When the injector is opened, fuel is sprayed into the airstream above the throttles, as shown in Figure 4–5. The ECM varies the open time of the injector to control the amount of fuel injected.

Complete Fuel System

A twin turbine fuel pump is mounted in the fuel tank with the gauge-sending unit. The fuel pump forces fuel through the filter under the vehicle to the throttle body injector assembly. Excess fuel from the injector assembly flows through the return fuel line to the fuel tank, as illustrated in Figure 4–6.

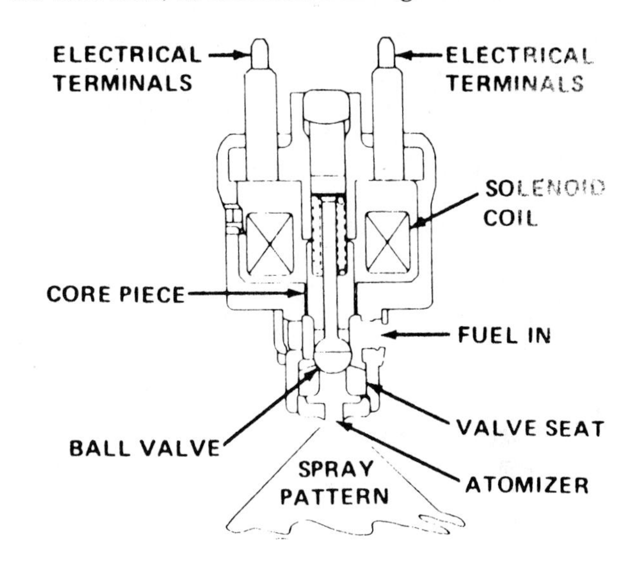

Figure 4–5 Injector Assembly. [Courtesy of General Motors Corporation]

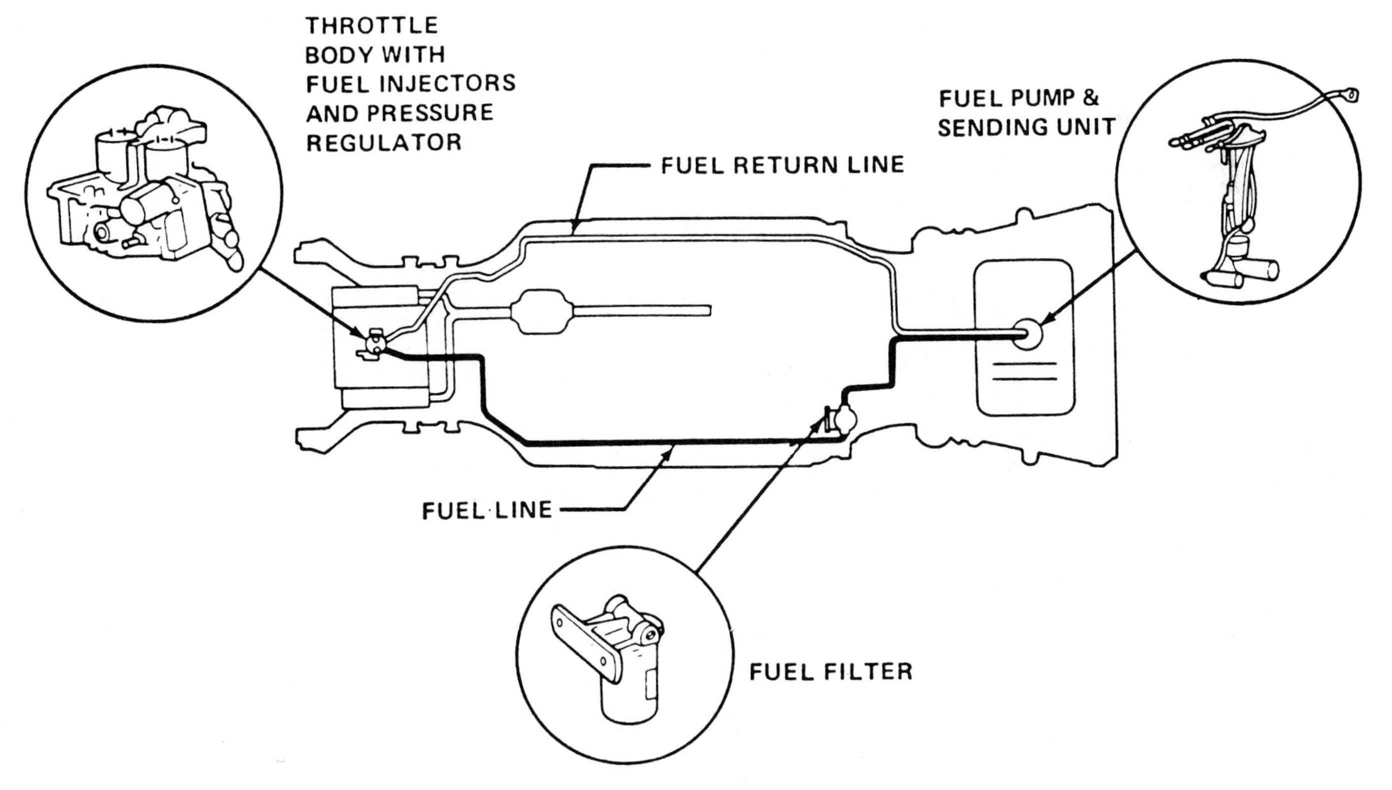

Figure 4–6 Complete Throttle Body Fuel System. [Courtesy of General Motors Corporation]

The ECM energizes the fuel pump relay winding when the ignition switch is turned on and the relay points close. Thus battery voltage is allowed to go through the relay points to the fuel pump, which will then supply fuel to the injectors. If no attempt is made to start the vehicle within 2 seconds from the time the ignition switch is turned on, the ECM will de-energize the fuel pump relay winding, and the relay points will open the circuit to the fuel pump. The ECM energizes the fuel pump relay winding as long as the engine is running. An alternate circuit is provided through a special set of oil pressure switch contacts to the fuel pump, as illustrated in Figure 4–7.

If the fuel pump relay becomes defective, power will still be supplied through the oil pressure switch contacts to the fuel pump. This makes it possible to drive the vehicle if the fuel pump relay is defective.

Throttle Body Injection (TBI) System Components

The electronic control module (ECM) is the "brain" of the TBI system. The ECM receives information from the sensors and performs the necessary control functions. A removable programmable read-only memory (PROM) is installed in the ECM, as pictured in Figure 4–8.

The PROM contains calibration data for each engine, transmission, and rear axle ratio. This information will always be retained in the permanent

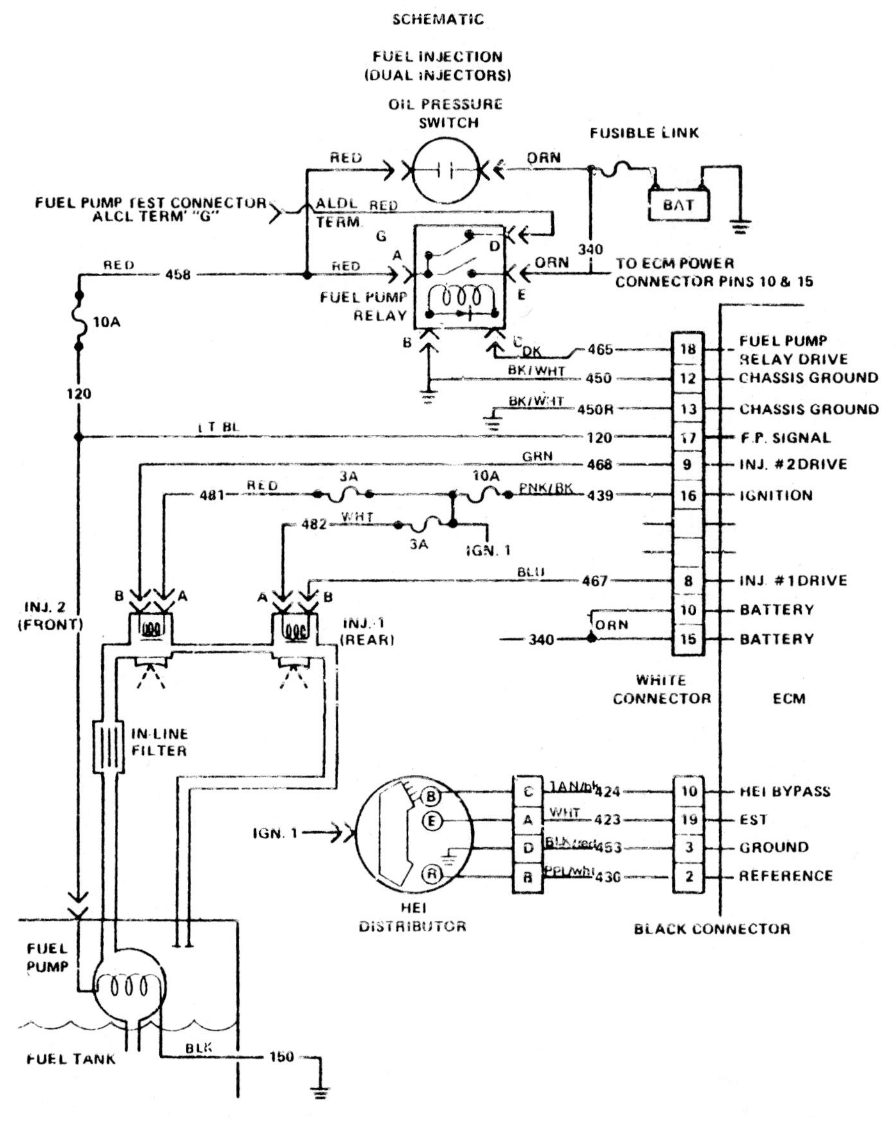

Figure 4–7 Wiring of Electronic Fuel-Injection System. (Courtesy of General Motors Corporation)

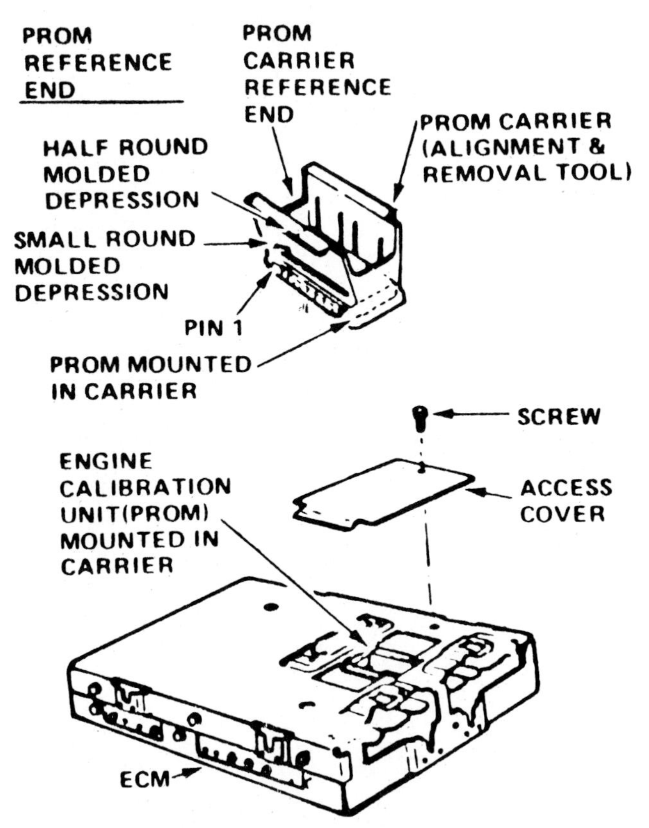

Figure 4–8 Electronic Control Module (ECM) with Programmable Read-Only Memory (PROM). (Courtesy of General Motors Corporation)

memory of the PROM. Special tools are available for PROM removal and replacement. The technician must be careful to install the PROM in the correct position and to be sure the pins on the PROM are not bent during installation. An alignment notch is used to position the PROM correctly, as shown in Figure 4–8.

The most important electronic components of TBI systems are:

1. Manifold absolute pressure (MAP) sensor.
2. Barometric pressure (BP) sensor.
3. Throttle position sensor (TPS).
4. Coolant temperature sensor (CTS).
5. Manifold air temperature sensor.
6. Oxygen (O_2) sensor.
7. Distributor reference pulses.
8. Vehicle speed sensor (VSS).
9. Park/neutral (P/N) switch.
10. Air conditioning clutch switch.

The input sensors are basically the same as the sensors used in the computer command control (3C) system. (See Chapter 3 for a complete description of the 3C system.) All the components listed above may not be used in every throttle body injection system.

Electronic Control Module (ECM) Control Functions

Fuel

The ECM maintains the correct air–fuel ratio of 14.7:1 by controlling the time that the injectors are open. The ECM varies the open time or on-time of the injectors from 1 to 2 ms at idle to 6 to 7 ms at wide-open throttle. As the on-time of the injector increases, the amount of fuel injected increases proportionately. The injectors are operated by the ECM in two different modes: synchronized and nonsynchronized. In the synchronized mode, the injector is energized by the ECM each time a distributor reference pulse is received by the ECM. On systems with dual throttle body injectors, the injectors are pulsed alternately. In the nonsynchronized mode of operation, the injectors are energized every 12.5 or 6.5 ms, depending on the application. This pulse time is controlled by the ECM, and it is independent of distributor reference pulses. The nonsynchronized pulses take place under the following conditions:

1. When the injector on-time becomes less than 1.5 ms.
2. During the delivery of prime pulses that are used to charge the intake manifold with fuel when the engine is being started or just before starting.
3. During acceleration enrichment.
4. During deceleration leaning.

If the engine coolant temperature is cold, the ECM will deliver "prime pulses" to the injector while the engine is being cranked and the throttle is less than 80 percent open. This system eliminates the need for a conventional choke. As the coolant temperature decreases, the ECM will increase the on-time or pulse width of the injector. At −33°F (−36°C) coolant temperature, the ECM will increase the injector pulse width to provide an air–fuel ratio of 1.5:1 for cold-start purposes. If dual throttle body injectors are used, the ECM may energize both injectors simultaneously during the prime pulses.

Should a cold engine become flooded, the ECM has the capability to clear this condition. To clear a flooded engine, the driver must depress the accelerator pedal to the wide-open position. The ECM will then reduce the injector pulse width to deliver an air–

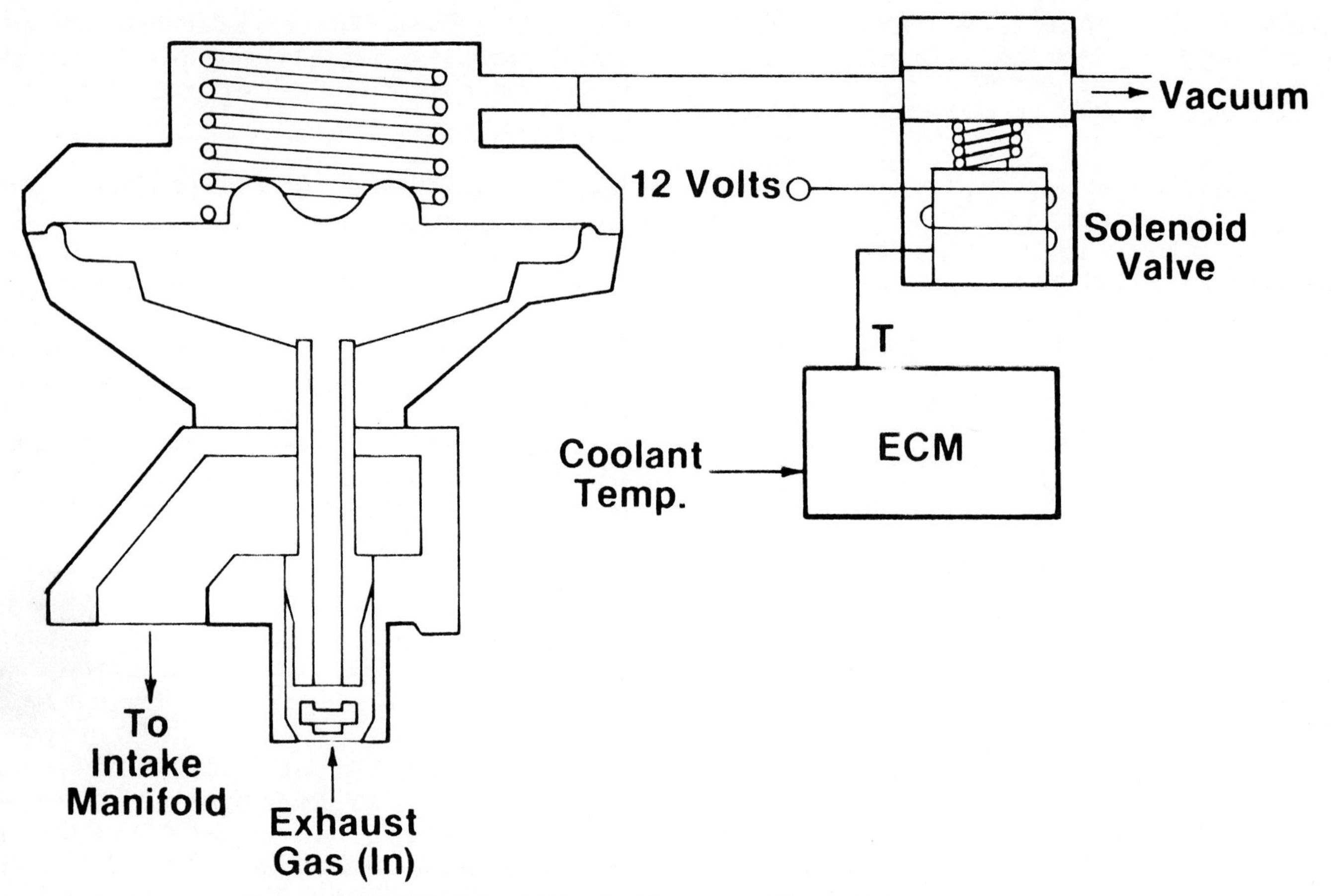

Figure 4-9 EGR Solenoid Circuit. [Courtesy of General Motors Corporation]

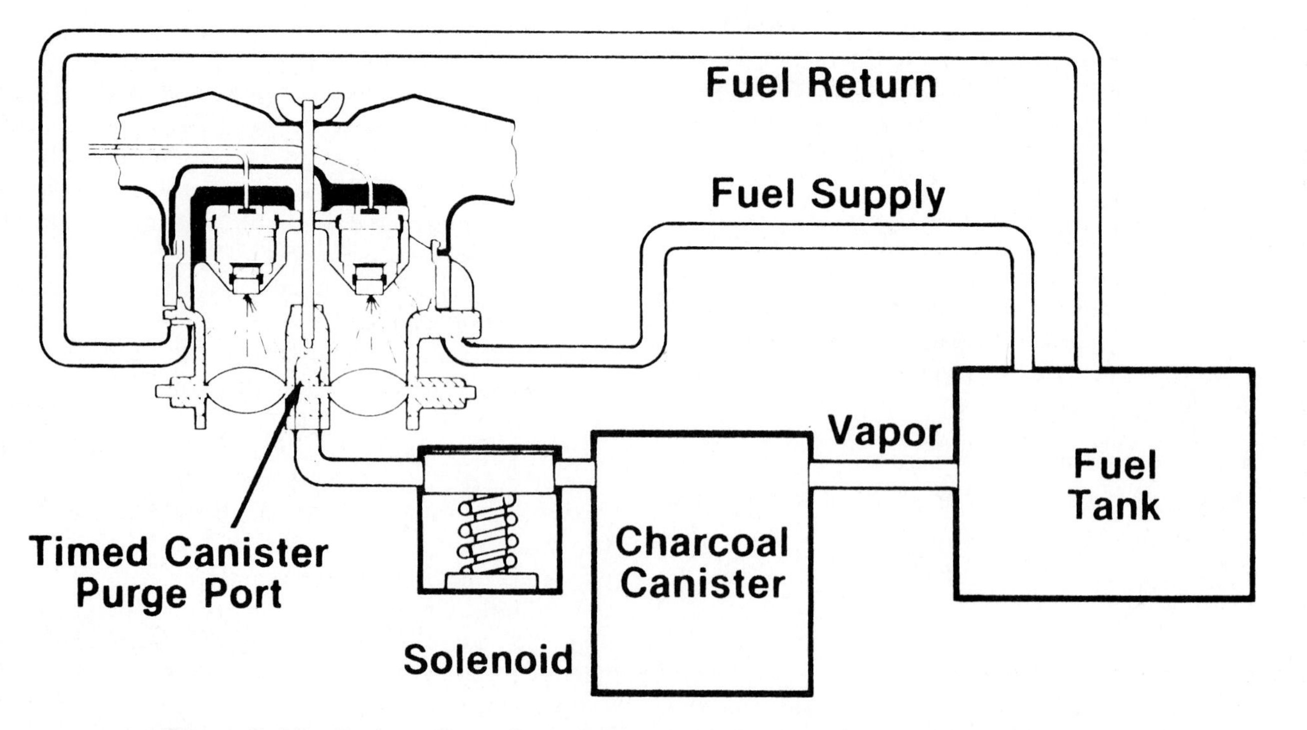

Figure 4-10 Canister Purge Solenoid Circuit. [Courtesy of General Motors Corporation]

fuel ratio of 20:1. This ratio will be maintained as long as the throttle is held wide open and the engine speed is below 600 rpm.

When the engine is running at a speed above 600 rpm, the system enters the open-loop mode. In this mode the ECM ignores the signal from the oxygen (O_2) sensor and calculates the injector pulse width on the basis of the input signals from the coolant temperature sensor (CTS) and the manifold absolute pressure (MAP) sensor. During the open-loop mode, the ECM analyzes the following information before it enters the closed-loop mode:

1. Oxygen sensor voltage output must be present.
2. Temperature of the CTS sensor (it must be at or above normal).
3. The time that has elapsed from the engine start-up (a specific period is required).

When the ECM is satisfied with the input signals, the system will enter the closed-loop mode. In this mode the ECM will modify the injector on-time on the basis of the signal from the O_2 sensor to provide the ideal air–fuel ratio of 14.7:1.

If the engine is accelerated suddenly, the manifold vacuum will decrease rapidly. The ECM senses the increase in throttle opening from the TPS sensor, and it senses the decrease in manifold vacuum from the MAP sensor. When the ECM receives input signals indicating sudden acceleration of the engine, it will increase the injector on-time to provide a slightly richer air–fuel ratio. This acceleration enrichment prevents a hesitation when the engine is accelerated.

If the engine is decelerated, a leaner air–fuel ratio is required to reduce carbon monoxide (CO) and hydrocarbon (HC) emission levels. A sudden increase in manifold vacuum occurs when the engine is decelerated. The ECM senses engine deceleration from the MAP sensor signal and the TPS sensor input. When the ECM receives signals indicating engine deceleration, it will reduce the injector on-time to provide a leaner air–fuel ratio and prevent high emissions of HC and CO. If a defect occurs in the system that makes it impossible for the ECM to operate in the normal modes, a throttle body backup (TBB) circuit in the ECM will take over operation of the injectors. When the TBB circuit is in operation, the on-time of the injectors will be increased, and the air–fuel ratio will be richer than normal, but the vehicle can be driven until the necessary repairs are made.

Electronic Spark Timing (EST)

The spark advance is controlled by the ECM in response to the input signals that it receives. (Refer to Chapter 5 for a complete description of ignition systems for General Motors fuel-injection systems.)

Exhaust-Gas Recirculation (EGR)

The ECM controls a vacuum solenoid that is used to open and close the vacuum circuit to the EGR valve. The ECM will energize the vacuum solenoid and shut off the vacuum to the EGR valve if the coolant temperature is below 150°F (64°C), or if the engine is operating at idle speed or under heavy load, wide-open throttle conditions. Under part-throttle conditions with the coolant temperature above 150°F (64°C), the ECM will deenergize the solenoid and allow vacuum to open the EGR valve. The EGR solenoid and related circuit are shown in Figure 4–9.

Canister Purge

Vapors from the fuel tank are collected in the charcoal canister. The ECM operates a solenoid in the purge hose between the canister and the intake manifold, as illustrated in Figure 4–10.

The ECM will energize the solenoid and close the purge hose from the canister if the engine is running and the coolant temperature is above 178°F (89°C). At all other times the solenoid will be deenergized, and the purge hose will be open. The purge line enters the throttle body assembly above the throttles.

Idle Speed

On some throttle body injection systems, the ECM operates an idle speed control (ISC) motor to control idle speed under slow or fast idle conditions. The same ISC motor is used in some computer command control (3C) systems. An ISC motor and the motor test procedure are illustrated in Figure 4–11.

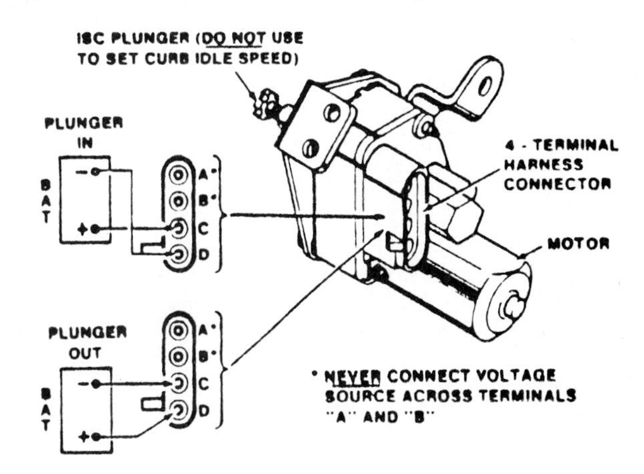

Figure 4–11 Idle Speed Control (ISC) Motor. (Courtesy of General Motors Corporation)

Some throttle body injection systems use an idle air control (IAC) motor to control idle speed. The IAC motor is operated by the ECM and it controls idle speed by opening or closing an air passage into the intake manifold. Air flow through the IAC passage bypasses the throttle, as shown in Figure 4–12.

The IAC motor is used to control fast idle and slow idle speeds. If the coolant temperature sensor (CTS) sends a signal to the ECM that indicates cold engine coolant, the ECM will operate the IAC motor and open the IAC passage to increase the idle speed. When the engine is at normal operating temperature, the IAC motor will provide a faster idle speed for a few seconds each time the engine is started. When dual throttle body assemblies are used, as in the crossfire injection system, an IAC motor is located in each throttle body assembly.

Air Injection

The air injection reactor (AIR) system is basically the same for throttle body injection systems or computer command control (3C) systems. (See Chapter 3 for a description of 3C AIR systems.) A complete AIR system is illustrated in Figure 4–13.

The ECM operates two electric solenoids in the air-control valve and air-switching valve to control air flow from the air pump to the air cleaner, exhaust ports, or catalytic converter. On many throttle body injection systems the ECM will operate the air-control valve to bypass air to the air cleaner for a few seconds each time the engine is started. Once the engine has been running for a brief period of time, the air-control valve will direct the air flow from the air pump to the air-switching valve. During the engine warmup time, the air-switching valve will direct air flow from the air pump to the exhaust ports. When the coolant reaches normal operating temperature and the system enters the closed-loop mode, the air-switching valve will direct the air flow from the pump to the catalytic converter.

Torque-Converter Clutch

The vehicle speed sensor (VSS) signal is used by the ECM to control the torque-converter lockup time. The VSS and the torque clutch system are the same on the throttle body injection systems and 3C systems. (See Chapter 3 for a description of the VSS and torque-converter clutch in the 3C system.) When the VSS on throttle body injection systems sends a speed signal above 35 mph (56 kph) to the ECM, the ECM will maintain the idle air control motor in the extended position. A defective VSS can cause erratic idling.

Diagnosis of Throttle Body Injection Systems

Fuel Pump Circuit

A defective fuel pump relay may cause a slightly longer cranking time when the engine is started because the oil-pressure switch contacts must close before the battery voltage can be supplied to the fuel pump. The fuel pump electrical circuit is shown in Figure 4–7.

When the ignition switch is turned on, it should be possible to hear the in-tank fuel pump run for approximately 2 seconds. If the fuel pump does not run, the circuit from the relay through the fuel pump may be tested by supplying battery voltage to terminal G in the 12-terminal assembly line communications link (ALCL), as shown in Figure 4–14.

If the fuel pump does not operate when battery voltage is supplied to terminal G, the fuel pump or the circuit from the fuel pump through the upper fuel pump relay contacts is defective. The fuel pump must

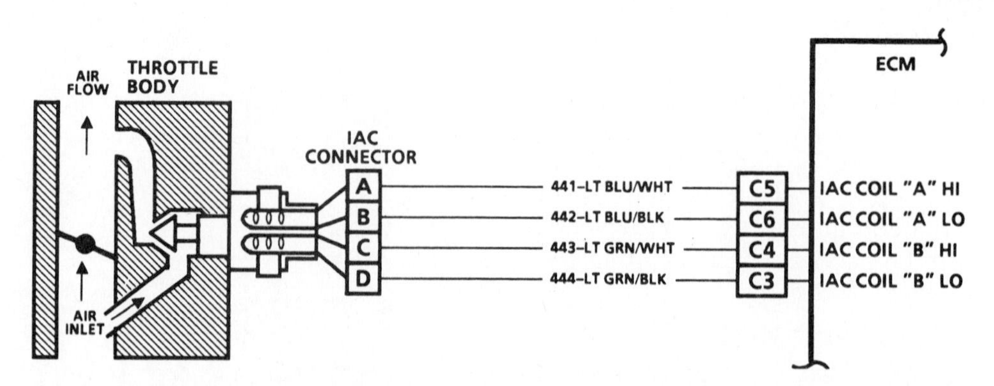

Figure 4–12 Idle Air Control (IAC) Motor. (Courtesy of General Motors Corporation)

be replaced if it fails to run when battery voltage is supplied to the fuel pump connector at the fuel tank. If the fuel pump runs when battery voltage is supplied to the connector at the tank but it did not run when battery voltage was supplied to terminal G, the defect exists in the circuit from the tank to the relay. If the fuel pump operates when battery voltage is supplied to terminal G but fails to operate when the ignition switch is turned on, the fuel pump relay, connecting wires, or the ECM is defective. In some vehicles, the fuel pump test connector is located in the engine compartment on the left fender shield rather than in the ALCL. A similar ALCL is used in TBI systems and 3C systems.

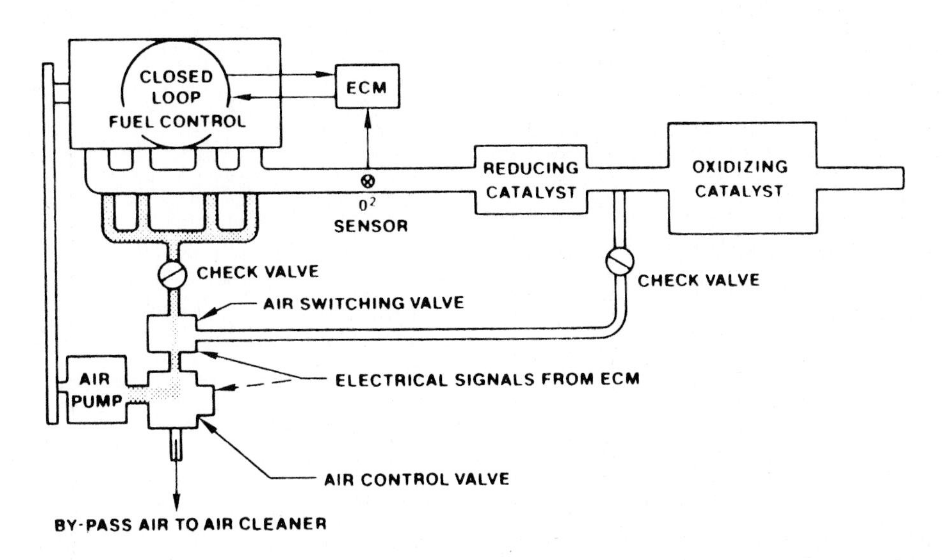

Figure 4–13 Air Injection Reactor [AIR] System. [Courtesy of General Motors Corporation]

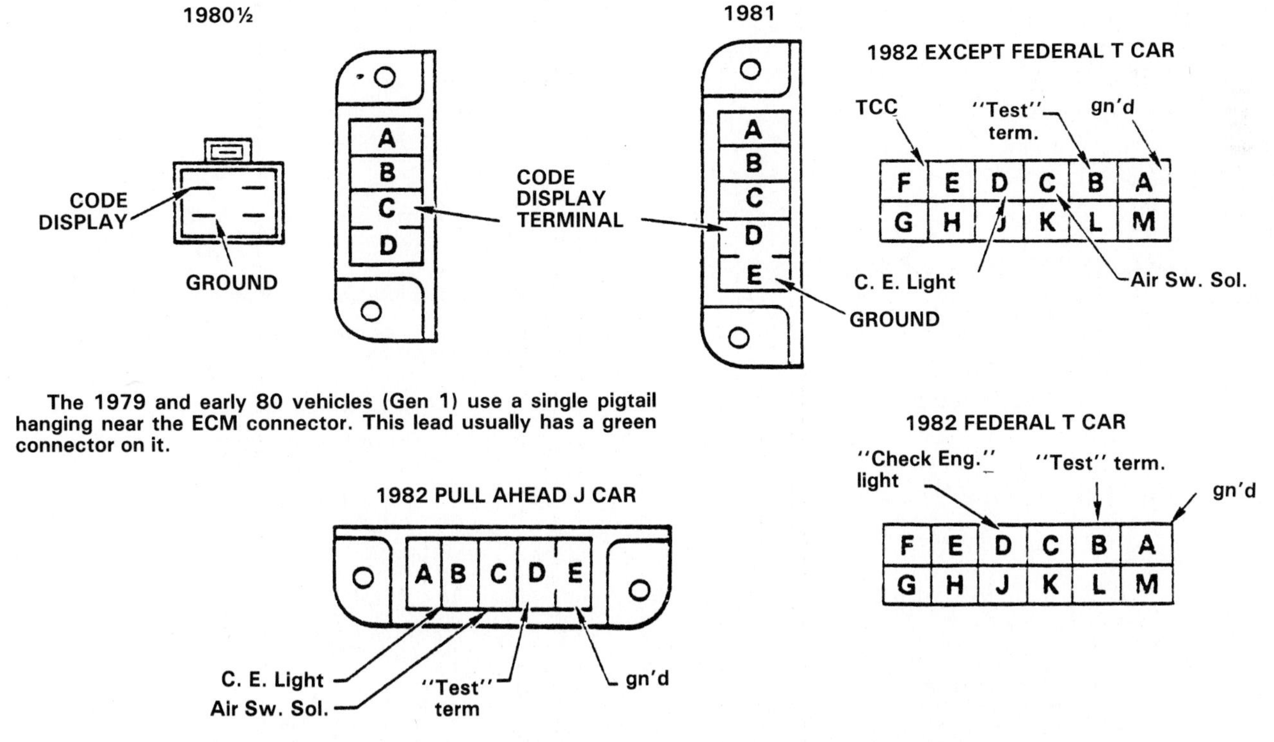

Figure 4–14 Assembly Line Communications Link [ALCL]. [Courtesy of General Motors Corporation]

Trouble Codes

When performance or economy complaints are received on a TBI system, the ignition system and the engine compression should be tested before the TBI system is diagnosed. A check-engine light in the instrument panel will come on if a defect occurs in the TBI system. When the ignition switch is turned on, the check-engine light should be illuminated and it should remain on for a few seconds after the engine is started. The check-engine light will be on all the time if a continuous defect exists in the TBI system. When an intermittent defect occurs, the check-engine light will be on only when the defect is present. However, the trouble code from the defect will be stored in the computer memory until the vehicle has been started 50 times. The check-engine light is illustrated in Figure 4–15.

When the test terminal is connected to the ground terminal in the ALCL with the ignition on, the system will enter the diagnostic mode. In this mode the check-engine light will flash out any trouble codes that are stored in the ECM memory. One flash followed by a pause and four flashes in a quick sequence indicates trouble code 14. Trouble code 12, which indicates that the diagnostic system is working, will be flashed first. The trouble codes will be given in numerical order, and they will be repeated three times. As long as the test connector is grounded, the check-engine light will keep repeating the trouble code sequence. When the system is in the diagnostic mode, the ECM will also energize all the relays that are controlled by the ECM. A trouble code indicates a defect in a specific circuit. For example, code 14 indicates that the engine coolant sensor or the connecting wires are defective, or that the ECM may not be able to receive the engine coolant sensor signal. The possible trouble codes and the operating requirements to set each code in the ECM memory are provided in Table 4–1.

Code 55 is interpreted as an ECM error. This indicates a defective ECM, oxygen sensor, or 5V supply wire from the ECM to some of the sensors. If the test terminal in the ALCL is grounded with the engine running, the system will enter the field service mode, where the frequency of check-engine light flashes indicates whether the system is in the open-loop mode or closed-loop mode. In the open-loop mode, the check-engine light will flash twice per second; it will flash once per second in the closed-loop mode.

When the necessary repairs have been completed, the trouble codes may be cleared from the computer memory by disconnecting the wire attached to the positive battery cable, as shown in Figure 4–16. A complete wiring diagram of the ECM used on various engines is provided in Figure 4–17.

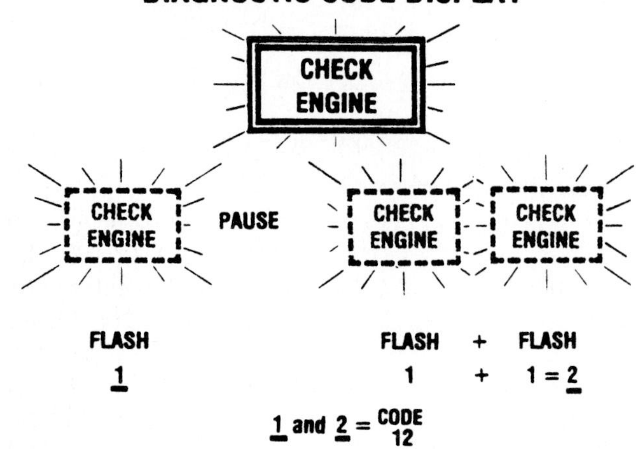

Figure 4–15 Check-Engine Light. (Courtesy of General Motors Corporation)

TABLE 4–1 TBI Trouble Codes.

Code 12	— Will flash three times to indicate that system diagnostics are working. If there are any stored fault codes, they will flash following Code 12. After flashing stored code(s), Code 12 will again flash to indicate that all codes have been displayed.
Code 13	— Oxygen Sensor Circuit — Engine running at normal operating temperature for about one minute (1200-1400 RPM in "neutral"). Oxygen sensor signal missing for 6 seconds.
Code 14	— High Coolant Temperature — Engine running 2 seconds with no coolant sensor signal voltage.
Code 15	— Low Coolant Temperature — Engine running for one minute Coolant sensor signal too high for 2 seconds.
Code 21	— High Throttle Position — Engine running below 1600 RPM, MAP less than 47 KPA (2.5 volts) and TPS is above 50% (2.5 volts) for 2 seconds.
Code 22	— Low Throttle Position — Engine running with TPS signal voltage zero for 2 seconds.
Code 24	— Vehicle Speed — Vehicle speed about 40-45 MPH steady throttle (decelerating 2.5 L) with no VSS signal for one minute.
Code 33	— MAP Too High — Engine idling and MAP signal is above 60 KPA (3 volts) for 5 seconds.
Code 34	— MAP Too Low — Engine running .025 seconds (250 m. seconds) with MAP signal voltage too low.
Code 42	— EST — Open or grounded EST line, open or grounded bypass line, and engine speed above 500 RPM'
Code 43	— ESC — Engine running and ESC circuit 485 is less than 6 volts for 4 seconds.
Code 44	— Lean Exhaust Indication — Engine running 2 minutes steady throttle above 2000 RPM with oxygen sensor signal less than 200 mv. (0.2 volts) for 60 seconds.
Code 45	— Rich Exhaust Indication — Engine running 3 minutes steady throttle above 2000 RPM with oxygen sensor signal above 750 mv.
Code 51	— PROM Error
Code 55	— ECM Error

(Courtesy of General Motors Corporation)

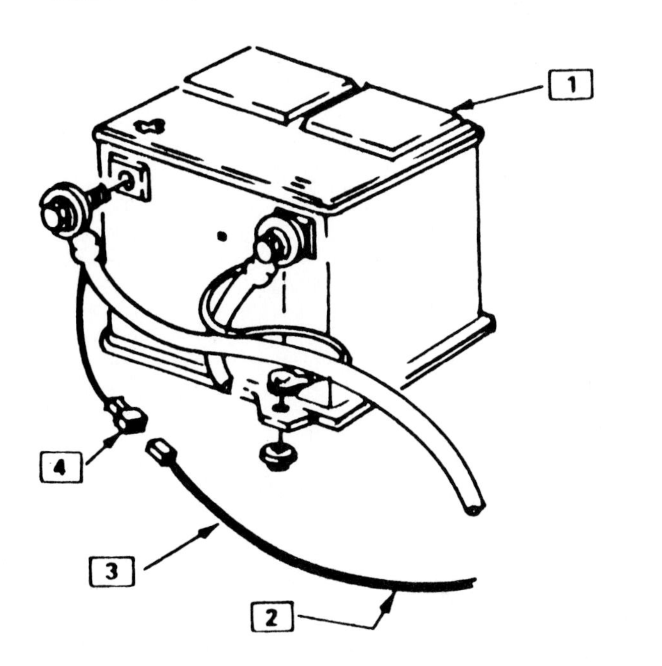

Figure 4–16 Trouble Code Clearing. (Courtesy of General Motors Corporation)

Idle Air Control (IAC) Motor Service

The IAC motor may be tested by removing the motor and reconnecting the wiring harness. With the ignition switch on and the test terminal in ALCL grounded, the motor pintle should pulse in and out. Erratic idle operation may be caused by an IAC motor that is sticking. When the IAC motor is reinstalled, the measurement from the pintle tip to the motor housing should not exceed 1.25 in. (32 mm), as shown in Figure 4–18.

When an IAC motor has been removed, the ECM will not know the position of the motor pintle until the vehicle speed reaches 35 mph (56 kph) and the ECM drives the motor to the fully extended position. Until the vehicle speed reaches this level, idling may be erratic. This problem can be overcome by reinstalling the IAC motor and disconnecting the speedometer cable from the speedometer head to the cruise control. With the engine idling, spin the speedometer cable by hand until the speedometer indicates more than 35 mph (56 kph). The ECM will then fully extend the IAC motor and begin controlling the motor in the normal manner.

Throttle Position Sensor (TPS) Adjustment

The TPS sensor may be adjusted in some applications, whereas in other systems no means of adjusting the TPS is provided. Before the rear TPS screw can be loosened, the spot weld on the screw must be drilled out, as illustrated in Figure 4–19.

With the ignition switch on and the engine stopped, connect a digital voltmeter to the TPS, as illustrated in Figure 4–20.

The voltmeter should read 0.450 to 0.600V. If the correct reading is not obtained on the voltmeter, loosen the TPS screws and rotate the TPS until the voltage is within specifications. Hold the TPS in this position and tighten the mounting screws.

Minimum Air Rate Adjustment

The minimum air rate should not have to be adjusted unless some throttle body parts have been replaced. It may be necessary to remove the throttle body unit and then remove the tamper-resistant plug on the minimum air adjustment screw, as outlined in Figure 4–21.

With the engine at normal operating temperature, plug the air passage to the IAC motor as shown in Figure 4–22.

With the transmission in park, adjust the minimum air rate screw to obtain an engine speed of 500 rpm. The injectors may be replaced in the throttle body assembly as shown in Figure 4–23.

The injector winding may be tested for continuity with an ohmmeter. Do not connect a 12V source to the injector winding for more than 5 seconds. The pressure regulator should not be disassembled.

Fuel Pump Pressure Test

The fuel pump pressure gauge should be connected in the fuel inlet at the throttle body assembly. Before the gauge is connected, the pressure in the system should be relieved. This may be accomplished by grounding one injector terminal and connecting the other terminal to a 12V source momentarily. Never loosen a fuel line without relieving the system pressure. The pressure gauge may be adapted to fit TBI systems with single or dual throttle body assemblies. With the ignition switch on, fuel pump pressure should be 10 to 12 psi (69 to 83 kPa). Never turn on the ignition switch with a fuel line disconnected.

WHITE POWER CONNECTOR

1.8L 2.5L	2.0L	5.0L 5.7L	Circuit	Pin	Pin	Circuit	1.8L 2.5L	2.0L	5.0L 5.7L
-	-	-	SPARE	1	24	SPARE	-	-	-
BRN	→	→	VEHICLE SPEED SENSOR	2	23	E-CELL	-	-	DK.BLU/ WHT
WHT/BLK	→	→	DIAGNOSTIC TEST ALCL	3	22	4TH GEAR SWITCH	-	-	5.7L ALT DK.GRN 5.0L AT
-	-	BLK	ELECTRONIC SPARK CONTROL	4	21	A/C CLUTCH	DK.GRN/ WHT	DK.GRN	→
ORN/BLK	→	→	PARK/NEUTRAL SWITCH	5	20	CHECK ENGINE LIGHT	BRN/WHT	→	→
-	-	LT.BLU	DUAL INJECTOR SELECT	6	19	CONVERTER CLUTCH	TAN/BLK	→	→
ORN	→	→	SERIAL DATA	7	18	FUEL PUMP RELAY DRIVE	DK.GRN/ WHT	→	→
BLU	2	2	INJECTOR #1	8	17	VOLTAGE MONITOR	TAN/WHT	LT.BLU	→
-	-	LT.GRN	INJECTOR #2	9	16	SWITCHED IGNITION	PNK/BLK	→	→
ORN	→	→	BATTERY	10	15	BATTERY	ORN	→	→
GRY	→	→	5 VOLT REFERENCE	11	14	MAP GROUND	BLK	→	1
BLK/WHT	→	→	ECM GROUND	12	13	ECM GROUND	BLK/WHT	→	→

1.8L 2.5L	2.0L	5.0L 5.7L	Circuit	Pin	Pin	Circuit	1.8L 2.5L	2.0L	5.0L 5.7L
PPL/WHT	PPL	PPL	CRANK SIGNAL	1	22	EGR	-	GRY	→
PPL/WHT	→	→	HEI REFERENCE	2	21	AMBIENT TEMP. SENSOR 1.8L	TAN	-	-
BLK/RED	→	→	HEI DIST. GROUND	3	20	MANIFOLD ABSOLUTE PRESSURE SIGNAL	LT.GRN	→	→
YEL	→	→	COOLANT SENSOR SIGNAL	4	19	EST SIGNAL	WHT	→	→
DK.BLU	→	→	T.P.S. SIGNAL	5	18	I.A.C. COIL "A" LO	LT.BLU/ BLK	LT.BLU/ RED	→
1.8L DK.GRN/ WHT	-	5.7L M.T. GRY/RED	3RD GEAR SIGNAL	6	17	I.A.C. COIL "A" HI	LT.BLU/ BLK	LT.BLU/ RED	→
DK.BLU	GRY/RED	5.0L M.T. GRY/RED	A/C RELAY OR HOOD LOUVRE CONTROL	7	16	AIR DIVERT SO.	-	DK.GRN/ YEL	5.7L M.T. BLK/PNK 5.0L ALL
PPL	→	→	OXYGEN SENSOR SIGNAL	8	15	OXYGEN SENSOR GROUND	TAN	→	→
1.8L LT.BLU/ BLK	DK.GRN/ YEL	→	COOLING FAN OR CANNISTER PURGE	9	14	AIR SWITCH SOLENOID	-	BRN	→
TAN/BLK	→	→	EST BYPASS	10	13	I.A.C. COIL "B" LO	LT.GRN/ BLK	→	→
BLK	→	→	COOLANT & TPS GROUND	11	12	I.A.C. COIL "B" HI	LT.GRN/ WHT	LT.GRN/ RED	→

→ = THE SAME

BLACK I/O CONNECTOR

Figure 4–17 Electronic Control Module (ECM) Terminals. (Courtesy of General Motors Corporation)

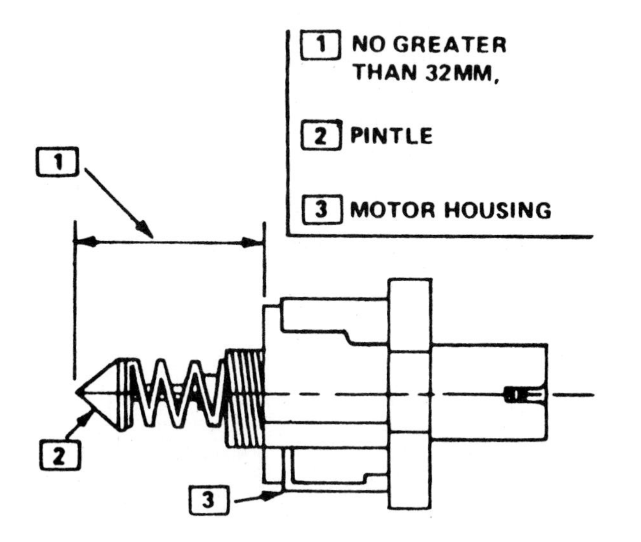

Figure 4-18 Idle Air Control Motor Installation. (Courtesy of General Motors Corporation)

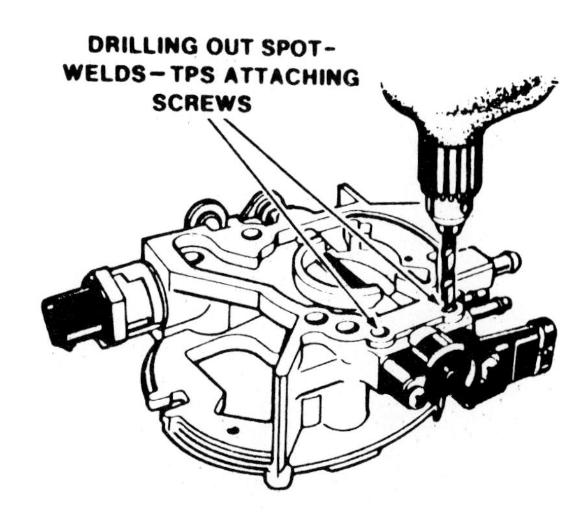

Figure 4-19 Throttle Position Sensor (TPS) Screw Removal. (Courtesy of General Motors Corporation)

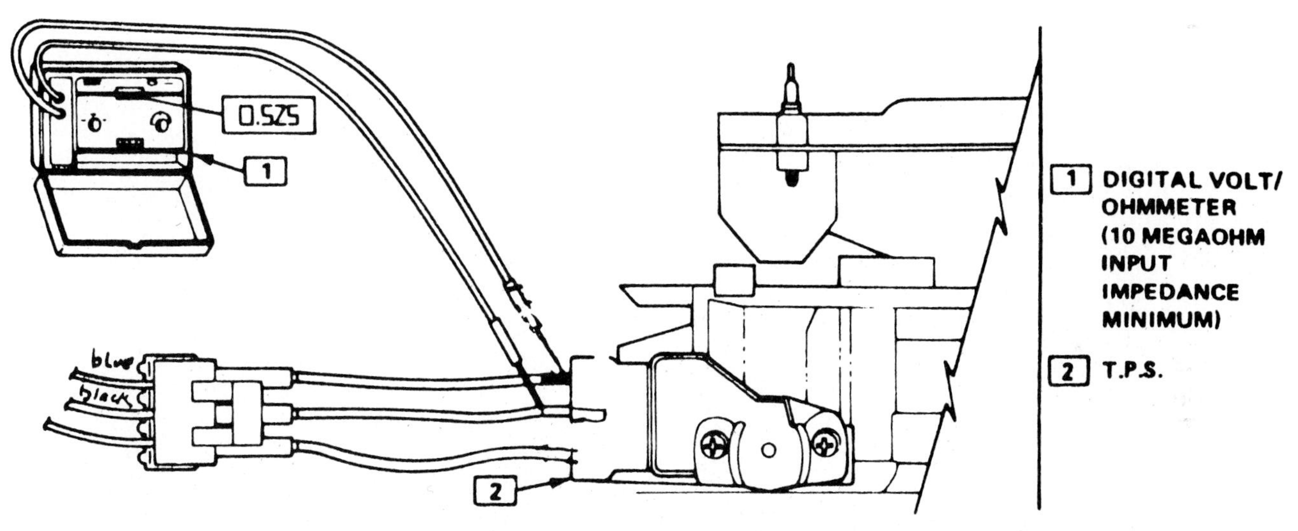

Figure 4-20 Throttle Position Sensor (TPS) Adjustment. (Courtesy of General Motors Corporation)

General Motors Port Fuel Injection

Fuel System

The port fuel injection (PFI) system has individual injectors located in the intake manifold ports. Air-injection pumps and heated air inlet systems are not required with PFI systems. When the PFI system is used on turbocharged engines, it may be referred to as a sequential fire injection (SFI) system. There are many similarities between the PFI and SFI systems. One of the most important differences between the two systems is that in the PFI system all the injectors are energized once for each crankshaft revolution by the electronic control module (ECM), whereas in the SFI system the injectors are energized individually each time the ignition system fires. A naturally aspired 186 CID 3.0L V6 engine with PFI is shown in Figure 4-24.

The electric fuel pump circuit is basically the same as the circuit used on General Motors throttle body injection systems. (An explanation of the fuel pump circuit was provided earlier in this chapter.) Fuel is forced through the fuel filter and accumulator

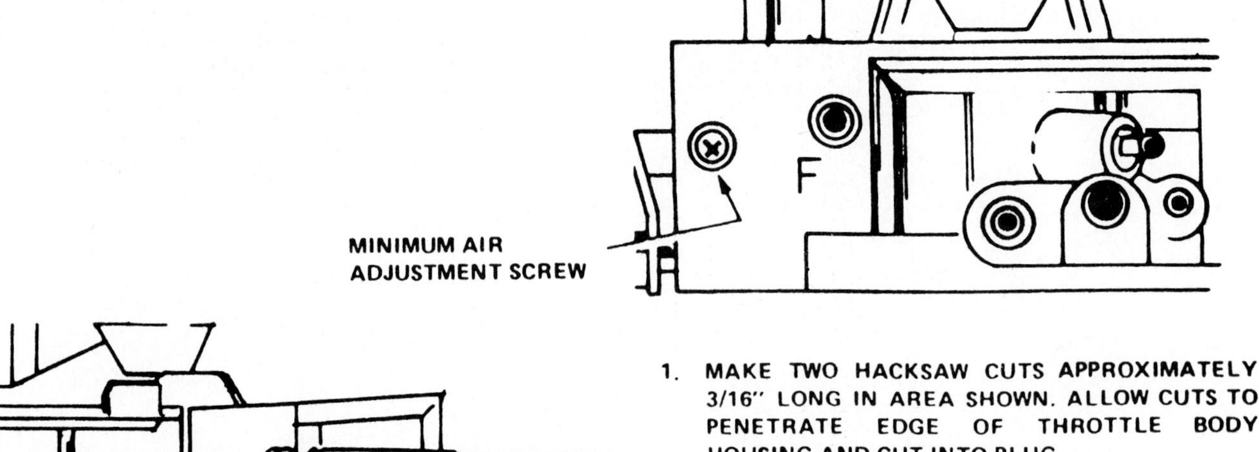
MINIMUM AIR
ADJUSTMENT SCREW
F

2
1
F

1. MAKE TWO HACKSAW CUTS APPROXIMATELY 3/16" LONG IN AREA SHOWN. ALLOW CUTS TO PENETRATE EDGE OF THROTTLE BODY HOUSING AND CUT INTO PLUG.
2. USING A SMALL PUNCH, KNOCK OUT PORTION OF CASTING CUT BY HACKSAW. HIT IN DIRECTION INDICATED BY ARROW 1.
3. KNOCK OUT STEEL PLUG WITH SMALL PUNCH BY HITTING PLUG FROM DIRECTION INDICATED BY ARROW 2.

Figure 4–21 Tamper-Resistant Plug Removal. (Courtesy of General Motors Corporation)

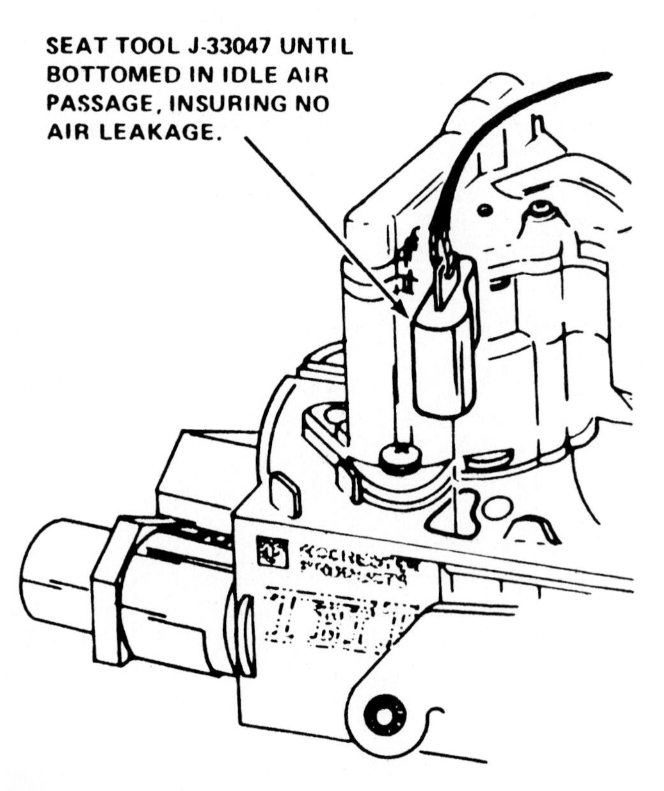
SEAT TOOL J-33047 UNTIL
BOTTOMED IN IDLE AIR
PASSAGE, INSURING NO
AIR LEAKAGE.

Figure 4–22 Minimum Air Rate Adjustment. (Courtesy of General Motors Corporation)

to the injector fuel rail by the electric pump. An "O" ring seal is located on each end of the injectors to seal the injectors to the fuel rail and the intake manifold.

The pressure regulator limits the fuel pressure to 26 to 46 psi (179 to 317 kPa), and excess fuel is returned from the regulator through the fuel return line to the tank. The fuel system components are shown in Figure 4–25.

A vacuum hose is connected from the lower side of the pressure regulator diaphragm to the intake manifold, as shown in Figure 4–26.

When the engine is idling, manifold vacuum will be 16 to 18 in hg (54 to 61 kPa). This low pressure is applied to the injector tips in the intake manifold. When the manifold vacuum is applied to the pressure regulator, less fuel pressure will be required to force the diaphragm and valve downward, and the fuel system pressure will be limited to 26 to 28 psi (179 to 193 kPa). As the throttle approaches the wide-open position, manifold vacuum will be 2 to 3 in hg (6.8 to 10 kPa), and this increase in pressure will be sensed at the injector tips. If the manifold vacuum applied to the regulator diaphragm is 2 to 3 in hg (6.8 to 10 kPa), fuel pressure will have to increase to force the diaphragm and valve downward, and fuel system pressure will increase to 33 to 38 psi (227 to 262 kPa).

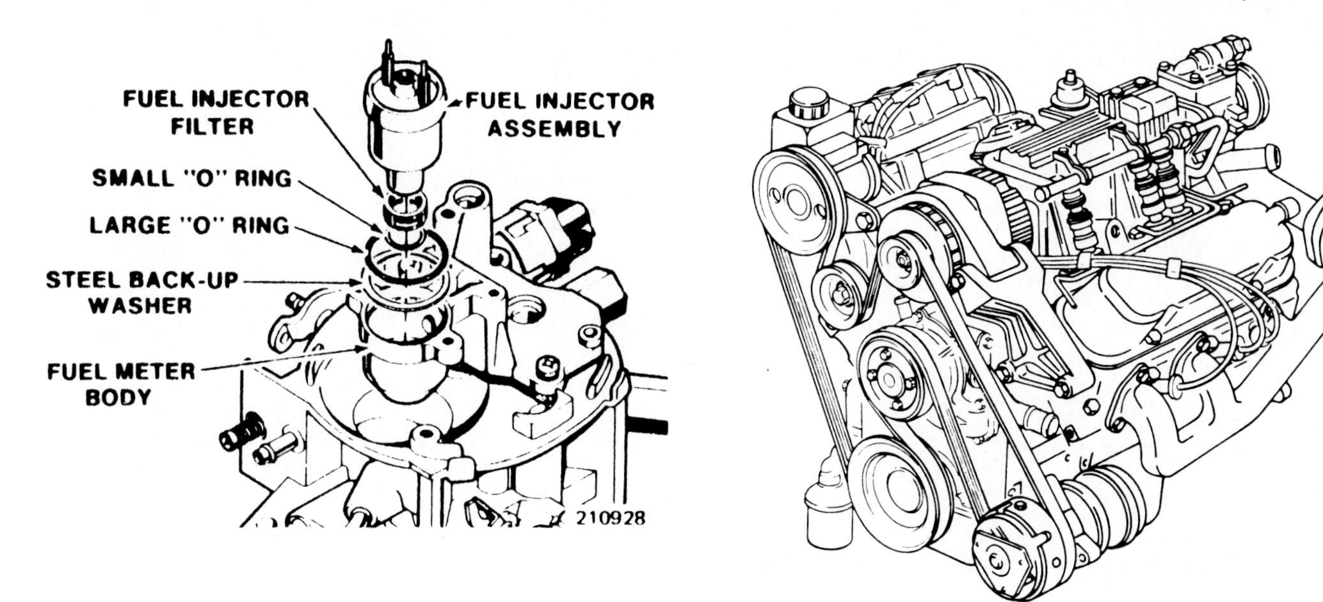

Figure 4–23 Injector Removal. (Courtesy of General Motors Corporation)

Figure 4–24 Nonturbocharged V6 with PFI. (Courtesy of General Motors Corporation)

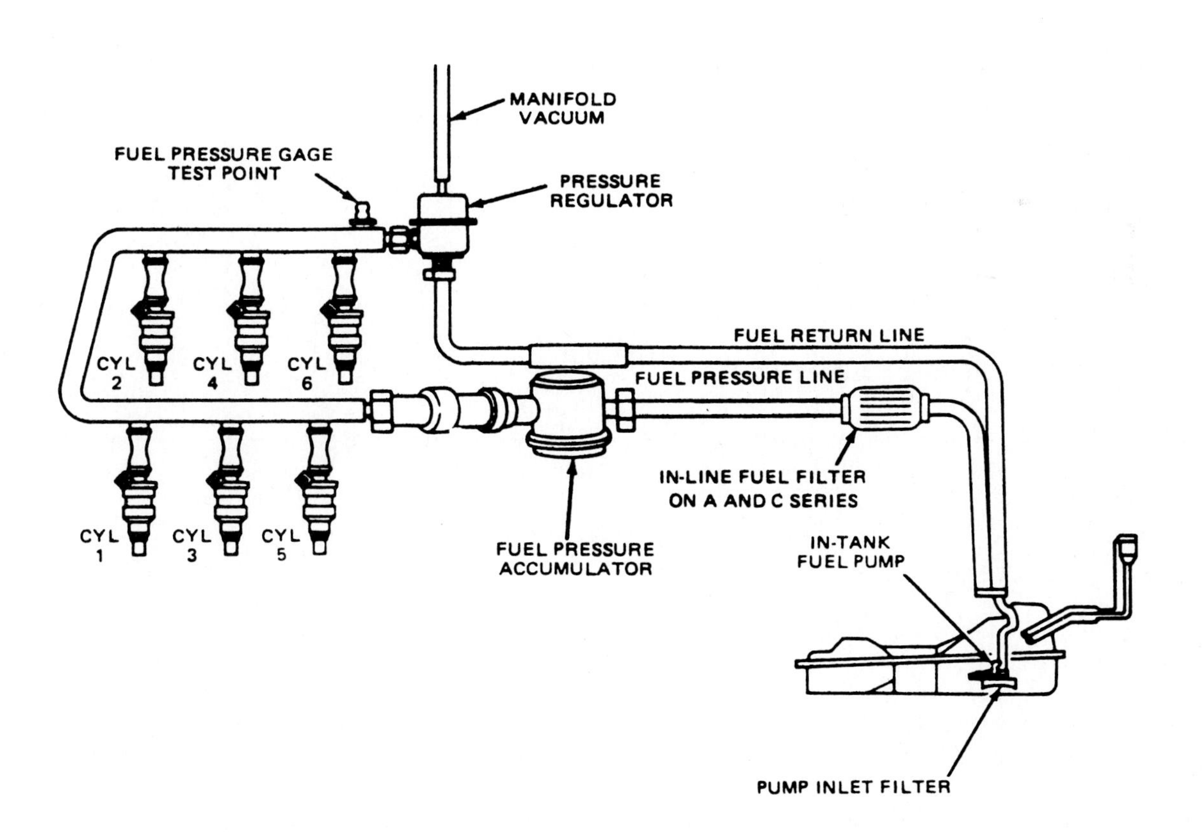

Figure 4–25 Fuel System Components. (Courtesy of General Motors Corporation)

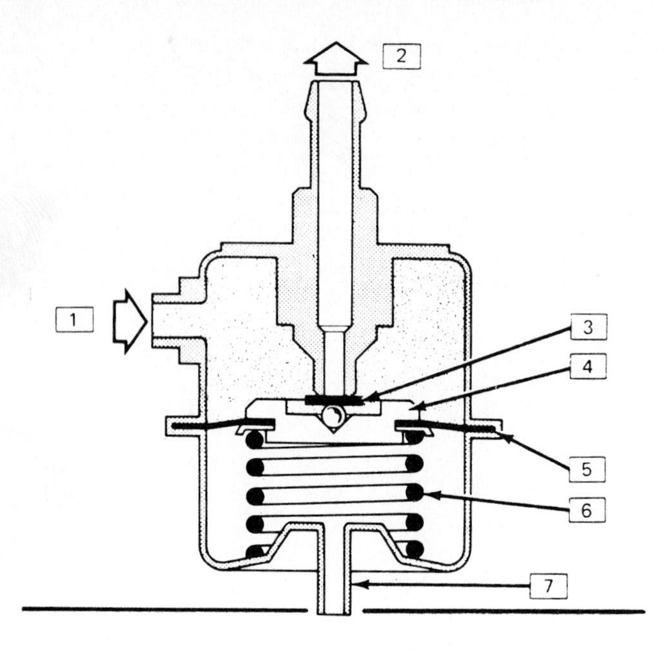

Figure 4-26 Pressure Regulator. [Courtesy of General Motors Corporation]

When the intake manifold pressure increases at wider throttle openings, the pressure regulator increases the fuel pressure at the injectors to maintain a constant pressure drop across the injectors and provide precise air–fuel ratio control.

Electronic Control Module (ECM)

The ECM is located under the instrument panel and is similar in appearance to those used with throttle body injection systems. Removable programmable read-only memory (PROM) and calibration package (CALPAC) units are located under an access cover in the ECM, as indicated in Figure 4–27.

The CALPAC unit allows fuel delivery if other parts of the ECM are damaged. When the PROM unit is replaced it must be installed in the original direction, and the connectors must not be bent. The removal and replacement procedure is pictured in Figure 4–28.

When the ECM connectors are being removed or installed, the ignition switch should be in the off position. The operating conditions sensed by the ECM and the system that it controls are outlined in Table 4–2.

Inputs

Similarities to Other Systems

The throttle-position sensor (TPS), oxygen (O_2) sensor, coolant temperature sensor, manifold absolute pressure (MAP) sensor, and the park/neutral switch are similar to the sensors used on General Motors 3C and throttle body injection systems. (Refer to Chapter 3 for an explanation of these sensors in the 3C systems.)

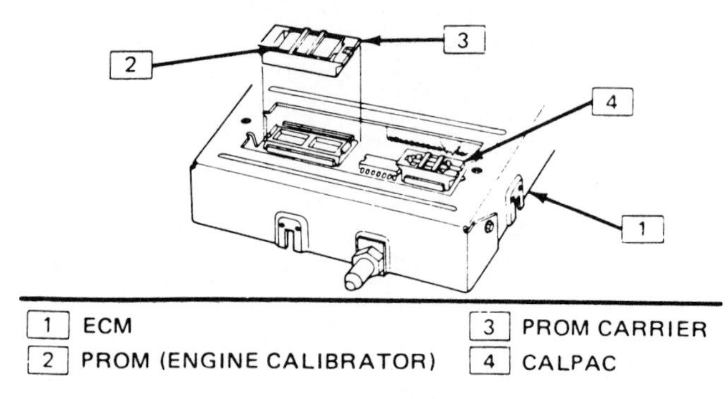

Figure 4-27 ECM with PROM and CALPAC Units. [Courtesy of General Motors Corporation]

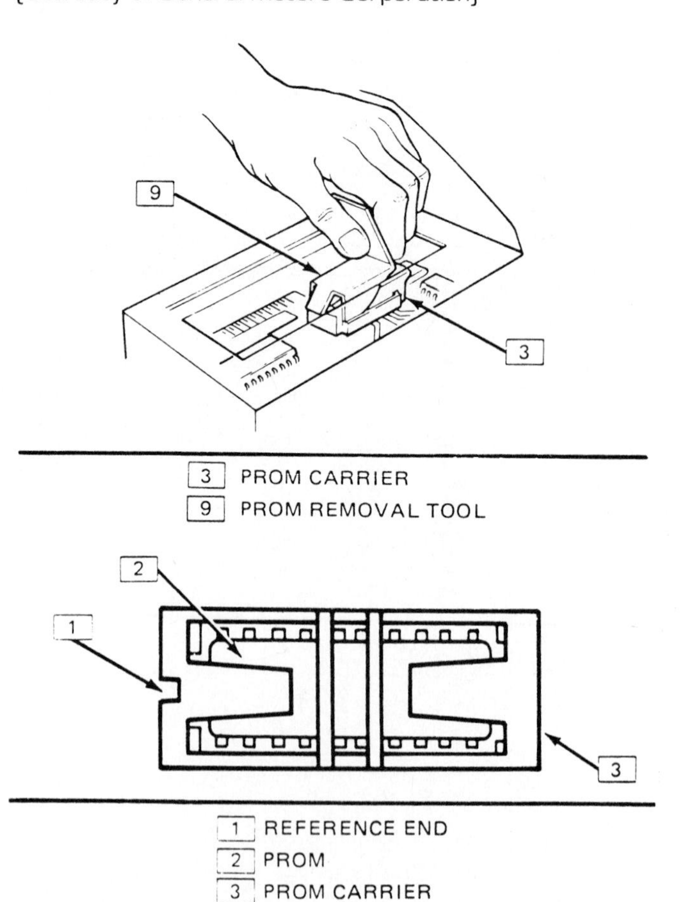

Figure 4-28 PROM or CALPAC Removal and Replacement. [Courtesy of General Motors Corporation]

TABLE 4–2 ECM Operating Conditions Sensed and Systems Controlled.

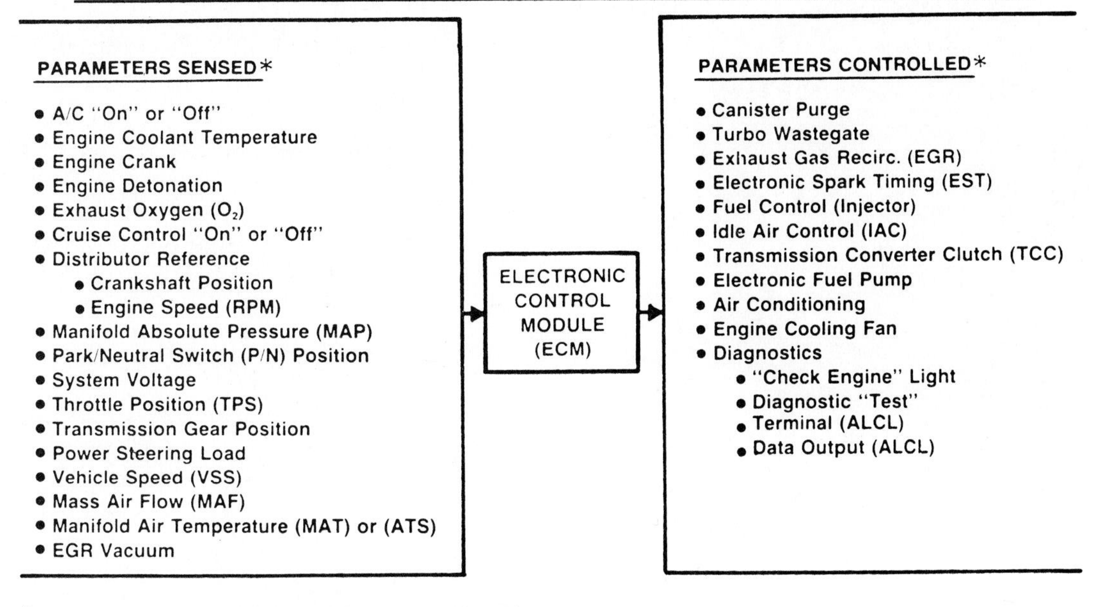

PARAMETERS SENSED*
• A/C "On" or "Off"
• Engine Coolant Temperature
• Engine Crank
• Engine Detonation
• Exhaust Oxygen (O_2)
• Cruise Control "On" or "Off"
• Distributor Reference
• Crankshaft Position
• Engine Speed (RPM)
• Manifold Absolute Pressure (MAP)
• Park/Neutral Switch (P/N) Position
• System Voltage
• Throttle Position (TPS)
• Transmission Gear Position
• Power Steering Load
• Vehicle Speed (VSS)
• Mass Air Flow (MAF)
• Manifold Air Temperature (MAT) or (ATS)
• EGR Vacuum
ELECTRONIC CONTROL MODULE (ECM)
PARAMETERS CONTROLLED*
• Canister Purge
• Turbo Wastegate
• Exhaust Gas Recirc. (EGR)
• Electronic Spark Timing (EST)
• Fuel Control (Injector)
• Idle Air Control (IAC)
• Transmission Converter Clutch (TCC)
• Electronic Fuel Pump
• Air Conditioning
• Engine Cooling Fan
• Diagnostics
• "Check Engine" Light
• Diagnostic "Test"
• Terminal (ALCL)
• Data Output (ALCL)

*NOT ALL SYSTEMS USED ON ALL ENGINES.

(Courtesy of General Motors Corporation)

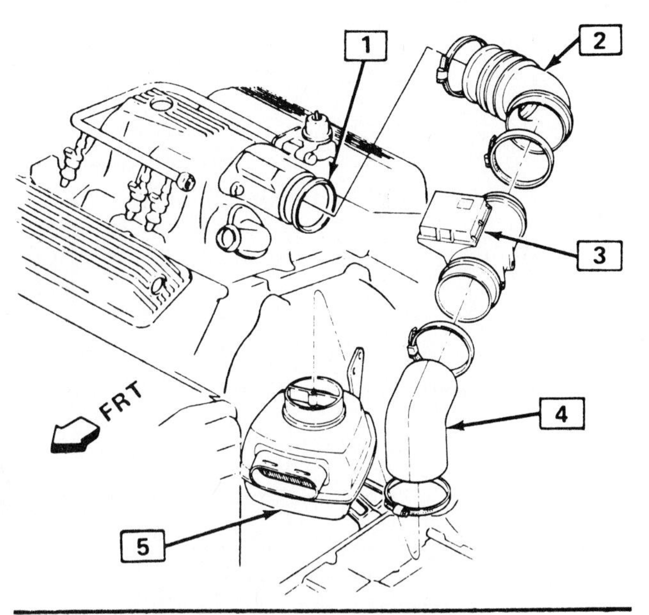

1 THROTTLE BODY ASM.
2 REAR AIR INTAKE DUCT
3 MASS AIR FLOW (MAF) SENSOR
4 INT. AIR INTAKE DUCT
5 AIR CLEANER ASM.

Figure 4–29 MAF Sensor Location. (Courtesy of General Motors Corporation)

Mass Air Flow (MAF) Sensor

The MAF sensor is used in place of the MAP sensor on many engines. Turbocharged engines always use the MAF sensor. The MAF sensor is located in the air intake between the air cleaner and the throttle body assembly, as indicated in Figure 4–29.

The MAF sensor contains a resistor that measures the temperature of the incoming air. A heated film in the sensor is a nickel grid coated with Kapton, a high temperature material. An electronic module in the top of the sensor maintains the temperature of the heated film at 167°F (75°C). If more energy is required to maintain the heated film at 167°F (75°C), the incoming mass of air has increased. This information is then sent to the ECM, and the ECM uses this input to provide a very precise air–fuel ratio control. The MAF sensor is shown in Figure 4–30.

Vehicle-Speed Sensor (VSS)

The VSS is a signal generator located in the transaxle. It is rotated mechanically and generates an electrical signal in relation to vehicle speed. This signal is sent through the VSS signal buffer to the ECM, as indicated in Figure 4–31.

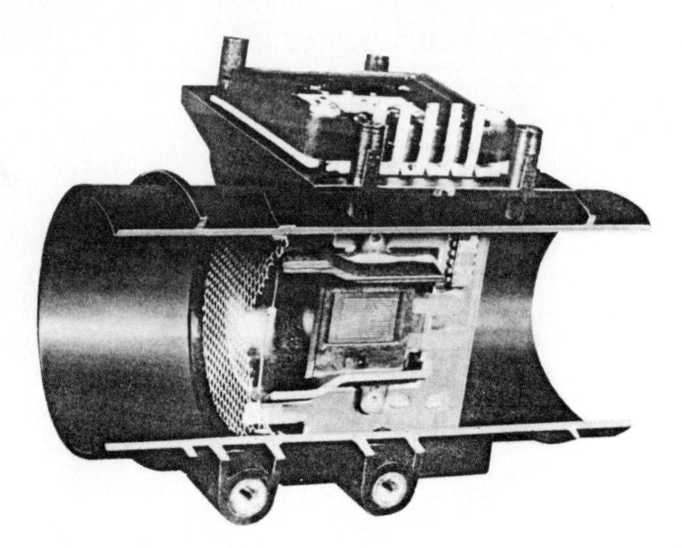

Figure 4–30 MAF Sensor. (Courtesy of General Motors Corporation)

Knock Sensor

A knock sensor is used on some of the PFI systems. The knock-sensor signal is sent through the electronic spark control (ESC) module to the ECM. When the engine detonates, this knock-sensor signal will cause the ECM to retard the spark advance. The knock-sensor circuit is shown in Figure 4–32.

Exhaust-Gas Recirculation (EGR) Vacuum Diagnostic Switch

The ECM operates a vacuum solenoid that applies vacuum to the EGR valve. When vacuum is applied to the EGR valve, the EGR vacuum diagnostic switch signals the ECM that the EGR valve is operating. If the EGR valve is operating when the necessary input signals are not present, the EGR vacuum diagnostic switch will signal the ECM and code 32 will be set in the ECM memory. The EGR vacuum diagnostic switch and the EGR solenoid circuits are illustrated in Figure 4–33.

Manifold Air Temperature (MAT) Sensor

Some PFI systems use a MAT sensor, which sends a signal to the ECM in relation to the temperature of the air in the intake manifold or in the air cleaner.

Distributor Reference Signal

Many PFI systems use the same ignition system as General Motors 3C systems. (These ignition systems are described in Chapter 5.) The distributor reference signal from the high-energy ignition (HEI) module to the ECM is used as a crankshaft position signal and engine speed signal.

ECM Outputs

Spark Advance

When the engine is running, the distributor pickup coil signal is sent through the HEI module and the distributor reference circuit to the ECM. This pickup signal travels through the electronic spark timing (EST) circuit in the ECM and then out to the HEI module through the EST wire. When the HEI module receives this pickup signal via the EST wire, it switches off the primary ignition circuit, which results in magnetic field collapse in the ignition coil and spark plug firing. The EST circuit in the ECM varies the pickup coil signal to provide the exact spark advance required by the engine. Figure 4–34 illustrates the spark advance control circuit.

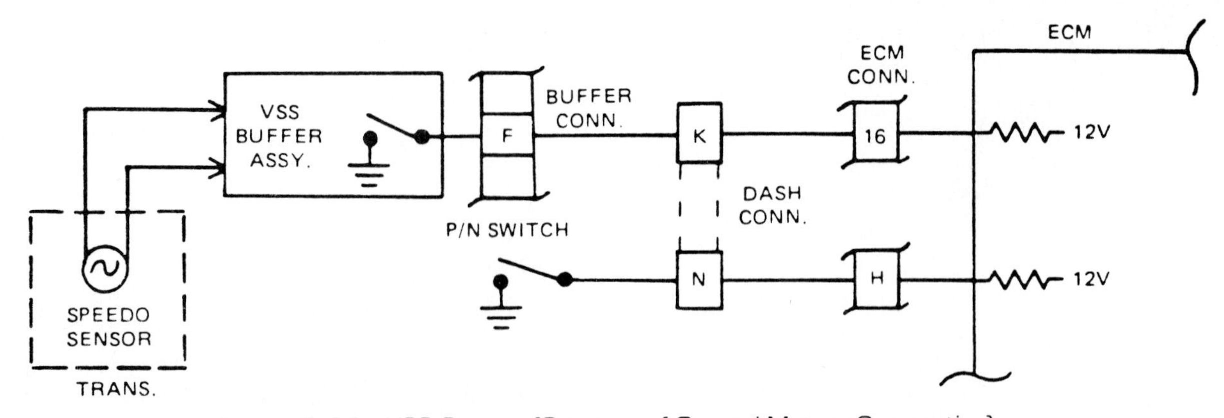

Figure 4–31 VSS Sensor. (Courtesy of General Motors Corporation)

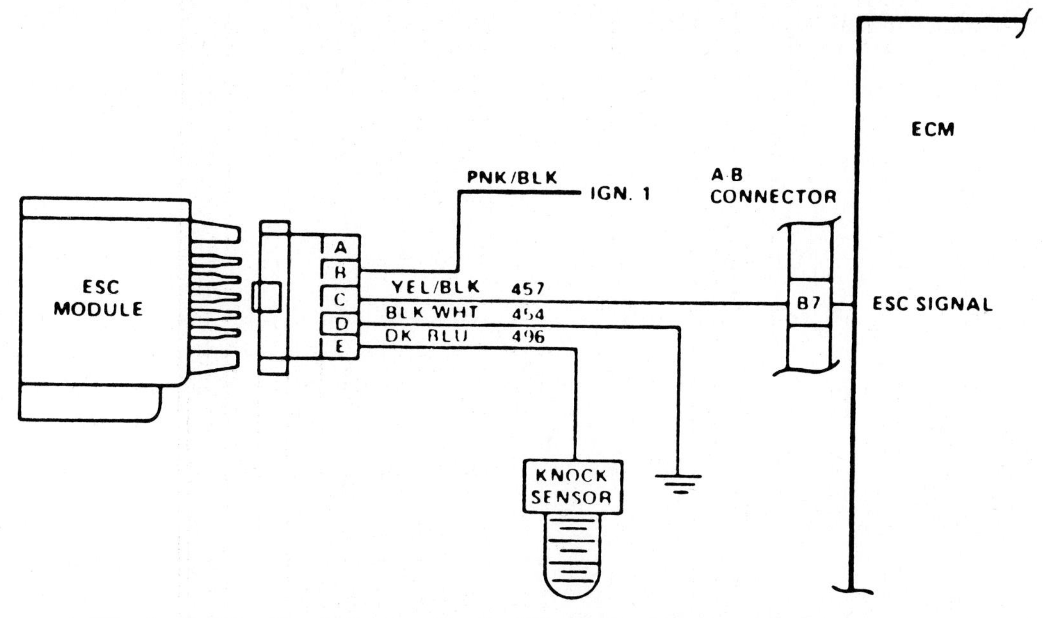

Figure 4–32 Knock-Sensor Circuit. [Courtesy of General Motors Corporation]

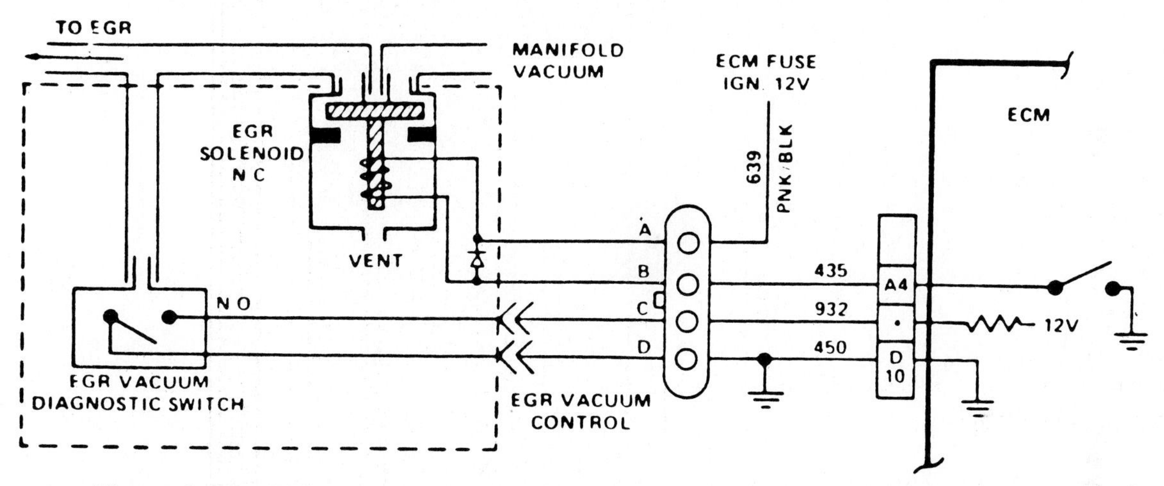

Figure 4–33 EGR Vacuum Diagnostic Switch and Solenoid. [Courtesy of General Motors Corporation]

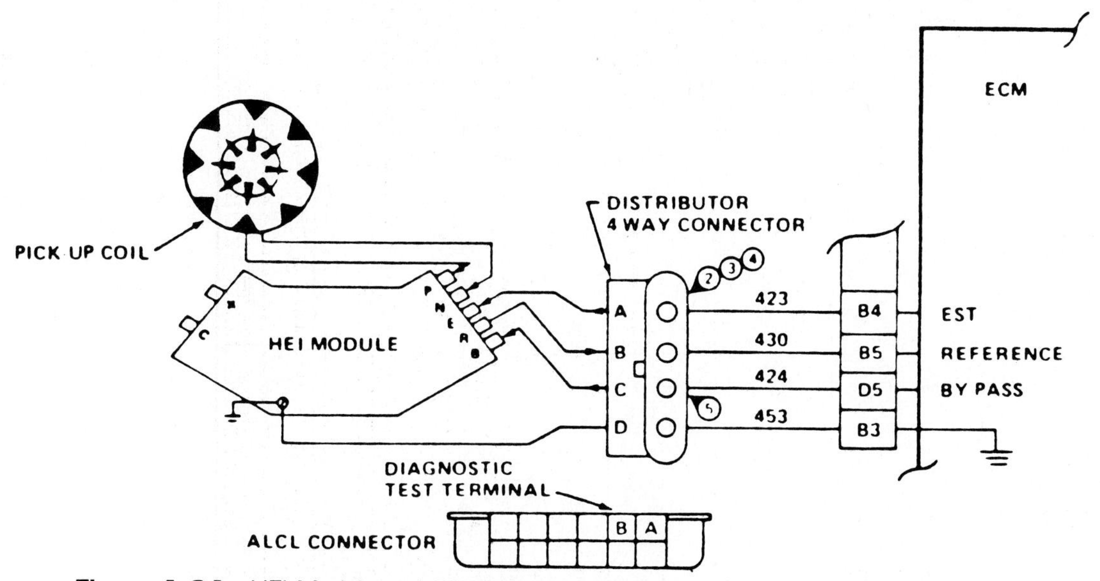

Figure 4–34 HEI Module and EST Circuit in ECM. [Courtesy of General Motors Corporation]

Coolant Fan

Several different coolant fan circuits are used on PFI-equipped vehicles. In these circuits the ECM will operate the coolant fan under certain conditions. One coolant fan circuit is outlined in Figure 4–35.

Some heavy-duty cooling systems have an optional cooling fan located beside the standard fan. The low-speed fan relay winding will be grounded by the ECM if the coolant temperature is above 208°F (98°C), the air conditioning (A/C) system pressure is below 260 psi (1793 kPa), and the vehicle speed is below 45 mph (72 kph). The "LO" contacts in the A/C head-pressure switch will ground the low-speed fan relay contacts if A/C is turned on and the pressure is above 260 psi (1793 kPa). When the low-speed fan

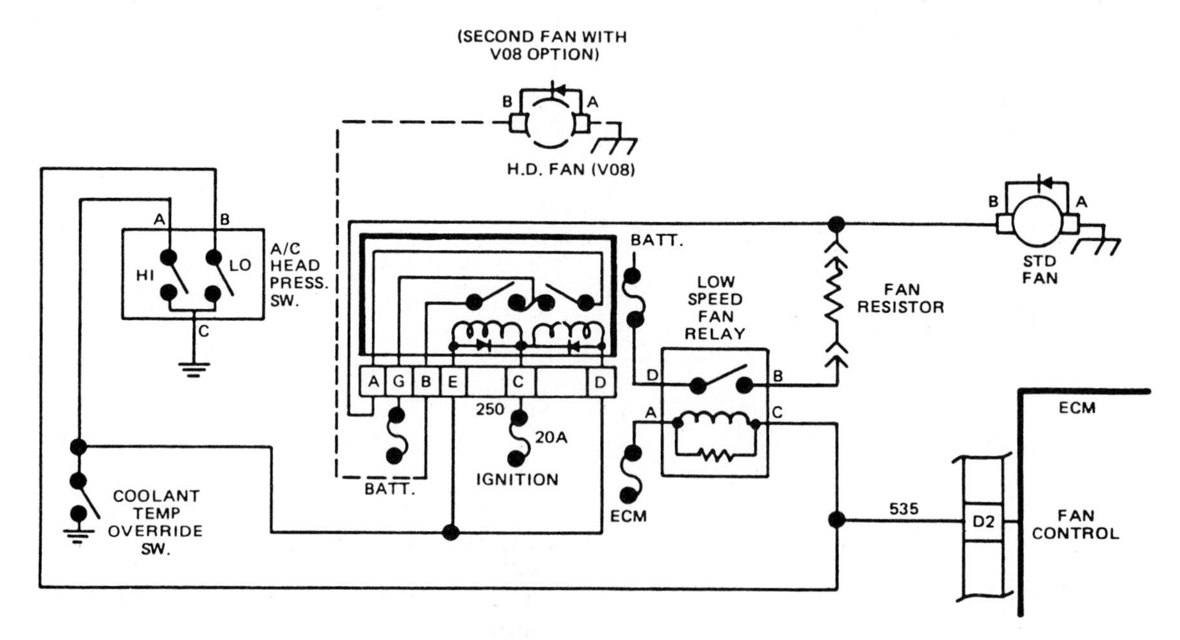

Figure 4–35 Coolant Fan Circuit without Timer. (Courtesy of General Motors Corporation)

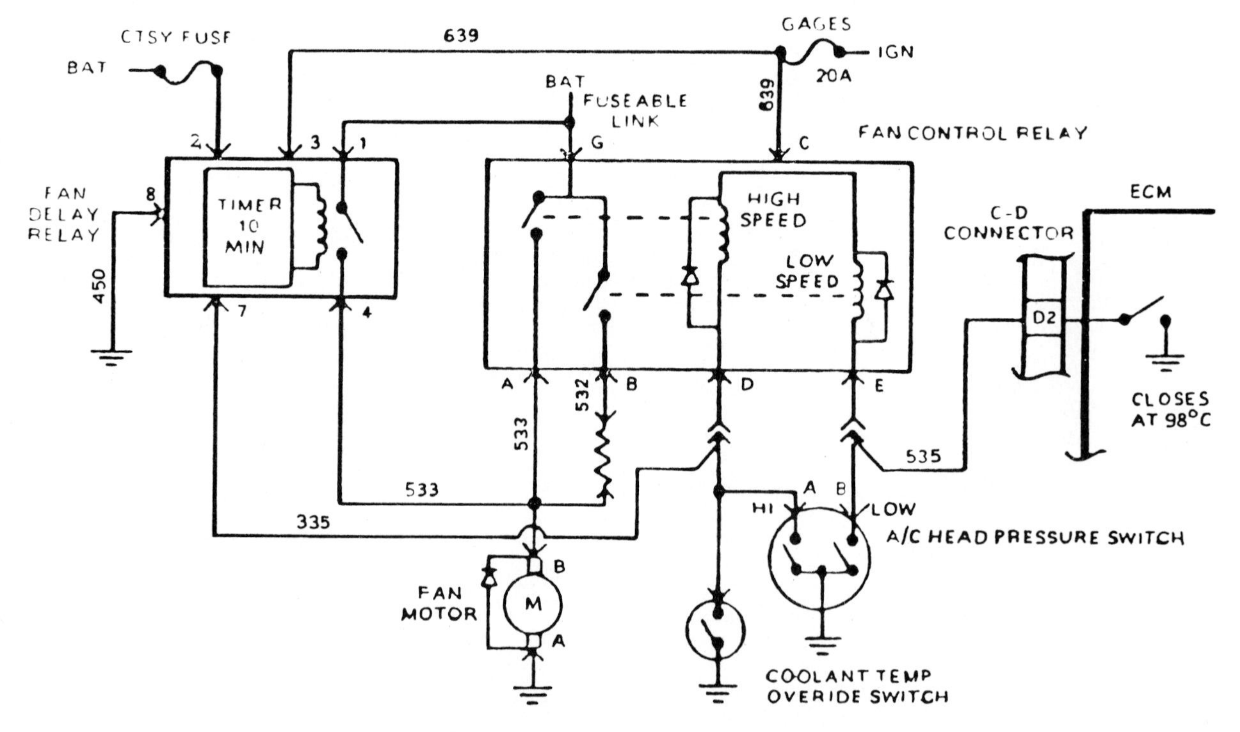

Figure 4–36 Coolant Fan Circuit with Timer. (Courtesy of General Motors Corporation)

relay contacts are closed, voltage will be supplied through the resistor to the standard fan motor, which causes the fan to run at low speed.

If the coolant temperature exceeds 223°F (160°C), the coolant temperature override switch will close, which grounds both windings in the dual contact relay. One set of points will supply voltage to the standard fan and the other contacts will complete the circuit to the heavy-duty fan, and both fans will operate at high speed. When the A/C is on and the pressure in the system exceeds 300 psi (2068 kPa), the "HI" contacts in the A/C head-pressure switch will close, which grounds both windings in the dual relay and causes both fans to operate at high speed. The standard fan will operate at low speed when terminals A and B are connected in the assembly line communications link (ALCL). The low-speed fan relay is mounted behind a plastic panel on the firewall with most of the other relays in the PFI system, and the dual relay is located near the brake booster.

Some coolant fan circuits have a timer that is used to operate the coolant fan after the engine is shut off. This type of circuit has one coolant fan that operates at two speeds. The coolant fan circuit is illustrated in Figure 4–36.

The ECM will operate the coolant fan at low speed if the following conditions are present.

1. Coolant temperature above 208°F (98°C).
2. Vehicle speed under 45 mph (72 kph).
3. A/C system pressure under 260 psi (1793 kPa).

The A/C head-pressure switch "LO" contacts will close and operate the coolant fan at low speed if the A/C system pressure exceeds 260 psi (1793 kPa). When the coolant temperature exceeds 223°F (106°C) the coolant temperature switch will close, which energizes the high-speed relay winding and supplies full voltage through the high-speed relay contacts to the coolant fan motor. This will operate the coolant fan at high speed. If the A/C system pressure exceeds 300 psi (2068 kPa) the "HI" contacts in the A/C head-pressure switch will close, which also causes the coolant fan to operate at high speed.

When the ignition is turned off and the coolant temperature is above 223°F (106°C), the coolant temperature override switch will remain closed, which grounds the timer relay winding. This will supply full voltage to the coolant fan motor through the timer relay contacts, and the coolant fan will operate at high speed. The coolant fan will be shut off after 10 minutes by the timer, or when the coolant temperature drops below 223°F (106°C), and the coolant temperature override switch contacts open.

A/C Compressor Clutch

When the A/C control switch is turned on and the cycling switch is closed, voltage will be supplied to terminal 88 on the ECM and to the A/C clutch control relay contacts. Under this condition the ECM will ground the A/C clutch control relay winding and the relay contacts will close, which supplies voltage to the A/C compressor clutch winding. If the power steering (P/S) pressure is high, such as when a full turn of the steering is completed, the P/S signal will cause the ECM to open the A/C clutch control relay winding circuit. When this occurs, the relay contacts will open and the compressor clutch will be disengaged. The ECM will also disengage the compressor clutch when the engine is operating at wide-open throttle (WOT). Figure 4–37 shows the compressor clutch circuit.

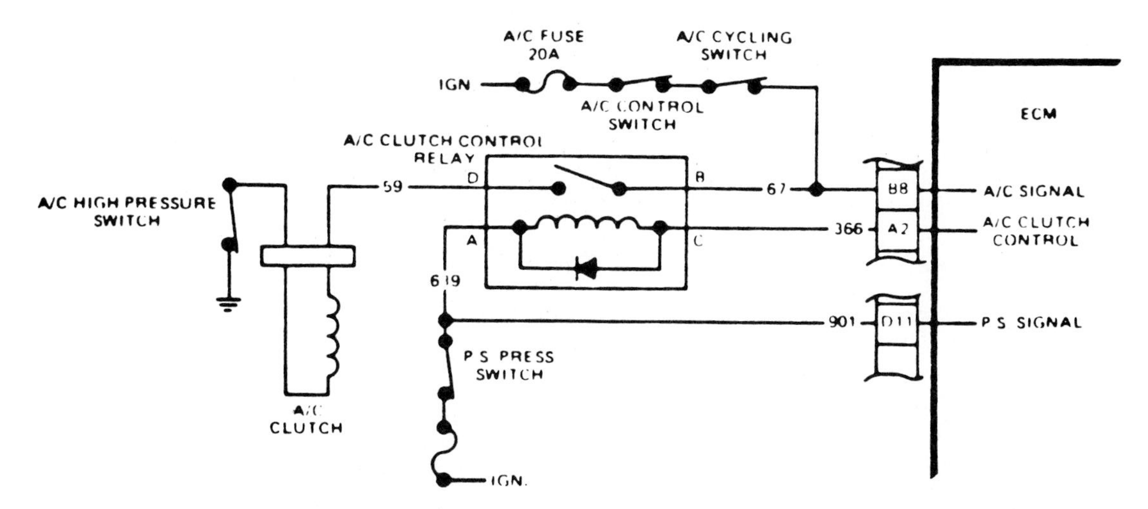

Figure 4–37 Compressor Clutch Circuit. (Courtesy of General Motors Corporation)

Idle Air-Control Motor

The ECM operates the IAC motor on the throttle body assembly to control idle speed under all operating conditions. (This motor was explained earlier in this chapter on General Motors throttle body injection systems.) The IAC motor connections to the ECM are indicated in Figure 4–38.

Transmission Converter Clutch (TCC)

The purpose of the TCC system is to eliminate torque-converter power loss when the vehicle is operating at moderate cruising speed. (The system was explained in Chapter 3.) The ECM will ground the apply solenoid winding and engage the TCC when the engine is operating at normal temperature and light load conditions above a specific road speed, and the transmission is in third or fourth gear. The TCC engagement speed may vary depending on the engine and transmission. Figure 4–39 illustrates the TCC circuit.

Canister Purge

The ECM operates a solenoid that allows fuel vapors to be purged from the canister into the intake manifold when the following conditions are present.

1. Engine running for more than 1 minute.
2. Coolant temperature above 165°F (80°C).
3. Vehicle speed above 10 mph (16 kph).
4. Throttle above idle speed.

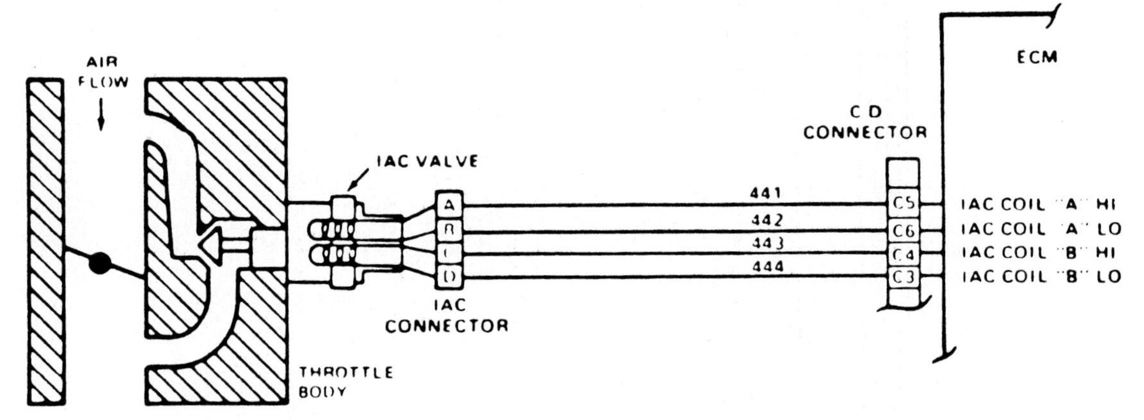

Figure 4–38 IAC Motor Circuit. (Courtesy of General Motors Corporation)

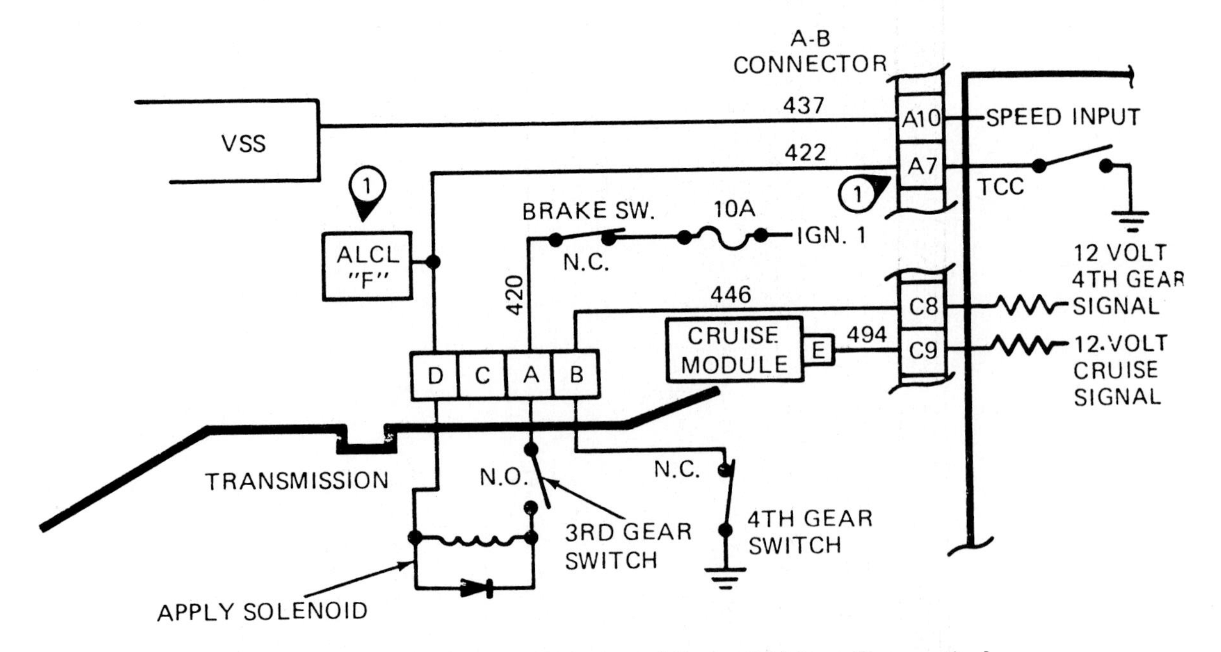

Figure 4–39 TCC Circuit. (Courtesy of General Motors Corporation)

The canister purge circuit is shown in Figure 4–40.

Fuel

The ECM controls the pulse width of the injectors to provide the precise air–fuel ratio required by the engine. When the engine is operating at normal temperature and idle speed, or moderate cruising speed, the ECM will control the air–fuel ratio at, or very close to, the stoichiometric value of 14.7 : 1. The ECM will provide mixture enrichment by increasing the injector pulse width when the engine coolant is cold, at wide-open throttle, and on sudden acceleration. The ECM will stop energizing the injectors when the engine is decelerated to lower emission levels and improve fuel economy.

Turbo Wastegate

The ECM operates a solenoid to limit the turbo boost pressure on turbocharged engines. When the ECM energizes the wastegate solenoid, some of the pressure supplied to the wastegate diaphragm is vented through the solenoid. This lowers the boost pressure, and the ECM will close the wastegate solenoid. The ECM uses the MAF sensor and engine speed signals to limit the boost pressure approximately 7.5 psi (51.7 kPa). (Turbocharger operation is explained in Chapter 8.) The wastegate solenoid circuit is illustrated in Figure 4–41.

General Motors Tuned Port Injection

Design

The tuned port injection (TPI) system is used on some 5.7L (350 CID) and 5.0L (305 CID) V8 engines. These engines have an air plenum and tubular runners in the intake manifold, which provide smooth unobstructed air flow into the cylinders. The TPI system is similar to the PFI system with a fuel injector located in each intake port. A manifold air temperature (MAT) sensor is used in the TPI system, and a separate cold-start injector is operated by the ECM to inject additional fuel when a cold engine is being started. An engine with a TPI system is illustrated in Figure 4–42.

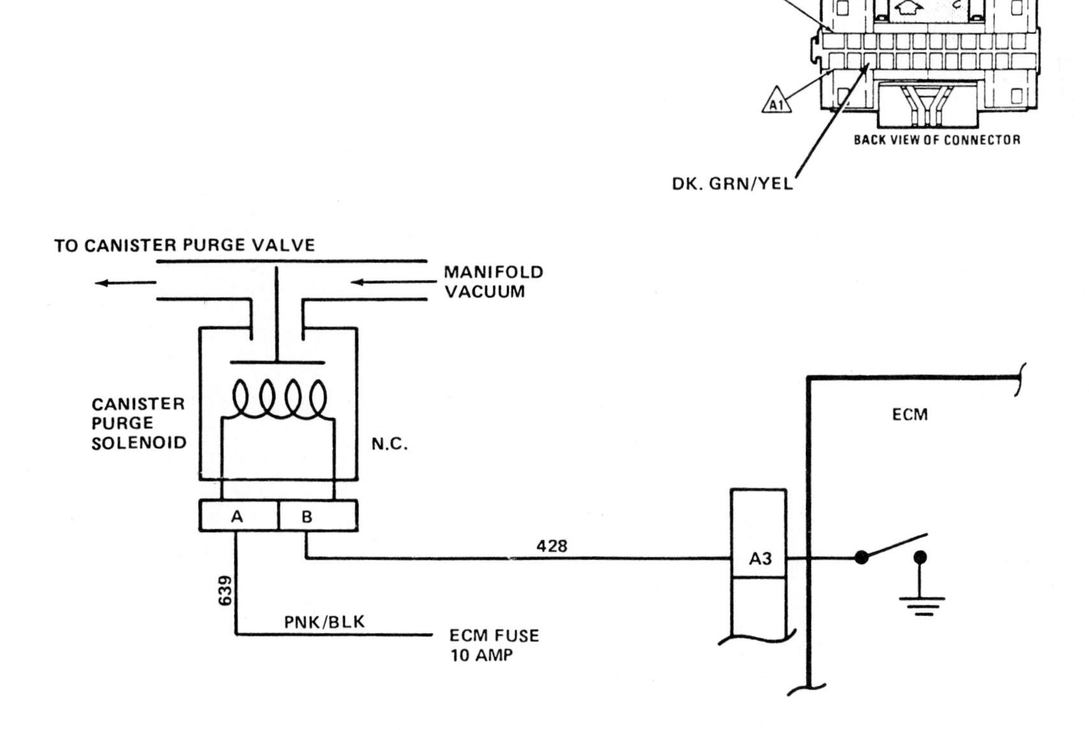

Figure 4–40 Canister Purge Circuit. (Courtesy of General Motors Corporation)

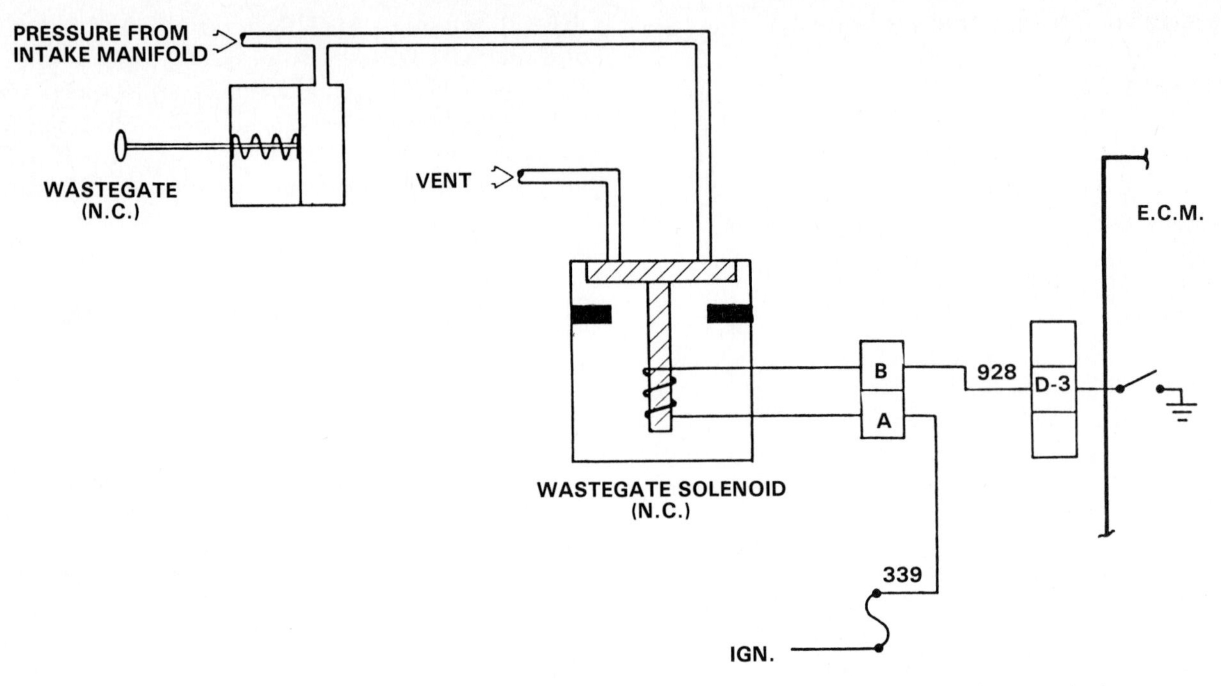

Figure 4-41 Wastegate Solenoid Circuit. [Courtesy of General Motors Corporation]

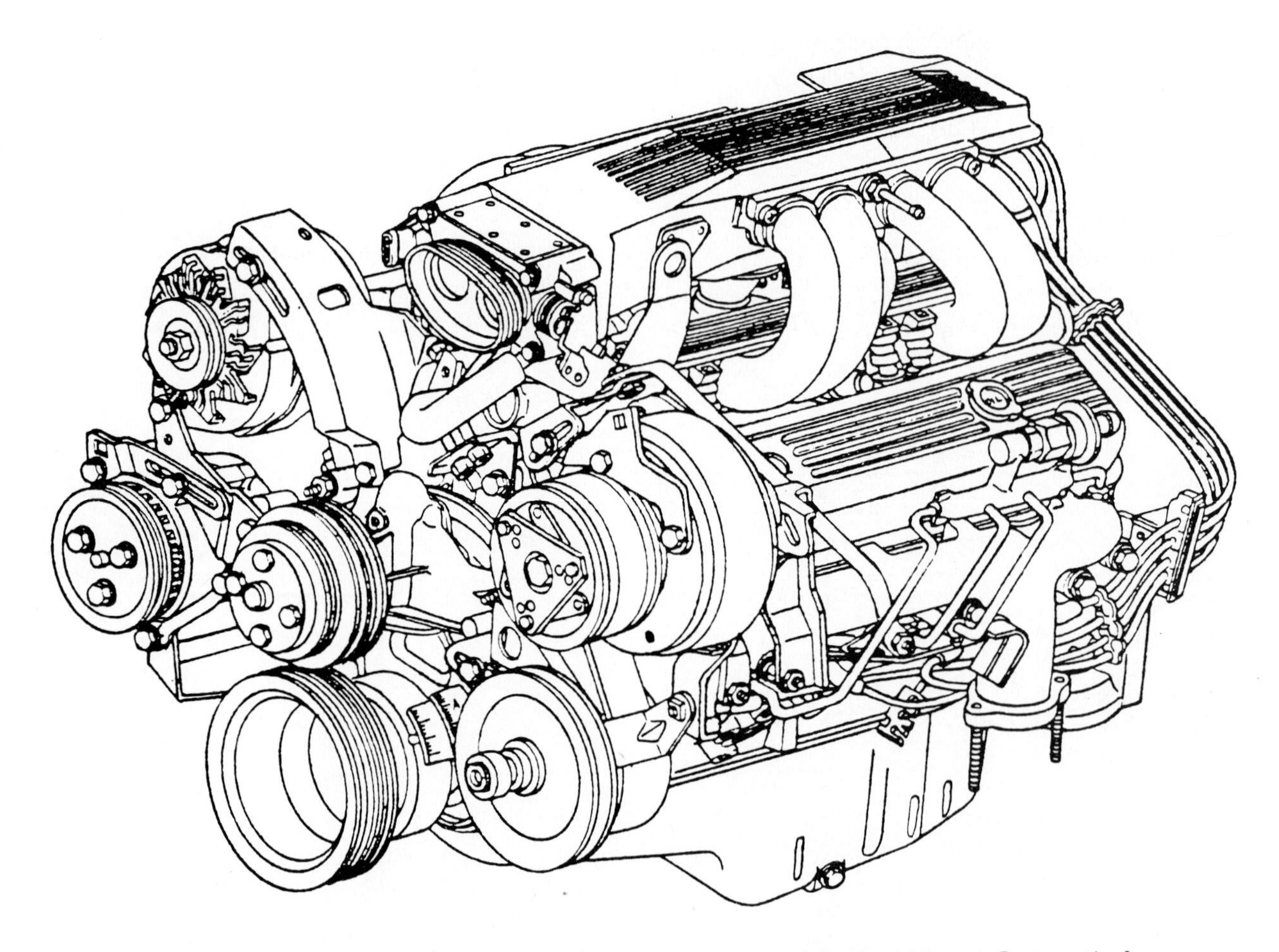

Figure 4-42 Tuned Port Injection System. [Courtesy of General Motors Corporation]

Diagnosis of PFI Systems

Diagnostic Codes

Many PFI systems use a ''service engine soon'' light in place of the check-engine light that is used on General Motors 3C and throttle body injection systems. The diagnostic code diagnosis of PFI and SFI systems is very similar to the diagnosis of 3C and throttle body injection systems. When the diagnostic test terminal is connected to the ground terminal in the assembly line communications link (ALCL), with the ignition switch on, the ''service engine soon'' light will flash out any diagnostic codes stored in the ECM. The ALCL is pictured in Figure 4–43.

When the diagnostic test terminal is grounded, the ECM will activate all the relays in the system and fully extend the IAC motor. The spark advance will also remain in a fixed position when the diagnostic test terminal is grounded. When the initial timing is being checked, the instructions on the underhood emission label should be followed. The diagnostic codes for PFI and SFI systems are shown in Table 4–3.

If the diagnostic test terminal is grounded with the engine running, the ECM will enter the field ser-

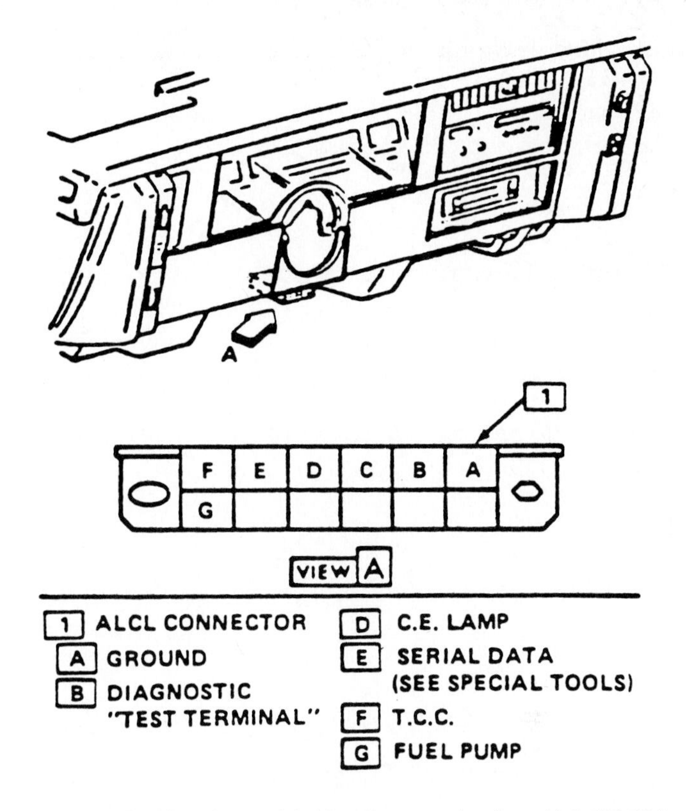

Figure 4–43 Assembly Line Communications Link (ALCL). (Courtesy of General Motors Corporation)

TABLE 4–3 PFI and SFI Diagnostic Codes.

Code	Description
12	NO REFERENCE PULSES TO ECM
13	OXYGEN SENSOR CIRCUIT FAILED
14	COOLANT READING TOO HIGH
15	COOLANT READING TOO LOW
21	THROTTLE POSITION SENSOR READING TOO HIGH
22	THROTTLE POSITION SENSOR READING TOO LOW
23**	MANIFOLD AIR TEMPERATURE LOW
24	VEHICLE SPEED SENSOR FAILED
25**	MANIFOLD AIR TEMPERATURE HIGH
31*	WASTEGATE ELECTRICAL SIGNAL OPEN OR GROUNDED
32	EGR ELECTRICAL OR VACUUM CIRCUIT MALFUNCTION
33*	MASS FLOW SENSOR READING TOO HIGH
33	MAP SENSOR HIGH
34*	MASS FLOW SENSOR READING TOO LOW OR NO SIGNAL FROM SENSOR
34	MAP SENSOR LOW
41*	CAM SENSOR FAILED
41**	CYLINDER SELECT ERROR
42*	ERROR IN DISTRIBUTOR OR C³ SYSTEM
43	ELECTRONIC SPARK CONTROL FAILURE
44	OXYGEN SENSOR LEAN TOO LONG
45	OXYGEN SENSOR RICH TOO LONG
51	CALIBRATION PROM ERROR
52*	CALPAK MISSING
53**	OVER VOLTAGE CONDITION
54**	LOW FUEL PUMP VOLTAGE
55	INTERNAL ECM ERROR (A/D CONVERTER)

NOTES

* = NEW CODES 1984
** = NEW CODES FOR 1985
NOT ALL CODES USED ON ALL APPLICATIONS

(Courtesy of General Motors Corporation)

vice mode in which the frequency of "service engine soon" flashes indicate whether the system is in open or closed loop. (The field service mode was explained earlier in this chapter.)

Checking Minimum Air Rate

The following procedure should be followed to check the minimum air rate in the throttle body.

1. With the engine at normal operating temperature and the ignition switch on, connect a jumper wire across terminals A and B in the ALCL.
2. Wait 30 seconds until the ECM extends the IAC motor.
3. Disconnect the IAC motor connector.
4. Connect a tachometer from the coil "tach" terminal to ground, and start the engine.
5. Idle speed should be 475 to 525 rpm. If the idle speed is not within specifications, the minimum air rate screw in the throttle body must be adjusted. (This procedure was explained earlier in this chapter.)

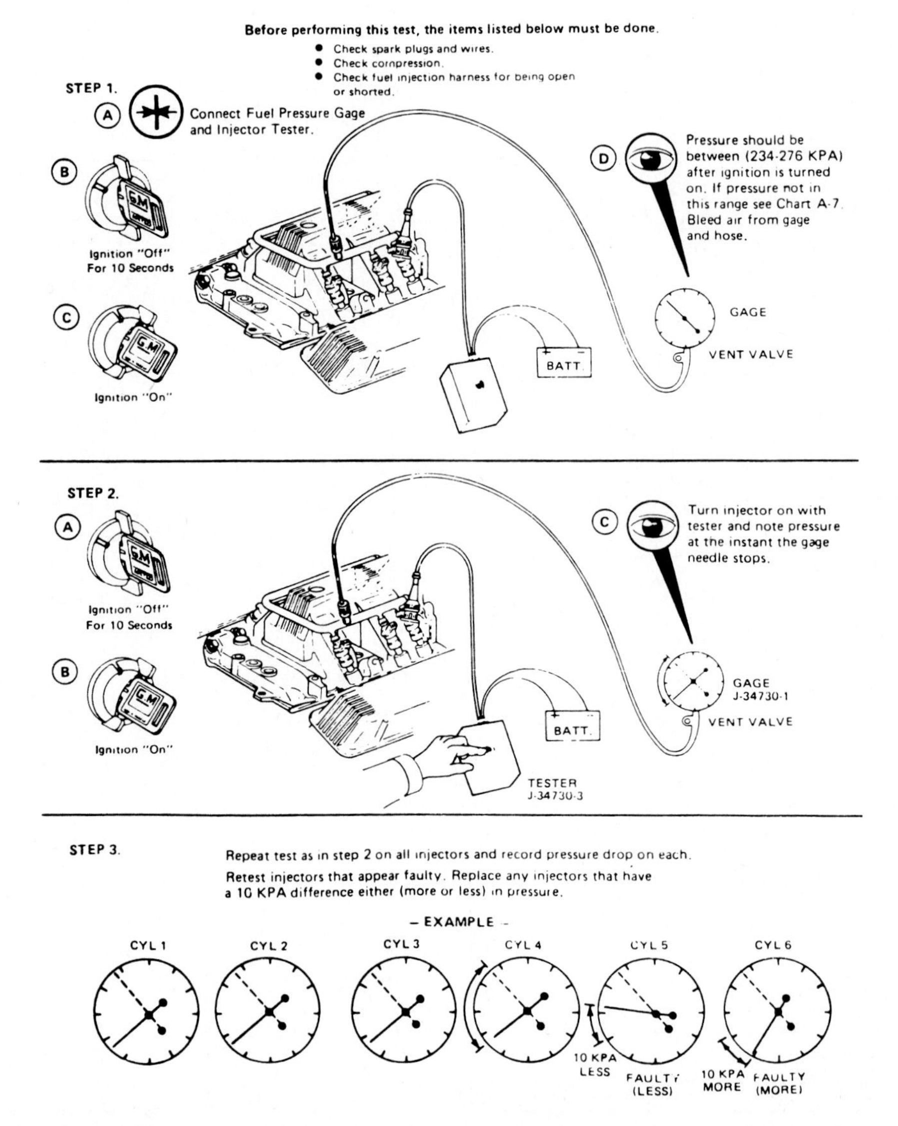

Figure 4-44 Injector Balance Test. (Courtesy of General Motors Corporation)

Injector Testing

The injectors may be tested as outlined in Figure 4–44.

The various sensors and circuits may be tested by measuring the voltage at the ECM terminals with a digital voltmeter. Voltage specifications at each ECM terminal on an SFI turbocharged engine are provided in Figure 4–45, and voltage specifications for the ECM terminals on a PFI system are shown in Figure 4–46.

Fuel Pump Pressure Test

The fuel pump pressure gauge must be connected to the test point on the fuel rail. (A complete fuel system diagram is shown in Figure 4–25.) Before the pressure gauge is connected, the fuel system pressure must be relieved. This may be done by disconnecting one injector terminal and connecting one injector terminal to ground while 12V is supplied momentarily to the other terminal. Never turn on the ignition switch while a fuel line is disconnected. Do

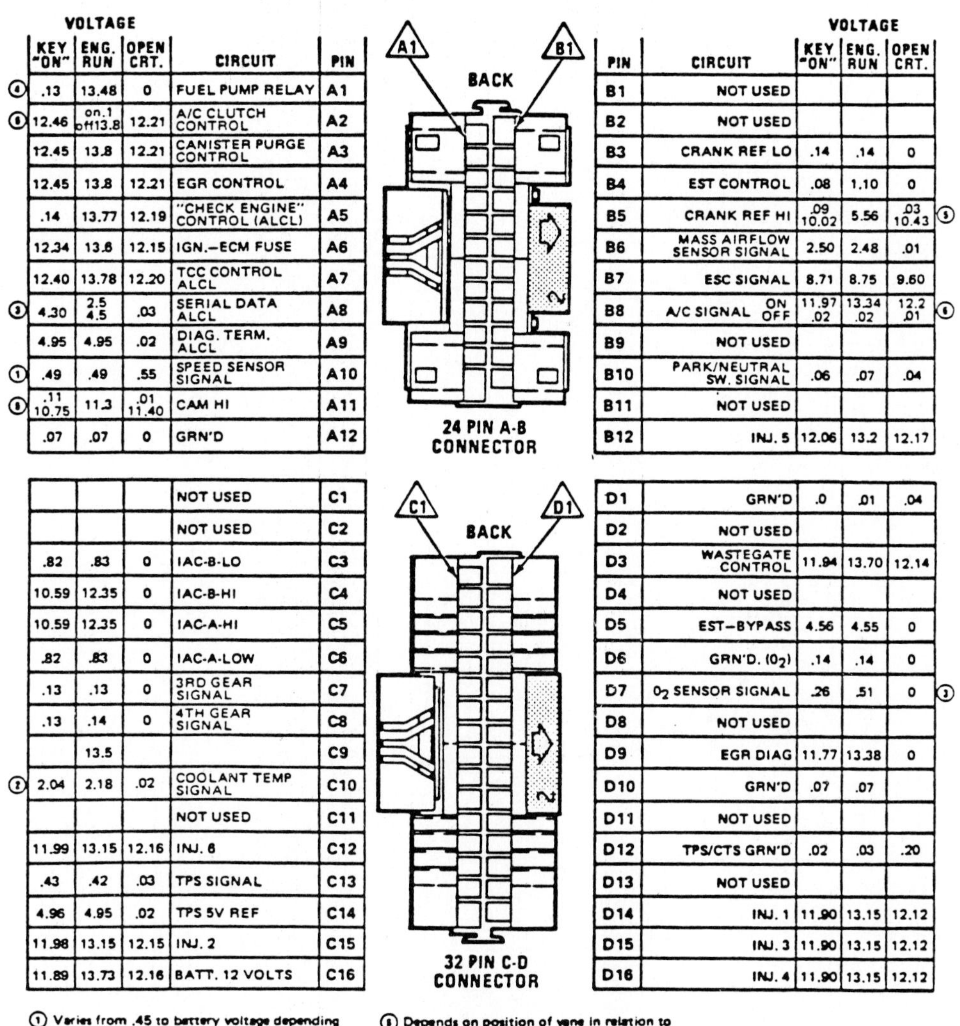

FUEL INJECTION ECM CONNECTOR IDENTIFICATION

THIS ECM VOLTAGE CHART IS FOR USE WITH A DIGITAL VOLTMETER TO FURTHER AID IN DIAGNOSIS. THE VOLTAGES YOU GET MAY VARY DUE TO LOW BATTERY CHARGE OR OTHER REASONS, BUT THEY SHOULD BE VERY CLOSE.
THE FOLLOWING CONDITIONS MUST BE MET BEFORE TESTING:
● ENGINE AT OPERATING TEMPERATURE ● ENGINE IDLING IN CLOSED LOOP (FOR "ENGINE RUN" COLUMN) ●
● TEST TERMINAL NOT GROUNDED ● ALCL TOOL NOT INSTALLED ●

Note	KEY "ON"	ENG. RUN	OPEN CRT.	CIRCUIT	PIN
4	.13	13.48	0	FUEL PUMP RELAY	A1
6	12.46	on .1 off 13.8	12.21	A/C CLUTCH CONTROL	A2
	12.45	13.8	12.21	CANISTER PURGE CONTROL	A3
	12.45	13.8	12.21	EGR CONTROL	A4
	.14	13.77	12.19	"CHECK ENGINE" CONTROL (ALCL)	A5
	12.34	13.6	12.15	IGN.–ECM FUSE	A6
	12.40	13.78	12.20	TCC CONTROL ALCL	A7
3	4.30	2.5 4.5	.03	SERIAL DATA ALCL	A8
	4.95	4.95	.02	DIAG. TERM. ALCL	A9
1	.49	.49	.55	SPEED SENSOR SIGNAL	A10
5	.11 10.75	11.3	.01 11.40	CAM HI	A11
	.07	.07	0	GRN'D	A12

PIN	CIRCUIT	KEY "ON"	ENG. RUN	OPEN CRT.	Note
B1	NOT USED				
B2	NOT USED				
B3	CRANK REF LO	.14	.14	0	
B4	EST CONTROL	.08	1.10	0	
B5	CRANK REF HI	.09 10.02	5.56	.03 10.43	5
B6	MASS AIRFLOW SENSOR SIGNAL	2.50	2.48	.01	
B7	ESC SIGNAL	8.71	8.75	9.60	
B8	A/C SIGNAL ON / OFF	11.97 / .02	13.34 / .02	12.2 / .01	6
B9	NOT USED				
B10	PARK/NEUTRAL SW. SIGNAL	.06	.07	.04	
B11	NOT USED				
B12	INJ. 5	12.06	13.2	12.17	

BACK
24 PIN A-B CONNECTOR

Note	KEY "ON"	ENG. RUN	OPEN CRT.	CIRCUIT	PIN
				NOT USED	C1
				NOT USED	C2
	.82	.83	0	IAC-B-LO	C3
	10.59	12.35	0	IAC-B-HI	C4
	10.59	12.35	0	IAC-A-HI	C5
	.82	.83	0	IAC-A-LOW	C6
	.13	.13	0	3RD GEAR SIGNAL	C7
	.13	.14	0	4TH GEAR SIGNAL	C8
		13.5			C9
2	2.04	2.18	.02	COOLANT TEMP SIGNAL	C10
				NOT USED	C11
	11.99	13.15	12.16	INJ. 6	C12
	.43	.42	.03	TPS SIGNAL	C13
	4.96	4.95	.02	TPS 5V REF	C14
	11.98	13.15	12.15	INJ. 2	C15
	11.89	13.73	12.16	BATT. 12 VOLTS	C16

PIN	CIRCUIT	KEY "ON"	ENG. RUN	OPEN CRT.	Note
D1	GRN'D	.0	.01	.04	
D2	NOT USED				
D3	WASTEGATE CONTROL	11.94	13.70	12.14	
D4	NOT USED				
D5	EST–BYPASS	4.56	4.55	0	
D6	GRN'D. (O_2)	.14	.14	0	
D7	O_2 SENSOR SIGNAL	.26	.51	0	3
D8	NOT USED				
D9	EGR DIAG	11.77	13.38	0	
D10	GRN'D	.07	.07		
D11	NOT USED				
D12	TPS/CTS GRN'D	.02	.03	.20	
D13	NOT USED				
D14	INJ. 1	11.90	13.15	12.12	
D15	INJ. 3	11.90	13.15	12.12	
D16	INJ. 4	11.90	13.15	12.12	

BACK
32 PIN C-D CONNECTOR

(1) Varies from .45 to battery voltage depending on position of drive wheels.
(2) Normal operating temperature.
(3) Varies.
(4) 12V first two seconds.
(5) Depends on position of vane in relation to "hall-effect" switch. Voltage will be low when vane is passing through switch.
(6) Engine running voltage will be high or low depending whether A/C is on or off.

Figure 4–45 ECM Terminal Identification and Voltage Specifications, SFI System with Turbocharger. (Courtesy of General Motors Corporation)

not attempt to loosen a fuel line, or connect the pressure gauge, until the fuel system pressure is relieved. With the ignition switch on, the fuel pump pressure should be 26 to 46 psi (179 to 317 kPa).

Chrysler Single-Point Electronic Fuel Injection (EFI) Used on 2.2L Engine

Modules

Power Module The power module is located in the left front fender well behind the battery. Adequate cooling for the electronic components in the module is supplied by intake air flowing through the module before it enters the air cleaner. The power module supplies an 8V signal to the logic module and the distributor pickup. A ground circuit for the automatic shutdown (ASD) relay is provided in the power module. When this ground circuit is completed, the ASD relay supplies voltage to the electric fuel pump, logic module, ignition coil positive terminal, and the injector and ignition coil drive circuits in the power module. The power module controls the operation of the fuel injector by opening and closing the injector ground circuit. Another function of the power module is to open and close the circuit from the coil neg-

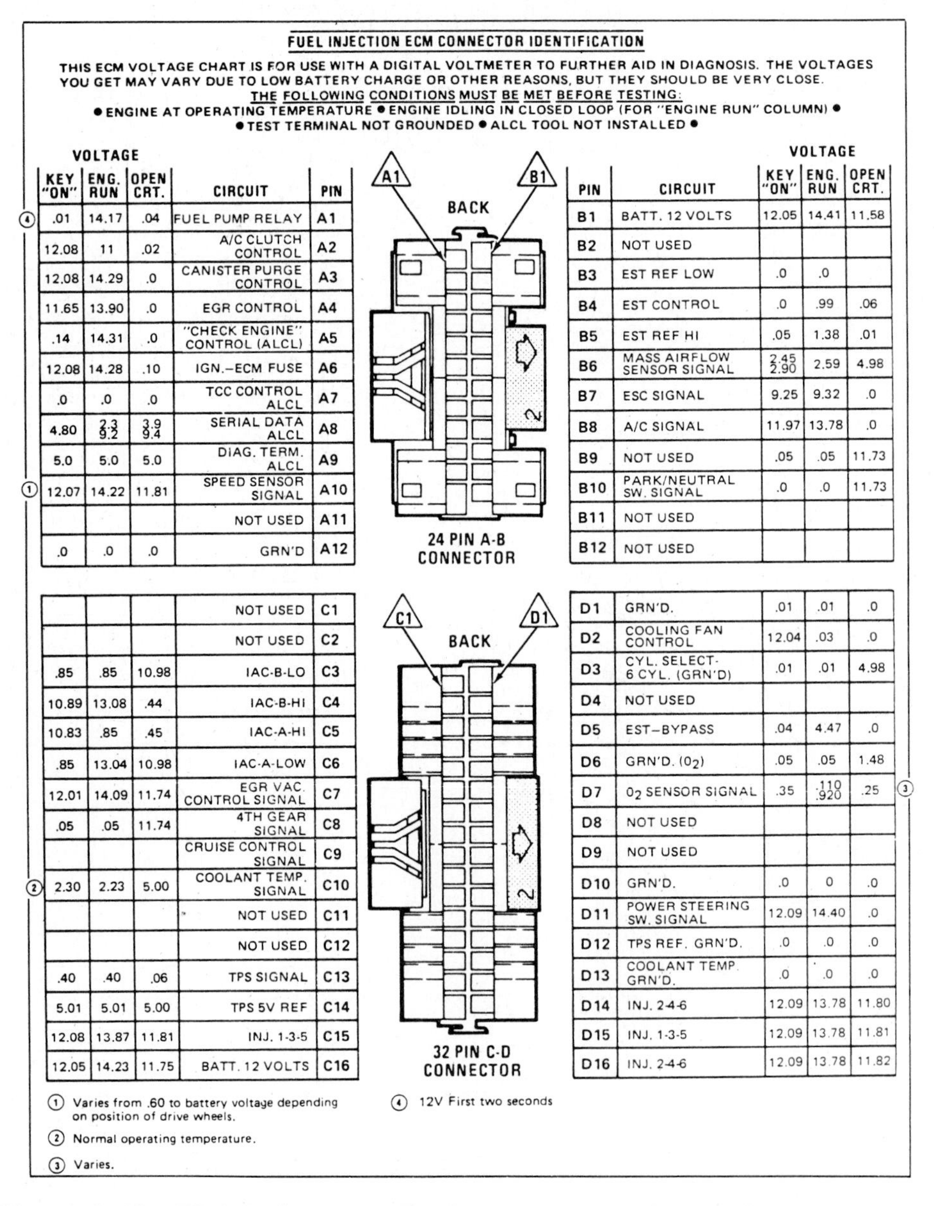

FUEL INJECTION ECM CONNECTOR IDENTIFICATION

THIS ECM VOLTAGE CHART IS FOR USE WITH A DIGITAL VOLTMETER TO FURTHER AID IN DIAGNOSIS. THE VOLTAGES YOU GET MAY VARY DUE TO LOW BATTERY CHARGE OR OTHER REASONS, BUT THEY SHOULD BE VERY CLOSE.

THE FOLLOWING CONDITIONS MUST BE MET BEFORE TESTING:

• ENGINE AT OPERATING TEMPERATURE • ENGINE IDLING IN CLOSED LOOP (FOR "ENGINE RUN" COLUMN) • • TEST TERMINAL NOT GROUNDED • ALCL TOOL NOT INSTALLED •

Note	VOLTAGE KEY "ON"	VOLTAGE ENG. RUN	VOLTAGE OPEN CRT.	CIRCUIT	PIN
(4)	.01	14.17	.04	FUEL PUMP RELAY	A1
	12.08	11	.02	A/C CLUTCH CONTROL	A2
	12.08	14.29	.0	CANISTER PURGE CONTROL	A3
	11.65	13.90	.0	EGR CONTROL	A4
	.14	14.31	.0	"CHECK ENGINE" CONTROL (ALCL)	A5
	12.08	14.28	.10	IGN.–ECM FUSE	A6
	.0	.0	.0	TCC CONTROL ALCL	A7
	4.80	2.3 / 9.2	3.9 / 9.4	SERIAL DATA ALCL	A8
	5.0	5.0	5.0	DIAG. TERM. ALCL	A9
(1)	12.07	14.22	11.81	SPEED SENSOR SIGNAL	A10
				NOT USED	A11
	.0	.0	.0	GRN'D	A12

PIN	CIRCUIT	VOLTAGE KEY "ON"	VOLTAGE ENG. RUN	VOLTAGE OPEN CRT.	Note
B1	BATT. 12 VOLTS	12.05	14.41	11.58	
B2	NOT USED				
B3	EST REF LOW	.0	.0		
B4	EST CONTROL	.0	.99	.06	
B5	EST REF HI	.05	1.38	.01	
B6	MASS AIRFLOW SENSOR SIGNAL	2.45 / 2.90	2.59	4.98	
B7	ESC SIGNAL	9.25	9.32	.0	
B8	A/C SIGNAL	11.97	13.78	.0	
B9	NOT USED	.05	.05	11.73	
B10	PARK/NEUTRAL SW. SIGNAL	.0	.0	11.73	
B11	NOT USED				
B12	NOT USED				

Note	VOLTAGE KEY "ON"	VOLTAGE ENG. RUN	VOLTAGE OPEN CRT.	CIRCUIT	PIN
				NOT USED	C1
				NOT USED	C2
	.85	.85	10.98	IAC-B-LO	C3
	10.89	13.08	.44	IAC-B-HI	C4
	10.83	.85	.45	IAC-A-HI	C5
	.85	13.04	10.98	IAC-A-LOW	C6
	12.01	14.09	11.74	EGR VAC. CONTROL SIGNAL	C7
	.05	.05	11.74	4TH GEAR SIGNAL	C8
				CRUISE CONTROL SIGNAL	C9
(2)	2.30	2.23	5.00	COOLANT TEMP. SIGNAL	C10
				NOT USED	C11
				NOT USED	C12
	.40	.40	.06	TPS SIGNAL	C13
	5.01	5.01	5.00	TPS 5V REF	C14
	12.08	13.87	11.81	INJ. 1-3-5	C15
	12.05	14.23	11.75	BATT. 12 VOLTS	C16

PIN	CIRCUIT	VOLTAGE KEY "ON"	VOLTAGE ENG. RUN	VOLTAGE OPEN CRT.	Note
D1	GRN'D.	.01	.01	.0	
D2	COOLING FAN CONTROL	12.04	.03	.0	
D3	CYL. SELECT- 6 CYL. (GRN'D)	.01	.01	4.98	
D4	NOT USED				
D5	EST–BYPASS	.04	4.47	.0	
D6	GRN'D. (O_2)	.05	.05	1.48	
D7	O_2 SENSOR SIGNAL	.35	.110 / .920	.25	(3)
D8	NOT USED				
D9	NOT USED				
D10	GRN'D.	.0	0	.0	
D11	POWER STEERING SW. SIGNAL	12.09	14.40	.0	
D12	TPS REF. GRN'D.	.0	.0	.0	
D13	COOLANT TEMP. GRN'D.	.0	.0	.0	
D14	INJ. 2-4-6	12.09	13.78	11.80	
D15	INJ. 1-3-5	12.09	13.78	11.81	
D16	INJ. 2-4-6	12.09	13.78	11.82	

(1) Varies from .60 to battery voltage depending on position of drive wheels.

(2) Normal operating temperature.

(3) Varies.

(4) 12V First two seconds

Figure 4–46 ECM Terminal Identification and Voltage Specifications, PFI System. (Courtesy of General Motors Corporation)

ative terminal to ground, which operates the ignition system. Commands from the logic module are used to control the power module illustrated in Figure 4–47.

Logic Module The logic module is located inside the vehicle behind the right front kick pad. This module supplies a 5V signal to the sensors in the system, and it also receives input signals from the sensors and the distributor pickup. On the basis of all the input signals received, the logic module will send the appropriate spark advance schedule to the power module under all engine operating conditions, and the power module opens the primary ignition circuit at the right instant to provide the correct spark advance. The logic module also commands the power module to supply the right injector pulse width, or on-time, to maintain engine performance, economy, and emission levels. Other functions of the logic module include the operation of the exhaust-gas recirculation (EGR) and canister purge solenoids, and the automatic idle speed (AIS) motor.

The logic module has the capability of testing many of its own input and output circuits. If a fault is found in a major system, the information is stored in the logic module for future reference. This fault can be displayed to the service technician by a flashing power-loss light on the instrument panel or by a digital reading on a tester that can be connected to the system. The logic module places the system in a "limp-in" mode if an unacceptable signal is received from the manifold absolute pressure sensor, throttle position sensor, or coolant sensor. In this mode the logic module will ignore some of the sensor signals and maintain the spark advance and injector pulse width to keep the engine running, but fuel economy and performance will decrease. In the limp-in mode, the logic module will illuminate the power-loss light. The logic module is shown in Figure 4–48.

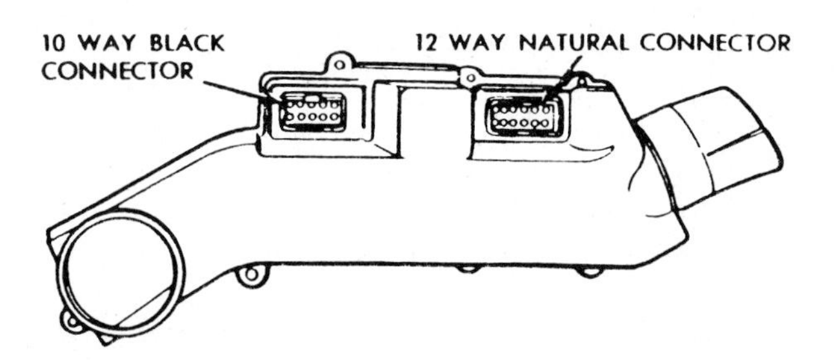

Figure 4–47 Power Module. (Courtesy of Chrysler Canada Ltd.)

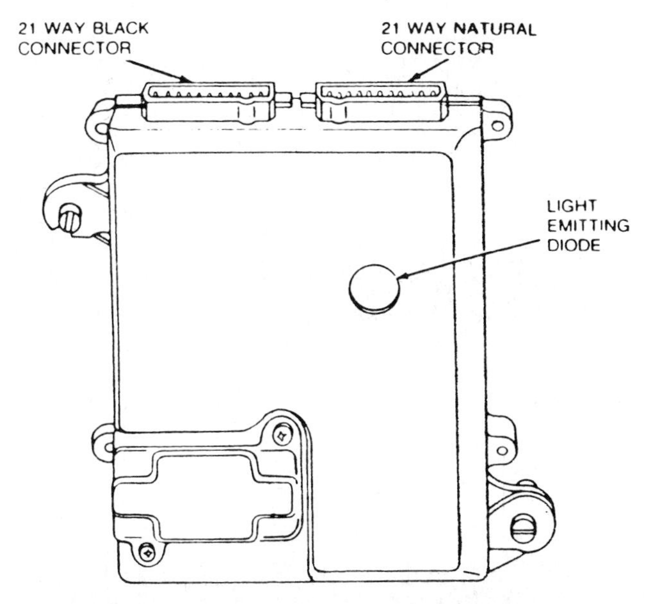

Figure 4–48 Logic Module. (Courtesy of Chrysler Canada Ltd.)

Switch Inputs

Park/Neutral Safety Switch This switch supplies information to the logic module and is used by the module to control the automatic idle speed (AIS) motor and provide the correct idle speed in all transmission selector positions. The park/neutral safety switch signal to the logic module may affect spark advance to some extent.

Electric Backlite (EBL) Switch When the EBL is turned on, the logic module will operate the AIS motor to increase throttle opening slightly to compensate for the additional alternator load on the engine.

Brake Switch In the event that the logic module does not receive a signal from the electric idle switch, the brake light switch will be used to sense idle throttle position.

Air Conditioning Switch If this switch is activated, the logic module will operate the AIS motor to increase idle speed.

Air Conditioning Clutch Switch The logic module will activate the AIS motor to give a one-time kick, which maintains engine speed when the clutch engages and prevents variations in idle speed. The switch inputs to the logic module are indicated in Figure 4–49.

Sensor Inputs

Manifold Absolute Pressure (MAP) Sensor The MAP sensor sends a signal to the logic module in

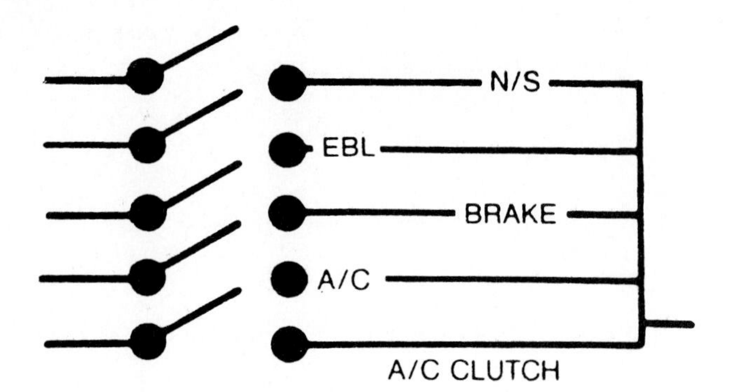

Figure 4–49 Switch Inputs to Logic Module. (Courtesy of Chrysler Canada Ltd.)

relation to manifold vacuum and barometric pressure. A reference voltage of 5V is applied from the logic module to the sensors. The MAP sensor sends a voltage signal of 0.3V to 4.9V to the logic module. This voltage signal is 4.9V at zero vacuum, and it may be as low as 0.3V at maximum vacuum. When engine load increases, manifold vacuum decreases and this signal is sent from the MAP sensor to the logic module. When this signal is received, the logic module increases the injector pulse width, which supplies the additional fuel requirements of the engine. As manifold vacuum increases the MAP sensor signal will cause the logic and power modules to shorten the injector pulse width and supply less fuel to the engine. The logic module uses the information from the MAP sensor and other inputs to determine the correct spark advance schedule under all engine operating conditions. A MAP sensor is pictured in Figure 4–50.

Throttle-Position Sensor The throttle-position sensor is a variable resistor connected to the throttle shaft on the throttle body assembly. As the throttle is

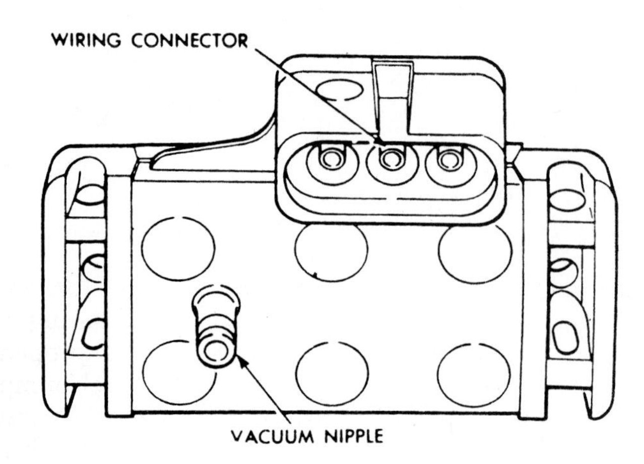

Figure 4–50 Manifold Absolute Pressure (MAP) Sensor. (Courtesy of Chrysler Canada Ltd.)

opened, a signal of 0.16V to 4.7V is sent from the sensor to the logic module. This signal and other sensor information is used by the logic module to adjust the air–fuel ratio to meet various conditions during acceleration, deceleration, wide-open throttle, and idle.

Oxygen Sensor The oxygen sensor is very similar to the oxygen sensor in the 3C system. (Refer to Chapter 3 for a description of this sensor.) This sensor generates a signal from 0V to 1V as the air–fuel ratio becomes richer. The oxygen sensor signal is used by the logic module along with other sensor data to provide the air–fuel ratio for optimum engine performance, economy, and emission levels. When there is a need for additional fuel enrichment, such as on sudden acceleration, the MAP sensor and throttle-position sensor signals may override the oxygen sensor signal.

Coolant Temperature Sensor The coolant temperature sensor is mounted in the thermostat housing. The resistance values of the resistive element vary from 11,000Ω at −4°F (−15°C) to 800Ω at 195°F (90.5°C). Along with other input data, the coolant sensor signal is used by the logic module in the scheduling of idle speeds, air–fuel ratio, and spark advance curves for all engine operating conditions. When the engine is cold, idle speed will increase, the air–fuel ratio will be enriched, and the spark advance curve will be altered to improve cold engine performance.

Vehicle-Speed Sensor The vehicle-speed sensor is located in the speedometer cable. This sensor contains an on/off micro switch that generates eight pulses per speedometer cable revolution. The vehicle-speed sensor signal and the throttle-position sensor signal are interpreted by the logic module to tell the difference between closed throttle deceleration and normal idle speed with the vehicle not in motion. During deceleration, the logic module operates the AIS motor to maintain a slightly higher idle speed and reduced emissions. The vehicle-speed sensor is illustrated in Figure 4–51.

Figure 4–51 Vehicle-Speed Sensor. (Courtesy of Chrysler Canada Ltd.)

Distributor Pickup Signal A Hall Effect switch unit is used in the distributor pickup assembly. Four metal shutter blades, one for each cylinder, are attached to the bottom of the distributor rotor. Each time one of these shutter blades rotates through the magnetic field of the Hall Effect switch, a pulse signal is generated by the switch. This pulse signal is sent to the logic module and the power module. When the distributor pickup signal is received, the power module grounds the automatic shutdown (ASD) relay winding, which supplies voltage to the fuel pump, ignition coil positive terminal, and the injector drive and coil drive circuits of the power module. If the distributor pickup signal is not present or is not correct, the ASD relay will not be energized by the power module. The logic module uses the distributor pulse signals as speed information to help determine the spark advance. Piston position is also determined by the logic module from the distributor pulse signal. The logic module signals the power module to operate the injector when the piston position signal is received from the distributor pickup. A distributor with the Hall Effect switch and shutter blade is shown in Figure 4–52.

(A complete explanation of Chrysler electronic ignition systems is provided in Chapter 5.) All the sensor inputs to the logic module are shown in Figure 4–53.

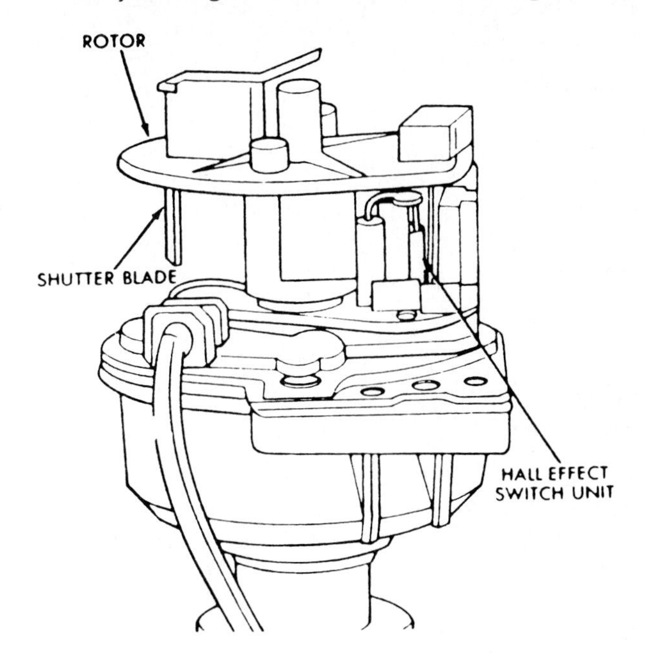

Figure 4–52 Hall Effect Distributor Pickup. (Courtesy of Chrysler Canada Ltd.)

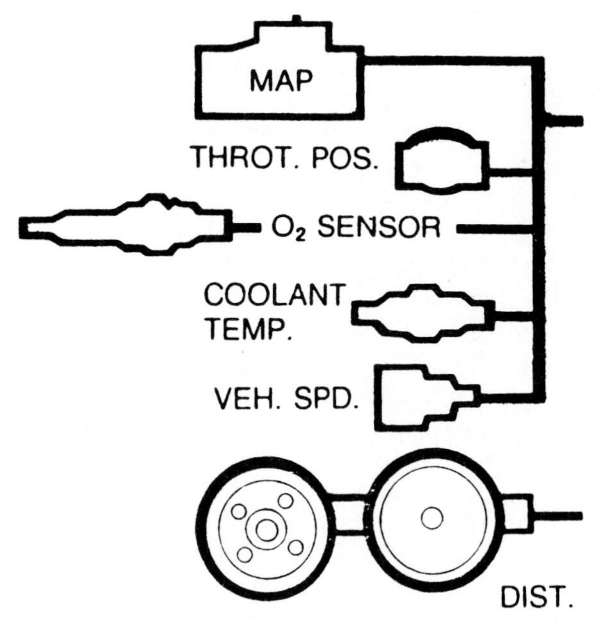

Figure 4–53 Sensor Inputs to Logic Module. (Courtesy of Chrysler Canada Ltd.)

Automatic Shutdown (ASD) Relay

Operation When the ignition switch is turned on, voltage is supplied through the J2 circuit to the power module, and the power module supplies 8V to the logic module and the distributor pickup. The power module also supplies voltage through the fused J2 (FJ2) circuit to the ASD relay. When this occurs, the power module grounds the ASD relay winding, and voltage is supplied through the relay contacts to the electric fuel pump, logic module, and coil positive terminal. When the ASD relay closes, the injector drive and ignition coil drive circuits are activated in the power module.

The logic module commands the power module to send a single "prime shot" of fuel to the engine when the ASD relay closes. This is accomplished by the power module grounding the injector winding for a brief instant. If the engine is not cranked within 1/2 second, the power module will open the circuit from the ASD relay winding to ground and the relay contacts will open the circuit to the fuel pump, logic module, and the coil positive terminal. If the power module does not sense battery voltage and distributor pulses at the rate of 60 rpm within 1/2 second after the first distributor pulse, the power module will open the circuit from the ASD relay winding to ground. This will cause the relay contacts to open, and the voltage supplied to the electric fuel pump, logic module, and coil positive terminal will be shut off. The power module will also open the circuit to the injector and deactivate the ignition coil drive circuit in the module.

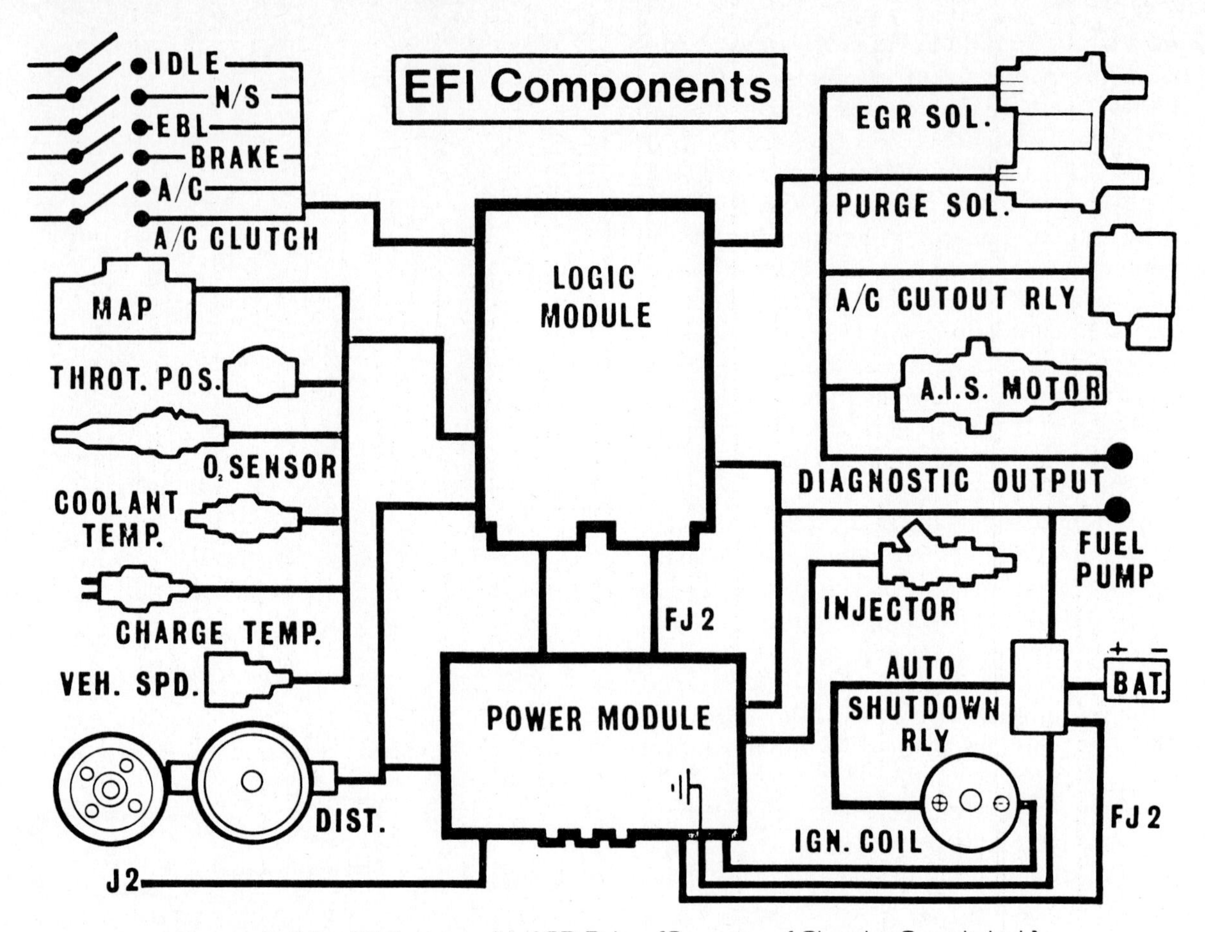

Figure 4–54 EFI System with ASD Relay. [Courtesy of Chrysler Canada Ltd.]

The ASD relay is a safety feature that shuts down the fuel pump and the ignition system if the vehicle is involved in an accident and the ignition switch is left on. The power module will ground the ASD relay winding continuously while the module receives distributor pulses at a rate above 60 rpm. This will keep the ASD relay contacts closed, and voltage will be supplied through the relay contacts to the circuits mentioned previously. The ASD relay is located under the right front kick pad with the logic module and MAP sensor. The ASD relay and the entire EFI system is illustrated in Figure 4–54.

Fuel System Components

Electric Fuel Pump The roller-vane type of fuel pump is driven by a permanent magnet electric motor. A check valve on the inlet side of the pump limits the maximum pump pressure to 120 psi (827 kPa) if the fuel system becomes completely plugged or restricted. Another check valve in the pump outlet prevents any movement of fuel in either direction when

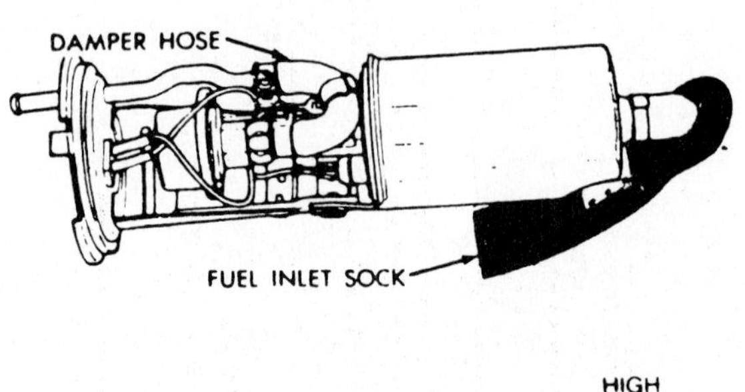

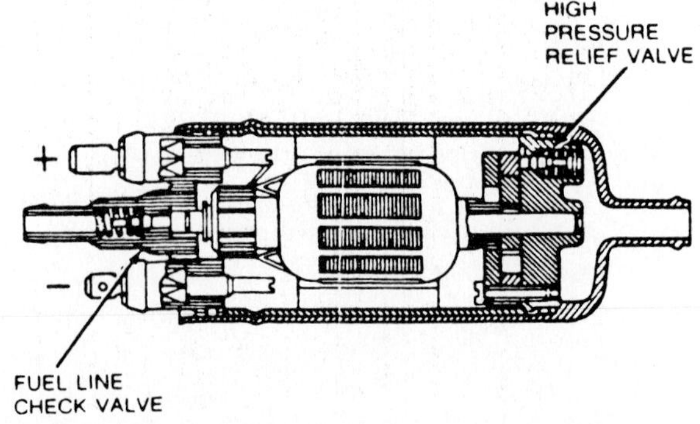

Figure 4–55 Electric Fuel Pump. [Courtesy of Chrysler Canada Ltd.]

the pump is not in operation. A 70-micron filter is provided by the fuel inlet sock on the pump inlet, which prevents water and other foreign particles from entering the fuel system. The fuel pump is shown in Figure 4–55.

The electric fuel pump and the fuel gauge sending unit are located in separate openings in the fuel tank. A swirl tank surrounds the fuel pump in the fuel tank. The return fuel line is connected from the throttle body assembly to the swirl tank. Excess fuel is being returned continuously from the throttle body assembly to the swirl tank while the engine is running. The swirl tank pictured in Figure 4–56 provides a supply of fuel at the pump inlet during all driving conditions.

The fuel flows through a low-pressure orifice in the end of return hose, and this creates a slight low-pressure area at the end of the hose, which causes increased fuel flow from the main fuel tank to the swirl tank. A return line check valve prevents fuel flow from the tank into the return line if the vehicle is rolled over in an accident.

In-Line Fuel Filter Besides the filter sock on the pump inlet, a 50-micron in-line fuel filter, as shown in Figure 4–57, is located near the fuel tank under the vehicle.

Throttle Body Assembly The throttle body assembly contains the injector, pressure regulator, automatic idle speed (AIS) motor, throttle position sensor, and throttle valve. The fuel pump supplies fuel through the in-line filter to the injector and pressure regulator. When pump pressure reaches 36 psi (248 kPa) the pressure regulator diaphragm will be forced downward by the fuel pressure, and excess fuel will be returned from the pressure regulator through the

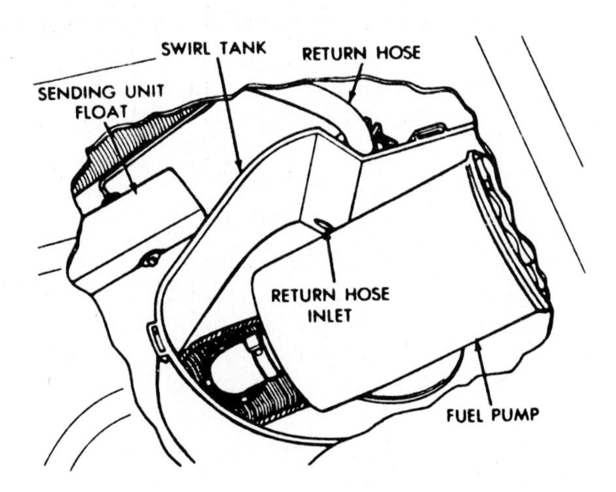

Figure 4–56 Swirl Tank. (Courtesy of Chrysler Canada Ltd.)

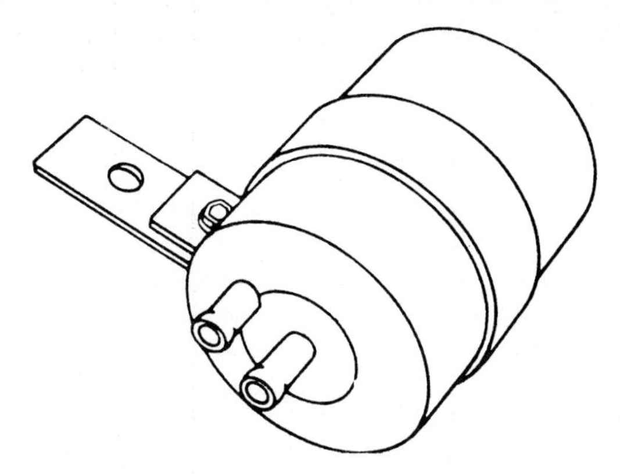

Figure 4–57 In-Line Fuel Filter. (Courtesy of Chrysler Canada Ltd.)

return line to the fuel tank, as illustrated in Figure 4–58.

Special high-pressure fuel hose and clamps are used in the EFI system. A vacuum hose is connected from the venturi area in the throttle body assembly to the spring chamber on the pressure regulator. During engine idle operation, venturi vacuum is low at the injector tip. Since this low vacuum is supplied to the pressure regulator diaphragm chamber, higher fuel pressure is required to push the diaphragm downward and limit fuel pressure. At wide-open throttle, venturi vacuum increases at the injector tip. An increase in vacuum results in a decrease in pressure. The increase in vacuum is also supplied to the pressure regulator diaphragm chamber. Therefore, less fuel pressure is required to force the diaphragm downward, and fuel pressure is reduced. This pressure regulator action provides a constant pressure drop of 36 psi (248 kPa) across the injector because fuel pressure is reduced when pressure is reduced at the injector tip.

The injector has a winding surrounding the movable plunger. Voltage is supplied to the injector winding from the power module, and the module also grounds the other end of the injector winding to energize the injector. The internal design of the injector is shown in Figure 4–59.

Air enters the throttle body assembly from the air cleaner. When the injector is energized, fuel is sprayed from the injector into the air stream above the throttle.

Output Control Functions

Fuel Injection Control The power module energizes the injector once for each piston intake stroke, and the amount of fuel delivered by the injector is determined by the pulse width, or on-time, of the injector. The pulse width is measured in milliseconds

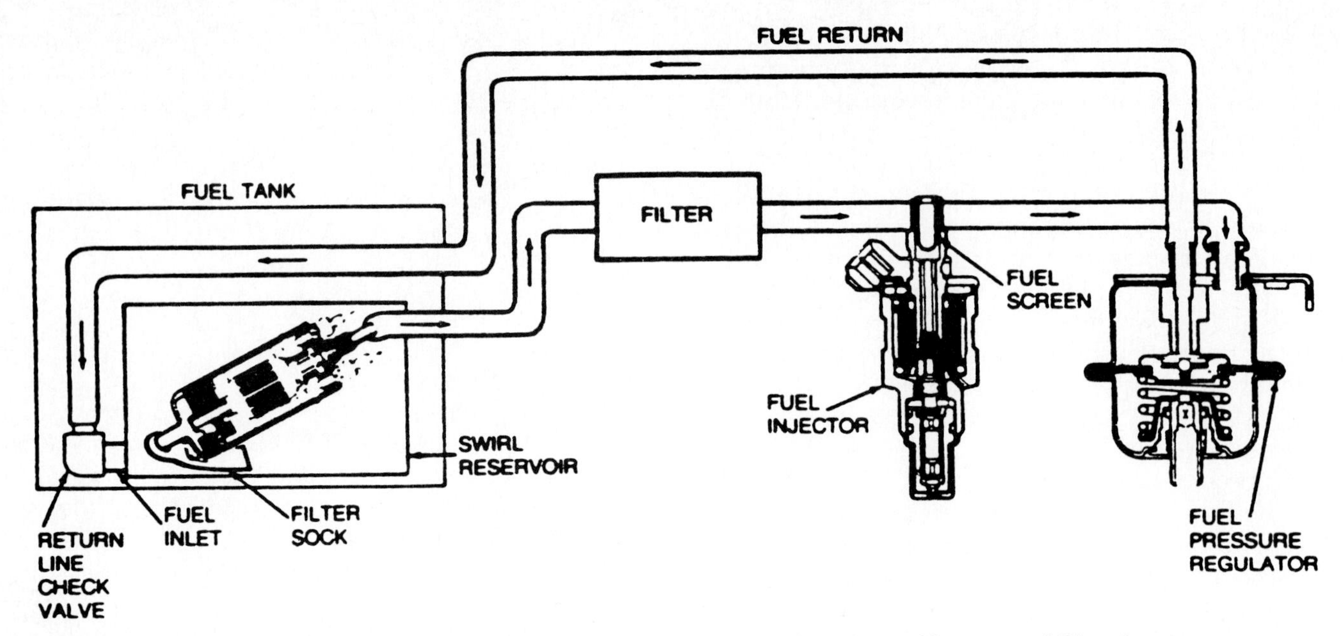

Figure 4–58 Fuel System with Injector and Pressure Regulator. [Courtesy of Chrysler Canada Ltd.]

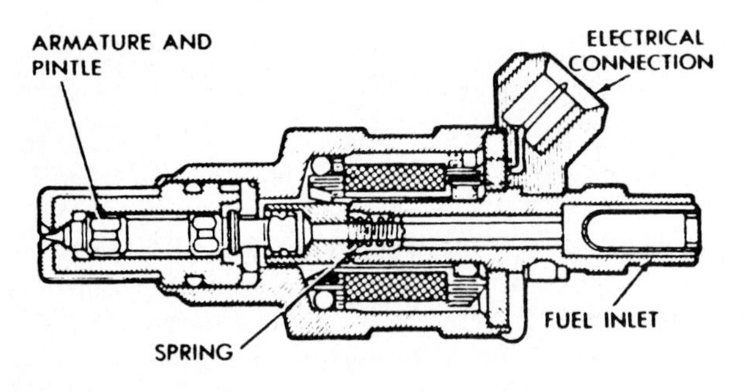

Figure 4–59 Injector Design. [Courtesy of Chrysler Canada Ltd.]

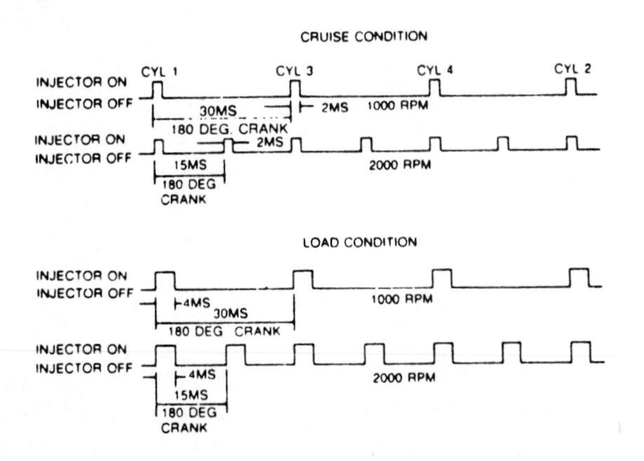

Figure 4–60 Injector Pulse Width. [Courtesy of Chrysler Canada Ltd.]

(ms), and the fuel flow from the injector increases in relation to the pulse width. Input data received by the logic module indicates the engine fuel requirements. The logic module commands the power module to supply the precise injector pulse width to meet these fuel requirements. As indicated in Figure 4–60, the pulse width will be 2 ms at 1000 rpm cruise conditions and the entire injector on/off time is 30 ms.

If the engine is operating at a 2000 rpm cruise condition, the entire injector on/off time is only 15 ms because the crankshaft is turning much faster. Under this condition the injector pulse width is still 2 ms, but more fuel is delivered by the injector because the injector is being energized every 15 ms. When the engine is operating under heavy load conditions at 1000 rpm or 2000 rpm, the injector pulse width is increased to 4 ms. The logic module and the power module will supply the correct pulse width to maintain engine performance, economy, and emission levels under all operating conditions. When the engine is being started, the injector is energized twice for each piston intake stroke. The injector pulse width will be increased when starting a cold engine to provide the necessary mixture enrichment.

Under certain operation conditions the system will operate in open-loop mode. In this mode the oxygen sensor signal is ignored by the logic module and the module maintains the air–fuel ratio at a predetermined value. The system will remain in open loop under the following conditions.

1. Cold engine (until the oxygen sensor generates a signal).
2. Park/neutral idle operation.
3. Wide-open throttle operation.
4. Deceleration conditions.
5. Oxygen sensor signal is not available for a specified period of time.

The system will operate in the closed-loop mode when the following conditions are present.

1. The coolant temperature is above a specified value.
2. The start-up delay timer has timed out in the logic module.
3. The oxygen sensor must be generating a valid signal to the logic module.
4. The vehicle must be operating under drive/idle or cruise conditions.

Ignition Spark Advance Control The logic module determines the precise spark advance requirements by interpreting data from the distributor rpm signal, MAP sensor, and coolant temperature sensor. When the engine is cold, the logic module will increase the spark advance for improved engine performance. The logic module commands the power module to open the primary ignition circuit at the right instant to provide the precise spark advance required by the engine.

Idle Speed Control While the engine is being started, the logic module will position the automatic idle speed (AIS) motor to provide easy starting without touching the accelerator pedal. When the engine is cold, the logic module will position the AIS motor to provide the correct cold fast idle speed. The AIS motor allows more air to flow past the motor plunger into the intake manifold to increase the idle speed. This air flow bypasses the throttle, as indicated in Figure 4–61.

The AIS motor will provide the correct idle speed when the air conditioner is on and the correct throttle opening when the engine is decelerating.

Exhaust-Gas Recirculation (EGR) Control The logic module energizes a vacuum solenoid in the EGR vacuum system. If the solenoid is energized, it shuts off the vacuum to the EGR valve, and a deenergized solenoid allows vacuum to pass through the solenoid to the EGR system. The solenoid will be energized at speeds below 1200 rpm, wide-open throttle operation, or when the coolant temperature is below 70°F (21°C). During all other engine operating conditions the solenoid is deenergized, and vacuum is supplied through the solenoid to the backpressure transducer. If the vehicle speed is below 30 to 35 mph (48 kph), the backpressure transducer will vent the vacuum in the EGR system. Above this speed the exhaust pressure will close the vacuum bleed in the transducer and the vacuum will be supplied to open the EGR valve. The EGR valve and the backpressure transducer are shown in Figure 4–62.

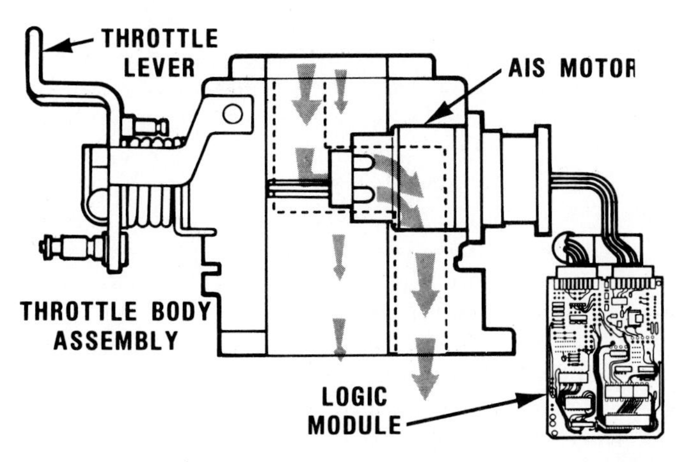

Figure 4–61 AIS Motor. [Courtesy of Chrysler Canada Ltd.]

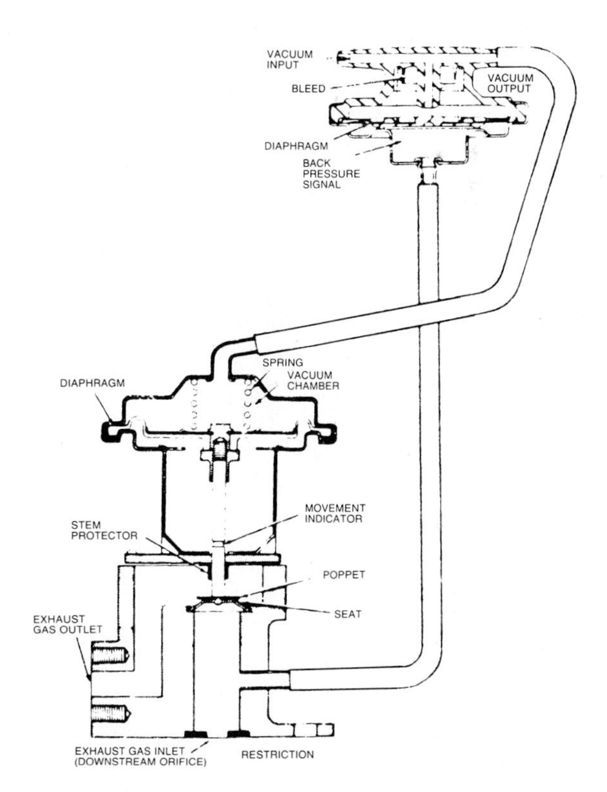

Figure 4–62 EGR Valve with Backpressure Transducer. [Courtesy of Chrysler Canada Ltd.]

Canister Purge Control The logic module operates a solenoid connected in the vacuum hose to the canister purge valve. When engine temperature is below 180°F (82°C), the logic module energizes the solenoid, which shuts off vacuum to the canister purge valve. Above this coolant temperature the solenoid is deenergized, which supplies vacuum through the solenoid to the purge control valve. Under this condition the canister is purged through a port in the throttle body. The canister purge solenoid and the EGR solenoid are located with the diagnostic connector under a cover on the fender well, as illustrated in Figure 4–63.

Air Conditioning Control When the throttle approaches the wide-open position, and the throttle position sensor voltage is above a specific value, the logic module will deenergize the air conditioning (A/C) wide-open throttle (WOT) cutout relay, and the relay contacts will open the circuit to the A/C compressor clutch. If the throttle is in the idle or cruising speed range, the logic module will energize the A/C WOT cutout relay, and the relay contacts will close, which supplies voltage to the A/C compressor clutch. If the engine speed drops below 500 rpm, the logic module will deenergize the A/C WOT cutout relay and open the circuit to the compressor clutch. This will prevent the engine from stalling under unusual conditions. If the engine is being cranked and the air conditioning is turned on, the logic module will not engage the compressor clutch until the engine speed exceeds 500 rpm. The A/C WOT cutout relay is located beside the starter relay near the battery. The complete wiring diagrams for the logic module and power module are shown in Figure 4–64 and 4–65.

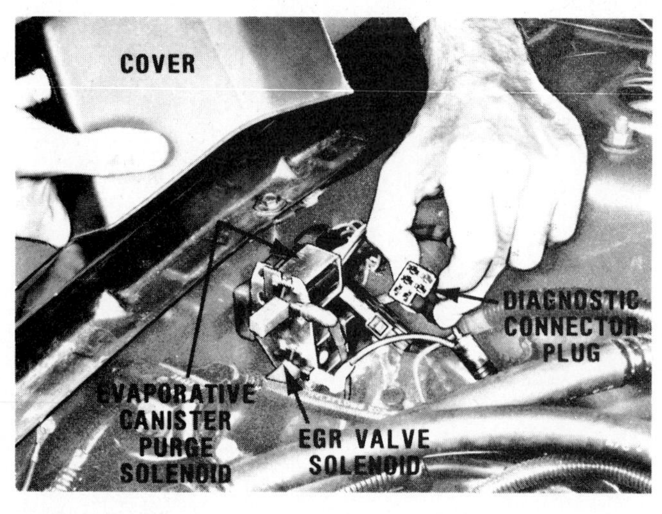

Figure 4–63 Canister Purge and EGR Solenoids with Diagnostic Connector. (Courtesy of Chrysler Canada Ltd.)

Later Model Chrysler Electronic Fuel Injection

System Changes The wiring diagrams in Figures 4–64 and 4–65 are for 1984 models. In the 1985 model year, several changes were made to the EFI system. For example, the automatic shutdown (ASD) relay is contained in the power module, and the logic module provides necessary signals to the power module for ASD relay control, charging system control, radiator fan control, and wastegate control on turbocharged engines. An external alternator voltage regulator is no longer required. A battery temperature sensor is located in the power module, and a battery charge signal is also sent to the power module. The power module operates a relay for radiator fan control and a solenoid for turbo wastegate control. Some new fault codes have been added, which apply to the battery and charging system. Diagnosis of the single-point EFI system is explained with the diagnosis of the multipoint EFI system later in this chapter. Wiring diagrams for 1985 single-point EFI systems are provided in Figures 4–66 through 4–70.

Chrysler Multipoint Electronic Fuel Injector (EFI) Used on 2.2L Turbocharged Engine

Modules and Inputs

Operation The logic module and the power module are very similar to the modules used in the single-point throttle body injection system, but they are not interchangeable. The modules perform basically the same functions as they did in the single-point EFI system; therefore we will describe the differences in the multipoint system.

Inputs The same inputs are used in the multipoint EFI system with the addition of a detonation sensor and a charge temperature sensor. If the engine detonates, the detonation sensor signals the power module to retard the spark advance until the detonation stops. The charge temperature sensor sends a signal to the logic module in relation to the temperature of the air–fuel mixture in the intake manifold. This signal acts as a backup for the coolant temperature sensor if the coolant temperature sensor fails. All the inputs and output control functions in the multipoint EFI system are illustrated in Figure 4–71.

The distributor in the multipoint EFI system has a reference pickup and a synchronizer, "sync," pickup. The reference pickup is the same as the Hall

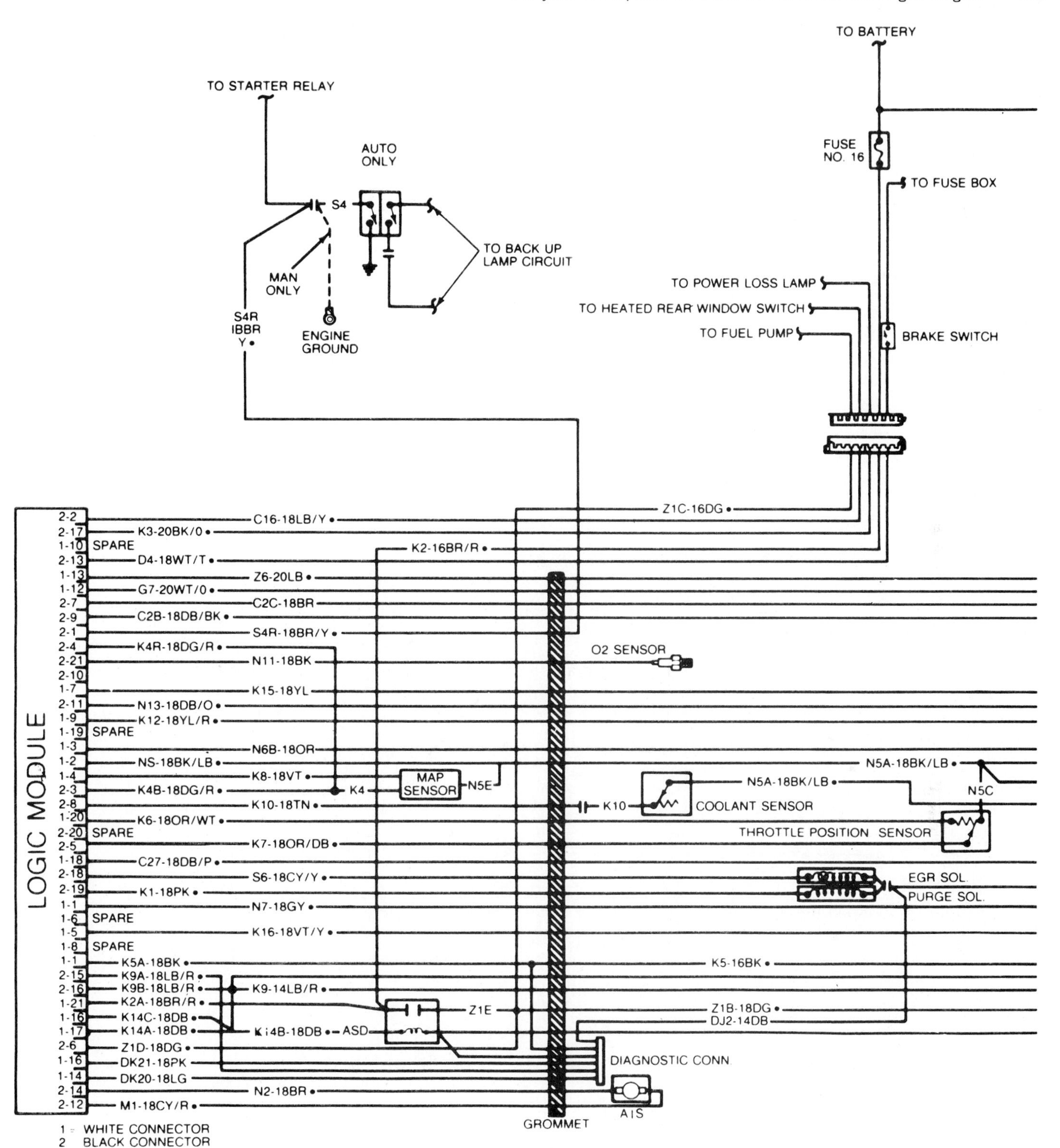

Figure 4–64 Logic Module Wiring Diagram. (Courtesy of Chrysler Canada Ltd.)

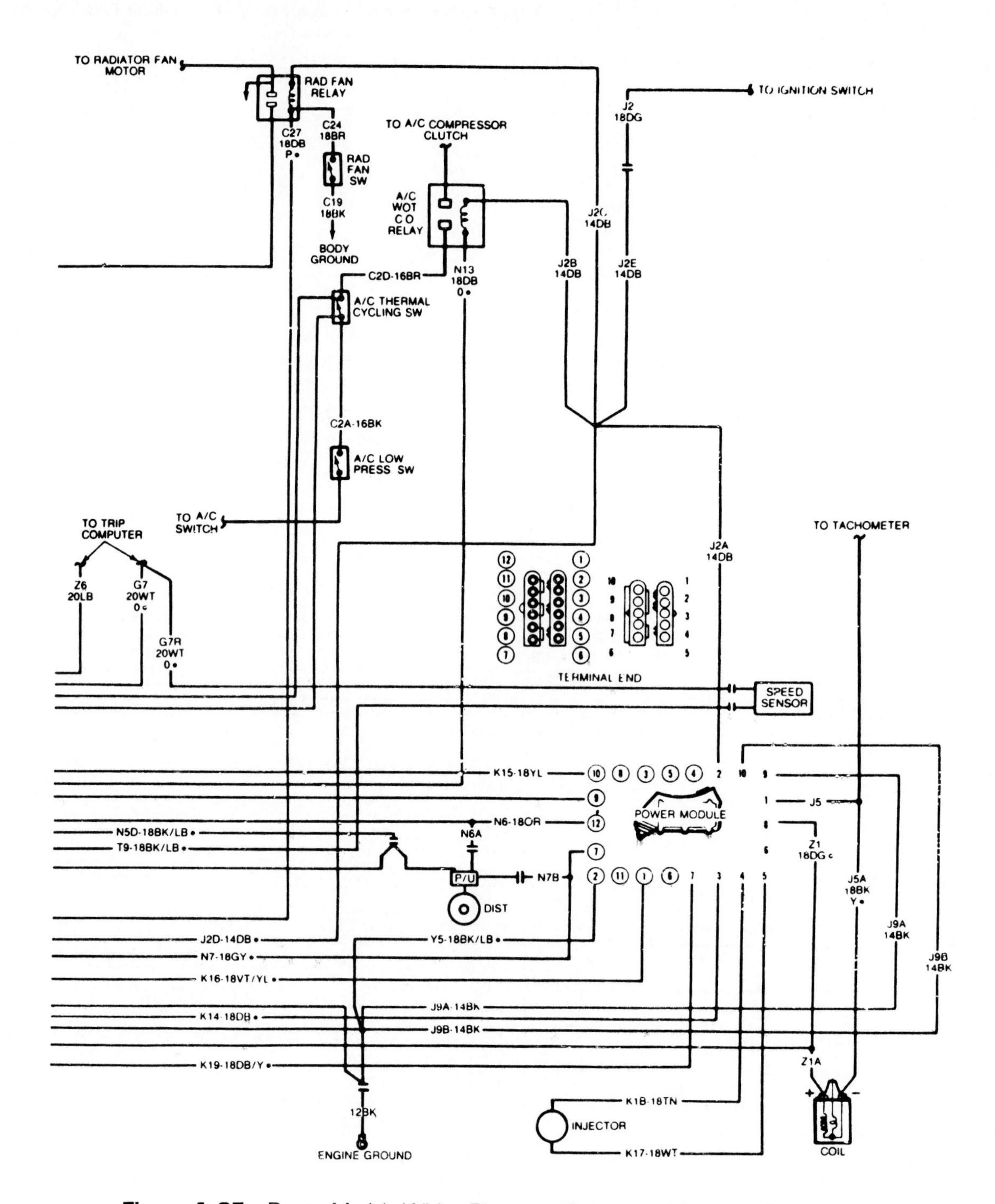

Figure 4–65 Power Module Wiring Diagram. (Courtesy of Chrysler Canada Ltd.)

Effect switch used in the single-point EFI system. Information data regarding engine speed and crankshaft position is supplied to the logic module by the reference pickup signal. The logic module uses this information with the other input data to determine the correct spark advance and injector pulse width. The sync pickup is mounted under the pickup plate, and a notched sync ring attached to the distributor shaft rotates through the pickup. This signal from the sync pickup tells the logic and power modules which pair of injectors to turn on and which pair of injectors to turn off. If the engine speed exceeds 6600 rpm, the logic module will shut off the fuel until the speed drops to a safe 6100 rpm. The reference pickup and sync pickup are pictured in Figure 4–72.

Output Control Functions

Fuel Injection Control All the output control functions in the multipoint EFI system are the same as the single-point EFI system except the fuel injection control. In the multipoint system, a single injector is placed in each intake port in the intake manifold rather than having one injector in the throttle body assembly. The injectors are similar to the injector in the single-point system.

The power module energizes injectors 1 and 2 and injectors 3 and 4 in pairs. Each pair of injectors is energized once for every two crankshaft revolutions. The four injectors located in the intake manifold are shown in Figure 4–73.

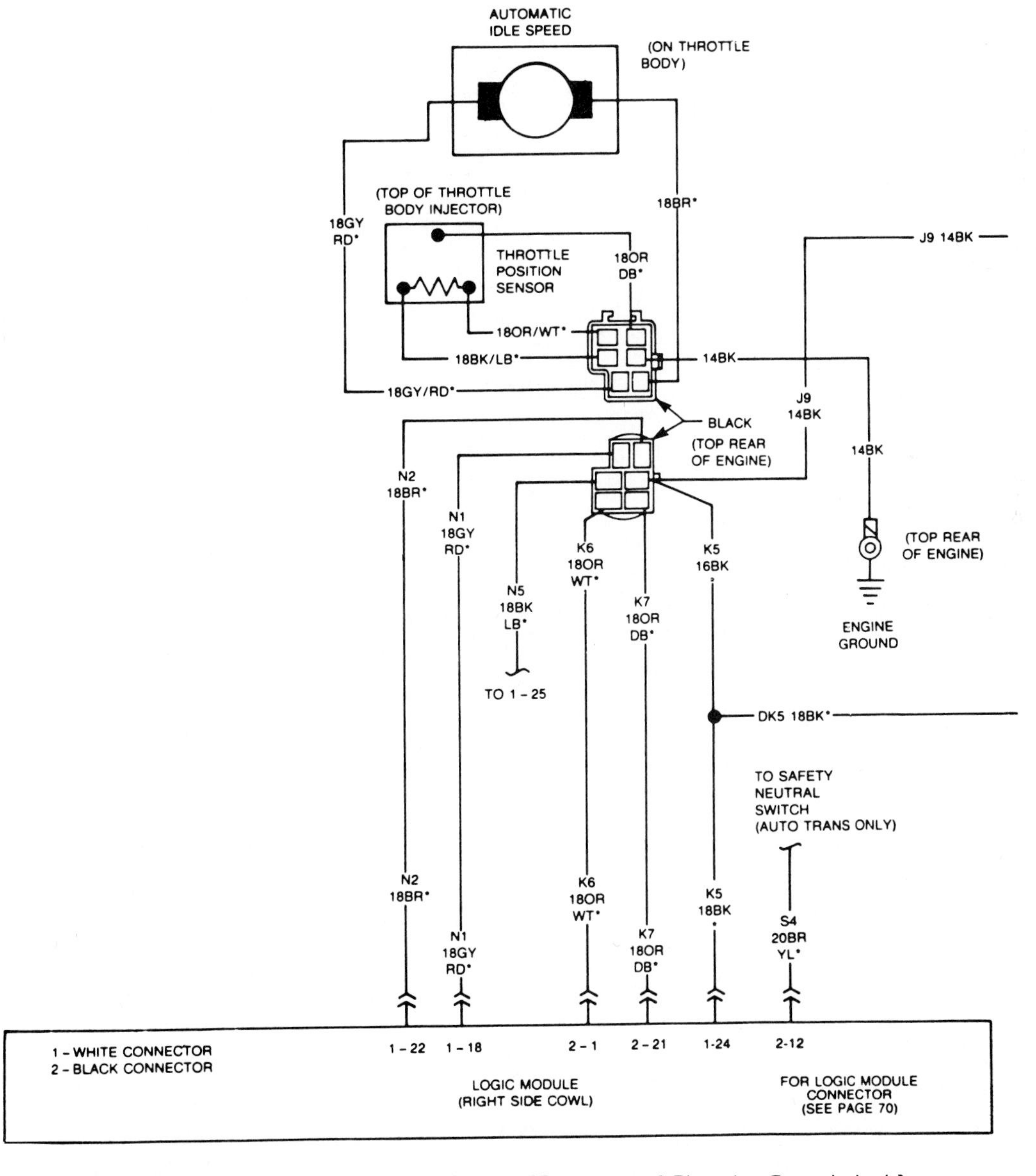

Figure 4–66 Logic Module Wiring. (Courtesy of Chrysler Canada Ltd.)

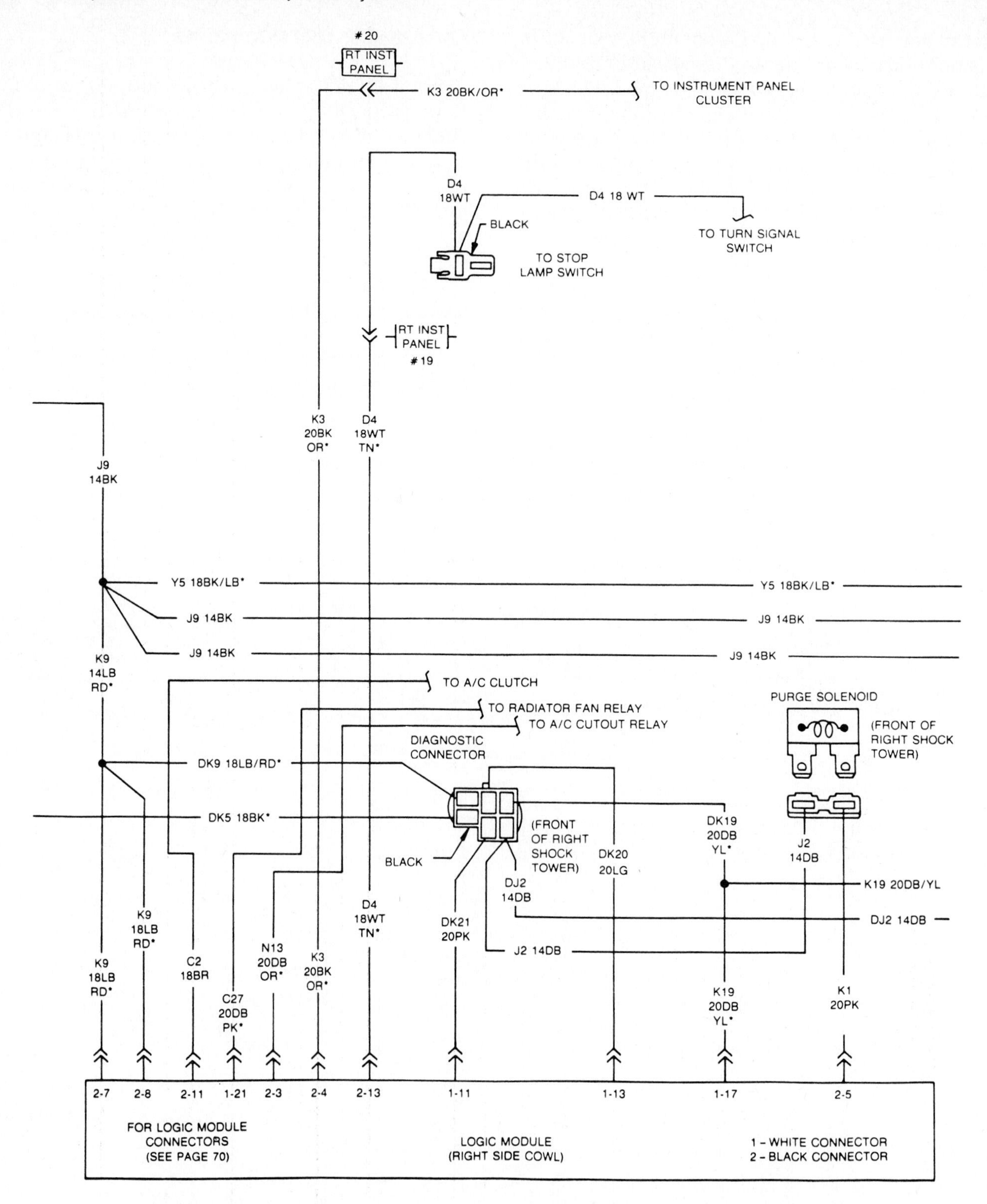

Figure 4-67 Logic Module Wiring Continued. (Courtesy of Chrysler Canada Ltd.)

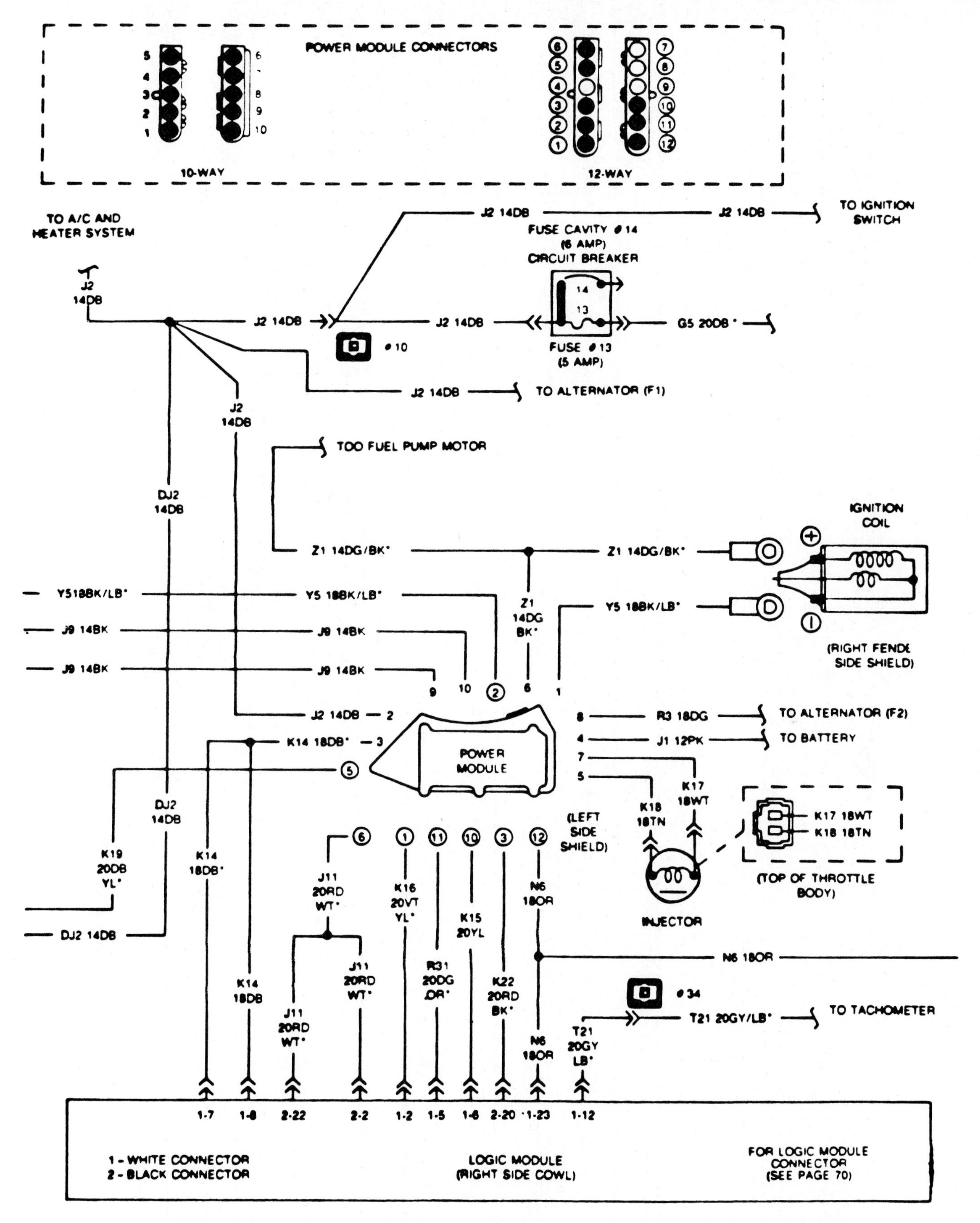

Figure 4-68 Power Module and Logic Module Wiring. (Courtesy of Chrysler Canada Ltd.)

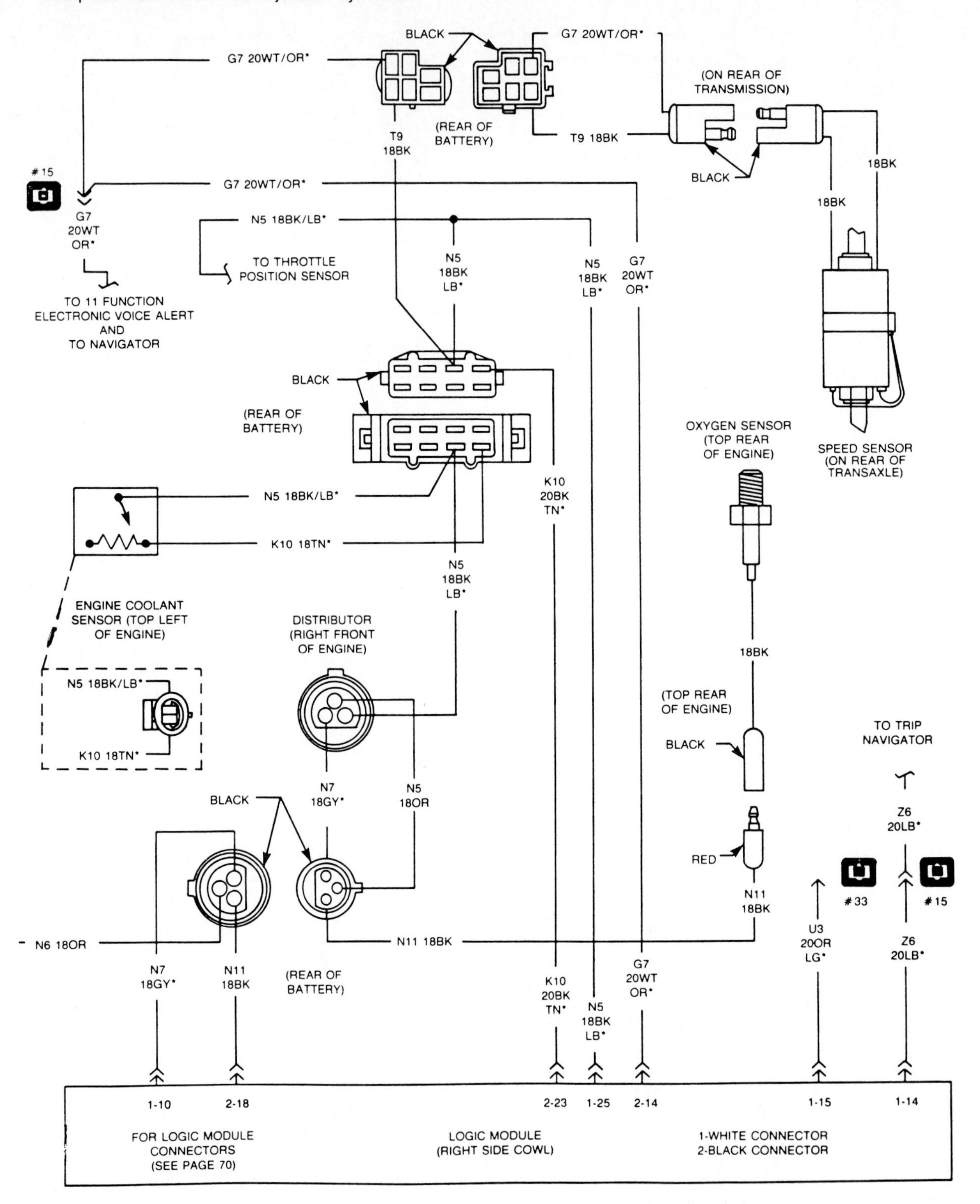

Figure 4-69 Logic Module Wiring Continued. [Courtesy of Chrysler Canada Ltd.]

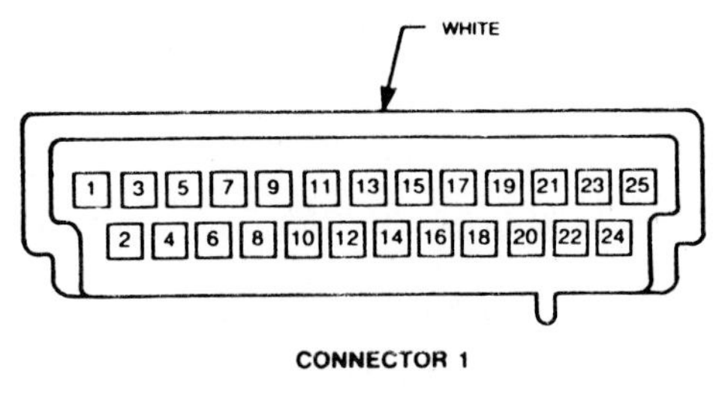

CONNECTOR 1

(CONNECTOR VIEWED FROM TERMINAL END)

CAV	CIRCUIT	GAUGE	COLOR	FUNCTION
1	————			
2	K16	20	VT/YL*	INJECTOR CONTROL
3	————			
4	————			
5	R31	20	DG/OR*	ALTERNATOR FIELD CONTROL
6	K15	20	YL	IGNITION CONTROL
7	K14	18	DB*	FUSED J2
8	K14	18	DB*	FUSED J2
9	————			
10	N7	18	GY*	DISTRIBUTOR PICK-UP SIGNAL
11	DK21	20	PK	DIAGNOSTIC CONNECTOR
12	T21	20	GY/LB*	TACHOMETER SIGNAL
13	DK20	20	LG	DIAGNOSTIC CONNECTOR
14	Z6	20	LB*	FUEL MONITOR SIGNAL
15	U3	20	OR/LG*	SHIFT INDICATOR LIGHT CONTROL
16	————			
17	K19	20	DB/YL*	AUTO SHUT DOWN RELAY CONTROL
18	N1	18	GY/RD*	AUTOMATIC IDLE SPEED OPEN SIGNAL
19	————			
20	————			
21	C27	20	DB/PK*	RADIATOR FAN RELAY CONTROL
22	N2	18	BR*	AUTOMATIC IDLE SPEED CLOSE SIGNAL
23	N6	18	OR	8 VOLT SUPPLY FROM POWER MODULE
24	K5	18	BK*	SENSOR GROUND
25	N5	18	BK/LB*	SENSOR GROUND

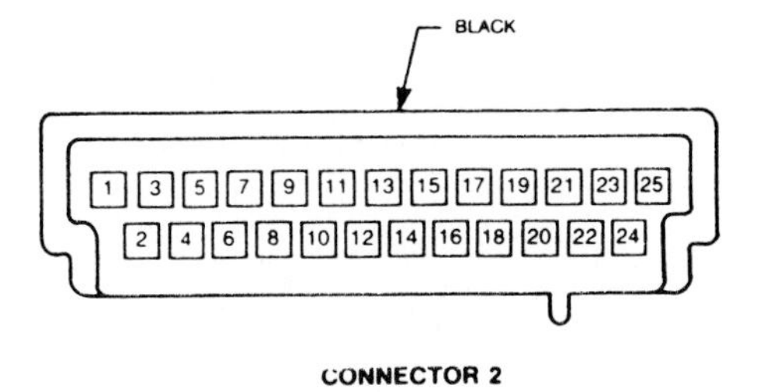

CONNECTOR 2

(CONNECTOR VIEWED FROM TERMINAL END)

CAV	CIRCUIT	GAUGE	COLOR	FUNCTION
1	K6	18	OR/WT*	5.0 VOLT SUPPLY FOR THROTTLE POT
2	J11	20	RD/WT*	DIRECT BATTERY FEED
3	N13	20	DB/OR*	A/C CUTOUT RELAY CONTROL
4	K3	20	BK/OR*	POWER LOSS LAMP CONTROL
5	K1	20	PK*	PURGE SOLENOID CONTROL
6				
7	K9	18	LB/RD*	POWER GROUND
8	K9	18	LB/RD*	POWER GROUND
9				
10				
11	C2	18	BR	AIR-CONDITIONING CLUTCH SIGNAL
12	S4	20	BR/YL*	PARK/NEUTRAL SWITCH SIGNAL
13	D4	18	WT/TN*	BRAKE SWITCH SIGNAL
14	G7	20	WT/OR*	SPEED SENSOR SIGNAL
15				
16				
17				
18	N11	18	BK	OXYGEN SENSOR SIGNAL
19				
20	K22	20	RD/BK*	BATTERY TEMPERATURE SIGNAL
21	K7	18	OR/DB*	THROTTLE POSITION SIGNAL
22	J11	20	RD/WT*	DIRECT BATTERY SENSE
23	K10	20	BK/TN*	COOLANT TEMPERATURE SENSOR SIGNAL
24				
25				

Figure 4–70 Logic Module Terminal Identification. (Courtesy of Chrysler Canada Ltd.)

The logic module commands the power module to provide the precise injector pulse width, which supplies the exact air–fuel ratio required by the engine. The pressure regulator maintains a constant pressure drop of 55 psi (379 kPa) across the injectors in the multipoint system. Many of the other components in the multipoint EFI system are similar to the components in the single-point system. The changes on the 1985 single-point system also apply to the multipoint system. These changes were explained under the single-point system, and they pertain to the alternator voltage regulator, battery charge and temperature signals, and the automatic shutdown relay. Fault codes and wiring diagrams vary depending on the year of vehicle. Wiring diagrams for 1985 multipoint EFI systems are provided in Figures 4–74 through 4–79.

Servicing and Diagnosis of EFI Systems

Service Precautions Personal injury or damage to the system components could result from improper service procedures. The following service precautions must be covered.

1. Using a 12V test light to test for continuity in EFI electrical circuits may damage expensive components in the system. Always use a voltmeter, ohmmeter, or diagnostic readout tester for the test procedure.

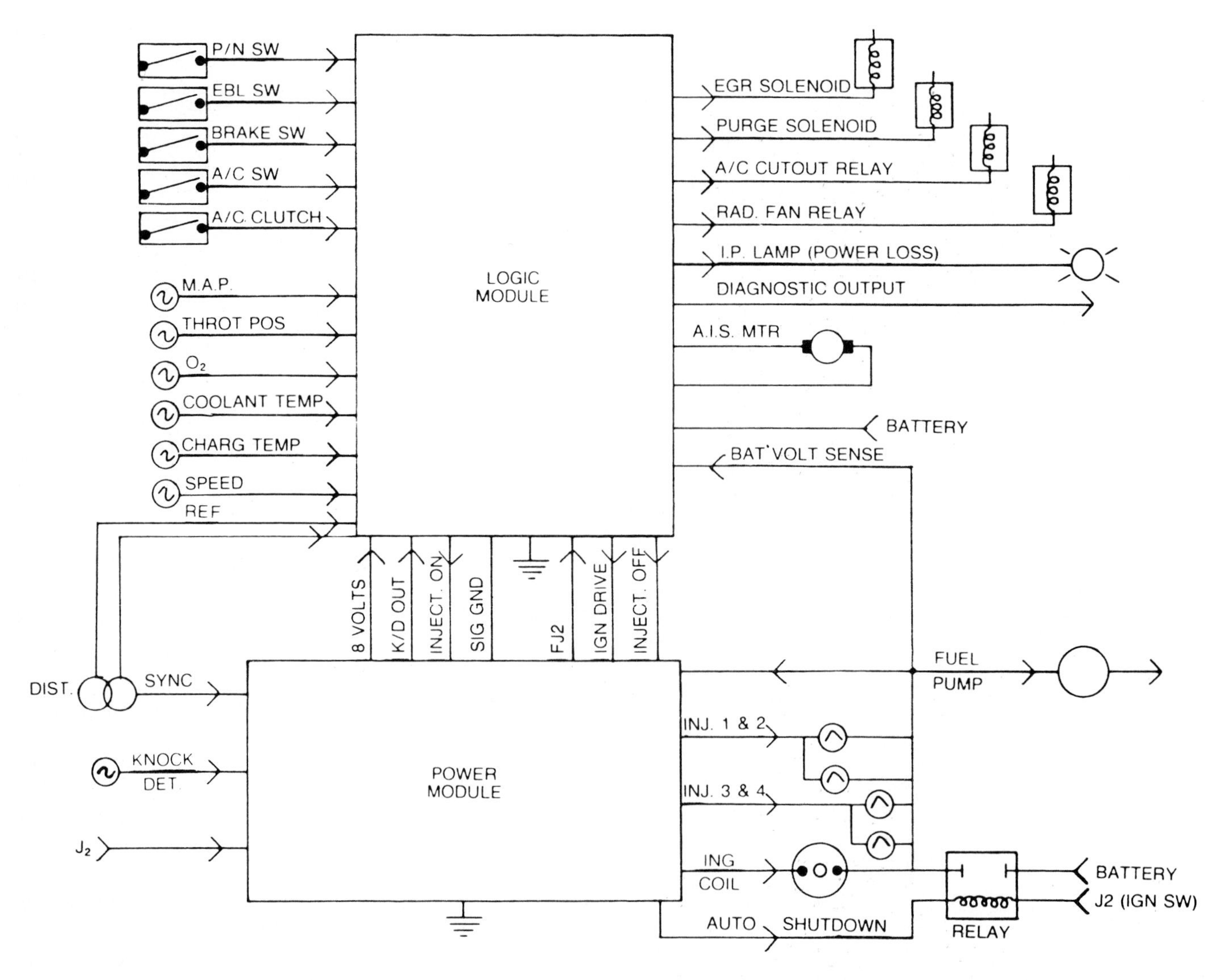

Figure 4–71 Multipoint EFI System Components. (Courtesy of Chrysler Canada Ltd.)

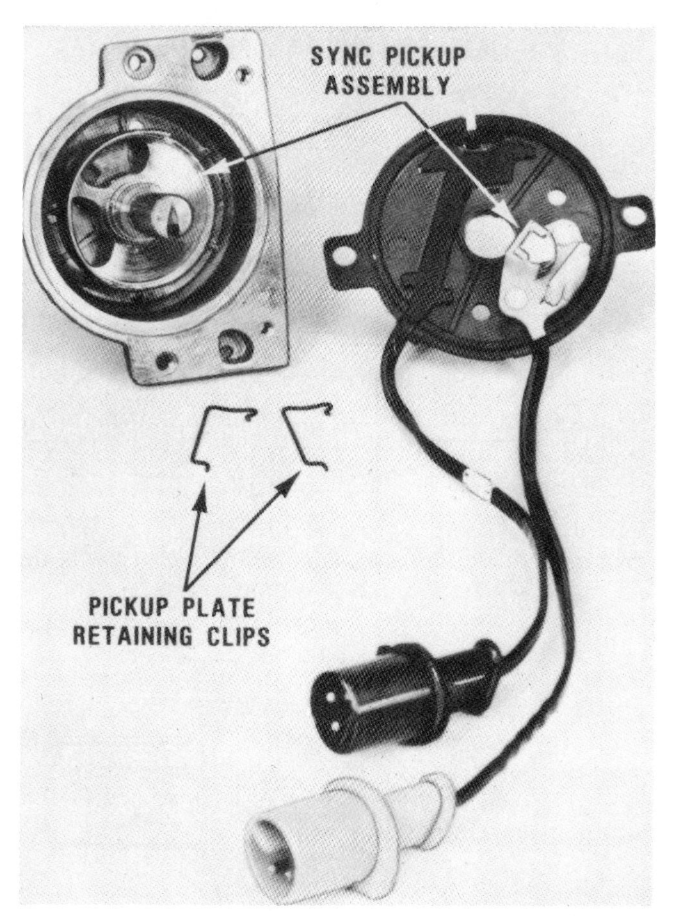

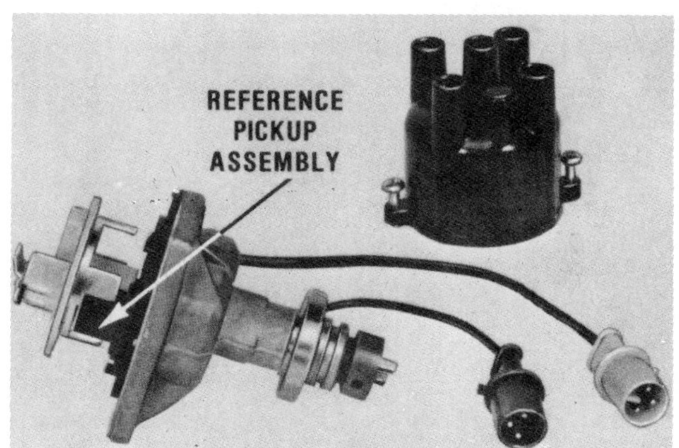

Figure 4-72 Reference and Sync Pickup Assemblies. (Courtesy of Chrysler Canada Ltd.)

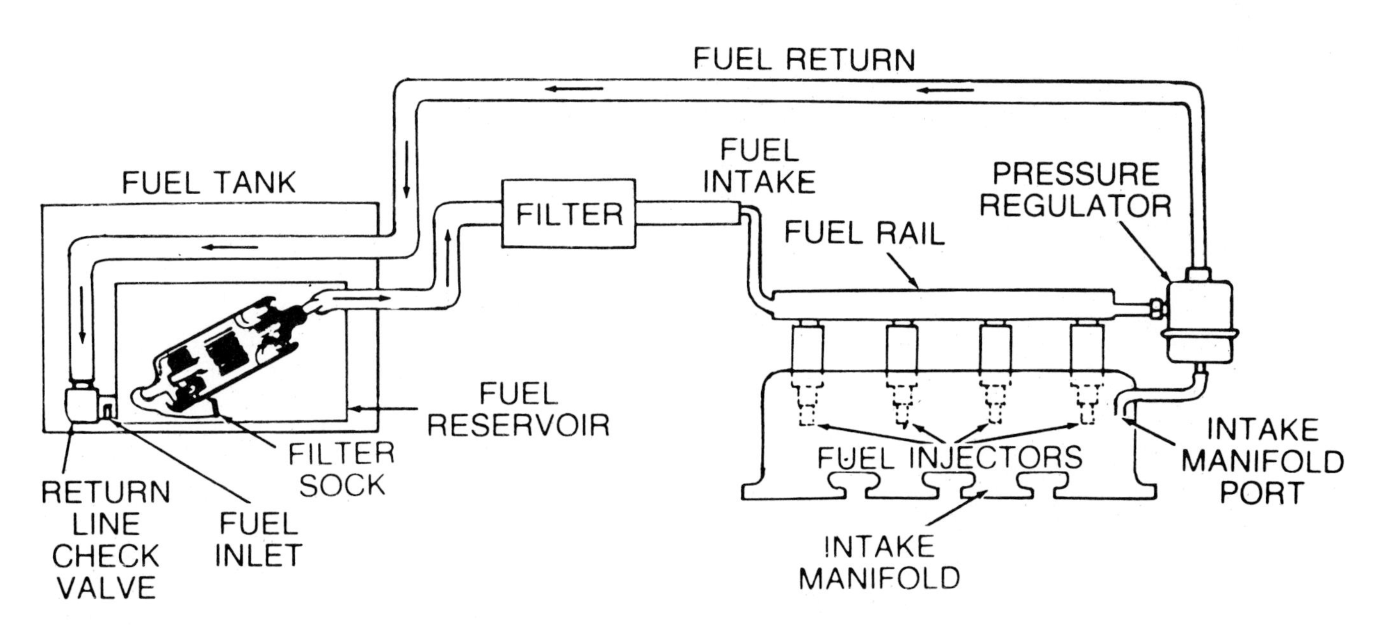

Figure 4-73 Multipoint EFI Fuel System. (Courtesy of Chrysler Canada Ltd.)

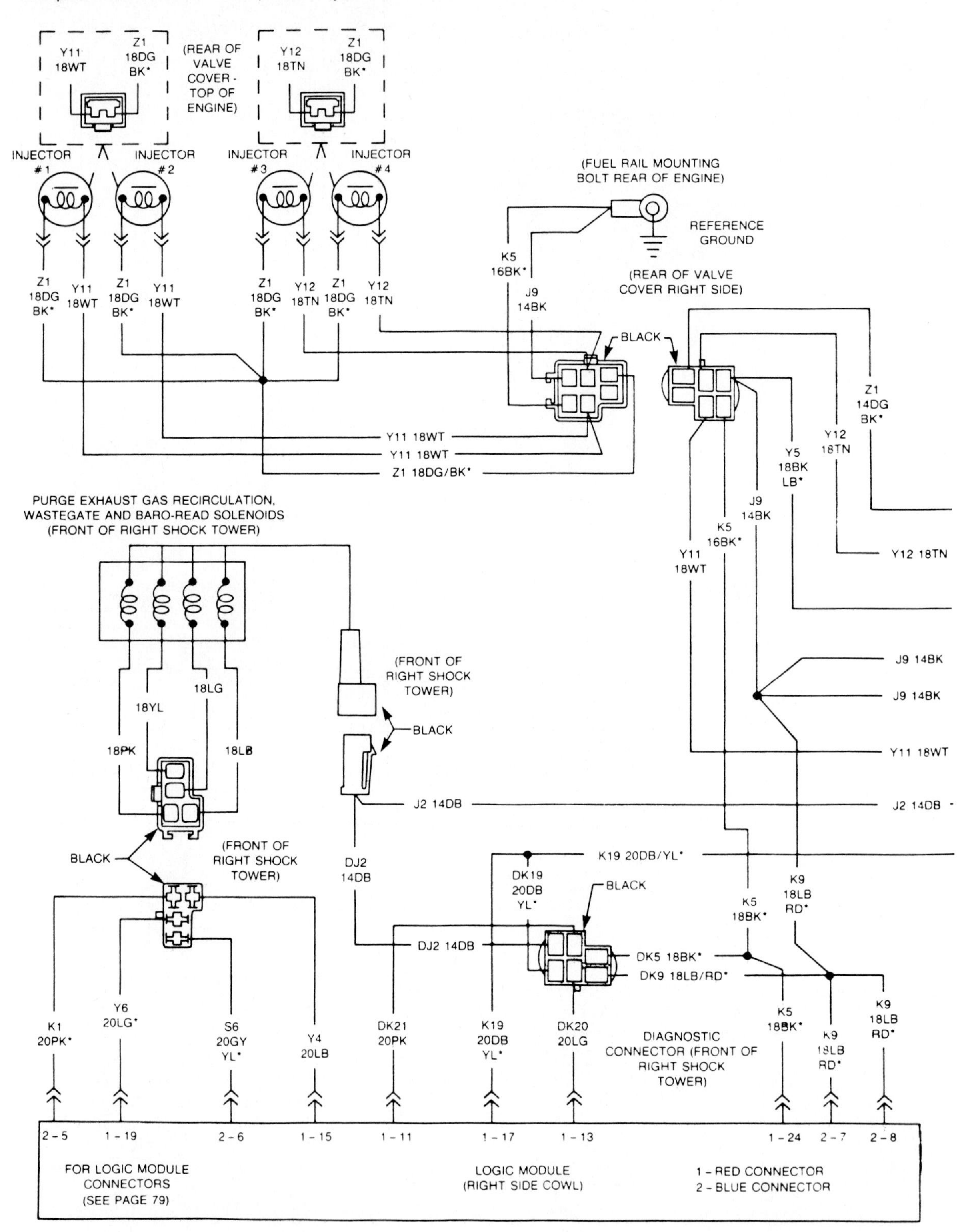

Figure 4–74 Logic Module Wiring, Chrysler Multipoint EFI Systems. [Courtesy of Chrysler Canada Ltd.]

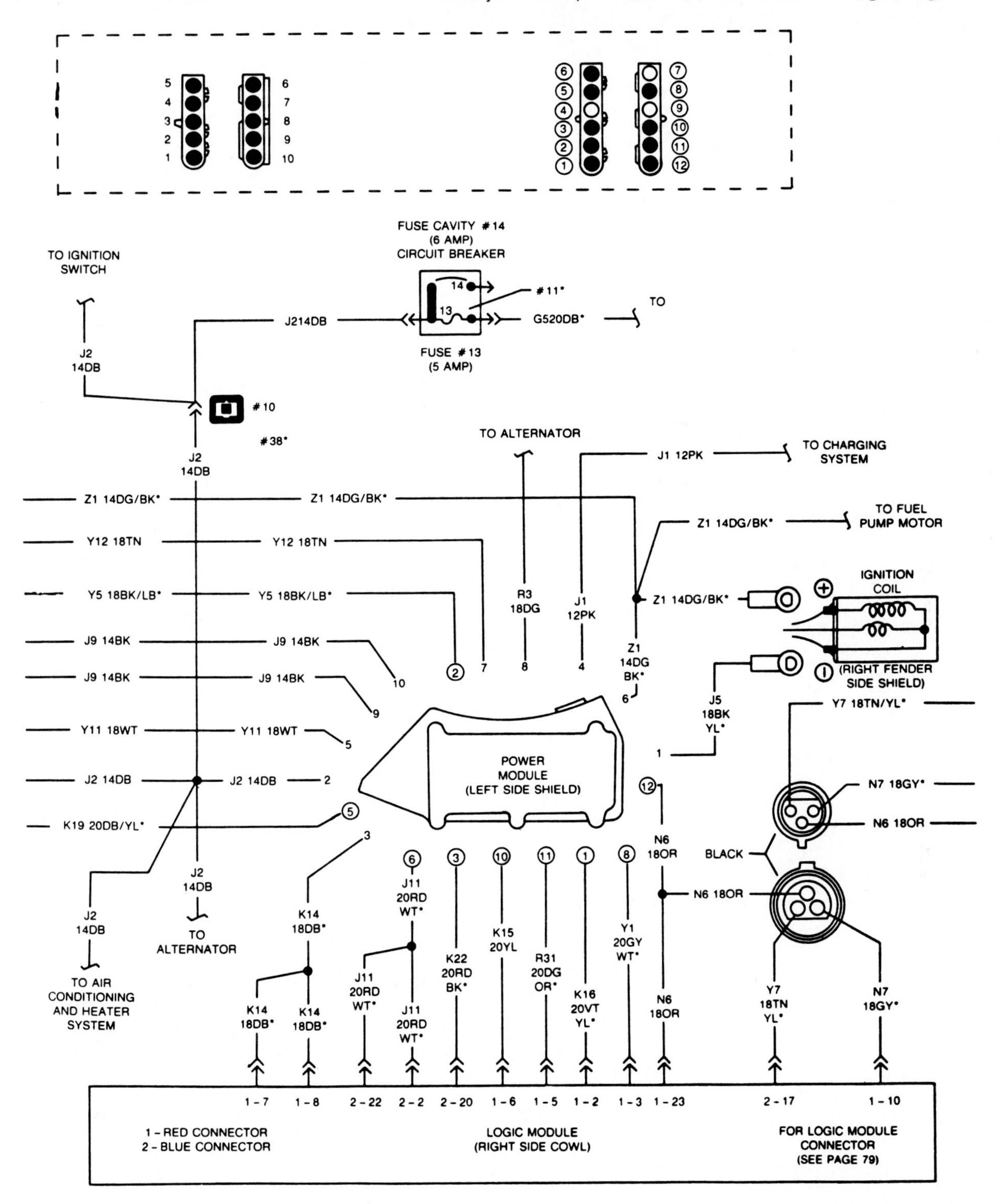

Figure 4–75 Logic Module and Power Module Wiring. (Courtesy of Chrysler Canada Ltd.)

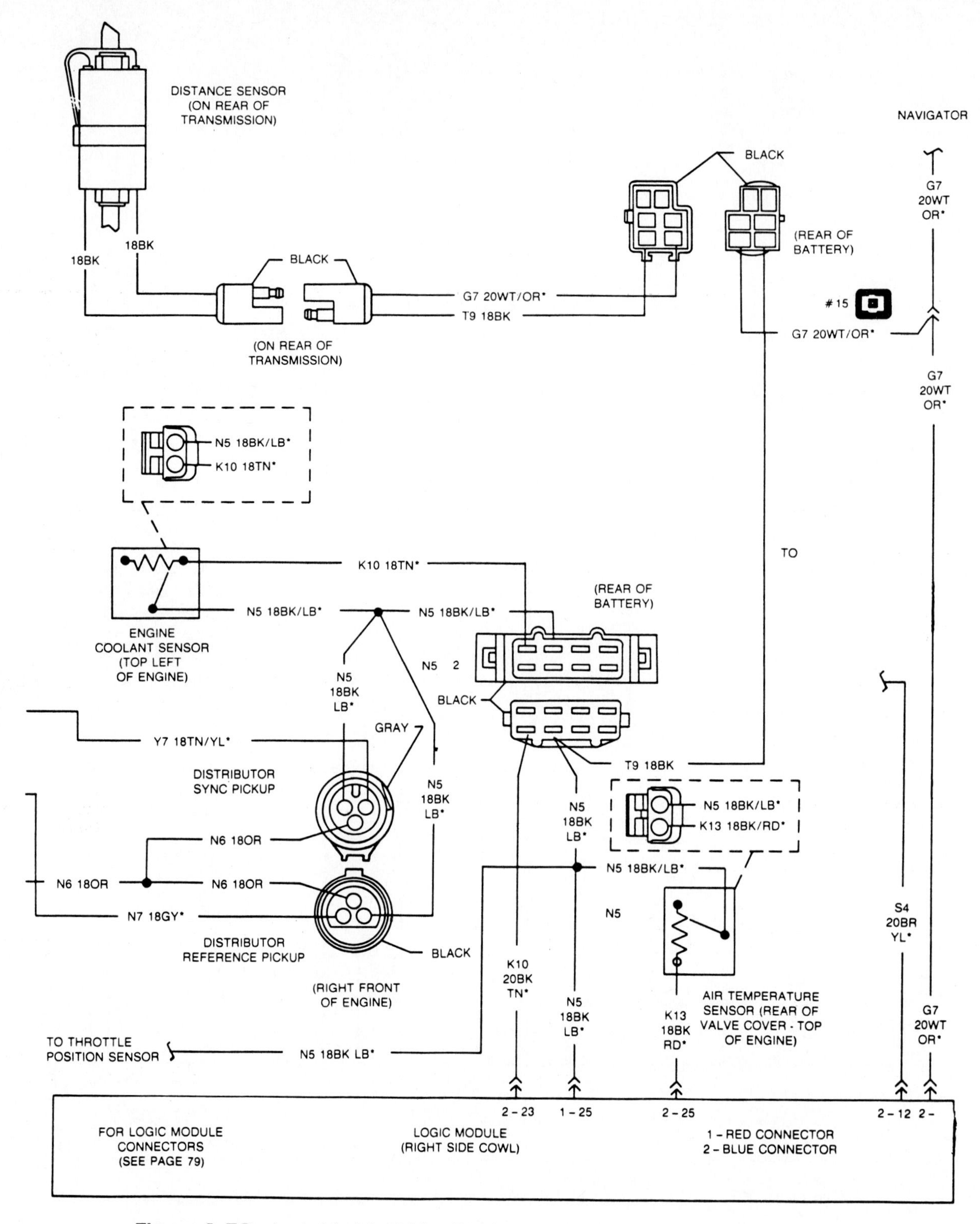

Figure 4–76 Logic Module Wiring Continued. [Courtesy of Chrysler Canada Ltd.]

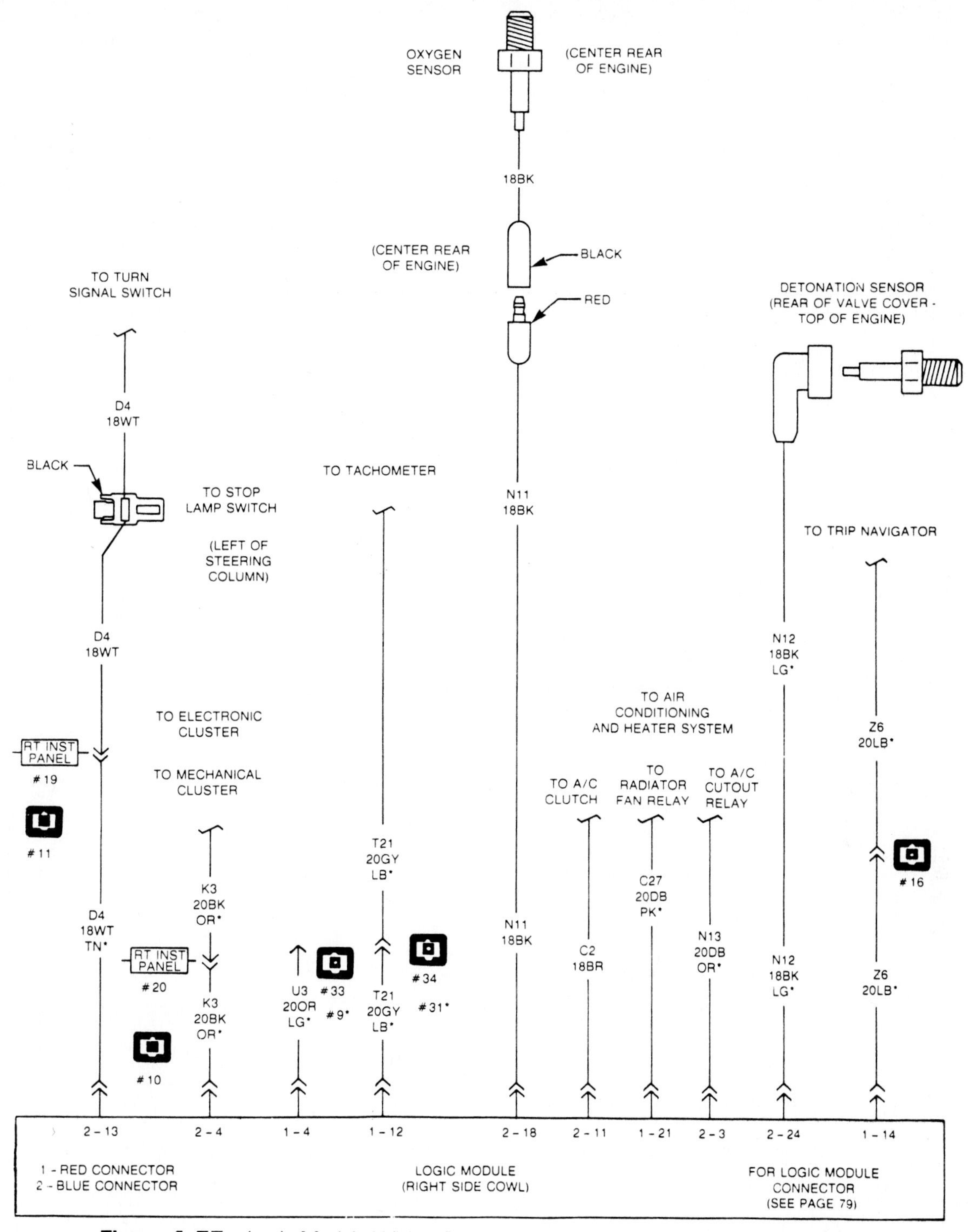

Figure 4–77 Logic Module Wiring Continued. [Courtesy of Chrysler Canada Ltd.]

2. Before any fuel line is loosened, the pressure in the fuel system must be relieved. High pressure in the fuel system could cause personal injury, or start a fire, if a fuel line connection is loosened without relieving the fuel pressure. The fuel pressure may be relieved by disconnecting the injector electrical connector and grounding one injector terminal with a jumper wire while the other terminal is connected to the battery positive terminal. In a multipoint system the fuel pressure can be relieved by activating one injector. The injector must not be activated for more than 10 seconds.
3. Never restrict a fuel line completely because this causes extremely high pump pressure.
4. Fuel line clamps must be torqued to 10 in lb (1 Nm).
5. Never substitute conventional gas line hose in place of the EFI hose. The EFI system also has special clamps that must not be replaced with other clamps. (Refer to Chapter 8 for precautions while servicing turbocharged engines.)

Adjusting Basic Ignition Timing When the basic timing is being checked, proceed as follows:

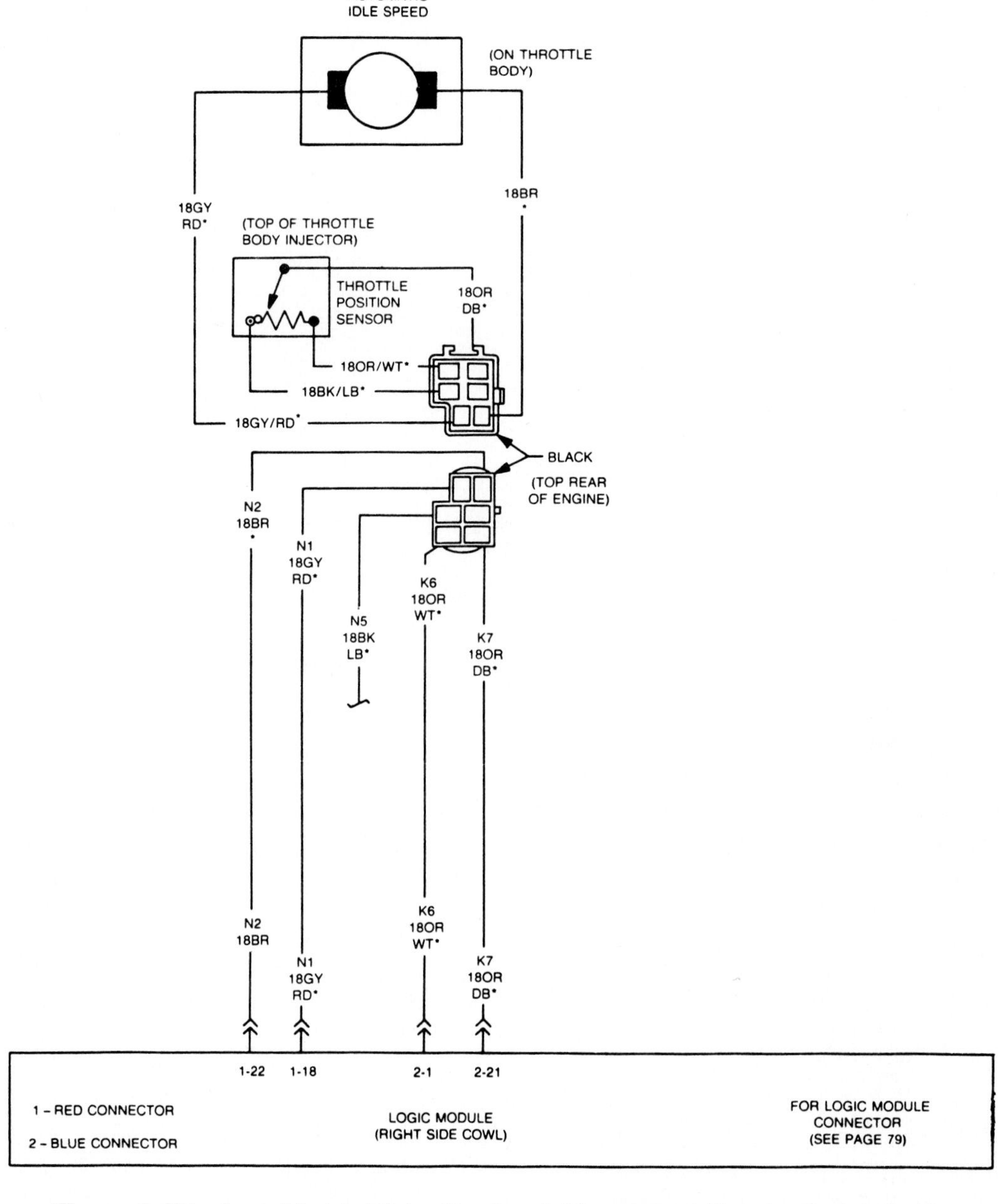

Figure 4–78 Logic Module Wiring Continued. [Courtesy of Chrysler Canada Ltd.]

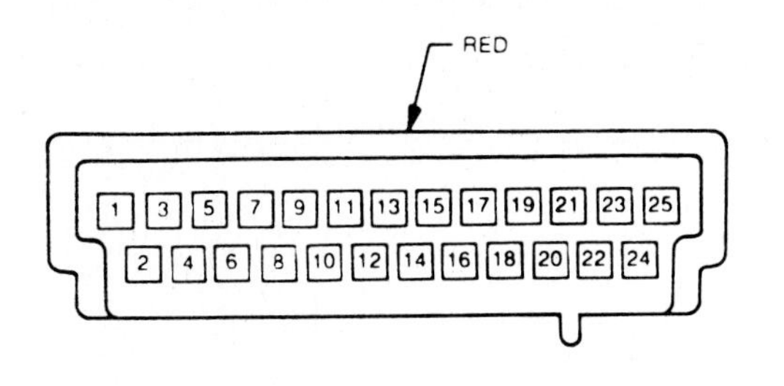

CONNECTOR 1

(CONNECTOR VIEWED FROM TERMINAL END)

CAV	CIRCUIT	GAUGE	COLOR	FUNCTION
1				
2	K16	20	VT/YL*	INJECTOR CONTROL (1-2)
3	Y1	20	GY/WT*	INJECTOR CONTROL (3-4)
4				
5	R31	20	DG/OR*	ALTERNATOR FIELD CONTROL
6	K15	20	YL	IGNITION CONTROL
7	K14	18	DB*	FUSED J2
8	K14	18	DB*	FUSED J2
9				
10	N7	18	GY*	DISTRIBUTOR REFERENCE PICKUP
11	DK21	20	PK	DIAGNOSTIC CONNECTOR
12	T21	20	GY/LB*	TACHOMETER SIGNAL
13	DK20	20	LG	DIAGNOSTIC CONNECTOR
14	Z6	20	LB*	FUEL MONITOR OUTPUT SIGNAL
15	Y4	20	LB	BARRO-READ SOLENOID
16				
17	K19	20	DB/YL*	AUTO SHUT DOWN RELAY
18	N1	18	GY/RD*	AUTOMATIC IDLE SPEED OPEN SIGNAL
19	Y6	20	LG*	WASTE GATE SOLENOID
20				
21	C27	20	DB/PK*	RADIATOR FAN RELAY
22	N2	18	BR*	AUTOMATIC IDLE SPEED CLOSE SIGNAL
23	N6	18	OR	8 VOLT SUPPLY FROM POWER MODULE
24	K5	18	BK*	SENSOR GROUND
25	N5	18	BK/LB*	SENSOR GROUND

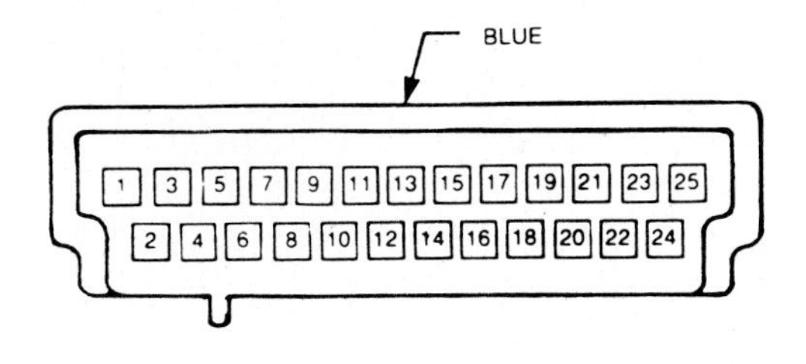

CONNECTOR 2

(CONNECTOR VIEWED FROM TERMINAL END)

CAV	CIRCUIT	GAUGE	COLOR	FUNCTION
1	K6	18	OR/WT*	5.0 VOLT SUPPLY FOR THROTTLE POT
2	J11	20	RD/WT*	DIRECT BATTERY FEED
3	N13	20	DB/OR*	A/C CUTOUT RELAY
4	K3	20	BK/OR*	POWER LOSS LAMP
5	K1	20	PK*	PURGE SOLENOID
6	S6	20	GY/YL*	EXHAUST GAS RECIRCULATOR SOLENOID
7	K9	18	LB/RD*	POWER GROUND
8	K9	18	LB/RD*	POWER GROUND
9				
10				
11	C2	18	BR	AIR-CONDITIONER CLUTCH SIGNAL
12	S4	20	BR/YL*	PARK/NEUTRAL SWITCH SIGNAL
13	D4	18	WT/TN*	BRAKE SWITCH SIGNAL
14	G7	20	WT/OR*	SPEED SENSOR SIGNAL
15				
16				
17	Y7	18	TN/YL*	DISTRIBUTOR SYNC PICKUP SIGNAL
18	N11	18	BK	OXYGEN SENSOR SIGNAL
19				
20	K22	20	RD/BK*	BATTERY TEMPERATURE SIGNAL
21	K7	18	OR/DB*	THROTTLE POSITION SIGNAL
22	J11	20	RD/WT*	DIRECT BATTERY SENSE
23	K10	20	BK/TN*	COOLANT TEMPERATURE SENSOR SIGNAL
24	N12	18	BK/LG*	DETONATION SENSOR SIGNAL
25	K13	18	BK/RD*	AIR TEMPERATURE SENSOR SIGNAL

Figure 4-79 Logic Module Terminal Identification. (Courtesy of Chrysler Canada Ltd.)

1. Operate the engine until normal operating temperature is reached.
2. Connect a timing light and tachometer to the engine.
3. Disconnect and reconnect the coolant temperature sensor connector at the thermostat housing. This will put the system in the limp-in mode, and the power-loss light will be on. If the system is working normally, the logic module varies the timing continuously at idle to provide smooth idle operation. In the limp-in mode, the timing is fixed by the logic module, and this condition is essential for checking basic timing.
4. Check the timing marks with the timing light. If the timing is not set at the specifications shown on the underhood emission label, loosen the distributor clamp and rotate the distributor to correct the timing. Retighten the distributor clamp.
5. Shut the engine off and remove the timing light and tachometer. Disconnect the quick disconnect terminal in the heavy red wire near the battery positive terminal for 10 seconds to clear the fault codes from the logic module's memory. Disconnecting the coolant sensor will store a fault code in the logic module's memory.
6. Reconnect the quick disconnect terminal at the battery and start the engine. Check to make sure the power-loss light goes out.

Checking Spark Advance The basic timing must be checked before the spark advance, and the power-loss light must be off. When the spark advance is being checked, there must not be any fault codes stored in the logic module's memory. (Fault-code diagnosis is discussed later in this chapter.) With an advance type of timing light and a tachometer connected to the engine, and the engine at normal operating temperature, increase the engine speed to 2000 rpm. Rotate the advance control on the timing light until the timing mark returns to the basic timing setting. Observe the spark advance on the timing light indicator. The spark advance should equal manufacturers' specifications. If the spark advance is not within specifications, the logic module should be replaced.

Testing Fuel Pump Pressure The fuel system pressure must be relieved before the pressure is tested. The pressure gauge must be connected in the fuel line at the throttle body assembly to test fuel pump pressure, as illustrated in Figure 4–80.

The pressure gauge can be connected to the shrader valve on the fuel rail to test fuel pump pressure on the multipoint system as shown in Figure 4–81.

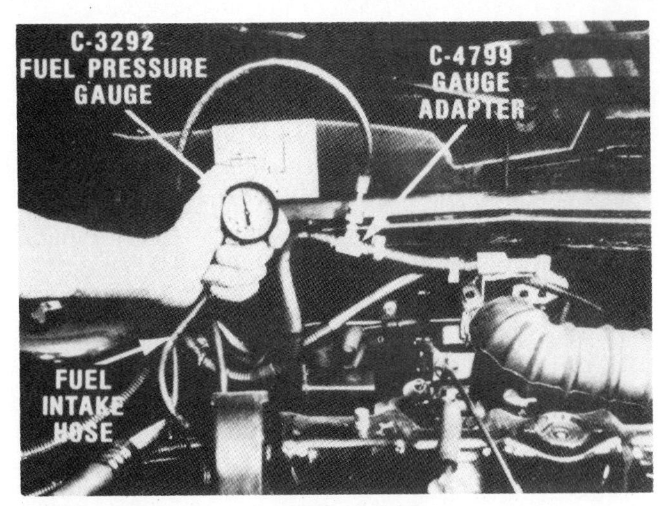

Figure 4–80 Testing Fuel Pump Pressure, Single-point EFI System. (Courtesy of Chrysler Canada Ltd.)

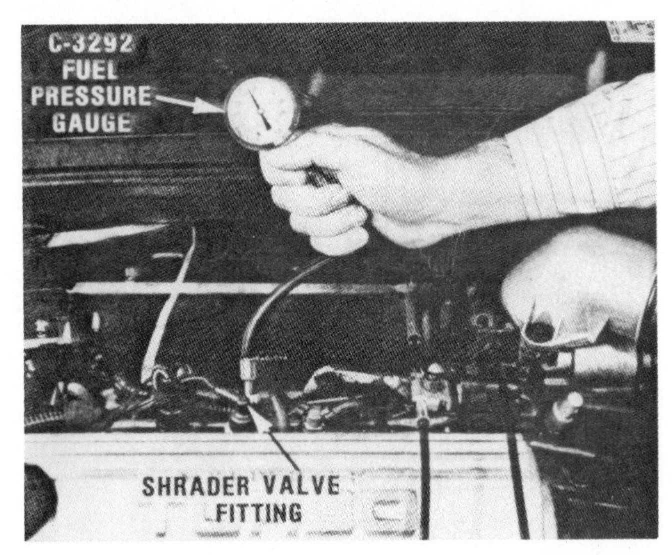

Figure 4–81 Testing Fuel Pump Pressure, Multipoint EFI System. (Courtesy of Chrysler Canada Ltd.)

Figure 4–82 Power-Limited Light. (Courtesy of Chrysler Canada Ltd.)

With the engine idling, the fuel pressure should be 48 to 62 psi (331 to 427 kPa) on multipoint systems and 29 to 43 psi (200 to 296 kPa) on single-point systems.

On-Board Diagnostics The logic module tests many of its input circuits and output control functions continuously. If an input signal is out of range or completely missing for a specific number of times, the logic module will consider this a "real" fault, and a fault code will be placed in the logic module memory. An output control function malfunctioning for a specific number of times will also cause a fault code to be stored in the logic module memory. The logic module will not store transient faults in its memory because they do not occur enough times. If the problem is repaired, or disappears, the logic module is programmed to erase the fault from its memory after 30 ignition switch cycles. The fault codes may be cleared from the logic module memory by momentarily disconnecting the quick disconnect in the heavy red wire connected to the battery positive cable.

Limp-In Mode If a defect occurs in the MAP sensor, throttle-position sensor, or coolant temperature sensor, the logic module places the system in a limp-in mode. In this mode the logic module substitutes other information for the missing or incorrect sensor input signal. For example, if the MAP sensor signal is not available, the logic module will use the engine speed signal and the throttle-position sensor signal to provide a "modified" MAP signal. In the limp-in mode, the logic module will keep the output control functions operational, but engine performance and economy will decrease, and exhaust emissions could increase. When the system is in the limp-in mode, the power-loss light on the instrument panel will be on. Some models use a power-limited light in place of the power-loss light as indicated in Figure 4–82.

If the fault is repaired and the fault codes are cleared from the logic module memory, and power-loss light will go out after the engine is started. When a fault that caused a limp-in mode corrects itself, the power-loss light will go out after the engine is stopped and restarted.

Diagnostic Readout Tester A diagnostic tester is available to read the fault codes in the EFI system, as shown in Figure 4–83.

The tester must be connected to the diagnostic connector illustrated in Figure 4–83. When the tester is connected it will perform six test functions:

1. A display check, code 88, should appear for a few seconds when the ignition switch is turned on, which indicates the tester is operational.
2. All fault codes in the logic module memory will be displayed on the tester.

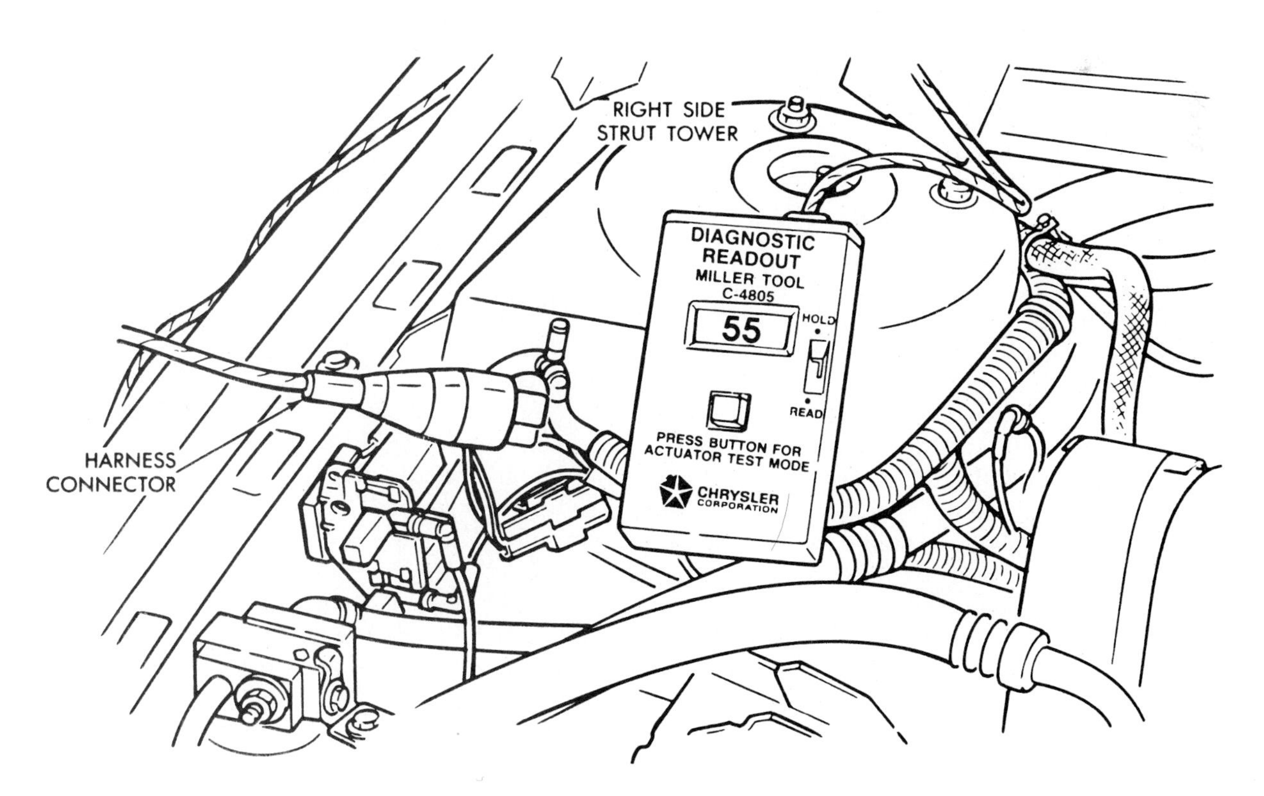

Figure 4–83 Diagnostic Readout Tester. (Courtesy of Chrysler Canada Ltd.)

3. A distributor signal should be seen on the tester when the engine is cranked. As long as the display changes while the engine is being cranked, a distributor signal is being received.
4. A switch test procedure may be performed by turning each switch input to the logic module on and off. It does not matter what code is on display before a switch is turned on and off, but the display must change when the switch is turned on and off, if the switch is working normally.
5. An oxygen sensor signal test may be performed with the engine running. If the oxygen sensor is generating a normal signal, the readout will continuously switch from 0 to 1 as the slight variations from lean to rich mixture occur.
6. An actuator test mode may be selected by depressing the actuator test mode (ATM) button on the tester with the ignition switch on. In this mode the ASD relay will be grounded, and the service technician may observe the following test sequence:
   (a) Three sparks from the coil wire when the coil wire is held ¼ in (6.3 mm) from a good engine ground.
   (b) Two automatic idle speed motor movements, one open, one closed.
   (c) One fuel pulse from the injector.

On 1985 EFI systems use the following procedure for fault-code diagnosis:

1. Connect the digital readout tester to the diagnostic connector and place the read/hold tester switch in the read position.
2. Turn the ignition switch on-off, on-off, on, within a 5-second interval.
3. Record all the fault codes displayed on the tester.

After the fault code display is completed, 55 will appear on the tester display. When this occurs, the actuation test mode (ATM) may be entered if the tester ATM button is pressed. In this mode ATM codes will be displayed on the tester. Each ATM code represents a specific solenoid, relay, or component in the circuit that will be cycled on and off if the ATM button is released when the code appears on the display representing the specific component. Cycling of a component will be stopped after 5 minutes, or when the ignition switch is turned off. The ATM tests allow the technician to listen to, observe, or perform voltage tests on specific components to locate various defects. ATM codes are provided with the fault codes later in this chapter.

After the fault-code diagnosis, a switch test procedure may be performed as follows:

1. Turn off all the input switch signals to the logic module.
2. With 55 indicated on the tester display at the end of the fault-code diagnosis, when each input switch is turned on and off the display reading should change if the switch signal is received by the logic module.

On the 1985 EFI systems, an oxygen sensor test may be performed with the engine running as outlined previously.

Power-Loss Light Diagnosis The power-loss light should come on when the ignition switch is turned on, and it should remain on for a few seconds after the engine is started. Once the light goes off it should remain off while the engine is running. The power-loss light may be used to perform three test functions:

1. A test of the distributor signal is performed while the engine is being cranked. If the power-loss light is on while the engine is being cranked, a distributor signal is being received.
2. Fault codes in the logic memory will be displayed by a flashing power-loss light if the ignition switch is turned off/on, off/on, off/on in a 5-second interval. For example, if the light flashes five times, pauses, and then flashes once, code 51 has been indicated.
3. A switch test may be performed after all the fault codes have been displayed. Each time a switch input to the logic module is turned on and off, the light should be illuminated momentarily.

A fault code indicates a problem in a certain area, not necessarily in a specific component. For example, a code 21 indicates a fault in the oxygen sensor, sensor wire, or the logic module is unable to receive an oxygen sensor signal. Code 12 indicates that the battery voltage has been disconnected from the logic module. This code can only be erased by 30 ignition switch cycles. The fault codes for the 1985 multipoint EFI system and the requirements to place the codes in memory are shown in Table 4–4, and the 1985 fault codes for the single-point EFI system are provided in Table 4–5.

TABLE 4–4 Fault Codes, Multipoint EFI System.

Code	Type	Power Loss/ Limit Lamp	Circuit	When Monitored By The Logic Module	When Put Into Memory	ATM Test Code
11	Fault	No	Distributor Signal	During cranking.	If no distributor signal is present since the battery was disconnected.	None
12	Indication	No	Battery Feed to the Logic Module	All the time when the ignition switch is on.	If the battery feed to the logic module has been disconnected within the last 20-40 engine starts.	None
13	Fault	Yes	M.A.P. Sensor (Vacuum)	When the throttle is closed during cranking and after the engine starts.	If the M.A.P. sensor vacuum level does not change between cranking and when the engine starts.	None
14	Fault	Yes	M.A.P. Sensor (Electrical)	All the time when the ignition switch is on.	If the M.A.P. sensor signal is below .02 or above 4.9 volts.	None
15	Fault	No	Vehicle Speed Sensor	Over a 7 second period during decel from highway speeds when the throttle is closed.	If the speed sensor signal indicates less than 2 mph when the vehicle is moving.	None
16	Fault	Yes	Battery Voltage Sensing (Charging System)	All the time after one minute from when the engine starts.	If the battery sensing voltage drops below 4 or between 7½ and 8½ volts for more than 20 seconds.	None
17	Fault	Yes	Detonation sensor	All the time when the engine is running.	If there is no knock signal above 5000 engine rpm for 3 seconds.	
21	Fault	No	Oxygen Sensor	All the time after 12 minutes from when the engine starts.	If there is no oxygen sensor signal for more than 22 seconds when in closed loop.	None
22	Fault	Yes	Engine Coolant Sensor	All the time when the ignition switch is on.	If the coolant sensor voltage is above 4.96 volts when the engine is cold or below .51 volts when the engine is warm.	None
23	Fault	Yes	Charge temperature sensor	All the time when the ignition switch is on.	If the Charge temperature sensor voltage is above 4.98 or below .06.	None
24	Fault	Yes	Throttle Position Sensor	All the time when the ignition switch is on.	If the throttle position sensor signal is below .16 or above 4.7 volts.	None
25	Fault	No	Automatic Idle Speed Motor (AIS)	Only when the AIS system is required to control the engine speed.	If proper voltage in the AIS system is not present. **NOTE:** Open circuit will not activate code.	03
26	Fault	No	Injector 1 and 2	During cranking.	If injectors 1 and 2 do not fire correctly.	02
27	Fault	No	Injectors 3 and 4	During cranking.	If injectors 3 and 4 do not fire correctly.	02
31	Fault	No	Canister Purge Solenoid	All the time when the ignition switch is on.	If the solenoid does not turn on and off when it should.	07
32	Fault	No	Power Loss/ Power Limit Lamp	All the time when the ignition switch is on.	If the lamp does not turn on and off when it should.	None
33	Fault	No	A/C Cutout Relay	All the time when the ignition switch is on.	If the relay does not turn on and off when it should.	05
34	Fault	No	E.G.R. Solenoid	All the time when the ignition switch is on	If the solenoid does not turn on and off when it should.	08
35	Fault	No	Radiator Fan Relay Circuit	All the time when the ignition switch is on.	If the relay does not turn on and off when it should.	04
36	Fault	Yes	Wastegate control solenoid	All the time when the ignition switch is on.	If the solenoid does not turn on and off when it should.	09

(Courtesy of Chrysler Canada Ltd.)

TABLE 4–4 (Continued) Fault Codes Multipoint EFI System.

Code	Type	Power Loss/ Limit Lamp	Circuit	When Monitored By The Logic Module	When Put Into Memory	ATM Test Code
37	Fault	No	Baro Read Solenoid	All the time when the ignition switch is on.	If the solenoid does not turn on and off when it should.	10
41	Fault	No	Alternator Field Control (Charging System)	All the time when the ignition switch is on.	If the field control fails to switch properly.	None
42	Fault	No	Auto Shutdown	All the time when the ignition switch is on.	If the control voltage of the relay pull in coil in the power module is not correct.	06
43	Fault	No	Spark Control	All the time when the ignition switch is on.	If the spark control interface fails to switch properly.	01
44	Fault	No	Battery Temperature Sensor (Charging System)	All the time when the ignition switch is on.	If the battery temperature sensor signal is below .04 or above 4.9 volts.	None
45	Fault	Yes	Overboost Monitor	All the time when the engine is running.	When M.A.P. sensor signal exceed a predetermine amount of boost indication.	None
46	Fault	Yes	Battery Voltage Sensing (Charging System)	All the time when the engine is running.	If the battery sense voltage is more than 1 volt above the desired control voltage for more than 20 seconds.	None
47	Fault	No	Battery Voltage Sensing (Charging System)	When the engine has been running for more than 6 minutes, engine temperature above 160°F and engine rpm above 1,500 rpm.	If the battery sense voltage is less than 1 volt below the desired control voltage for more than 20 seconds.	None
51	Fault	No	Oxygen Feedback System	During all closed loop conditions.	If the system stays lean for more than 2 minutes.	None
52	Fault	No	Oxygen Feedback System	During all closed loop conditions.	If the system stays rich for more than 2 minutes.	None
53	Fault	No	Logic Module	All the time in the diagnostic mode.	If the logic module fails.	None
54	Fault	Yes	Distributor sync. pickup	All the time when the engine is running.	If there is no distributor sync. pickup signal.	None
55	Indication	No			Indicates end of diagnostic mode.	None
88	Indication	No			Indicates start of diagnostic mode. **NOTE:** This code must appear first in the diagnostic mode or fault codes will be inaccurate.	
0	Indication	No			Indicates oxygen feedback system is lean with the engine running.	None
1	Indication	No			Indicates oxygen feedback system is rich with the engine running.	
8	Indication	No	Knock Circuit		Indicates knock sensor system is detecting knock.	None

(Courtesy of Chrysler Canada Ltd.)

TABLE 4–5 Fault Codes Single-Point EFI System.

Code	Type	Power Loss/ Limit Lamp	Circuit	When Monitored By The Logic Module	When Put Into Memory	ATM Test Code
11	Fault	No	Distributor Signal	During cranking.	If no distributor signal is present since the battery was disconnected.	None
12	Indication	No	Battery Feed to the Logic Module	All the time when the ignition switch is on.	If the battery feed to the logic module has been disconnected within the last 20-40 engine starts.	None
13	Fault	Yes	M.A.P. Sensor (Vacuum)	When the throttle is closed during cranking and after the engine starts.	If the M.A.P. sensor vacuum level does not change between cranking and when the engine starts.	None
14	Fault	Yes	M.A.P. Sensor (Electrical)	All the time when the ignition switch is on.	If the M.A.P. sensor signal is below .02 or above 4.9 volts.	None
15	Fault	No	Vehicle Speed Sensor	Over a 7 second period during decel from highway speeds when the throttle is closed.	If the speed sensor signal indicates less than 2 mph when the vehicle is moving.	None
16	Fault	Yes	Battery Voltage Sensing (Charging System)	All the time after one minute from when the engine starts.	If the battery sensing voltage drops below 4 or between 7½ and 8½ volts for more than 20 seconds.	None
21	Fault	No	Oxygen Sensor	All the time after 12 minutes from when the engine starts.	If there is no oxygen sensor signal for more than 22 seconds when in closed loop.	None
22	Fault	Yes	Engine Coolant Sensor	All the time when the ignition switch is on.	If the coolant sensor voltage is above 4.96 volts when the engine is cold or below .51 volts when the engine is warm.	None
24	Fault	Yes	Throttle Position Sensor	All the time when the ignition switch is on.	If the throttle position sensor signal is below .16 or above 4.7 volts.	None
25	Fault	No	Automatic Idle Speed Motor (AIS)	Only when the AIS system is required to control the engine speed.	If proper voltage in the AIS system is not present. **NOTE:** Open circuit will not activate code.	03
26	Fault	No	Fuel Injector	During cranking.	If the current through the fuel injector does not reach its proper peak level.	02
27	Fault	No	Fuel Control	All the time when the ignition switch is on.	If the fuel control interface fails to switch properly.	None
31	Fault	No	Canister Purge Solenoid	All the time when the ignition switch is on.	If the solenoid does not turn on and off when it should.	07
32	Fault	No	Power Loss/ Power Limit Lamp	All the time when the ignition switch is on.	If the lamp does not turn on and off when it should.	None
33	Fault	No	A/C Cutout Relay	All the time when the ignition switch is on.	If the relay does not turn on and off when it should.	05
35	Fault	No	Radiator Fan Relay Circuit	All the time when the ignition switch is on.	If the relay does not turn on and off when it should.	04
37	Fault	No	Shift Indicator Lamp (Manual Trans Only)	All the time when the ignition switch is on.	If the lamp does not turn on and off when it should.	None
41	Fault	No	Alternator Field Control (Charging System)	All the time when the ignition switch is on.	If the field control fails to switch properly.	None
42	Fault	No	Auto Shutdown	All the time when the ignition switch is on.	If the relay does not turn on and off when it should.	06
43	Fault	No	Spark Control	All the time when the ignition switch is on.	If the spark control interface fails to switch properly.	01

(Courtesy of Chrysler Canada Ltd.)

TABLE 4–5 (Continued) Fault Codes Single-Point EFI System.

Code	Type	Power Loss/ Limit Lamp	Circuit	When Monitored By The Logic Module	When Put Into Memory	ATM Test Code
44	Fault	No	Battery Temperature Sensor (Charging System)	All the time when the ignition switch is on.	If the battery temperature sensor signal is below .04 or above 4.9 volts.	None
46	Fault	Yes	Battery Voltage Sensing (Charging System)	All the time when the engine is running.	If the battery sense voltage is more than 1 volt above the desired control voltage for more than 20 seconds.	None
47	Fault	No	Battery Voltage Sensing (Charging System)	When the engine has been running for more than 6 minutes, engine temperature above 160°F and engine rpm above 1,500 rpm.	If the battery sense voltage is less than 1 volt below the desired control voltage for more than 20 seconds.	None
51	Fault	No	Oxygen Feedback System	During all closed loop conditions.	If the system stays lean for more than 2 minutes.	None
52	Fault	No	Oxygen Feedback System	During all closed loop conditions.	If the system stays rich for more than 2 minutes.	None
53	Fault	No	Logic Module	All the time in the diagnostic mode.	If the logic module fails.	None
55	Indication	No			Indicates end of diagnostic mode.	None
88	Indication	No			Indicates start of diagnostic mode. **NOTE:** This code must appear first in the diagnostic mode or fault codes will be inaccurate.	
0	Indication	No			Indicates oxygen feedback system is lean with the engine running.	None
1	Indication	No			Indicates oxygen feedback system is rich with the engine running.	None

(Courtesy of Chrysler Canada Ltd.)

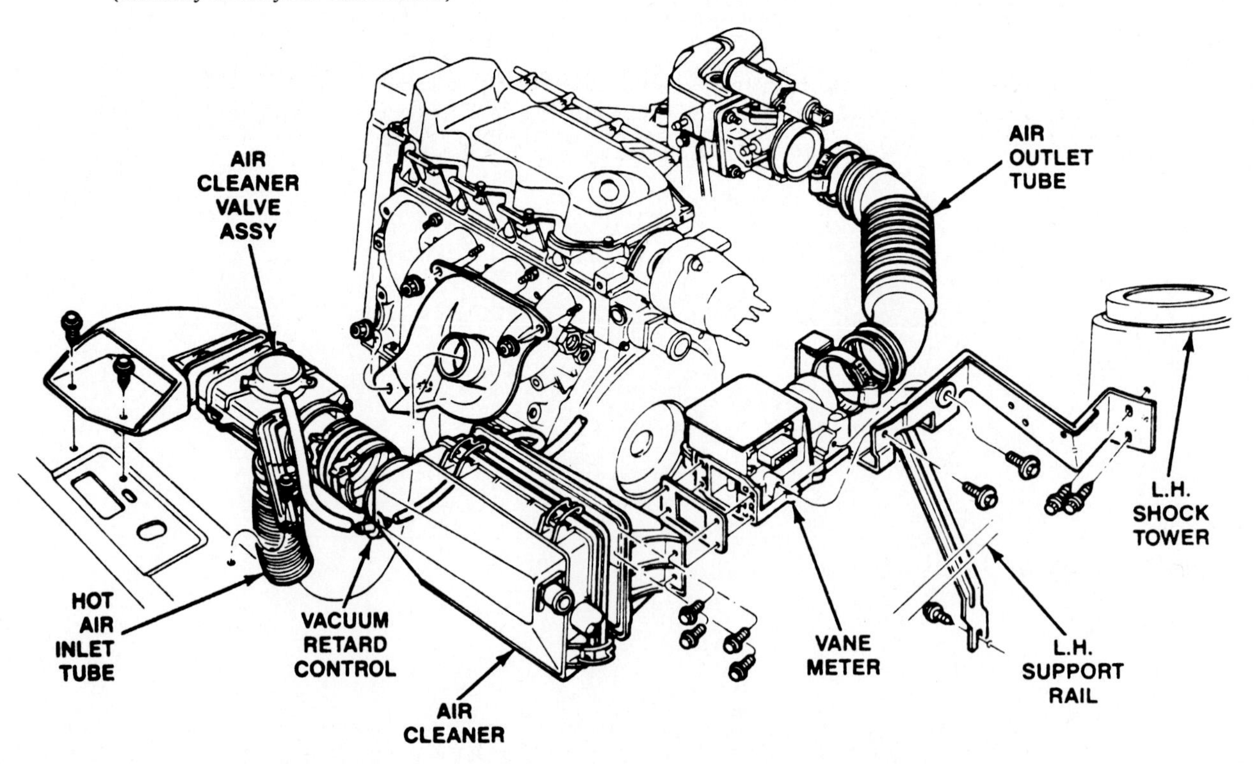

Figure 4–84 Vane Meter Location. (Courtesy of Ford Motor Co.)

Ford Electronic Engine Control (EEC IV) Systems Used with 2.3L Turbocharged Engine

Input Sensors

The 2.3L turbocharged engine has an EEC IV system with electronic fuel injection (EFI). This system uses many of the same input sensors that are used in other EEC IV systems. The vane meter is only used on the EEC IV system with EFI. Two sensors are located in the vane meter. These sensors are referred to as the "vane air flow" (VAF) sensor and the "vane air temperature" (VAT) sensor. The vane meter is located in the engine air intake system as illustrated in Figure 4–84.

All the air flow into the engine must travel through the vane meter. The air flow rotates a vane that is mounted on a pivot pin in the meter body, as pictured in Figure 4–85.

The movement of the vane will be proportional to the volume of air flow through the air intake system. The VAF sensor is a variable resistor that has a sliding contact that is attached to the vane shaft. The ECA applies a 5V reference voltage to the VAF sensor. As the vane shaft rotates, it moves the sliding contact on the variable resistor. The voltage output signal from the VAF sensor to the ECA will vary between 0V and 5V. A higher volume of air flow will produce a higher voltage output signal from the VAF sensor. The VAF sensor in the vane meter is shown in Figure 4–86.

A vane air temperature (VAT) sensor is also located in the vane meter. The ECA calculates a mass air flow value from the VAT and VAF sensor signals. This value is used by the ECA to provide the correct air–fuel ratio of 14.7:1. The VAT sensor in the vane meter is shown in Figure 4–87.

Some EEC IV systems have a three-wire exhaust-gas oxygen (EGO) sensor in place of the single-wire sensor. This three-wire sensor has an internal electric heater to improve sensor response time. An air charge temperature sensor is used in some systems on 2.3L engines. The input sensors and output controls in the EEC IV systems on the 2.3L turbocharged engine are pictured in Figure 4–88.

Output Controls

An exhaust-gas recirculation (EGR) solenoid controls the vacuum that is applied to the EGR valve. The EGR solenoid shuts off the vacuum to the EGR valve if the engine coolant is cold, or when the engine is operating at idle speed or wide-open throttle (WOT). At all other operating conditions the ECA will energize the EGR solenoid, and this will apply vacuum to the EGR valve.

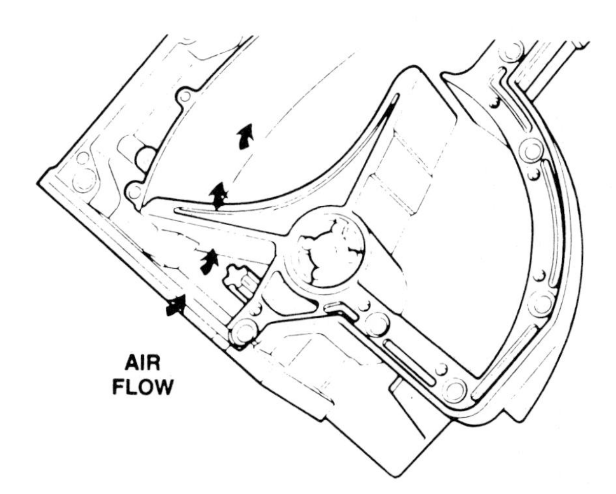

Figure 4–85 Vane Meter Air Vane. (Courtesy of Ford Motor Co.)

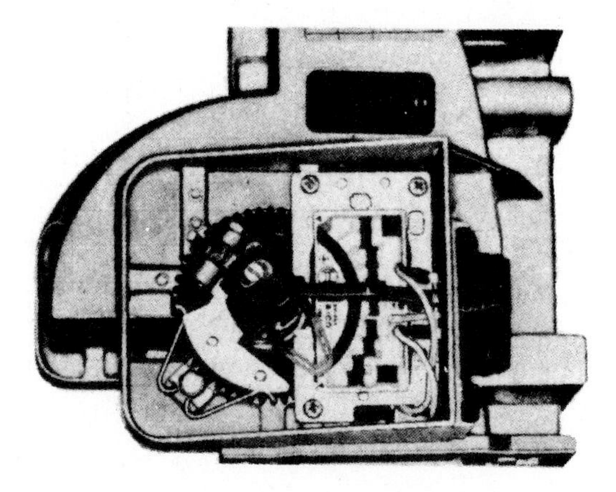

Figure 4–86 Vane Air Flow (VAF) Sensor. (Courtesy of Ford Motor Co.)

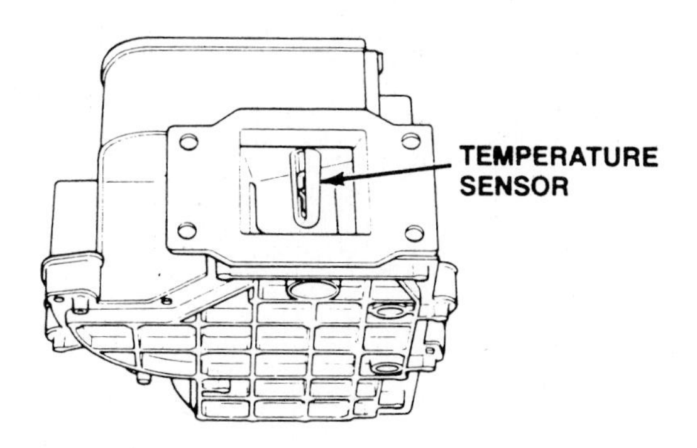

Figure 4–87 Vane Air Temperature Sensor. (Courtesy of Ford Motor Co.)

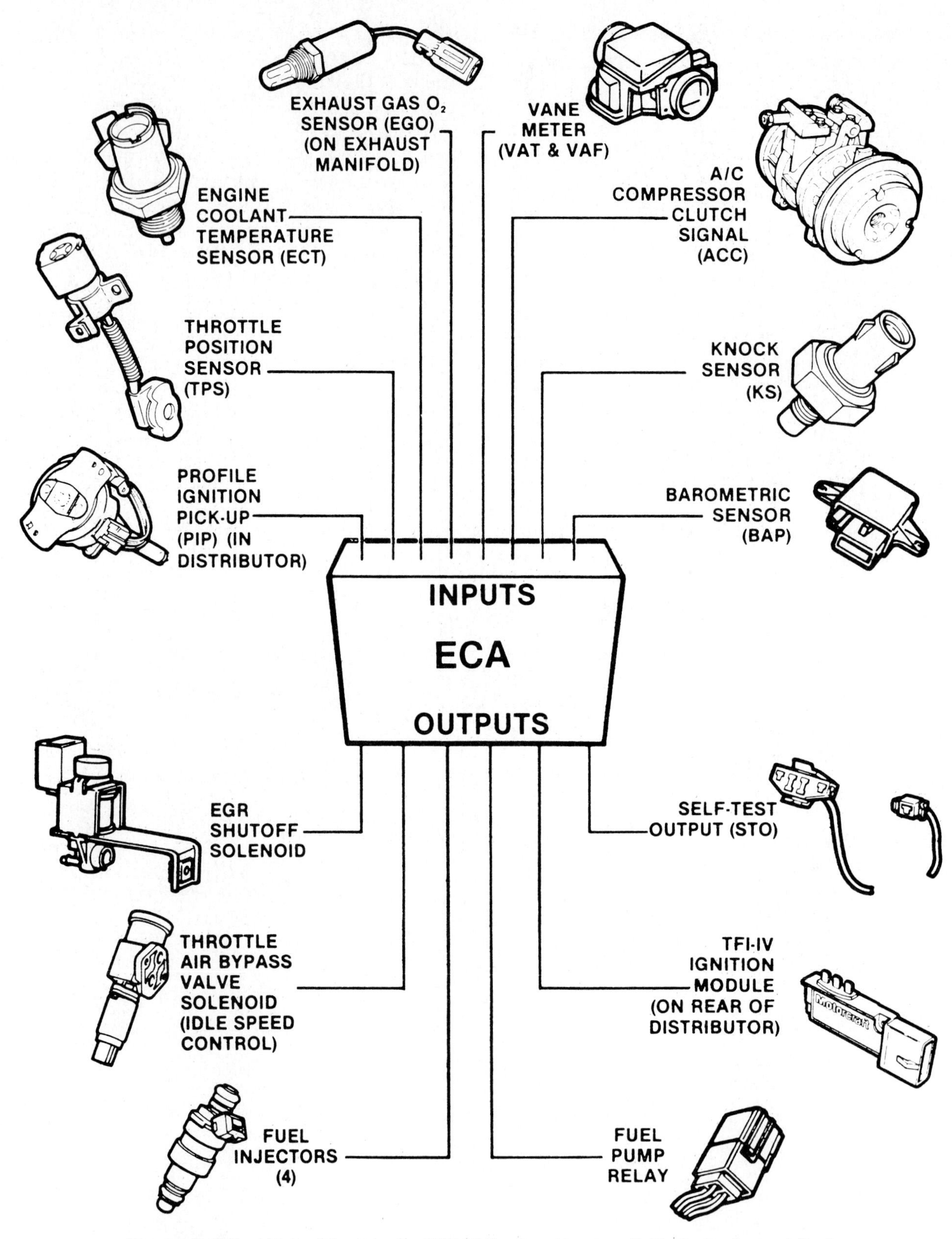

Figure 4–88 Wiring Diagram for EEC IV System Used on 2.3L Turbocharged Engine. (Courtesy of Ford Motor Co.)

The throttle body assembly contains the throttle valve, throttle-position sensor (TPS), and the throttle air bypass valve solenoid is illustrated in Figure 4–89.

The fuel injectors are mounted in the intake ports, and there is no fuel delivered to the throttle body assembly. The throttle air bypass valve solenoid is operated by the ECA. When the ECA increases the voltage that is supplied to the solenoid, the throttle air bypass valve opening will increase, and thus allow additional air to bypass the throttle valve and increase the engine idle speed. The throttle air bypass valve controls both the cold fast idle speed and the idle speed at normal engine temperatures. There are no curb idle or fast idle adjustments on the throttle air bypass valve solenoid. While the engine is being cranked, and ECA moves the throttle air bypass valve to the fully open position. This allows "no touch starting," which means the engine can be started at any temperature without touching the throttle. Each time the engine is started, the throttle air bypass valve will provide fast idle speed for a period of time. The time period will increase as engine coolant temperature decreases. The throttle air bypass valve solenoid provides the same function as the throttle kicker or idle speed control motor on other systems. Air flow through the throttle air bypass valve is illustrated in Figure 4–90.

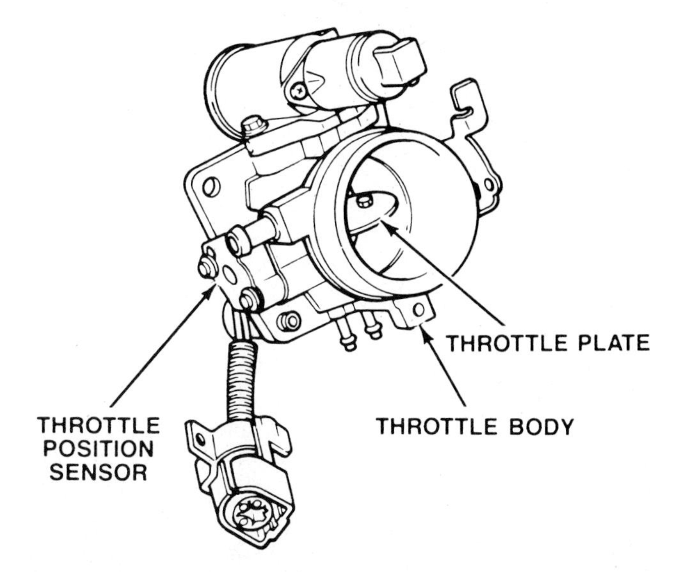

Figure 4–89 Throttle Body Assembly. [Courtesy of Ford Motor Co.]

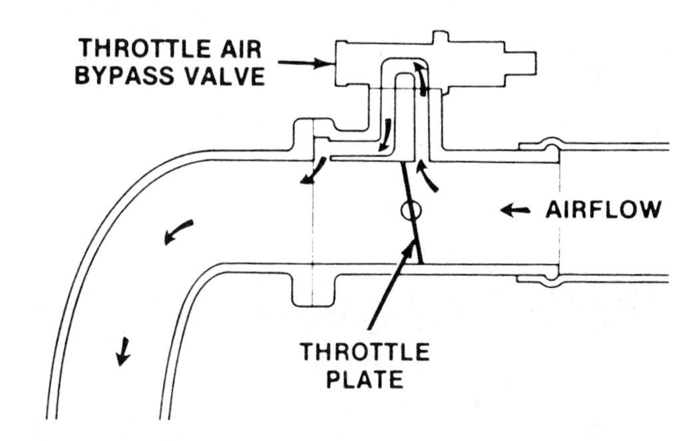

Figure 4–90 Throttle Air Bypass Valve Operation. [Courtesy of Ford Motor Co.]

Fuel System

The EEC IV system with electronic fuel injection (EFI) has a low-pressure electric pump mounted in the fuel tank and a high-pressure electric pump located on the right frame rail, as shown in Figure 4–91.

When the ignition switch is turned on, the ECA will ground the fuel pump relay winding, and the relay contacts will close to provide power to the fuel pumps, as shown in Figure 4–92.

If the engine is not cranked within one second, the ECA will open the circuit from the fuel pump relay to ground, and the relay contacts will open the circuit to the fuel pumps. The dual fuel pumps are capable of supplying fuel to 100 psi (690 kPa). A fuel filter is mounted near the high-pressure pump in the frame rail.

Fuel is delivered from the fuel pumps to the fuel rail and the pressure regulator. The fuel rail, pressure regulator, and the injectors are mounted on the intake manifold as shown in Figure 4–93.

When the pressure in the fuel rail reaches 39 psi (269 kPa), the pressure regulator diaphragm will be

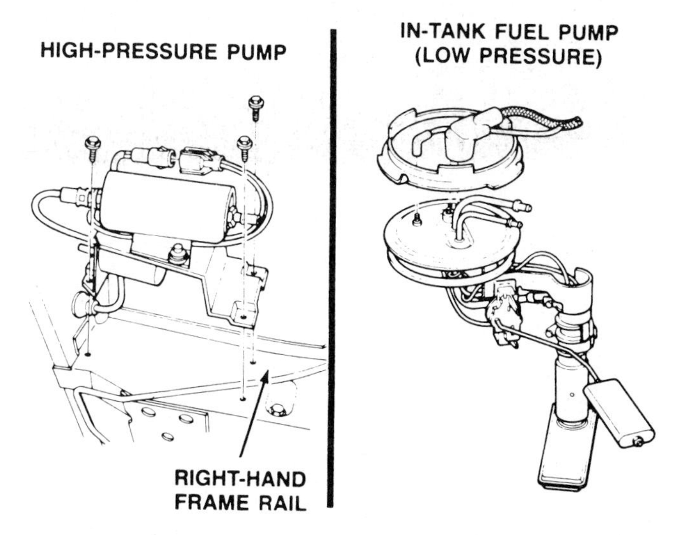

Figure 4–91 EEC IV with EFI Fuel Pumps. [Courtesy of Ford Motor Co.]

forced upward, which will open the valve, and excess fuel will return to the fuel tank as illustrated in Figure 4–94.

A vacuum hose is connected from the intake manifold to the top of the pressure regulator. This vacuum signal maintains a constant pressure drop across the injectors during all intake manifold vacuum conditions. When the engine is idling, high manifold vacuum will be applied to the upper side of the pressure regulator diaphragm, and the diaphragm can be forced upward at 39 psi (269 kPa). At high engine speeds the turbocharger will pressurize the intake manifold to 8 psi (55 kPa), and this pressure is applied to the upper side of the pressure regulator diaphragm. Under this condition 50 psi (345 kPa) will be required to move the diaphragm upward, and fuel pressure to the injectors will be maintained at this higher value. When the intake manifold pressure increases at high speeds, the increase in fuel pressure will maintain the same pressure difference across the injectors.

The injectors are mounted in the intake ports of the intake manifold. Each injector contains an electric solenoid that is energized by the ECA. The amount of fuel that each injector delivers is determined by the on-time of the injector because there is a constant pressure across the injector. A cutaway view of an injector is illustrated in Figure 4–95.

The ECA controls the on-time of the injectors to provide the precise air–fuel ratio that is required by the engine under all operating conditions. When the ignition switch is turned on, battery voltage is supplied to the injectors from the power relay. The injectors for cylinders 1 and 2 and the injectors for cylinders 3 and 4 are connected in parallel electrically as shown in the wiring diagram in Figure 4–96.

The ECA energizes each pair of injectors when it completes the circuit from the injectors to ground through the ECA. When the engine is decelerated in the closed throttle mode, the ECA will not energize the injectors. This provides improved emission levels and fuel economy during deceleration. The injectors are energized again when the engine speed drops to a predetermined rpm. Enrichment of the air–fuel ratio on a cold engine is provided by increasing the on-time of the injectors, and this eliminates the need for a conventional choke. If the engine becomes flooded, the condition may be cleared by

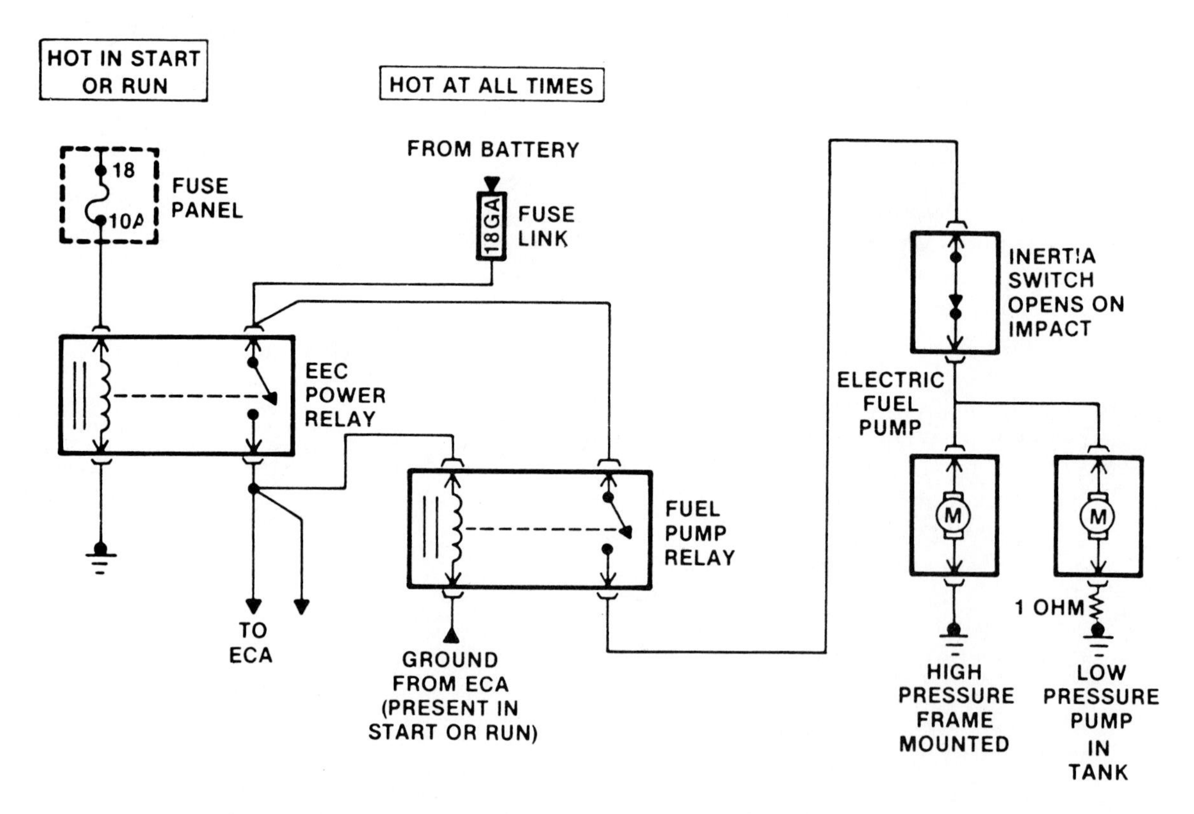

Figure 4–92 Electric Fuel Pump Circuit. (Courtesy of Ford Motor Co.)

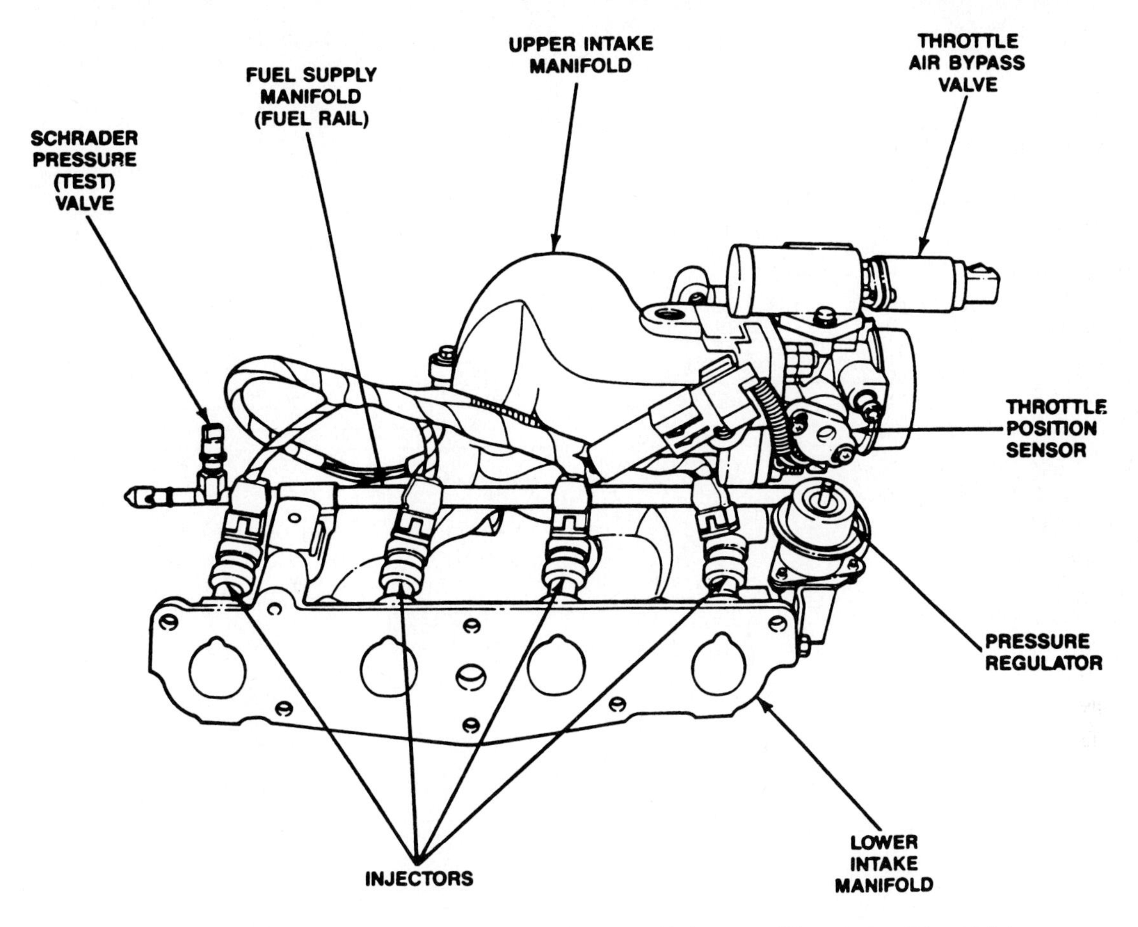

Figure 4–93 Fuel Rail with Pressure Regulator and Injectors. (Courtesy of Ford Motor Co.)

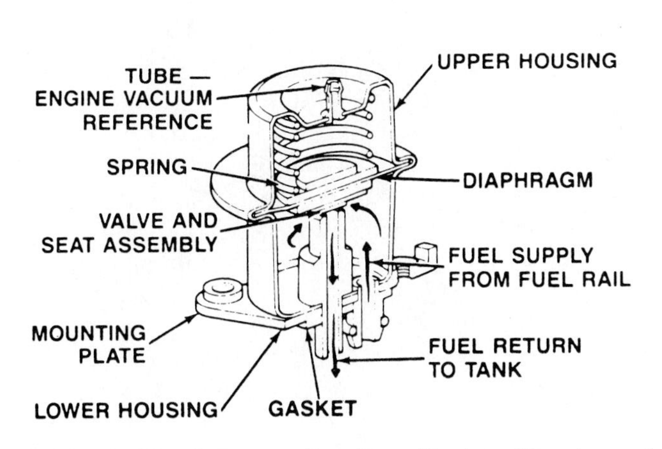

Figure 4–94 Pressure Regulator Design. (Courtesy of Ford Motor Co.)

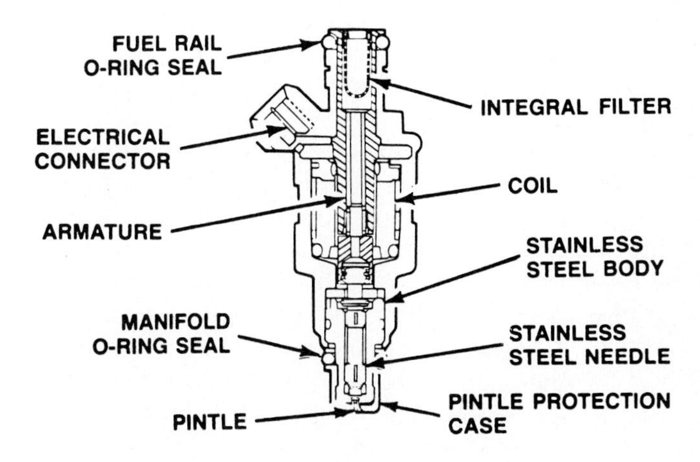

Figure 4–95 Injector Design. (Courtesy of Ford Motor Co.)

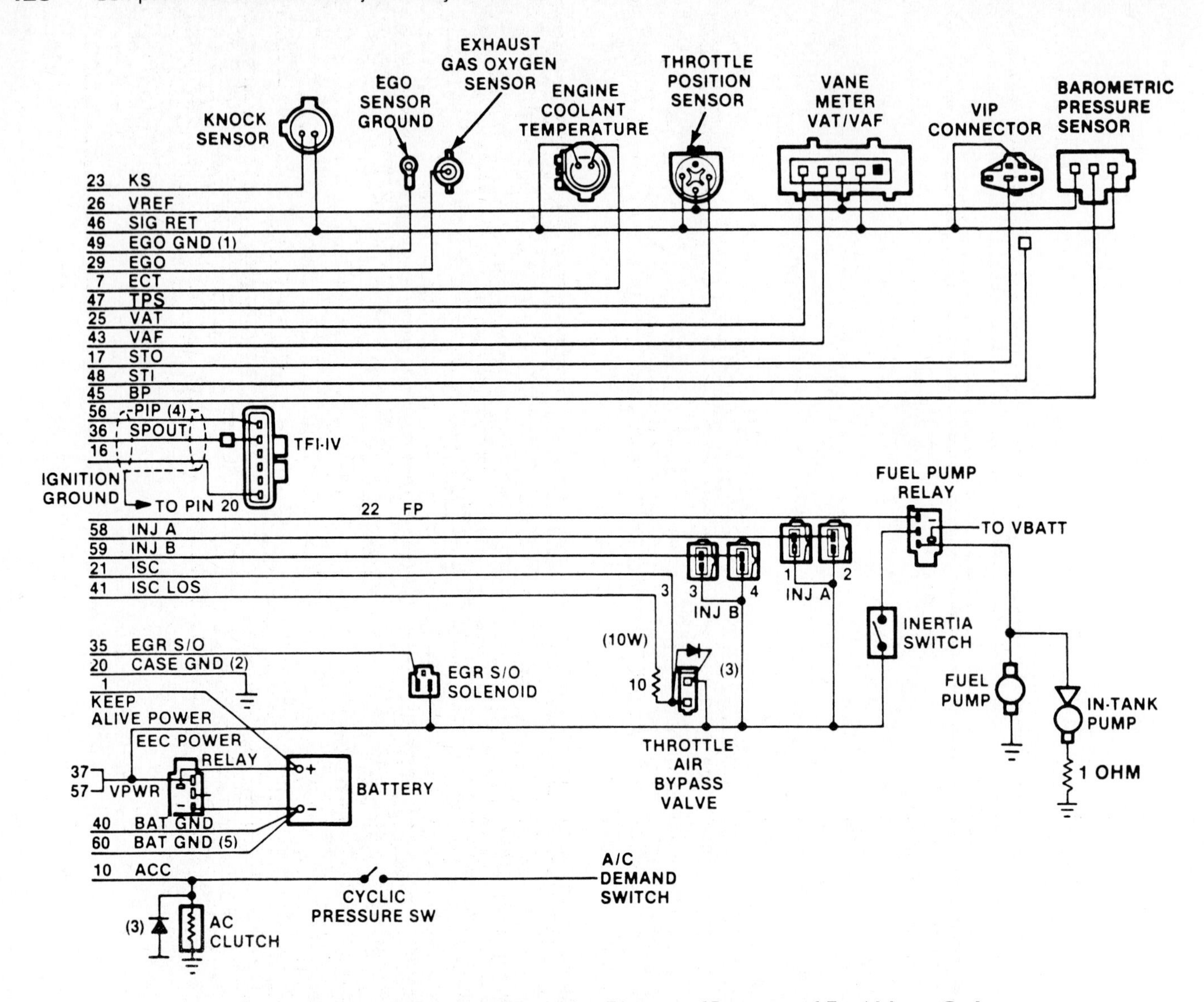

Figure 4–96 EEC IV with EFI Wiring Diagram. (Courtesy of Ford Motor Co.)

depressing the throttle to the wide-open position while cranking the engine.

When the ECA receives a wide-open throttle signal from the throttle-position sensor (TPS), the ECA will stop energizing the injectors while the engine is being cranked. This will clear the flooded condition, and as soon as the engine starts on the excess fuel in the intake manifold, the ECA will begin energizing the injectors again. The modes of operation on the EEC IV system with EFI are similar to the operating modes of other EEC IV systems with feedback carburetors that were discussed in Chapter 3.

Diagnosis of EEC IV Systems

Fuel Pump Pressure Test The fuel pump pressure in the EEC IV system with EFI may be diagnosed in the same way as the fuel pump pressure in the General Motors port injection system that was described earlier in this chapter.

Self-Test Initiation The same test procedure may be used to diagnose the EEC IV systems with fuel injection, or the EEC IV systems with feedback carburetors, that were described in Chapter 3. When defects occur in the EEC IV system, service codes will be stored in the ECA memory. The service codes may be obtained from the ECA memory by connecting a voltmeter or STAR tester to the self-test connector that is located in each EEC IV wiring harness. The voltmeter or STAR tester must be connected to the self-test connector on 1.6L engines, as shown in Figure 4–97.

If a 2.3L engine is being diagnosed, connect the STAR tester or voltmeter as outlined in Figure 4–98.

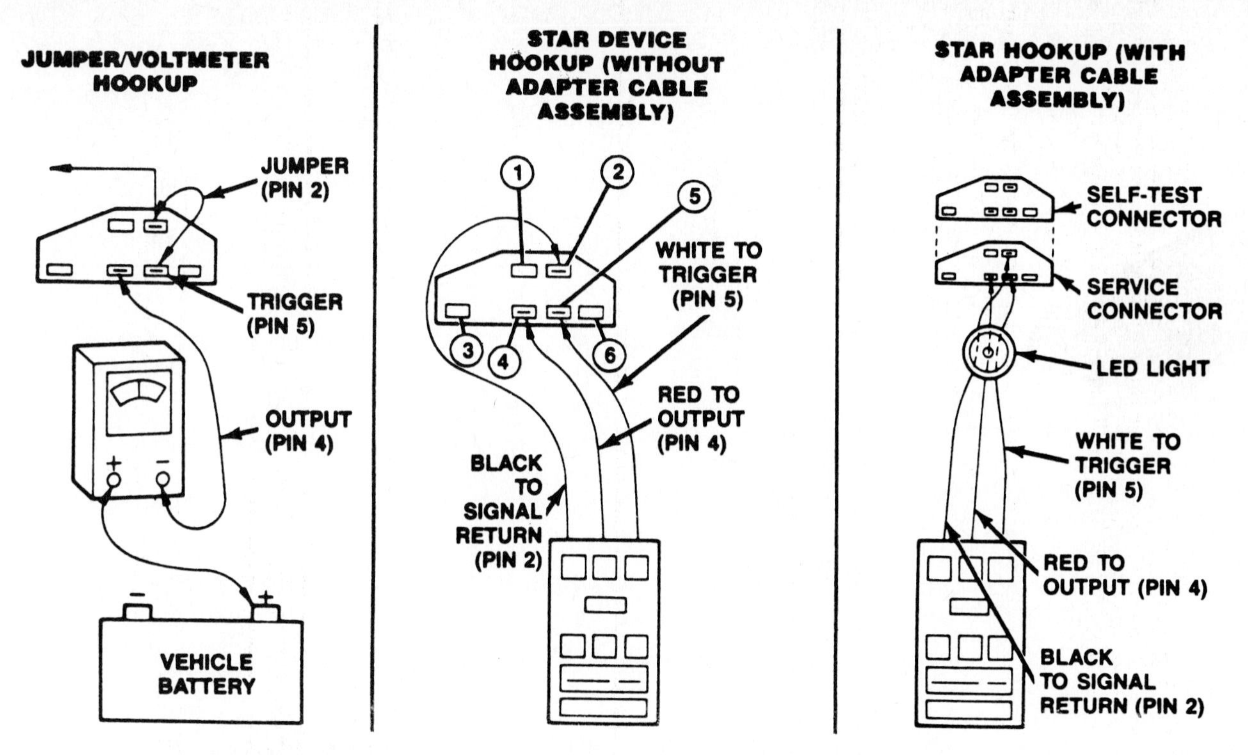

Figure 4-97 Voltmeter and STAR Tester Connections to the Self-Test Connector on 1.6L Engines. [Courtesy of Ford Motor Co.]

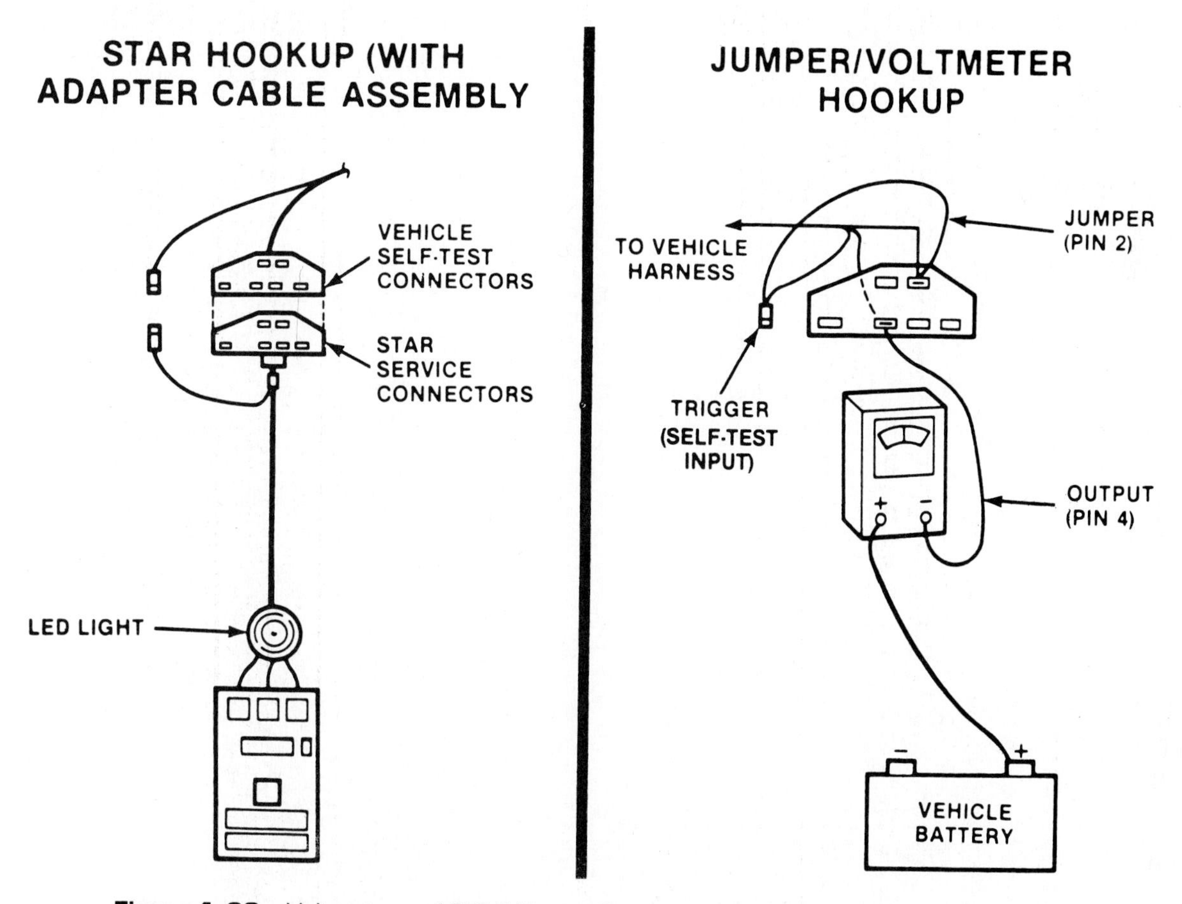

Figure 4-98 Voltmeter and STAR Tester Connections to Self-Test Connector on 2.3L Engines. [Courtesy of Ford Motor Co.]

When the STAR tester power switch is turned on, 88 will be displayed in the tester window, which will be followed by a 00 display. These displays indicate that the tester is ready to begin the self-test, which is initiated by depressing a pushbutton on the tester. The button will latch down, and a colon must appear beside the 00 display before any service codes will be displayed. If the LO BAT indicator is displayed steadily at any time, the internal battery in the tester must be replaced. The initial STAR tester displays are indicated in Figure 4–99.

When a voltmeter and a jumper wire are connected to the self-test connector, the service codes will appear as pulsations on the voltmeter needle. Two pulsations of the voltmeter needle followed by a 2-second pause and three more needle pulsations indicate service code 23, as pictured in Figure 4–100.

Key-On/Engine-Off Test When the STAR tester or the voltmeter is connected to the self-test connector and the ignition switch is turned on, service codes caused by hard failures will be displayed followed by a code 10 and service codes caused by soft failures, which are the result of intermittent faults. Hard failures are defects that are present at the time of testing.

The key-on/engine-off service code format is outlined in Figure 4–101.

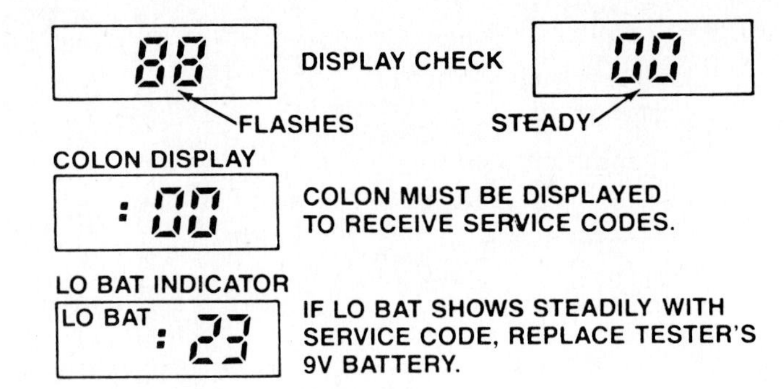

Figure 4–99 Initial STAR Tester Displays. (Courtesy of Ford Motor Co.)

Output Cycling Test This test is performed in the key-on/engine-off test after the service codes have been displayed. Without disabling the self-test connections, momentarily depress the throttle to the wide-open position and then release the throttle to initiate the output cycling test. This test procedure allows the technician to force the ECA to activate the ECA output controls one after the other. The technician may listen to, or observe, each output control such as relays or motors to determine if they are operating normally. A second throttle depression will stop the output cycling tests, or the test will shut off automatically after 10 minutes.

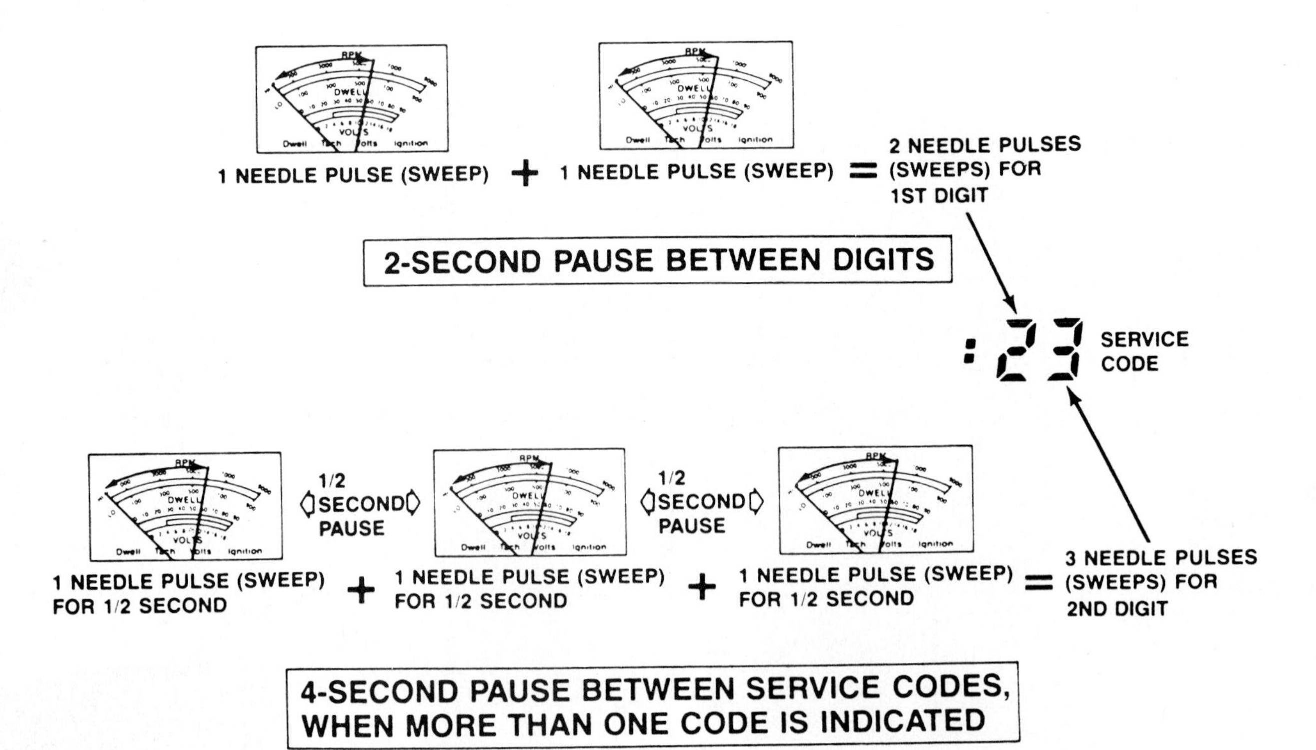

Figure 4–100 Voltmeter Reading of Service Codes. (Courtesy of Ford Motor Co.)

Engine Running Test Before the engine running test is attempted, complete these procedures:

1. Run the engine at 1500 rpm for 15 minutes to warm up the oxygen sensor.
2. Turn the engine off and depress the self-test button on the STAR tester. If a voltmeter is used for test purposes, connect the jumper wire to the self-test connector.
3. Wait 10 seconds and start the engine.

With the STAR tester or the voltmeter connected to the self-test connector and the engine running, the ECA will monitor all the sensors for proper operation, and it will energize all the output actuators and check the corresponding results. An engine identification code will be displayed first in the engine running test code format. The engine identification code pulses will be equal to half the number of engine cylinders. For example, three pulses on the voltmeter would be the engine identification code for a six-cylinder engine. Three engine identification pulses will be displayed as 30 on the STAR tester. The engine running service code format is shown in Figure 4–102.

The service code sequence for the key-on/engine-off test and the engine running test with a STAR tester is outlined in Table 4–6, and the service code sequence when these tests are performed with a voltmeter is shown in Table 4–7.

Wiggle Test If a voltmeter is connected to the self-test connector, the jumper wire must be disconnected from the self-test connector during this test.

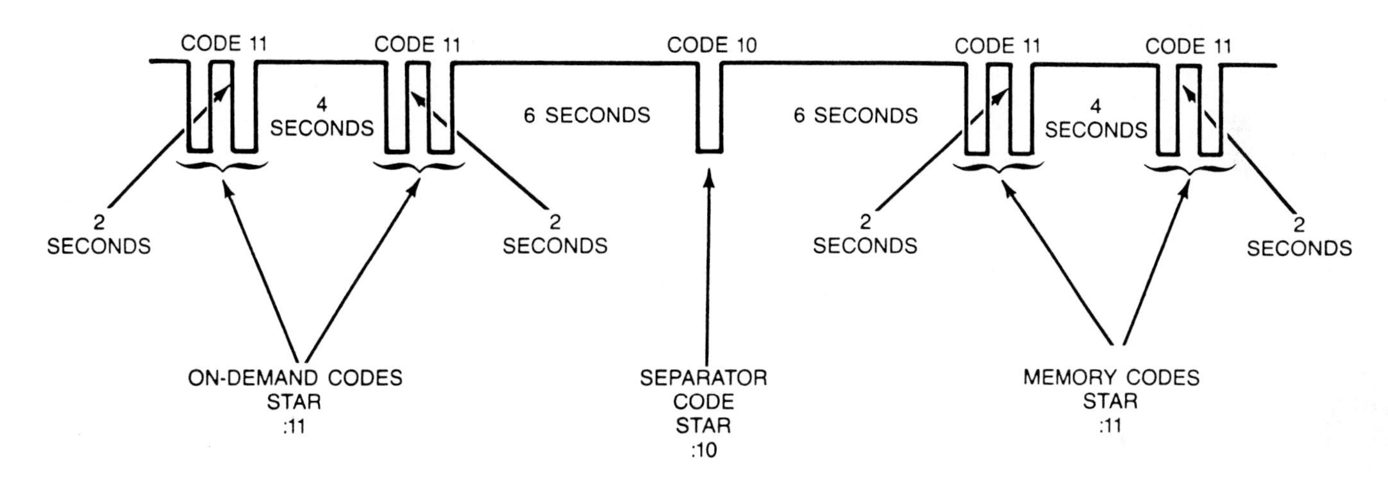

Figure 4–101 Key-On/Engine-Off Service Code Format. [Courtesy of Ford Motor Co.]

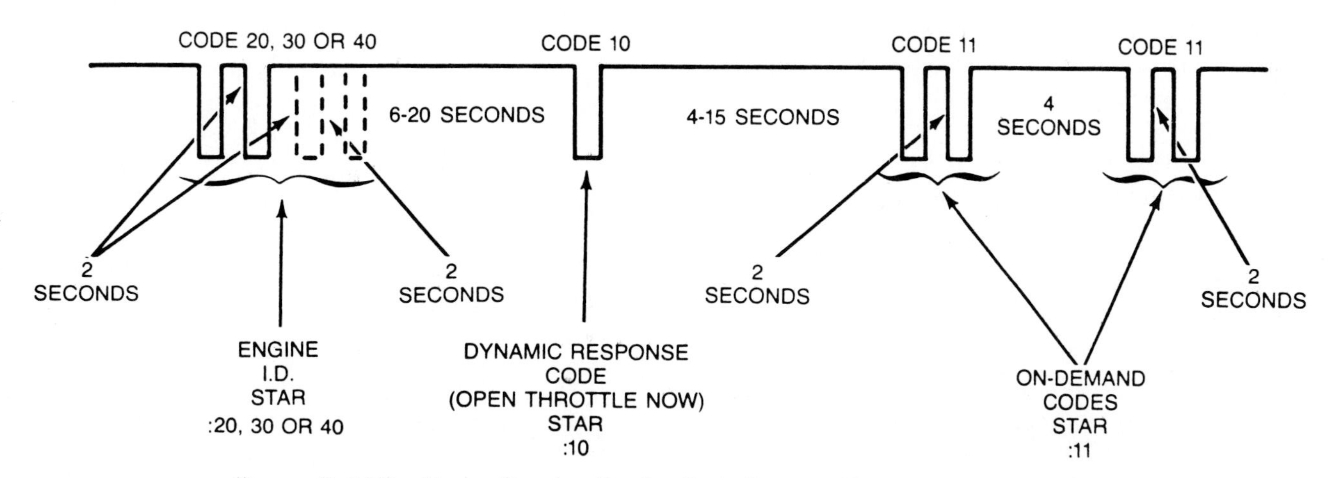

Figure 4–102 Engine Running Service Code Format. [Courtesy of Ford Motor Co.]

TABLE 4–6 Service Code Sequence with STAR Tester.

KEY-ON/ENGINE-OFF SEGMENT	*ENGINE-RUNNING SEGMENT*
1. FAST CODES—LED LIGHT FLICKERS	1. ENGINE I.D. CODE—NUMERALS 20, 30 OR 40
2. ON-DEMAND CODES—HARD FAULTS	2. DYNAMIC RESPONSE CODE—NUMERAL 10 (GOOSE THE THROTTLE)
3. SEPARATOR CODE—NUMERAL 10	3. FAST CODES—LED LIGHT FLICKERS
4. MEMORY CODES—INTERMITTENT FAULTS	4. ON-DEMAND CODES—HARD FAULTS

(Courtesy of Ford Motor Co.)

TABLE 4–7 Service Code Sequence with Voltmeter.

KEY-ON/ENGINE-OFF SEGMENT	*ENGINE-RUNNING SEGMENT*
1. FAST CODES—NEEDLE FLUCTUATES	1. ENGINE I.D. CODE—TWO, THREE OR FOUR NEEDLE SWEEPS
2. ON-DEMAND CODES—HARD FAULTS	2. DYNAMIC RESPONSE CODE—NEEDLE SWEEPS ONCE (GOOSE THE THROTTLE)
3. SEPARATOR CODE—NEEDLE SWEEPS ONCE	3. FAST CODES—NEEDLE FLUCTUATES
4. MEMORY CODES—INTERMITTENT FAULTS	4. ON-DEMAND CODES—HARD FAULTS

(Courtesy of Ford Motor Co.)

When the ignition is on and the wiring harness or connectors are wiggled, the voltmeter needle will deflect each time a defect occurs, and a service code will be stored. A STAR tester with a light emitting diode (LED) light as illustrated in Figure 4–96 may be used for this test. With the tester pushbutton not depressed and the ignition on, the LED light will go out each time a defect occurs as the wiring harness and connectors are wiggled. The LED light will stay off as long as the defect exists. The wiggle test may be repeated with the engine running and the self-test mode activated with the jumper wire connected to the self-test connector or the self-test button depressed on the STAR tester. The key-on/engine-running tests may be repeated to read any service codes that were stored in the ECA during the wiggle test.

Erase Code Procedure The service codes may be erased from the ECA if the following procedure is used:

1. Turn the ignition switch off.
2. Depress the self-test button to deactivate the STAR tester.
3. Turn the ignition switch on.
4. Depress the self-test button to reactivate the STAR tester.
5. When the STAR tester begins to display the first code, depress the self-test button to deactivate the tester. All codes should be erased from the ECA.

Service Procedure and Codes Intermittent codes should not be erased from the ECA until the necessary repairs have been completed. Hard faults should be repaired before intermittent faults. Fast codes are indicated by a slight flutter of the voltmeter needle or flash of the LED light on the STAR tester. These codes should be ignored, because they are used in factory test procedures. (Refer to Table 4–6 for fast codes.) If the basic timing is being checked, the in-line timing connector near the distributor must be disconnected. (See Chapter 5 for a complete description of EEC IV ignition.) The computed timing may be checked by depressing the self-test button on the STAR tester or connecting the jumper wire to the self-test connector as in the voltmeter diagnosis, with the engine running. The computed timing should be advanced 20° more than basic timing. Service codes should be disregarded during this test. The service codes are listed in Table 4–8.

The 90 series codes were introduced because dual oxygen sensors are used on later model 3.8L engines. Various Ford engines have different sensors or components in the EEC IV system. Therefore some service codes will only be available on a specific engine. The service codes may vary depending on the model year. When repairs are made to correct the defects

TABLE 4–8 EEC IV Service Codes.

Code	Description	Code	Description
11	System "pass"	55	Electrical charging under voltage
12	Rpm out of spec (extended idle)	56	MAF (VAF) input too high
13	Rpm out of spec (normal idle)	58	Idle tracking switch input too high (engine running test)
14	PIP was erratic (continue test)	61	ECT input too low
15	ROM test failed	63	TPS input too low
16	Rpm too low (fuel lean test)	64	ACT (VAT) input too low
17	Rpm too low (upstream/lean test)	65	Electrical charging over voltage
18	No tach	66	MAF (VAF) input too low
21	ECT out of range	67	Neutral drive switch—drive or accelerator on (engine off)
22	MAP out of range	68	ITS open or AC on (engine-off test)
23	TPS out of range	72	No MAP change in "goose test"
24	ACT out of range	73	No TPS change in "goose test"
25	Knock not sensed in test	76	No MAF (VAF) change in "goose" test
26	MAF (VAF) out of range	77	Operator did not do "goose test"
31	EVP out of limits	81	Thermactor air bypass (TAB) circuit fault
32	EGR not controlling	82	Thermactor air diverter (TAD) circuit fault
33	EVP not closing properly	83	EGR control (EGRC) circuit fault
34	No EGR flow	84	EGR vent (EGRV) circuit fault
35	Rpm too low (EGR test)	85	Canister purge (CANP) circuit fault
36	Fuel always lean (at idle)	86	WOT A/C cut-off circuit fault (all 3.8L and 5.0L Continental)
37	Fuel always rich (at idle)	87	Fuel pump circuit fault
41	System always lean	88	Throttle kicker circuit fault (5.0L)
42	System always rich	89	Exhaust heat control valve circuit fault
43	EGO cooldown occurred	91	Right EGO always lean
44	Air management system inoperative	92	Right EGO always rich
45	Air always upstream	93	Right EGO cooldown occurred
46	Air not always bypassed	94	Right secondary air inoperative
47	Up air/lean test always rich	95	Right air always upstream
48	Injectors imbalanced	96	Right air always not bypassed
51	ECT input too high	97	Rpm drop (with fuel lean) but right EGO rich
53	TPS input too high	98	Rpm drop (with fuel rich) but right EGO lean
54	ACT (VAT) input too high		

(Courtesy of Ford Motor Co.)

that are indicated by service codes, always begin repairing the defect that was indicated by the lowest service code number and always begin the repair procedure with the defect indicated by hard failures. In many cases the defective sensor can cause more than one service code to be stored in the ECA.

Test Questions

Questions on General Motors Throttle Body Injection Systems and Port Fuel Injection Systems

1. The pressure regulator limits the fuel pressure at the injectors to __________ psi.
2. The amount of fuel that is injected is determined by the injector __________.
3. When the ignition switch is on and no attempt is made to start the engine, the fuel pump will run until the battery is discharged. T F
4. In the open-loop mode the electronic control module ignores the oxygen sensor signal. T F
5. Cold fast idle speed is controlled by the idle air control (IAC) motor. T F
6. A code 13 would indicate a defect in the __________ circuit.
7. A defective coolant sensor could result in an engine flooding complaint. T F
8. A defective vehicle speed sensor (VSS) could result in:
   (a) erratic idle operation.
   (b) detonation.
   (c) reduced maximum speed.
9. The minimum air rate should be adjusted each time an engine tune-up is performed. T F

10. On a port-type injection system, trouble code 33 indicates a defect in the ______________ circuit.

Questions on Chrysler Electronic Fuel Injection Systems

1. The fuel hose from a conventional carburetor fuel system may be used on an EFI system. T F
2. A 12V test light should be used to test for power at the logic module terminals. T F
3. When the basic timing is being checked, the system must be in the ______________ mode.
4. If the power-loss light is being used to diagnose the EFI system, two light flashes followed by a pause and one light flash indicates a defect in the ______________ circuit.
5. The automatic shutdown (ASD) relay will remain closed when the ignition switch is on and the engine is not running. T F
6. When the power-loss light is on, the EFI system is in the ______________ mode.
7. Battery voltage is supplied to the electrical fuel pump through the ______________ relay contacts.
8. In the single-point EFI system the pressure regulator limits the fuel pump pressure to ______________ psi.
9. An in-line fuel filter is located in the engine compartment. T F
10. The power module increases the injector pulse width when engine load is increased. T F
11. The automatic idle speed AIS motor increases the idle speed by allowing more ______________ to bypass the throttle.
12. In the multipoint EFI system the power module energizes the injectors individually. T F

Questions on Ford Electronic Engine Control (EEC IV) Systems

1. Hard faults may be defined as defects that are present when the sytem is being diagnosed. T F
2. When repairs are being made to correct defects that were indicated by service codes, the defect indicated by the service code with the highest number should be repaired first. T F
3. Defects that are indicated by intermittent service codes should be replaced before defects indicated by hard fault codes. T F
4. When the basic timing is being checked, disconnect the:
   (a) coolant sensor connector.
   (b) in-line timing connector.
   (c) PIP wire at the distributor.
5. The vane air temperature (VAT) sensor sends a signal to the ECA in relation to:
   (a) intake air temperature.
   (b) coolant temperature.
   (c) fuel–air mixture temperature.
6. While the engine is being cranked, the ECA moves the throttle air bypass valve to the ______________ position.
7. The high-pressure fuel pump is located in the fuel tank. T F
8. When a cold engine is flooded, the condition can be corrected by:
   (a) turning off the ignition switch and waiting 10 seconds.
   (b) pushing the throttle wide open while the engine is cranked.
   (c) leaving the throttle in the idle position while the engine is cranked.

5

Electronic Ignition Systems

Practical Completion Objectives

1. Perform voltmeter and ohmmeter tests on General Motors high-energy ignition systems.
2. Diagnose General Motors Computer-Controlled Coil Ignition (C^3I) systems.
3. Diagnose a no-start complaint using a 12V test lamp and a spark tester.
4. Perform voltmeter and ohmmeter tests on Ford Duraspark II ignition systems and Thick Film Integrated (TFI) ignition systems.
5. Diagnose Ford Electronic Engine Control IV (EEC IV) ignition systems when a no-start complaint is received.
6. Diagnose Duraspark III ignition systems when the engine fails to start.
7. Diagnose Chrysler electronic spark-advance systems with a Hall Effect switch, in a no-start situation.
8. Check and adjust ignition timing on General Motors, Ford, and Chrysler ignition systems.

General Motors High-Energy Ignition Systems

Design

Point ignition systems are subject to dwell and timing changes because of point-rubbing block wear. High exhaust emissions are caused by dwell and timing changes. Electronic ignition systems provide more stable operation because the dwell is determined by a solid-state module. Exhaust emissions are reduced by the increased stability of electronic ignition systems.

The high-energy ignition (HEI) system is self-contained in the distributor. Early model HEI systems were equipped with conventional vacuum and centrifugal advance mechanisms. On later model HEI systems, the spark advance is controlled by the electronic control module (ECM), and the advance mechanisms are not required. A later model HEI system is shown in Figure 5–1.

Many HEI coils are mounted on top of the distributor cap while some models have an externally mounted coil. A spring contact connects the secondary coil terminal to the center cap terminal. High-voltage leakage is prevented by the seal around the center cap terminal as illustrated in Figure 5–2.

The ground terminal dissipates induced voltages from the coil frame to ground on the distributor housing. Early model HEI coils have the secondary winding connected from the center cap terminal to the primary winding. Secondary windings are connected from the center cap terminal to the coil frame on later model coils. Four screws attach the coil to the distributer cap. The primary coil leads, identified as ''bat'' and ''tach'' terminals, are mounted in the distributor cap. A double connector is used on each primary terminal. One connection on the bat terminal is connected to the ignition switch. The bat terminal is also connected to the HEI module. Tach inner terminals are connected to the module, and the outer tach connection extends to the dash-mounted tachometer or diagnostic connector. Discussion of the diagnostic connector is included under the heading High-Energy Ignition Diagnosis.

A timer core is attached to the distributor shaft. The number of teeth on the timer core matches the

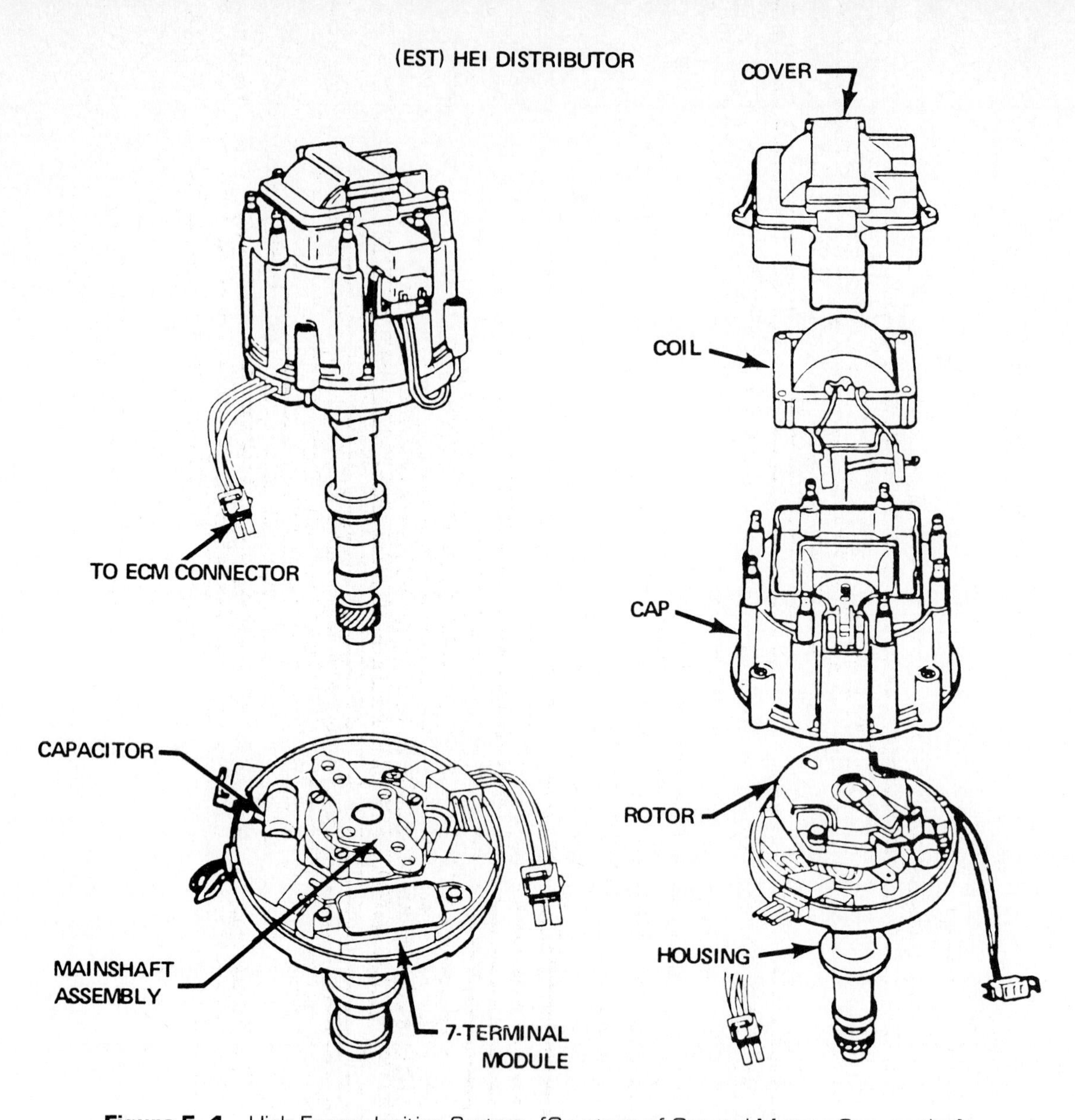

Figure 5-1 High-Energy Ignition System. [Courtesy of General Motors Corporation]

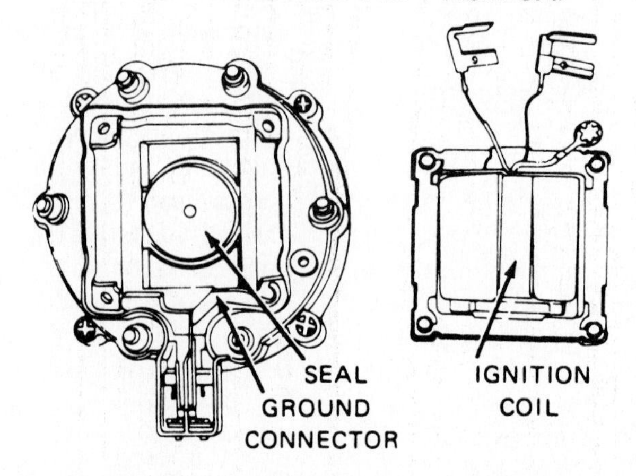

Figure 5-2 High-Energy Ignition Coil. [Courtesy of General Motors Corporation]

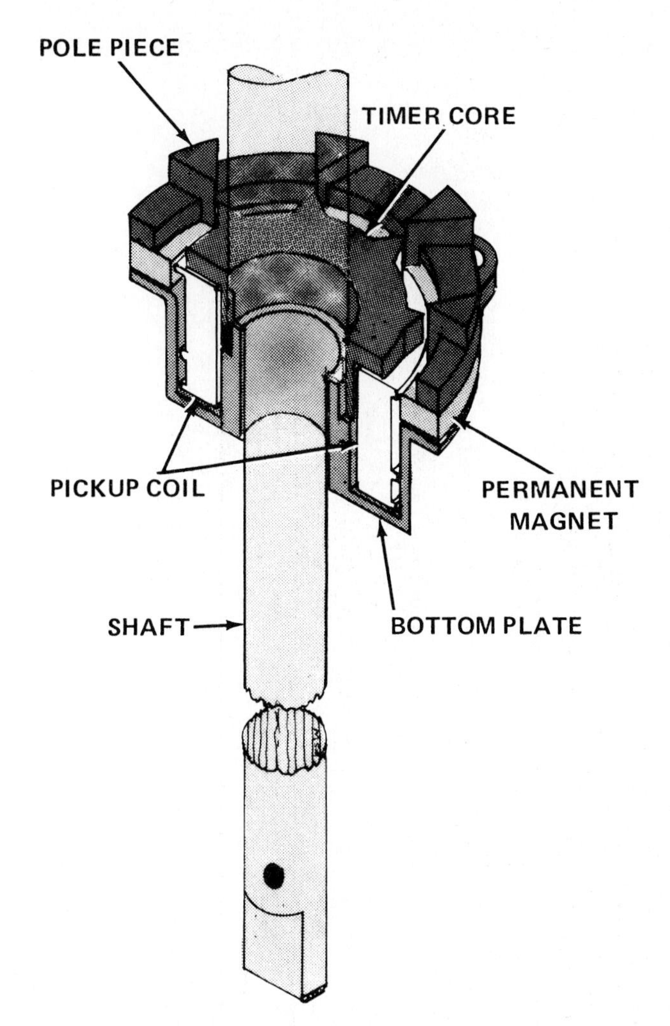

Figure 5-3 High-Energy Ignition Timer Core. (Courtesy of Sun Electric Corporation)

number of engine cylinders. A pickup coil assembly surrounds the timer core, and the number of teeth on the pickup pole piece match the timer core teeth as pictured in Figure 5-3.

Operation

As the timer core teeth approach alignment with the teeth on the pole piece teeth, the magnetic field builds up around the pickup coil. The resulting induced voltage in the pickup coil signals the module to turn on the primary current as illustrated in Figure 5-4. With the ignition switch on and the distributor shaft not turning, there is no primary current flow.

When the timer-core teeth begin moving out of alignment with the pole-piece teeth, the magnetic field suddenly collapses across the pickup coil. The module opens the primary circuit when the pickup

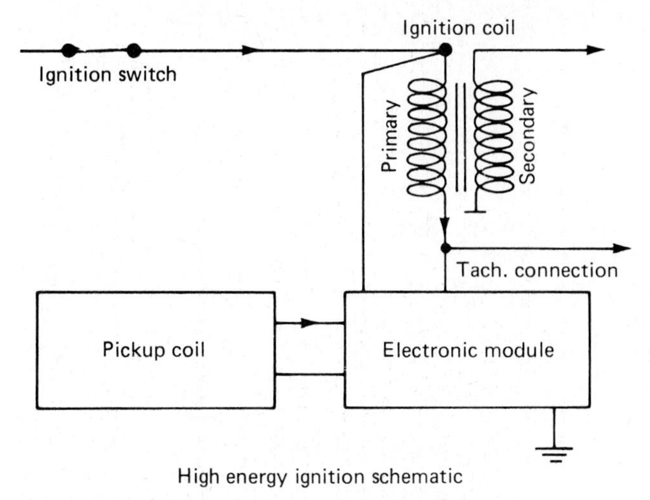

Figure 5-4 High-Energy Ignition Primary Current Flow. (Courtesy of General Motors Corporation)

coil induced voltage signal is received. Magnetic collapse occurs across the ignition coil windings, when the module opens the primary circuit. High voltage required to fire the spark plug is induced in the secondary winding, when the primary circuit is opened by the module. Primary dwell time is the length of time that the primary circuit remains closed by the module. The module extends primary dwell time as engine speed increases, providing higher primary magnetic strength and maximum secondary voltage. Other electronic ignition systems operate on similar basic principles, but many systems provide a constant dwell time regardless of engine rpm. The HEI system with the primary circuit open is shown in Figure 5-5.

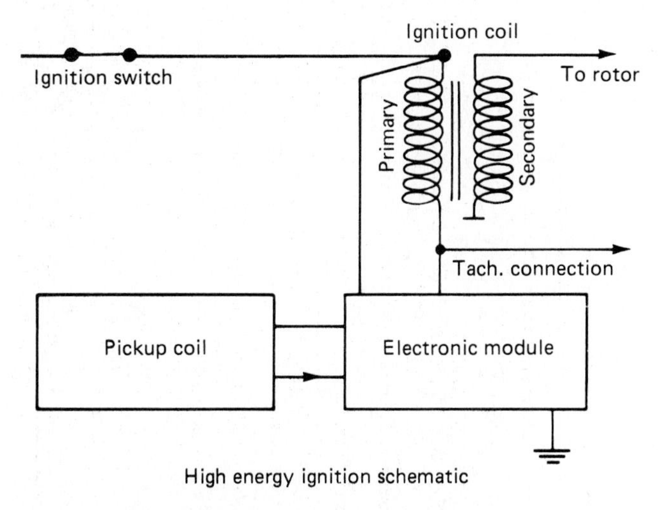

Figure 5-5 High-Energy Ignition Primary Circuit Open. (Courtesy of General Motors Corporation)

High-Energy Ignition Used with Computer Command Control System

Design and Operation

A seven-wire ignition module is used in the computer command control (3C) HEI system. The additional three-module terminals are connected to the electronic control module (ECM). Conventional vacuum and centrifugal advance mechanisms are discontinued, and the correct spark advance is supplied by the ECM in relation to the input sensor signals that it receives. Input sensor signals from the barometric pressure sensor, manifold absolute pressure sensor, coolant temperature sensor, and the crankshaft position sensor are used by the ECM to determine the correct spark advance, as shown in Figure 5–6.

While the engine is cranking, the pickup coil signal goes directly through the module signal converter and bypass circuit. The ECM does not affect spark advance when the engine is being started. Approximately 5 seconds after the engine is started, a 5V disable signal is sent from the ECM to the module. This signal opens the module bypass circuit and completes the compensated ignition spark timing circuit from the ECM to the module, as illustrated in Figure 5–7.

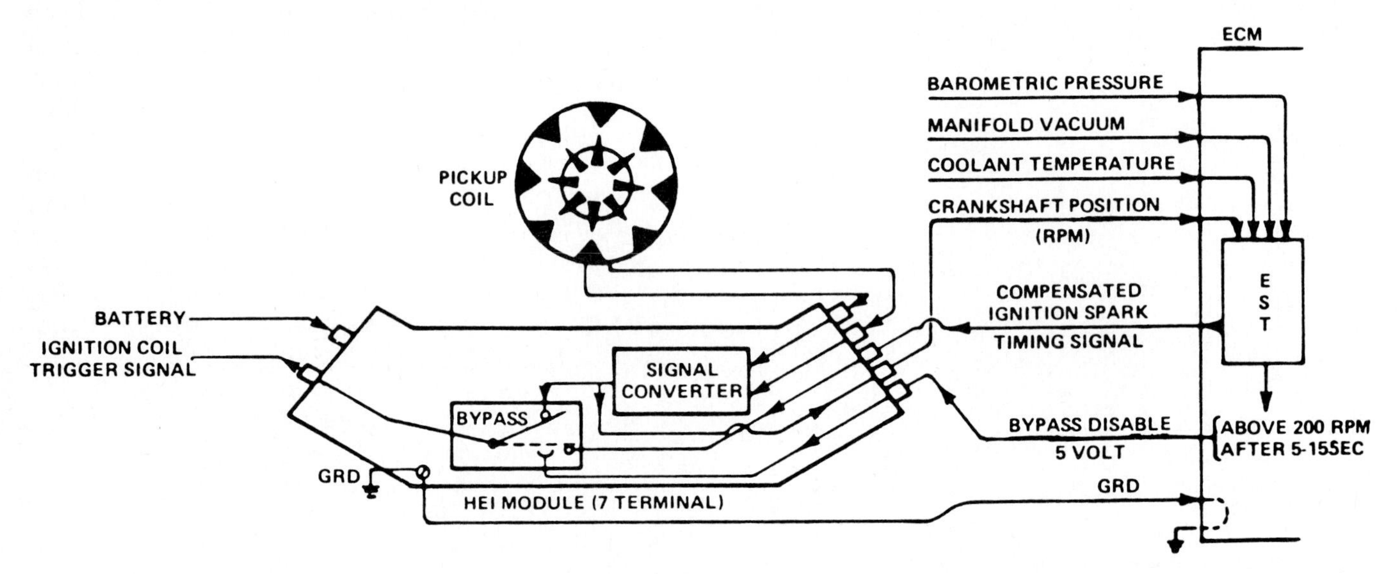

Figure 5–6 Computer Command Control High-Energy Ignition System, Start Mode. [Courtesy of General Motors Corporation]

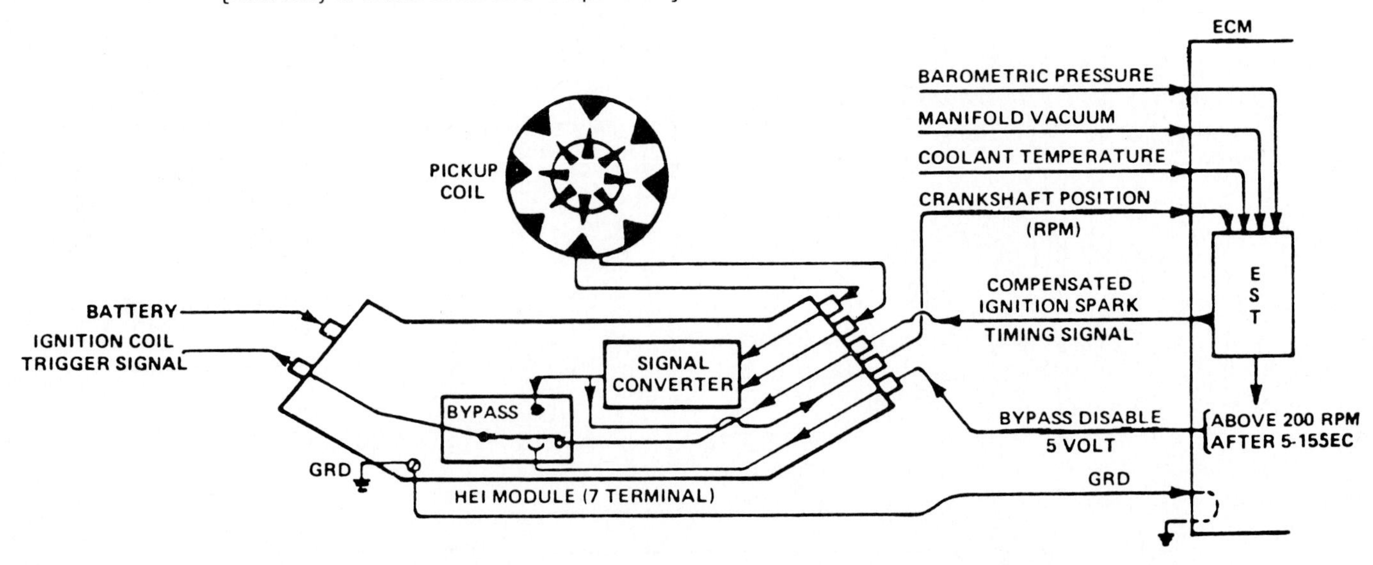

Figure 5–7 Computer Command Control High-Energy Ignition System, Run Mode. [Courtesy of General Motors Corporation]

The pickup coil signal now travels through the module signal converter, the ECM, and the compensated ignition spark timing circuit to the module. Ignition spark advance is controlled by the electronic spark-timing (EST) circuit in the ECM under this condition. The distributor pickup coil signal that is sent through the module to the ECM is referred to as a crankshaft-position rpm signal because it provides the ECM with an engine-speed signal. (The complete 3C system is explained in Chapter 3.)

High-Energy Ignition Used with Fuel Injection Systems

Design and Operation

Some throttle body injection systems have a conventional pickup coil and a Hall Effect switch in the distributor. The conventional pickup coil is only used while the engine is being cranked. A signal is sent from the Hall Effect switch through the ECM to the HEI module while the engine is running. The Hall Effect switch sends an rpm signal to the ECM, and the R terminal on the HEI module that was used on other systems for this purpose is no longer connected to the ECM, as shown in Figure 5–8.

When the engine is being cranked, the conventional pickup coil signal is sent directly to the transistor in the HEI module. A 5V disable signal is sent from the ECM through the tan/black wire and the winding in the HEI module to ground. This 5V disable signal changes the HEI module circuit into the EST mode as shown in the illustration in the lower left side of Figure 5–8. The signal from the Hall Effect switch can now travel through the ECM and the HEI module circuit to the transistor in the module. The time-variable circuit in the ECM will vary the signal from the Hall Effect switch to provide the precise spark advance that is required by the engine. Each time a signal is received by the transistor in the HEI module, it will open the primary ignition circuit to provide magnetic collapse in the ignition coil and the high induced voltage in the secondary winding to fire the spark plug. In the EST mode the circuit is open from the conventional pickup coil to the transistor in the HEI module. The Hall Effect switch provides a signal each time a reflector blade that is attached to the distributor shaft rotates past the switch. The HEI module circuit shown in Figure 5–8 is for illustration purposes: the actual module circuit would be much more complex.

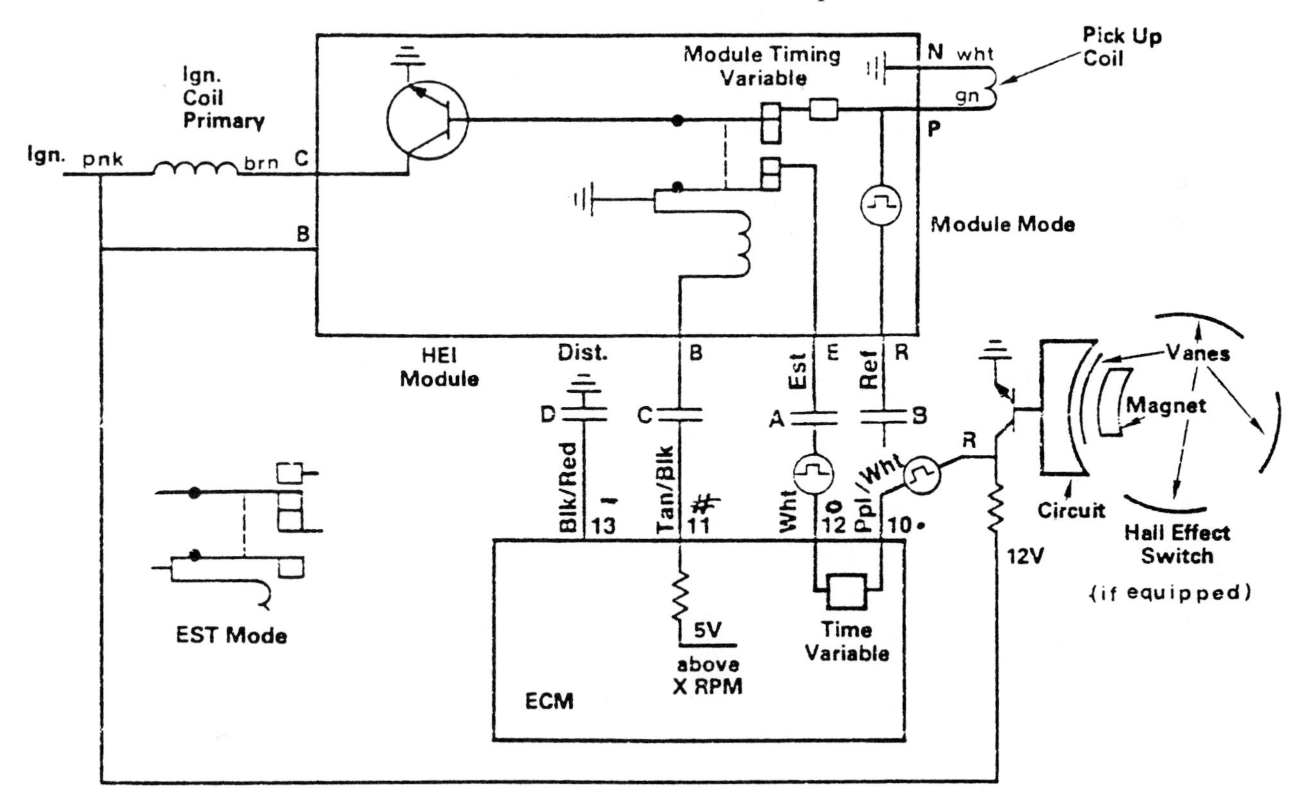

Figure 5–8 High-Energy Ignition with Hall Effect Switch. (Courtesy of General Motors Corporation)

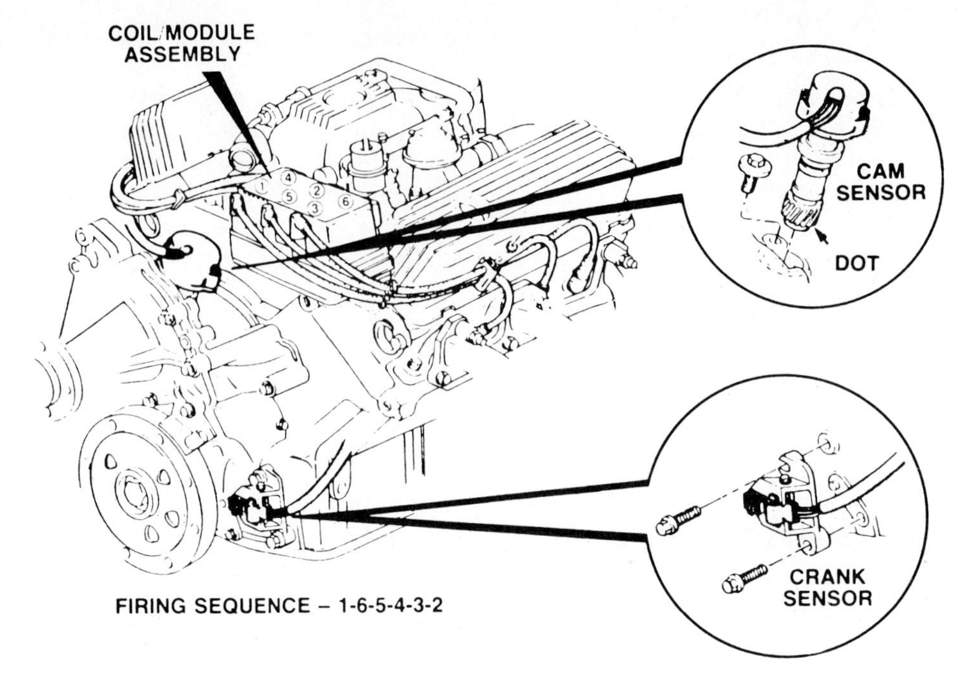

Figure 5–9 Computer-Controlled Coil Ignition System. [Courtesy of General Motors Corporation]

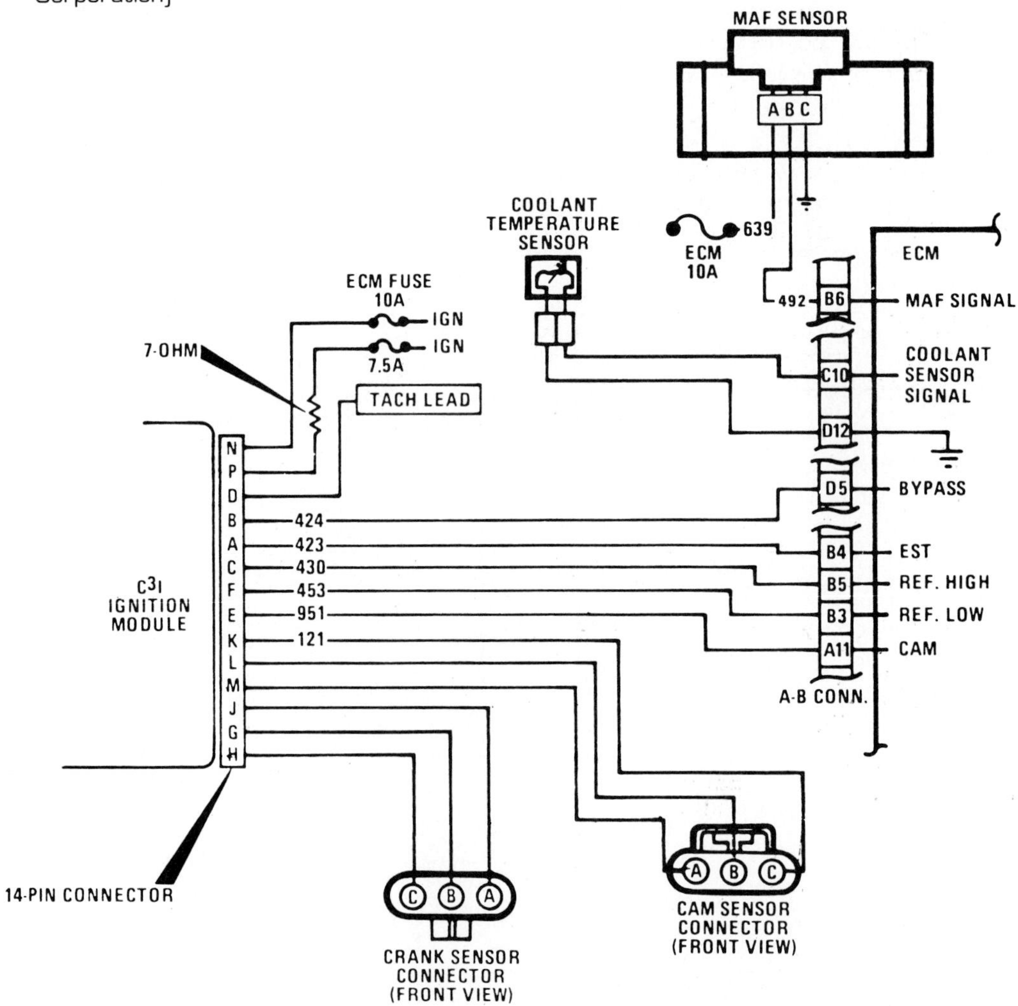

Figure 5–10 Computer-Controlled Coil Ignition System Wiring Diagram. [Courtesy of General Motors Corporation]

Computer-Controlled Coil Ignition Systems Used with Port Fuel Injection

Design and Operation

The main components in the computer-controlled coil ignition C³I system are the cam sensor, crank sensor, and the electronic coil module, as pictured in Figure 5–9.

The cam sensor and the crank sensor both contain Hall Effect switches. A conventional distributor is not required with the C³I system, and the cam sensor is mounted in the engine block in place of the distributor. This sensor is driven by the camshaft. The crank sensor is mounted at the front of the crankshaft, and this sensor is operated by an interruption ring attached to the crankshaft pulley. Signals from the crank sensor inform the electronic coil module and the ECM when each piston is at top dead center (TDC). This signal is used by the ECM and the coil module for correct timing and spark advance. A signal is also sent from the cam sensor to the coil module and the ECM. This sensor generates one signal for each sensor revolution. The ECM uses this signal to time the injector opening. For example, if the engine has sequential fuel injection (SFI), the ECM will begin to energize the injectors in the correct sequence when it receives the cam sensor signal.

Three ignition coils are mounted in the electronic coil module assembly, and an electronic module is located underneath the coils. The C³I system has two spark plug wires connected to each coil. On some V6 engines, plug wires 1 and 4, 5 and 2, and 6 and 3, are paired. With this type of system each coil will fire both spark plugs at the same time. The wires are paired so that one spark plug is firing when the piston is on the compression stroke and the other piston is on the exhaust stroke. If a spark plug fires when the piston is on the exhaust stroke, it has no effect. When the engine is being cranked, the crank sensor signal goes directly to the coil module, which opens the primary circuit of each coil. Once the engine starts, the crank sensor signal goes through the reference low or reference high circuit to the ECM. This signal is sent from the ECM through the EST circuit to the coil module. On the basis of the input signals, the ECM will vary the crank sensor signal to provide the precise spark advance required by the engine. An initial timing adjustment is not required on the C³I system. The wiring diagram for a C³I system is provided in Figure 5–10.

With the ignition switch on, voltage is supplied from terminal P on the coil module through the module circuit to the blue input wires on each coil primary winding. The coil module opens and closes the circuit from each primary winding to ground. When the ignition switch is turned on, voltage is supplied from terminal M on the coil module through terminal H to terminal A on the crank and cam sensor. Some engines have a combined crank and cam sensor mounted at the front of the crankshaft. The wiring diagram for this circuit is shown in Figure 5–11.

A tachometer lead is connected to the coil module, and the terminal on this wire is usually located near the coil module. Only digital tachometers should be connected to the tachometer lead.

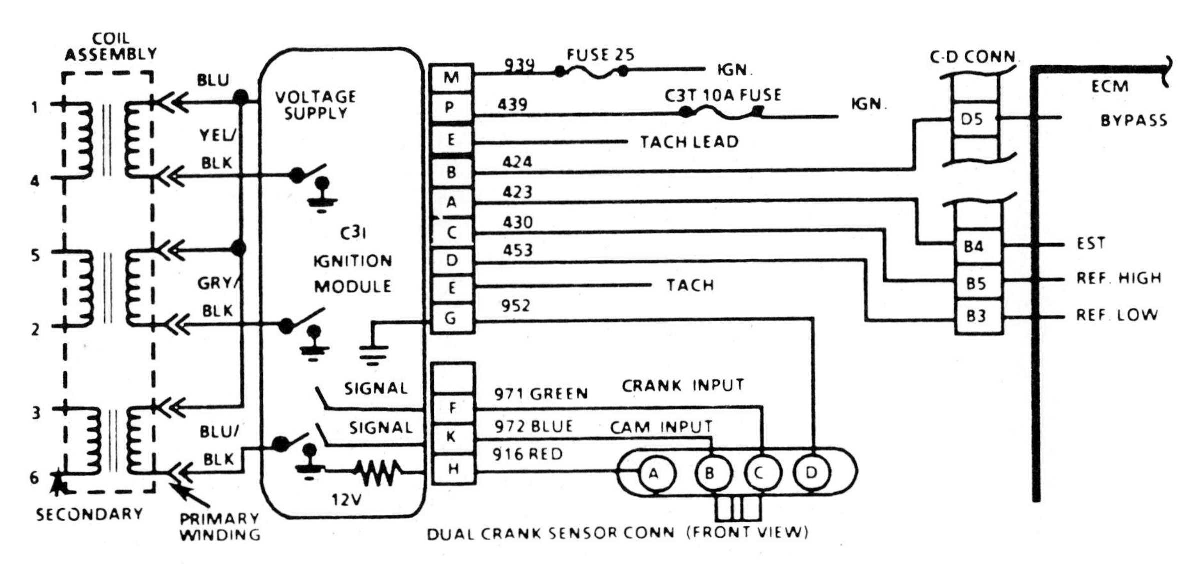

Figure 5–11 Computer-Controlled Coil Ignition System with Combined Crank and Cam Sensor. [Courtesy of General Motors Corporation]

Diagnosis of HEI Systems

Ohmmeter, Voltmeter Diagnosis

With the ignition switch on, connect a voltmeter from the coil bat terminal to ground. The voltmeter reading should exceed 12V. A zero voltage reading indicates an open circuit between the ignition switch and the coil bat terminal. When the voltmeter reads less than 12V, the battery is discharged or a high-resistance problem exists between the ignition switch and the coil bat terminal. The pickup coil may be checked for open circuits and shorts by connecting an ohmmeter across the disconnected pickup coil leads as indicated by ohmmeter 2 in Figure 5–12.

An infinite reading indicates an open pickup coil. The pickup coil is shorted if the ohmmeter reading is below the specification range of 650 to 850Ω. Ohmmeter 1 in Figure 5–12 is testing the pickup coil for a grounded condition. A low ohmmeter reading indicates a grounded pickup coil. The pickup coil is satisfactory if ohmmeter 1 displays an infinite reading.

Pull lightly on the pickup leads when testing for open circuits. A fluctuating ohmmeter reading indicates an intermittent open circuit. Severe engine surging usually occurs if an intermittent open circuit exists in the pickup coil.

If the distributor has a conventional pickup coil and a Hall Effect switch, this switch must be tested separately. With the Hall Effect switch removed from the distributor, connect a 12V battery and a voltmeter to the switch as indicated in Figure 5–13.

When a knife blade is inserted in the Hall Effect switch, the voltmeter reading should be within 0.5V of battery voltage. If the knife blade is removed from the switch, the voltmeter reading should be less than 0.5V. The Hall Effect switch must be replaced if the voltmeter readings are not within specifications.

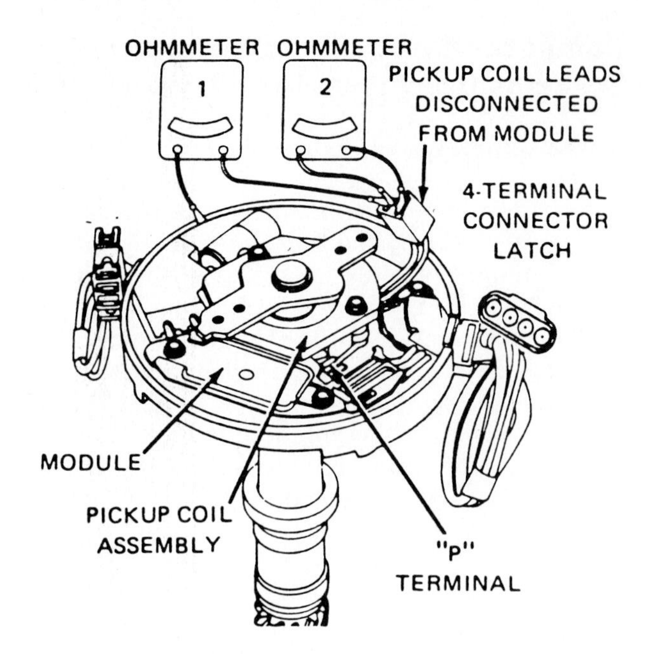

Figure 5–12 High-Energy Ignition Pickup Coil Tests. [Courtesy of General Motors Corporation]

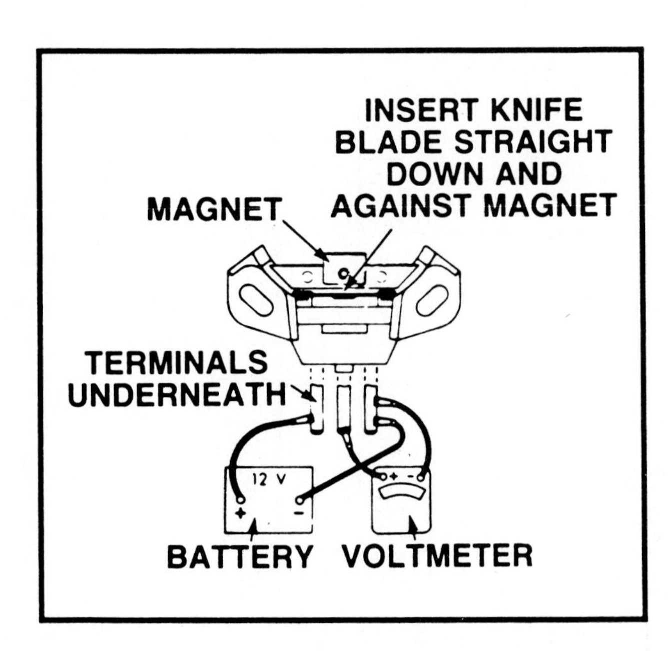

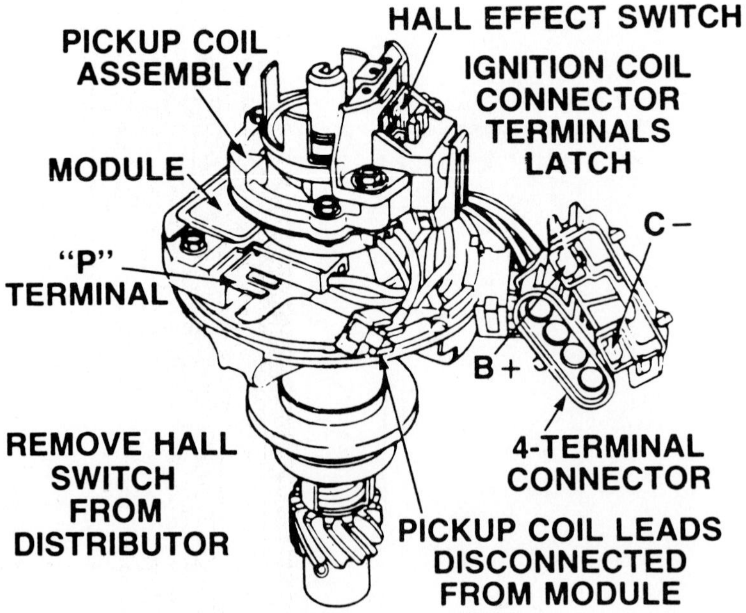

Figure 5–13 Hall Effect Switch Tests. [Courtesy of General Motors Corporation]

The primary ignition coil winding should be tested by connecting an ohmmeter across the primary terminals as outlined by ohmmeter 1 in Figure 5–14.

A reading below the specified value of .5Ω indicates a shorted winding. The primary winding is open if an infinite reading is obtained. Ohmmeter 2 is connected from the center cap terminal to the ground terminal to test the secondary winding. The secondary winding should have 12,000 to 20,000Ω. Early model HEI secondary windings are connected from the center cap terminal to one of the primary terminals. When testing early model coils, the ohmmeter leads must be connected from the center cap terminal to one of the primary terminals.

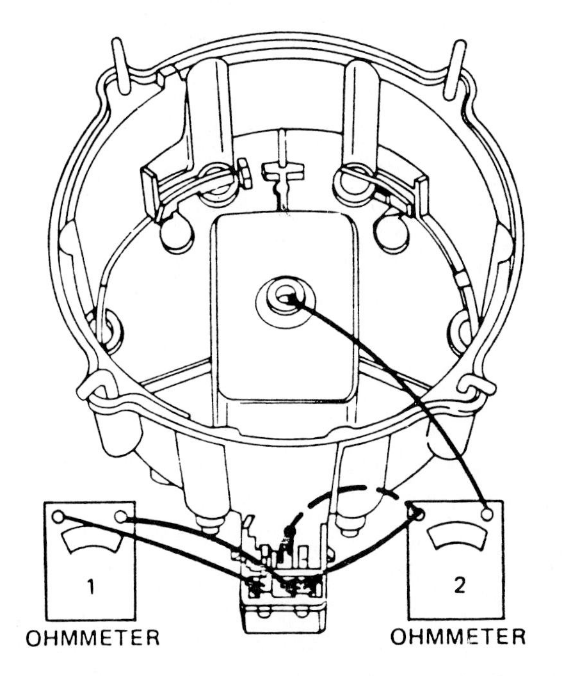

Figure 5–14 High-Energy Ignition Coil Tests. [Courtesy of General Motors Corporation]

The distributor cap and rotor should be inspected for cracks, any sign of leakage, and corroded terminals. Inspect the underside of the rotor for burn marks, which indicate leakage problems. If all the voltmeter and ohmmeter tests are satisfactory, the module or the ignition coil is defective if there is no spark while cranking the engine. Ohmmeter tests will not indicate insulation leakage defects in the ignition coil. Computer testing of modules and coils will be detailed in Chapter 16.

Test-Lamp Diagnosis

When the engine fails to start, a 12V test lamp may be connected from the tach terminal to ground. On vehicles with a diagnostic connector, the test lamp may be connected from the #6 terminal in the diagnostic connector to ground as pictured in Figure 5–15.

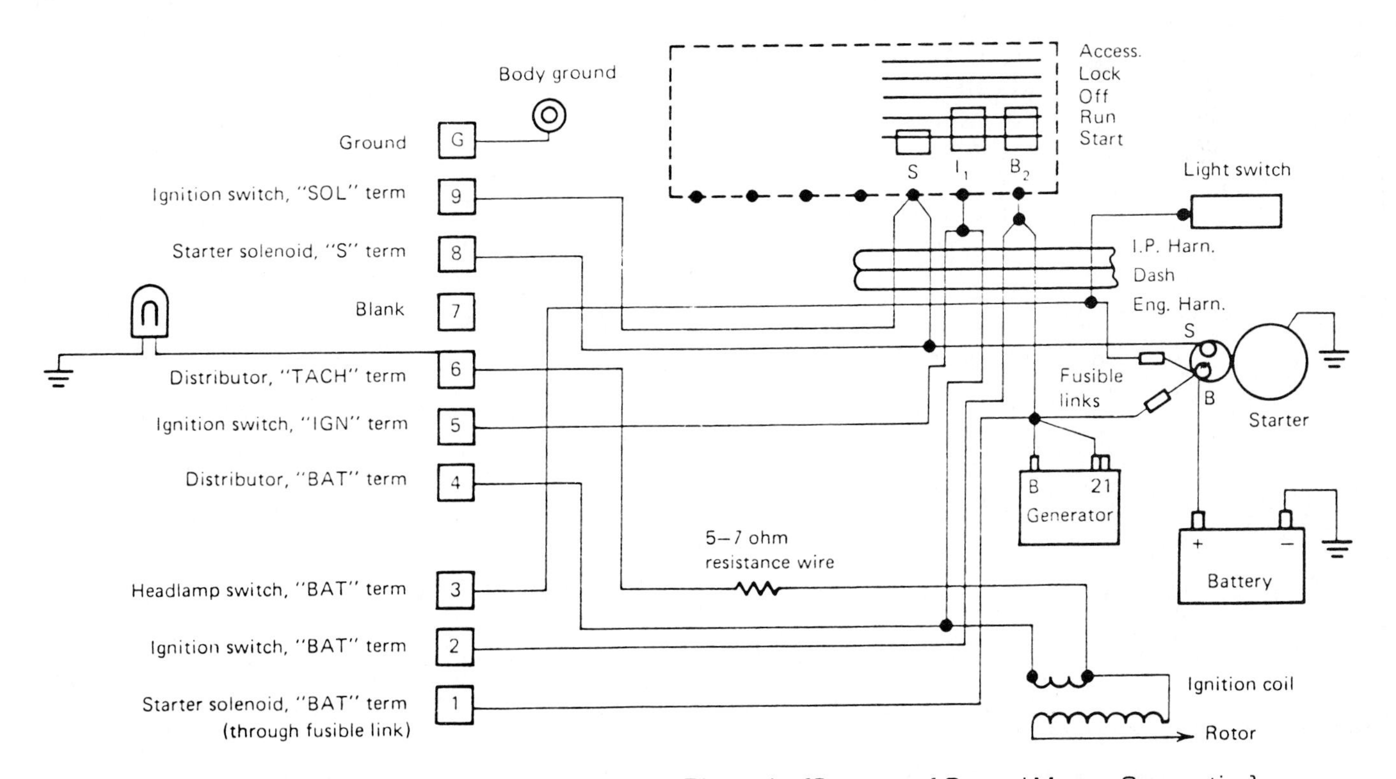

Figure 5–15 High-Energy Ignition Test-Lamp Diagnosis. [Courtesy of General Motors Corporation]

Failure of the test lamp to light with the ignition switch on indicates an open circuit in the coil primary or in the circuit from the ignition switch to the coil bat terminal. While cranking the engine, the test lamp should flash on and off. The module or pickup coil is defective if the test lamp is on continuously while cranking the engine.

If the test-lamp diagnosis indicates a satisfactory pickup coil and module, the no-start condition is caused by a defect in the secondary circuit. The HEI ST 125 tester may be used to test the secondary circuit. A spark-plug boot is used to connect the ST 125 tester to the center distributor cap terminal. The ST 125 tester is a specially designed spark plug with the ground electrode and part of the center electrode removed. Voltage requirement of the tester is 25,000V. The spark-plug wires must be removed from the cap, and the primary coil leads must be connected, as illustrated in Figure 5–16, while the tester is connected to the center cap terminal.

Failure of the ST 125 tester to fire while cranking the engine indicates a defective ignition coil, assuming the module and pickup coil were satisfactory in the test-lamp diagnosis. With the distributor cap and wires installed, the ST 125 tester may be connected from one of the spark plug wires to ground. If the tester will not fire while cranking the engine, a leakage defect exists in the rotor, cap, or plug wire, assuming that the tester fired when connected to the center cap terminal.

Various tests may be performed by connecting a 12V test lamp from the diagnostic connector terminals to ground. The circuit or component that may be tested at each diagnostic connector terminal is identified in Figure 5–15.

When the basic timing is being adjusted on the ignition systems used with General Motors 3C ignition, or with throttle body injection, one of the following procedures will be necessary.

1. On some systems a set timing connector must be grounded.
2. The four-wire connector near the distributor must be disconnected.
3. The tan wire in the four-wire distributor connector must be disconnected.
4. In other systems the test terminal in the assembly line communications link (ALCL) must be grounded. (The ALCL is explained in Chapters 3 and 4.) The underhood emission label should provide the correct procedure for checking the basic timing on each vehicle.

Diagnosis of Computer-Controlled Coil Ignition Systems

Engine Misfires

If the engine is misfiring, connect an ST 125 HEI test spark plug to a plug wire from each coil to ground while the engine is cranked. If one or two coils are not firing, check those coil windings and spark plug wires. An ohmmeter may be connected across the two secondary terminals on each coil to test the secondary winding. This winding should have less than 15,000Ω.

If the coil secondary winding and spark plug wires are satisfactory, remove the six assembly screws

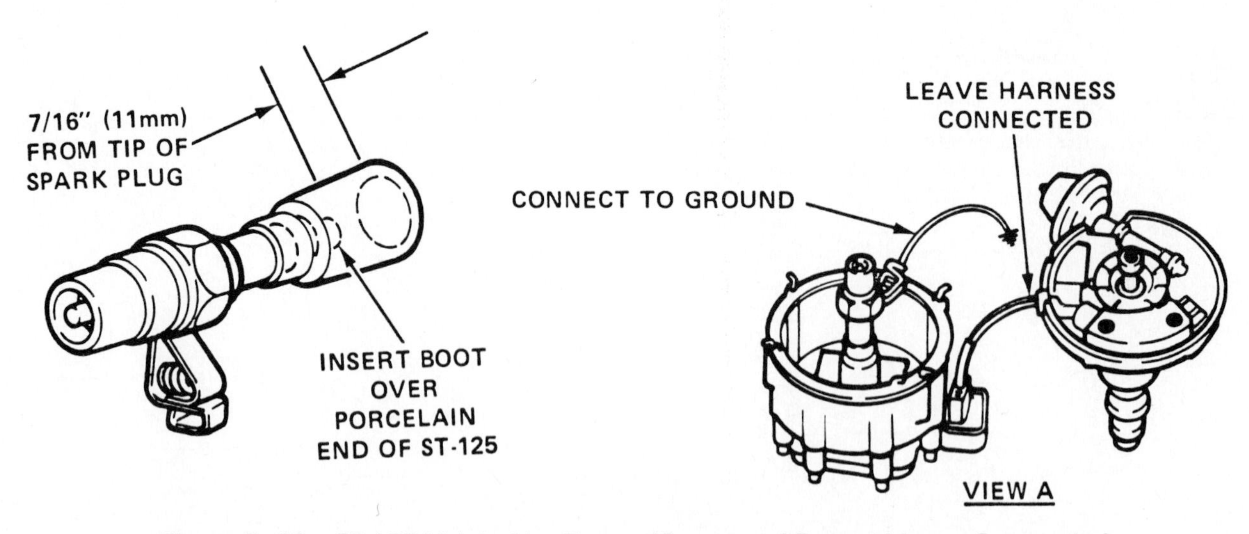

Figure 5–16 ST 125 High-Ignition Tester. (Courtesy of General Motors Corporation)

and detach the coils from the module. Connect a 12V test light across the primary coil terminals while the engine is cranked. If the test light blinks, but the test spark plug did not fire in the previous test, replace the coil assembly. When the light remains on or fails to come on, the coil module is defective.

If the test spark plug fires on each plug wire while the engine is cranked, the ignition system is satisfactory. Check the engine compression and the fuel system to locate the cause of the engine misfiring.

Engine Fails to Start

If the test spark plug does not fire when it is connected to any of the spark plug wires, proceed as follows:

1. Connect a digital voltmeter from coil module terminals M and P to ground with the ignition switch on. If 12V is not available at these terminals, check the fuses and connecting wires. Refer to Figure 5–11 for terminal identification.
2. With the ignition switch off, disconnect the ECM A–B connector and connect a digital voltmeter from ECM terminal B5 to ground while the engine is cranked. This reading should be 1 to 7V and varying. If this reading is not within specifications, proceed to the sensor diagnosis. When the reading is normal at terminal B5, the crank and cam sensor signal is being generated; therefore proceed to step 3.
3. Remove the coils from the module and connect a 12V test light from the blue primary coil leads to ground with the ignition on. If the light does not come on and the readings in step 1 are satisfactory, replace the coil module.
4. Connect a 12V test light across the primary leads on each coil while the engine is cranked. The test light should blink. If the test light remains on, replace the coil module. If the test light blinks on all three coils, connect the ST 125 test spark plug to a plug wire from each coil and ground while the engine is cranked. If no spark occurs, the coils are defective. It is unlikely that all three coils would be unsatisfactory.

Sensor Diagnosis if Engine Fails to Start

This sensor diagnosis will apply to the C^3I system with a combination cam and crank sensor, which is illustrated in Figure 5–11. Diagnosis would be similar on the C^3I system with separate cam and crank sensors, which is shown in Figure 5–10. If the voltmeter reading is not satisfactory in step 2 of the engine-fails-to-start diagnosis, proceed with the sensor diagnosis as follows:

1. Check voltage from coil module terminals M and P to ground as in step 1 of engine-fails-to-start diagnosis.
2. Disconnect the crank and cam sensor connector and connect a digital voltmeter across terminals A and B. With the ignition switch on, the voltmeter should read 5 to 11V. If the reading is not within specifications, check the wires from terminals A and B to coil module terminals H and K for open circuits and grounds. If the wires are satisfactory, replace the coil module.
3. Repeat the procedure in step 2 on terminals A and C in the sensor connector, and the same voltmeter reading should be obtained. When the reading is not satisfactory, check the wires from terminals A and C to coil module terminals H and F for open circuits and grounds. If the wires are satisfactory, replace the coil module.
4. With the ignition switch on, connect a digital voltmeter across terminals A and D in the sensor connector. This voltmeter reading should be 5 to 11V. When the reading is not within specifications, check the wire from terminal D to coil module terminal G for open circuits and grounds. If the wire is satisfactory but the voltmeter reading is out of specifications, replace the coil module.
5. Connect four short jumper wires between the cam and crank sensor terminals and the terminals in the wiring harness connector to allow meter connections to these terminals with the sensor connected.
6. Connect a digital voltmeter from terminals B and C to ground while the engine is cranked. The voltmeter reading at each terminal should be .7 to 9V and varying. If either voltmeter readings are not within specifications, the crank and cam sensor is defective.

Sensor Clearance

The dual sensor bracket cannot be moved to adjust basic timing. However, the clearance should be 0.030 in (0.762 mm) between each side of the crankshaft pulley discs and the sensors. This clearance may be adjusted by loosening the clamp bolt and moving the sensor up or down in the bracket.

Cam Sensor Installation

If a separate cam sensor is mounted in the engine block and the sensor is removed, it must be reinstalled correctly. The cam-sensor drive gear has a dot on one side that must face opposite the cam-sensor disc window. When the cam sensor is installed, this dot must face away from the timing chain with number 1 piston TDC on the compression stroke. The cam sensor harness must face toward the timing chain. When the cam sensor is being installed, use the following procedure.

1. Rotate the engine until number 1 piston is at TDC on the compression stroke.
2. Rotate the crankshaft until the timing mark on the harmonic balancer is 25° after TDC.
3. Remove plug wires from the coil assembly.
4. Disconnect cam-sensor connector and connect three short jumper wires between the sensor terminals and the wiring-harness connector.
5. Connect a digital voltmeter from terminal B to ground. Refer to Figure 5–10 for terminal identification.
6. With the ignition switch on, rotate the cam sensor counterclockwise until the sensor switch closes. This action will be indicated by a decrease on the voltmeter reading.
7. Tighten the cam-sensor clamp in this position. Reinstall the plug wires and reconnect the sensor connector.

Ford Duraspark Ignition Systems

Design

Duraspark II ignition systems have a six-wire module, and the distributor is equipped with conventional centrifugal and vacuum-advance mechanisms. A Duraspark II ignition wiring diagram is illustrated in Figure 5–17.

Some Duraspark II ignition systems have a universal ignition module (UIM), which has an additional wiring-harness connector that contains three wires. This extra wiring-harness connector may be connected to a vacuum switch, a barometric pressure switch, or to the microprocessor control unit (MCU) system depending on the engine application. The vacuum switch signals the module to retard the spark advance 3° to 6° if the manifold vacuum drops below 6 in hg. If a barometric pressure switch is used, the signal from the switch to the module will retard the timing 3° to 6° if the vehicle is operating at elevations below 2400 ft (731 m). When the UIM is used in the MCU system, the additional module connector is connected to a knock sensor, which signals the UIM to retard the timing if the engine begins to detonate. When the basic timing is being checked, the additional three-wire connector on the UIM should be disconnected.

Duraspark III ignition systems have a five-wire ignition module. This ignition system is used with the Electronic Engine Control (EEC) system. The distributor advances are not required in this system, be-

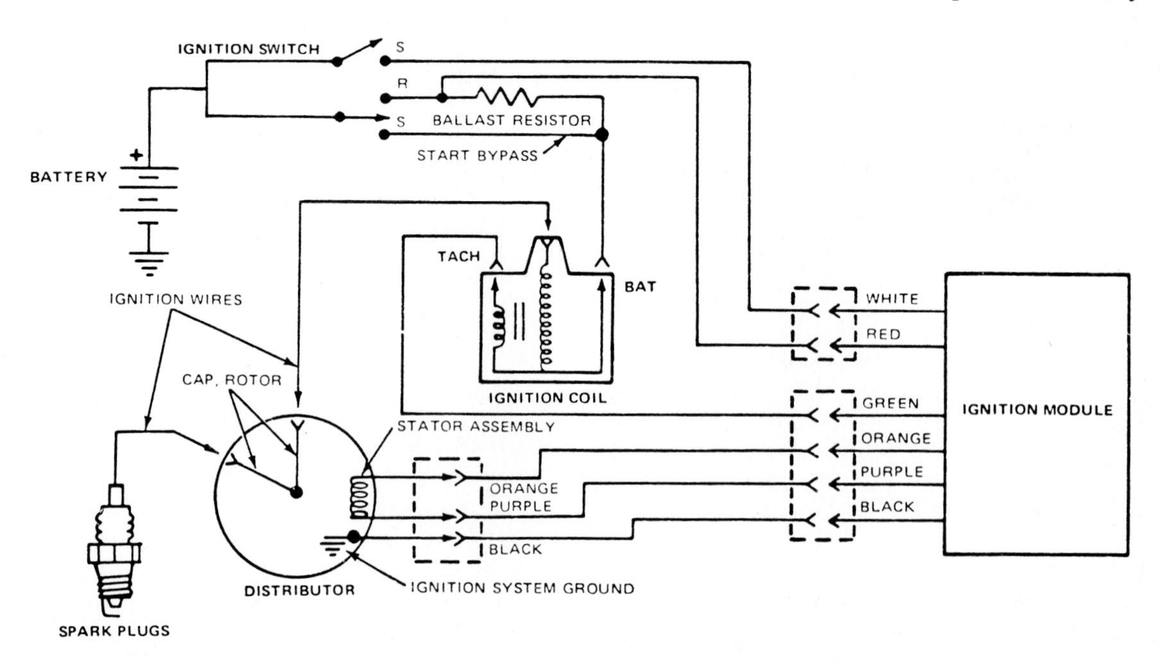

Figure 5–17 Duraspark II Ignition Wiring Diagram. (Courtesy of Ford Motor Co.)

cause the spark advance is determined by the microprocessor in the EEC system.

Many Duraspark III ignition systems have a pickup assembly located at the front on the crankshaft. A cam ring with four high points is located on the crankshaft pulley. When these high points rotate past the pickup, the pickup signal is sent through the microprocessor to the ignition module. The instant this signal is received, the module opens the primary ignition circuit. The microprocessor can vary the pickup signal to provide the precise spark advance required by the engine. This type of pickup is a crankshaft position (CP) sensor. The basic timing cannot be adjusted on a Duraspark III ignition system with a CP sensor, because the sensor cannot be rotated. A Duraspark III system with a CP sensor is shown in Figure 5–18, and a Duraspark III ignition wiring diagram is illustrated in Figure 5–19.

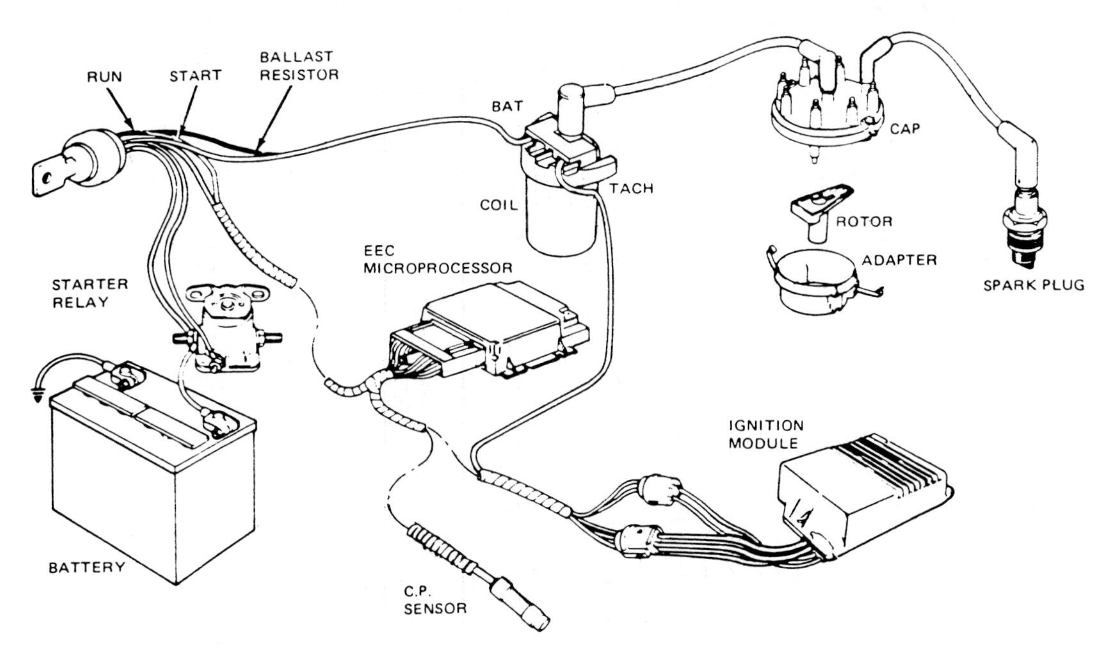

Figure 5–18 Duraspark III Ignition System with Crankshaft Pickup Assembly. (Courtesy of Ford Motor Co.)

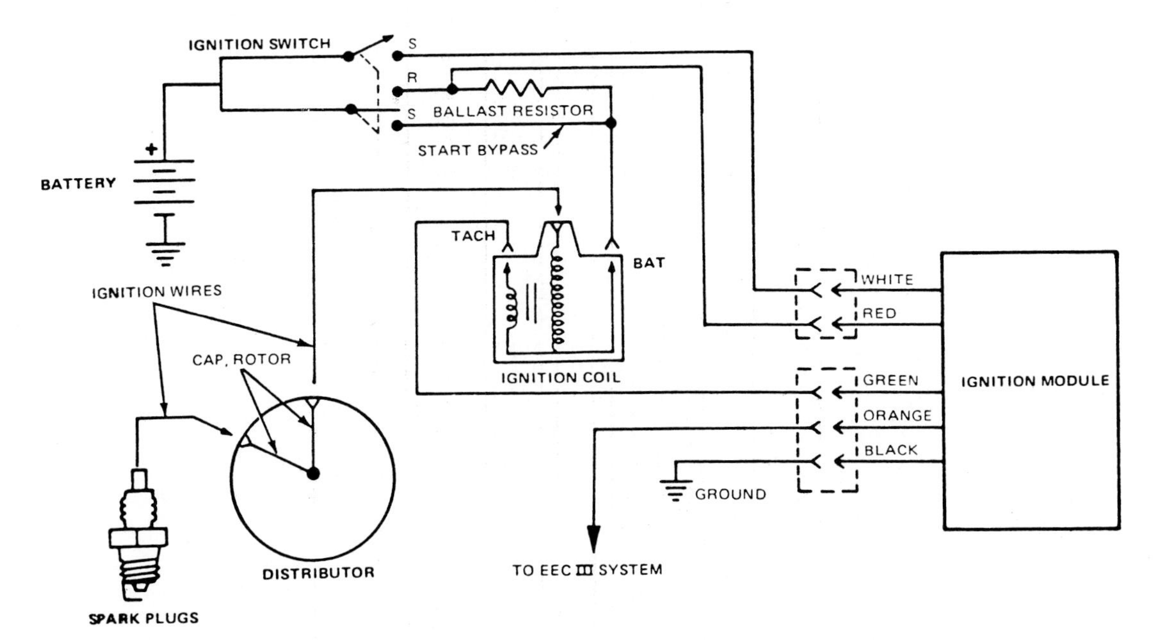

Figure 5–19 Duraspark III Ignition System Wiring Diagram. (Courtesy of Ford Motor Co.)

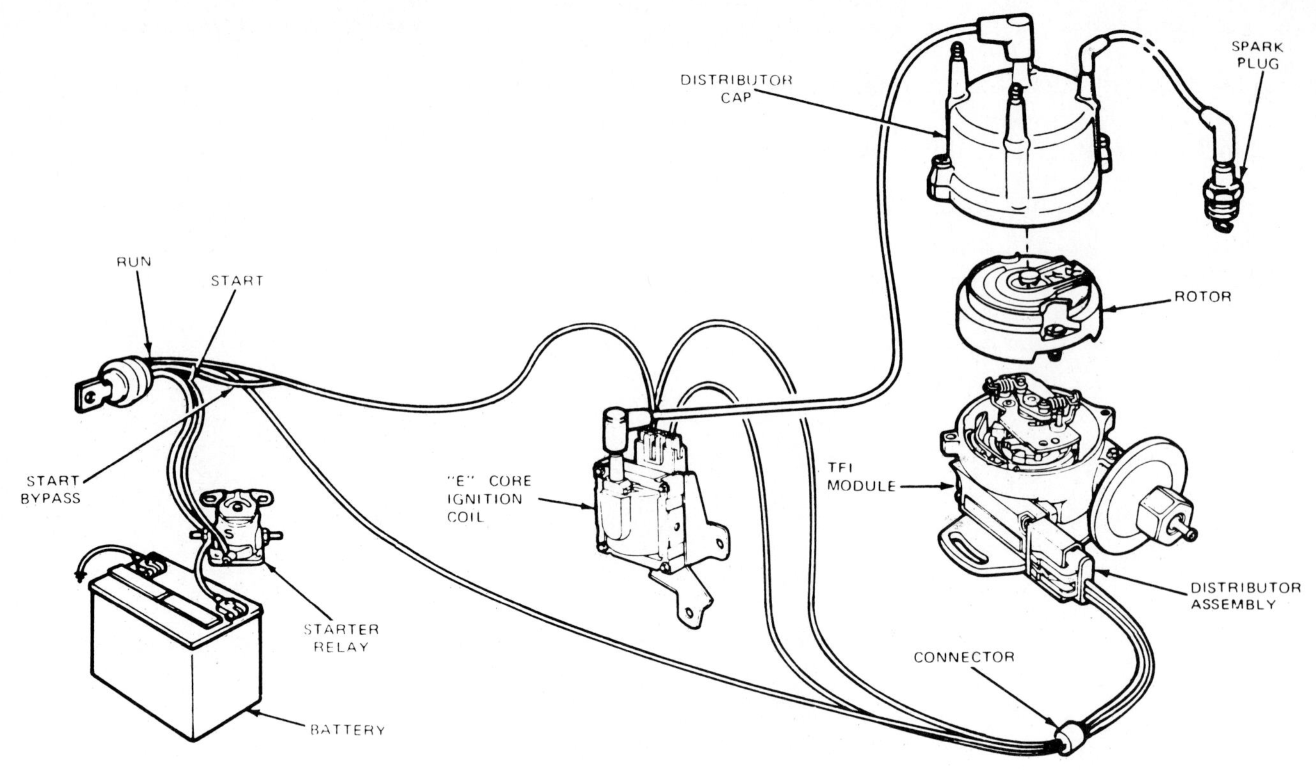

Figure 5-20 TFI Ignition System. [Courtesy of Ford Motor Co.]

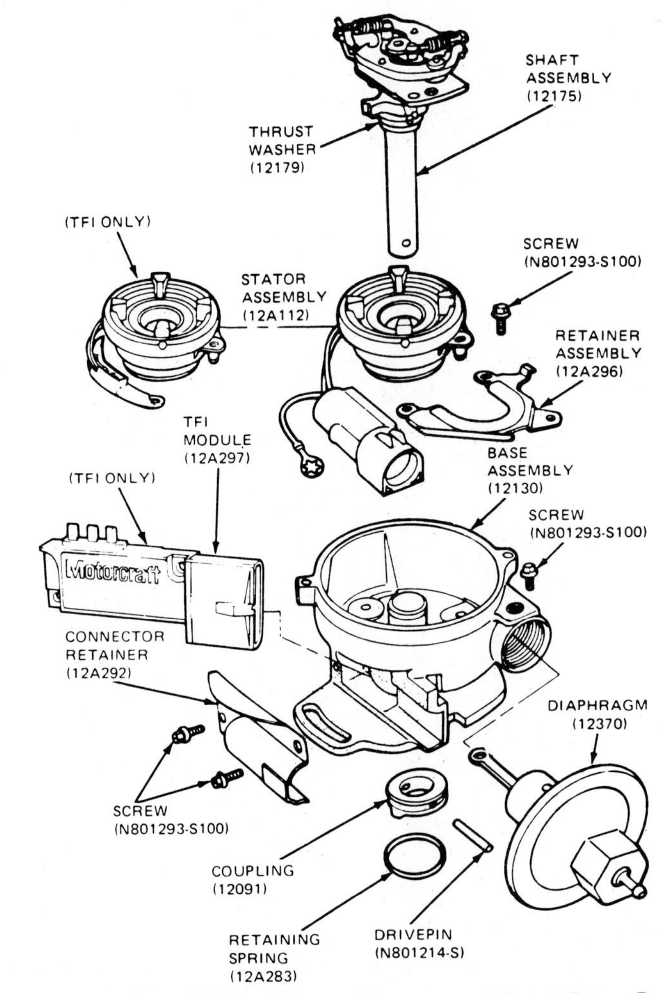

Figure 5-21 TFI Distributor. [Courtesy of Ford Motor Co.]

Thick Film Integrated Ignition Systems

The main differences in thick film integrated (TFI) systems are listed below.

1. A TFI module is attached to the distributor housing. Module circuitry increases dwell time as engine speed increases.
2. TFI ignition coil windings are set in epoxy. An iron frame surrounds the coil windings.
3. The primary circuit-resistance wire normally connected between the ignition switch and the coil primary winding is eliminated on TFI systems.
4. Pickup coil leads are connected to the TFI module inside the distributor housing. A TFI ignition system is shown in Figure 5–20, and a TFI distributor is pictured in Figure 5–21.

Thick Film Integrated Ignition Systems Used with Electronic Engine Control IV Systems

Design and Operation

TFI ignition systems are used with all Electronic Engine Control (EEC IV) systems. The distributor in the EEC IV TFI ignition system is similar to the distributor that is used with the conventional TFI ignition system. A Hall Effect device is used in place of the conventional pickup coil in the distributor. This device is referred to as a "profile ignition pickup" (PIP). The EEC IV TFI distributor is illustrated in Figure 5–22.

EEC IV TFI ignition systems use a six-wire TFI module. The primary ignition circuit is connected to the module in the same way as it is on conventional TFI ignition systems. However, the EEC IV TFI ignition system has three extra wires that are connected from the module to the electronic control assembly (ECA), as shown in Figure 5–23.

The electrical ground circuit between the ECA and the TFI module is provided by circuit number 16. A PIP signal is sent from the Hall Effect device to the ECA through circuit number 56. This signal informs the ECA when each piston is at 10° before top dead center (BTDC). The ECA provides a spark output (SPOUT) signal through circuit number 36 to the TFI module, which provides the precise spark advance that is required by the engine under all operating conditions. This eliminates the need for conventional advance mechanisms on the EEC IV TFI distributor.

Diagnosis of Ford Electronic Ignition Systems

Diagnosis of Duraspark II Ignition Systems

Duraspark II systems and earlier model seven-wire solid-state ignition systems may be diagnosed using the chart in Table 5–1.

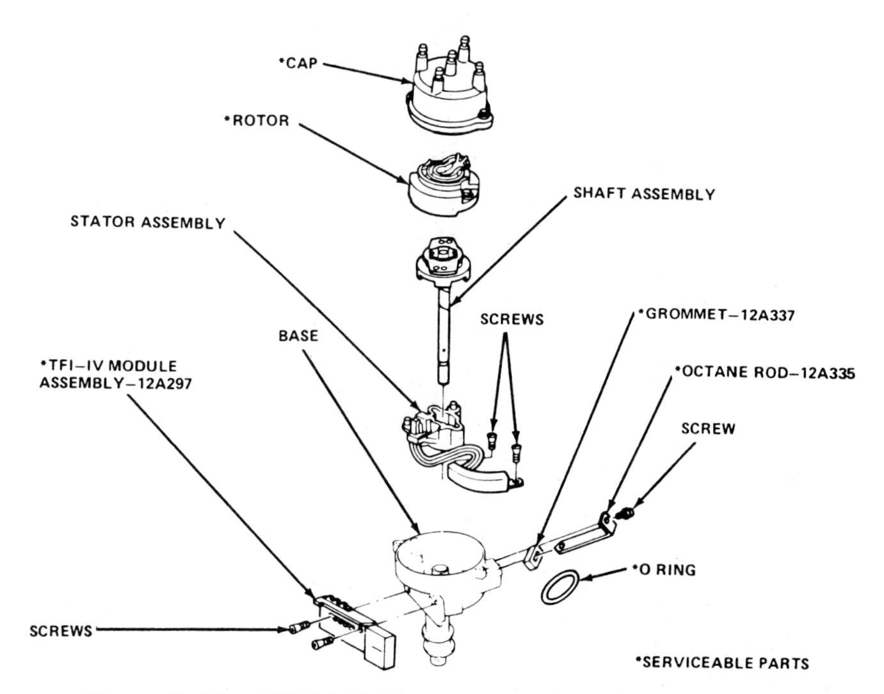

Figure 5–22 EEC IV TFI Distributor. (Courtesy of Ford Motor Co.)

TFI Diagnosis, Test-Lamp Method

Connect a 12V test lamp to the ignition coil negative terminal as pictured in Figure 5–24.

Failure of the test lamp to flash while cranking the engine indicates a defective pickup coil or module. A flashing test lamp while cranking the engine indicates a satisfactory pickup coil and module.

Connect the spark tester from the coil secondary wire to ground as indicated in Figure 5–25.

If the spark tester fails to fire while cranking the engine, the ignition coil is defective. Connect the spark tester from a spark plug wire to ground. A defective distributor cap, rotor, or plug wire is indicated if the spark tester does not fire, assuming it fired when connected to the secondary coil wire.

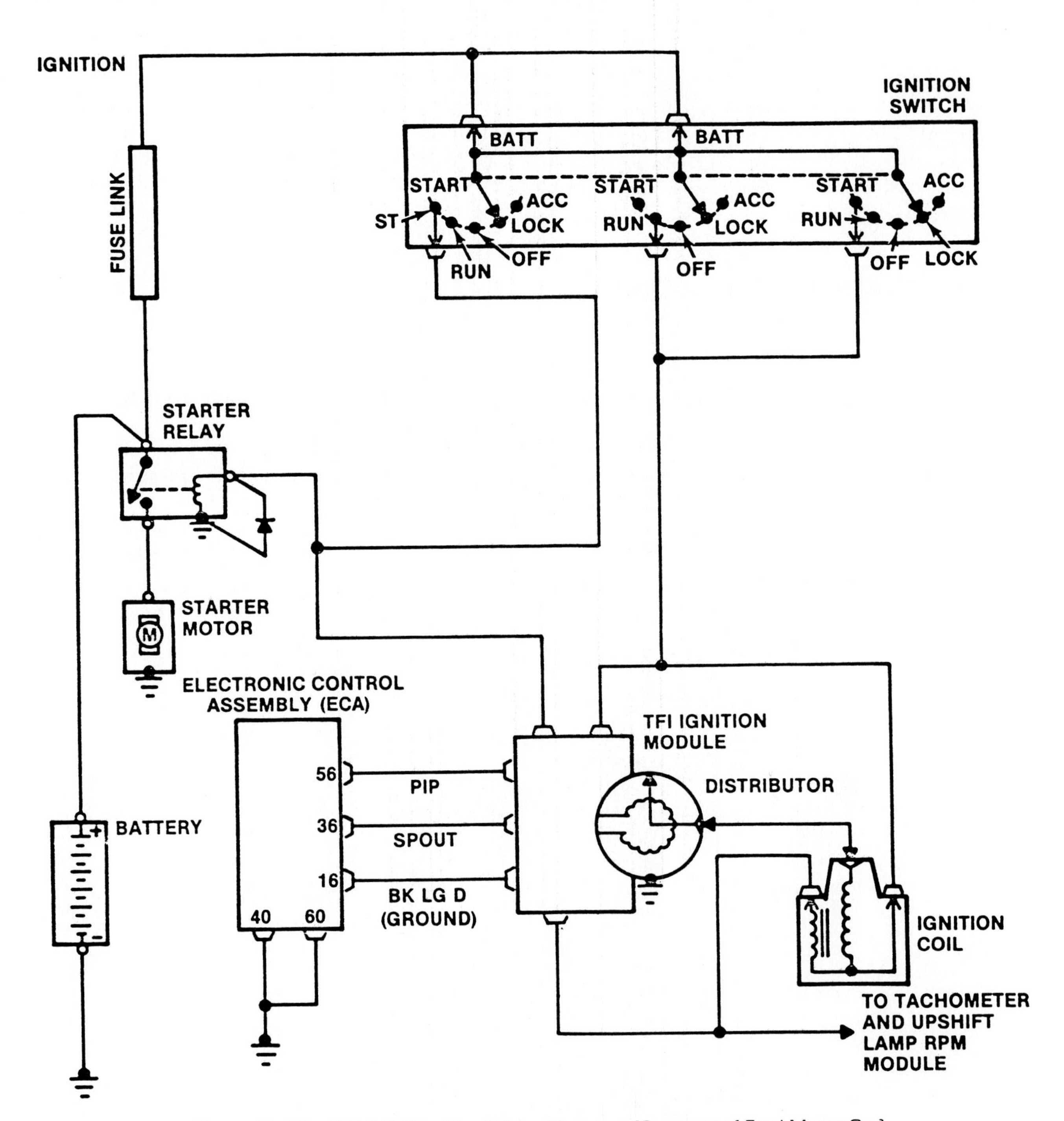

Figure 5–23 EEC IV TFI Ignition Wiring Diagram. (Courtesy of Ford Motor Co.)

TABLE 5–1 Diagnosis of Ford Duraspark II Ignition Systems

	Test Voltage Between	*Specifications*	*Testing*
Key on	Red wire and ground	over 11.5V	voltage to module
	Blue wire and ground omit 76 and up	over 11.5V	ignition wire
	Green wire and ground	over 11.5V	resistor wire and primary winding
Key on	Battery terminal of coil and ground	5–7V	ignition
	Ground tach terminal of coil when making above test		resistor wire
Key on cranking	Battery terminal of coil and ground	over 9V	bypass wire
	Ground tach terminal of coil when making above test		
	Test Resistance Between	*Specifications*	*Testing*
Key off	Purple and orange wire	400–800Ω	pickup coil open
	Purple or orange wire and ground	inf. × 100 scale	pickup coil grounds
	Black wire and ground	0Ω	ground wire
	Primary coil terminals	1–2Ω	primary winding
	Primary coil terminal to coil tower	7000–13000Ω	secondary winding

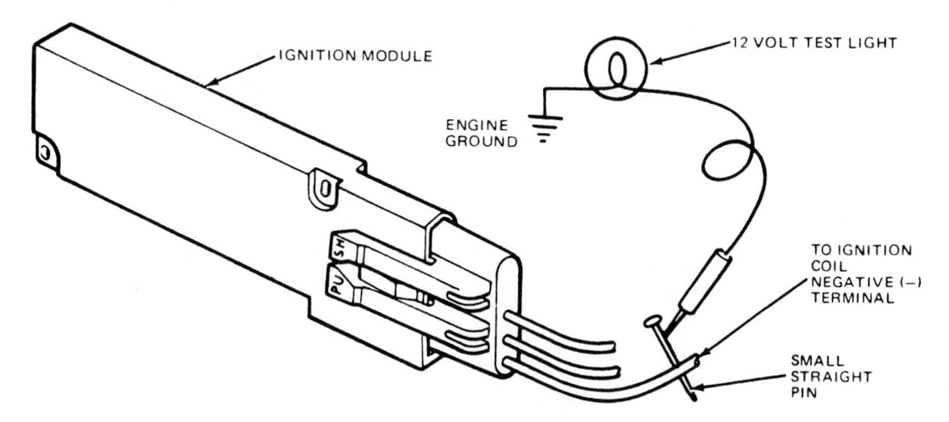

Figure 5–24 TFI Test-Lamp Diagnosis. [Courtesy of Ford Motor Co.]

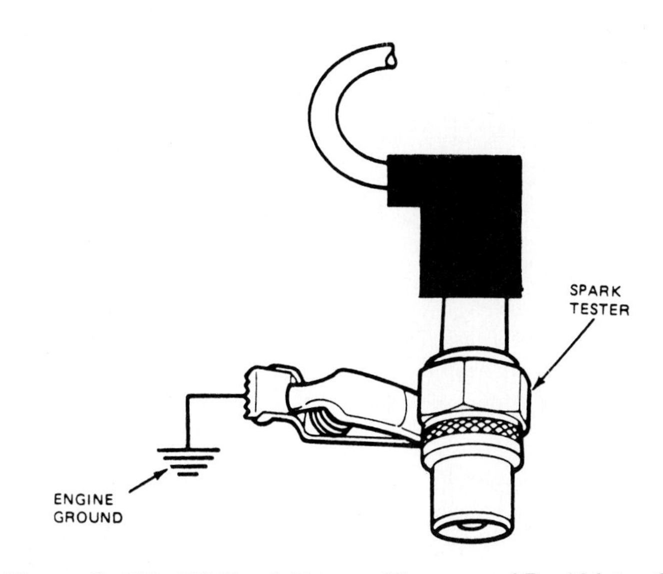

Figure 5–25 TFI Spark Tester. [Courtesy of Ford Motor Co.]

TFI Voltmeter Ohmmeter Diagnosis

Connect a voltmeter from the positive primary coil terminal to ground as shown in Figure 5–26.

With the ignition switch on, the voltmeter reading should exceed 12V. A zero voltmeter reading indicates an open circuit between the ignition switch and the positive coil terminal.

Disconnect the module connector, and connect a voltmeter from terminals 1 and 2 to ground as pictured in Figure 5–27.

With the ignition switch on, the voltmeter should indicate over 12V at both terminals. A zero reading at terminal 1 indicates an open circuit in the coil primary winding, or in the wire from the coil negative terminal to terminal 1. An open circuit exists in the wire from the coil positive terminal to terminal 2, if the voltmeter reading is zero at terminal 2.

With the ignition switch in the start position, connect a voltmeter from module connector terminal 3 to ground. The voltmeter should indicate over 9V while cranking the engine. A zero voltmeter reading indicates an open circuit in the ignition switch start terminal, or in the wire from the start terminal to module terminal 3.

Connect the ohmmeter leads across the primary coil terminals, as shown in Figure 5–28.

A resistance reading between .3 and 1Ω indicates a satisfactory primary winding. The primary winding is shorted if the resistance is below .3Ω. An ohmmeter reading above 1Ω indicates excessive primary resistance.

Connect the ohmmeter from the coil secondary coil terminal to one of the primary terminals as pictured in Figure 5–29.

A resistance reading below 9,000Ω indicates a shorted winding. Secondary resistance is excessive if the resistance reading exceeds 11,500Ω.

Connect the ohmmeter to the pickup coil leads, as illustrated in Figure 5–30.

A resistance below 800Ω indicates a shorted pickup coil. Pickup coil resistance is excessive if the ohmmeter reading exceeds 975Ω.

Diagnosis of EEC IV TFI Ignition

When checking the initial ignition timing, the spark output (SPOUT) circuit number 36 must be disconnected at the distributor connector. A separate con-

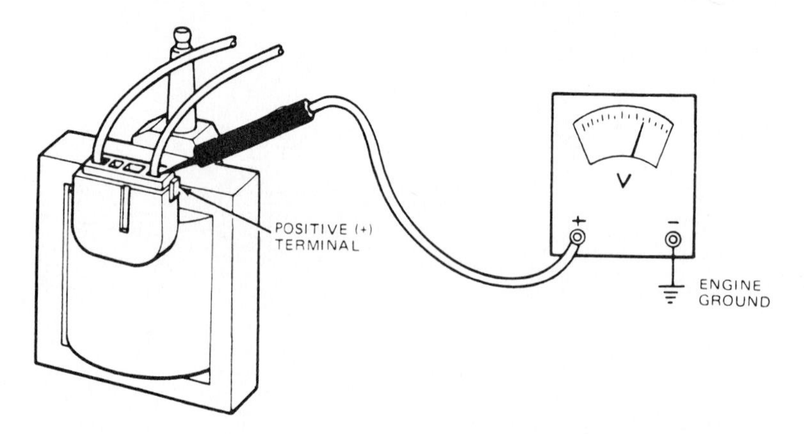

Figure 5–26 TFI Voltmeter Diagnosis. (Courtesy of Ford Motor Co.)

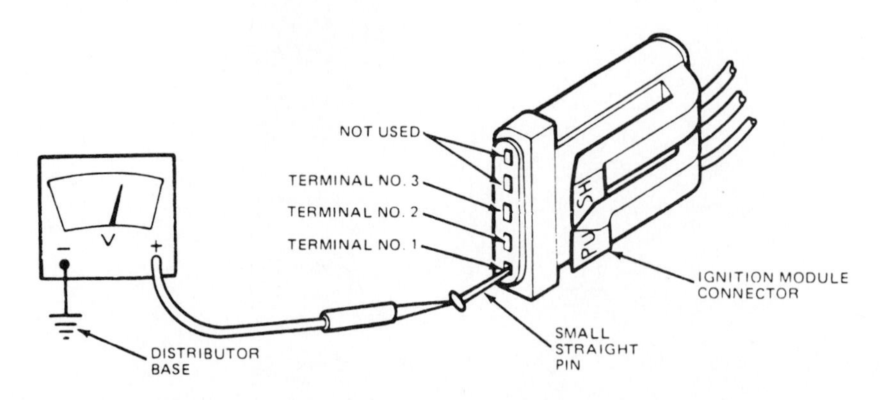

Figure 5–27 TFI Voltmeter Diagnosis. (Courtesy of Ford Motor Co.)

nector is provided for this wire. Refer to Figure 5–23 for circuit identification. The ignition timing is fixed by the TFI module in a limited operation strategy (LOS) mode when circuit number 36 is disconnected. The distributor may be rotated to adjust the initial timing. If the engine fails to start, the following diagnostic procedure should be followed.

1. Check for spark with a spark tester at the coil secondary wire and the spark plug wires while the engine is being cranked. If spark occurs at the coil secondary wire, but no spark is available at the spark plug wires, the distributor cap or rotor is defective.
2. If spark is not available at the coil secondary wire while the engine is being cranked disconnect the ECA connector. Connect a digital voltmeter across circuits 56 and 36 in the wiring harness connector. While the engine is being cranked, the voltmeter should indicate 3–6V.
3. If the voltmeter reading in step 2 is not correct, the signal from the TFI module or the profile ignition pickup (PIP) is defective.
   (a) Disconnect the distributor connector and connect an ohmmeter across circuit 56 at the ECA connector and the distributor connector. The ohmmeter should indicate zero ohms. Connect the ohmmeter across circuits 36 and 16 at the ECA connector and the distributor connector. Each wire should indicate zero ohms. If an infinite reading is obtained, that wire has an open circuit. With the distributor connector disconnected, connect the ohmmeter from circuits 56, 36, and 16 at the ECA connector to ground. Each wire should provide an infinite reading. If a low ohmmeter reading is obtained, that wire is grounded.
   (b) Perform the voltmeter and ohmmeter tests at the coil and the module connector as outlined

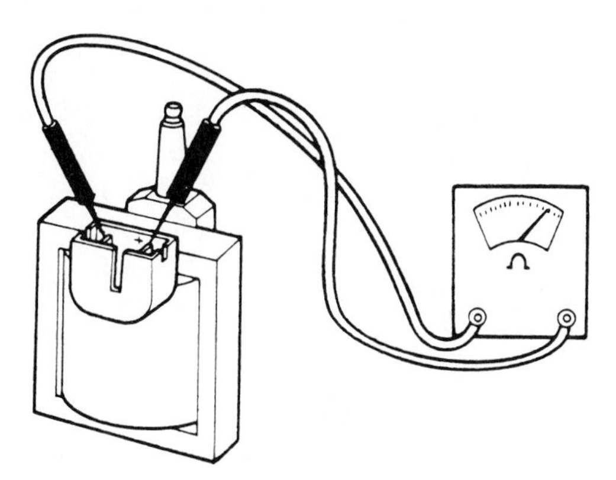

Figure 5–28 Testing TFI Coil Primary Winding. (Courtesy of Ford Motor Co.)

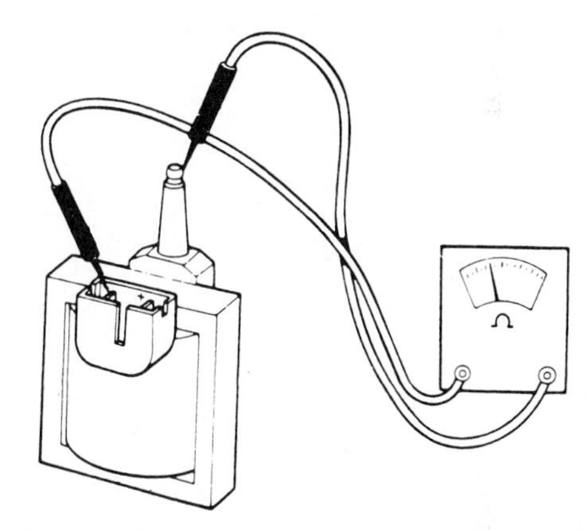

Figure 5–29 Testing TFI Coil Secondary Winding. (Courtesy of Ford Motor Co.)

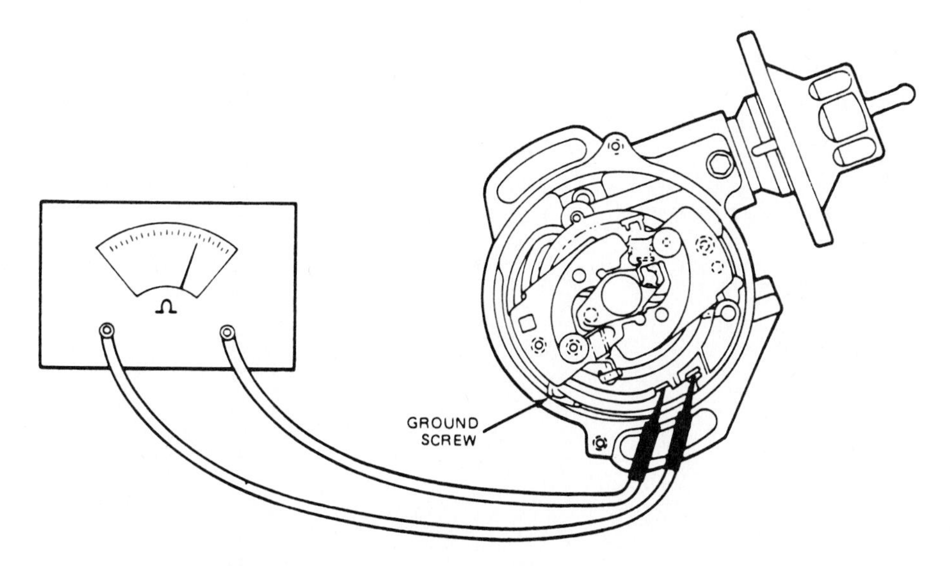

Figure 5–30 Testing TFI Pickup Coil. (Courtesy of Ford Motor Co.)

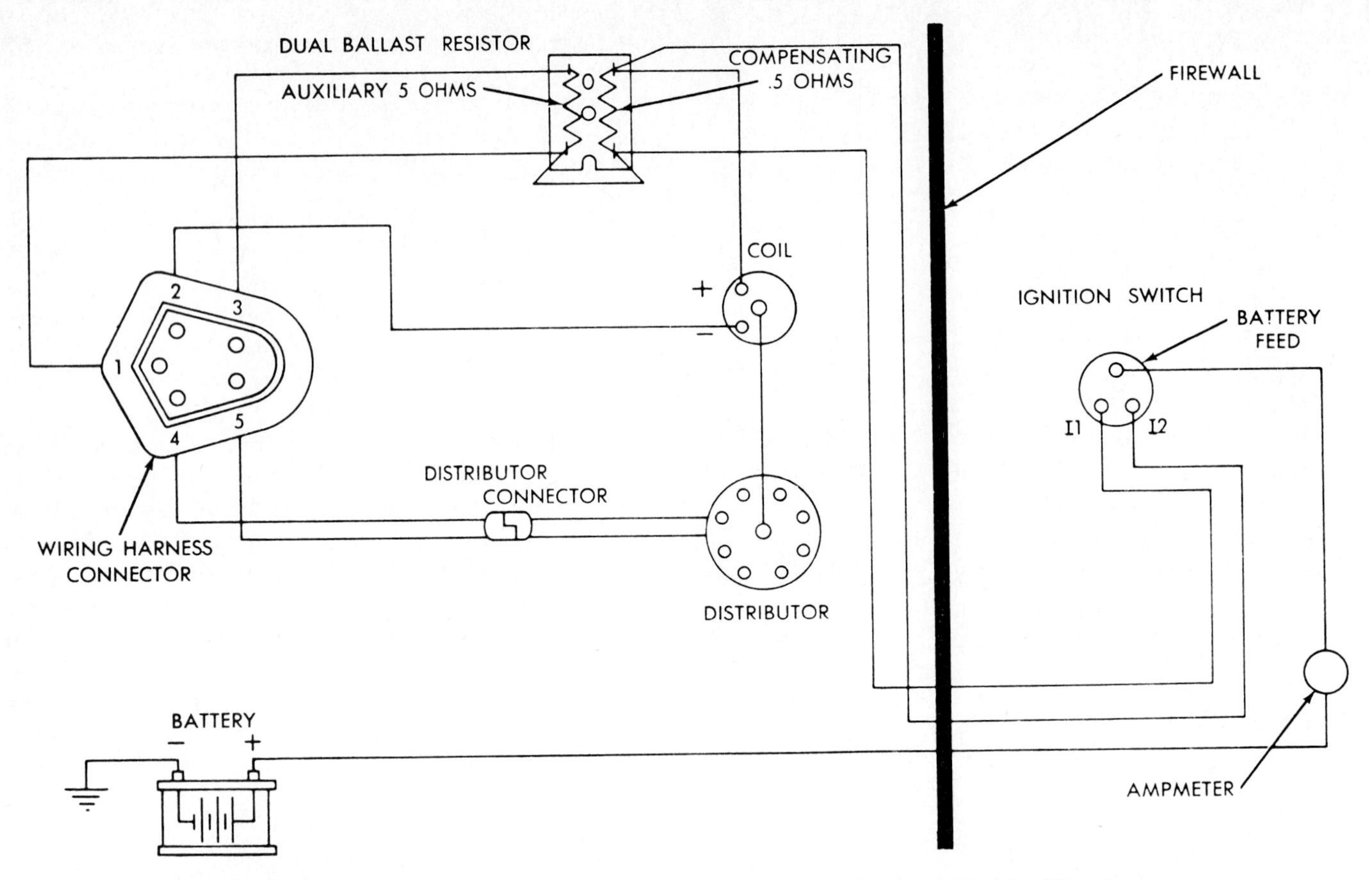

Figure 5-31 Chrysler Electronic Ignition System with Five-Terminal Module. (Courtesy of Chrysler Canada Ltd.)

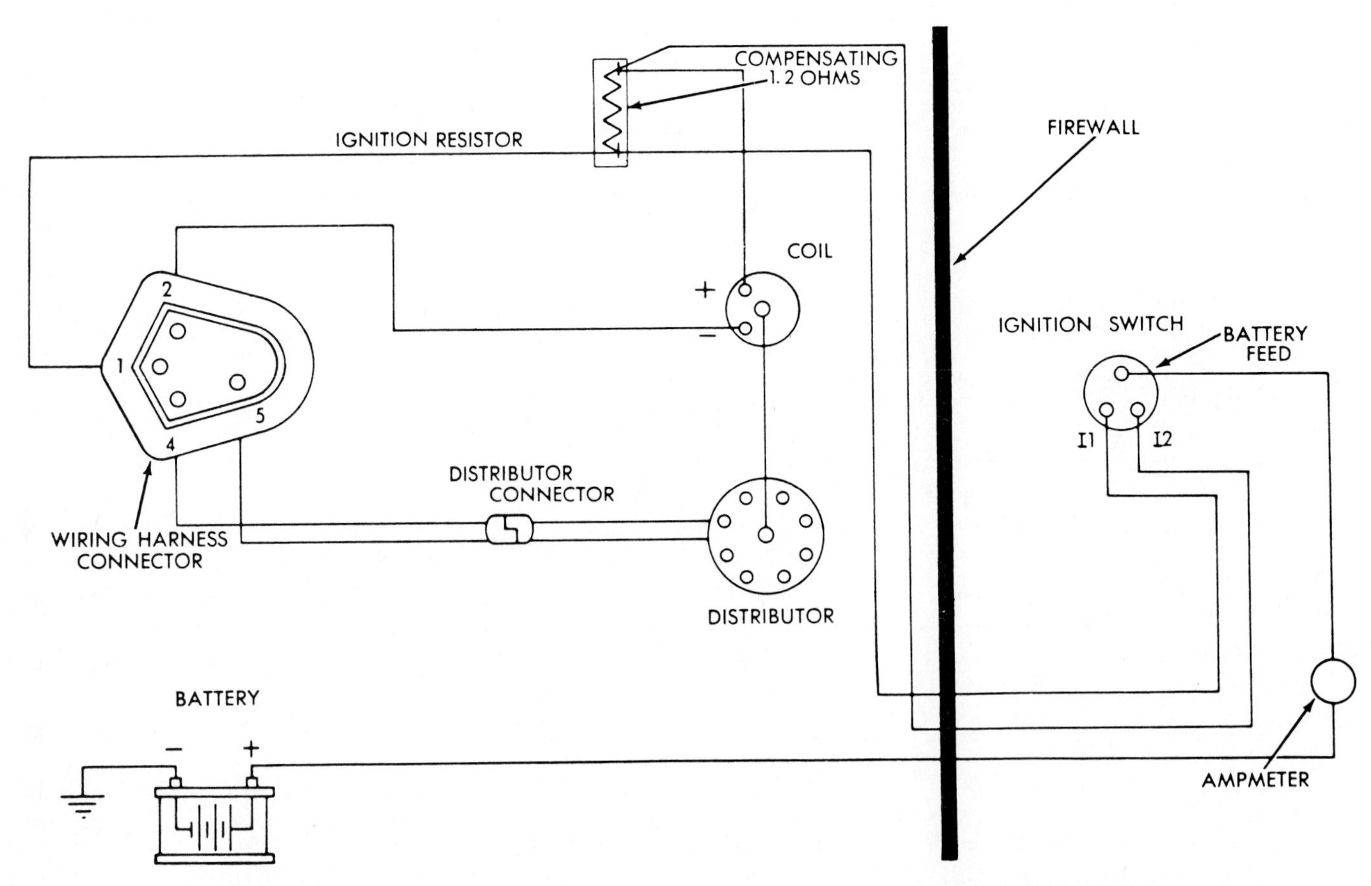

Figure 5-32 Chrysler Electronic Ignition System with Four-Terminal Module. (Courtesy of Chrysler Canada Ltd.)

previously in this chapter. The Hall Effect pickup cannot be tested with a voltmeter or ohmmeter.

(c) If no defects are located in steps 3(a) or 3(b), the module or Hall Effect pickup is defective. Replace the module and pickup individually and repeat step 2.

4. If the voltmeter reading in step 2 is normal but no spark was available in step 1, test or replace the ignition coil. Repeat step 1 with the ECA connector installed.
5. When spark does not occur at the spark tester in step 1, connect a voltmeter from ECA circuit 36 to ground with the ECA connector installed. While the engine is being cranked, the voltmeter should read 3–6V. The ECA is defective if the voltmeter reading is not within specifications.

Duraspark III Diagnosis

A Duraspark III ignition system may be diagnosed with a 12V test lamp and a test spark plug as explained previously on the TFI system. While the engine is being cranked, if a test lamp connected from the coil tach terminal to ground flutters but the test spark plug connected to the coil secondary wire fails to fire, the coil is defective. If the test lamp does not flutter while the engine is cranked, the following procedure may be used to locate the defective component.

1. Probe the module orange and yellow wire with a pin and connect an AC voltmeter from the orange and yellow wire to ground. Use the lowest scale on the AC voltmeter.
2. A pulsating voltmeter reading while the engine is being cranked indicates the crankshaft position (CP) sensor and the electronic control assembly (ECA) are satisfactory. When the AC voltage signal is available at the orange and yellow wire connected to the module, but the test lamp connected from the tach terminal to ground does not flutter while the engine is being cranked, the module is defective.
3. If there is no AC voltmeter signal at the module orange and yellow wire while the engine is being cranked, check the AC voltage signal from the orange and yellow wire at the ECA terminal to ground while the engine is being cranked. When an AC voltage signal is available at this ECA terminal but there is no voltage signal available at the module orange and yellow wire, the orange and yellow wire has an open circuit or a grounded condition.
4. If an AC voltage signal is not available from the orange and yellow wire at the ECA while the engine is being cranked, check the AC voltage signal at ECA terminals connected to the CP sensor. If a voltage signal is present at the CP sensor leads while the engine is being cranked, but there is no voltage signal from the orange and yellow wire at the ECA, the ECA is defective.
5. When the AC voltage signal is not available at the CP sensor leads at the ECA while the engine is being cranked, connect the AC voltmeter leads at the CP sensor. If an AC voltage signal is available at the CP sensor, but there is no AC voltage signal at the CP sensor leads at the ECA, the CP sensor leads from the sensor to the ECA have an open circuit or a grounded condition. When there is no AC voltage signal from the CP sensor while the engine is being cranked, the CP sensor is defective.

Chrysler Electronic Ignition Systems

Electronic Ignition Circuits

Chrysler-built vehicles with electronic ignition have four or five terminal modules, and these systems are equipped with conventional distributor advances. An electronic ignition system with a five-terminal module is shown in Figure 5–31, and the ignition circuit with the four-terminal module is illustrated in Figure 5–32.

Electronic Lean Burn and Electronic Spark Advance Ignition Systems

In the electronic lean burn (ELB) systems or electronic spark advance (ESA) systems, the spark advance is computer controlled, and the conventional distributor advances are discontinued. A centrifugal advance was used in some ELB distributors. In many ELB systems, two pickup coils were located in the distributor. The start pickup signal provided a signal to the computer while the engine was being cranked, and the run pickup signal was used once the engine started.

A diagram of an ELB system is provided in Figure 5–33.

In the ESA systems, single or dual pickup coils may be used in the distributors on V8 engines depending on the year of the vehicle. On four-cylinder ESA systems, a single Hall Effect pickup is used in the distributor. In early model ESA systems, the computer only controlled the spark advance, whereas in later systems the air–fuel ratio and other functions are computer controlled. An early model ESA system with a single pickup coil and a single ignition resistor is shown in Figure 5–34.

A later model ESA system with a computer-controlled oxygen (O_2) feedback carburetor system from a 2.2L four-cylinder engine is outlined in Figures 5–35 and 5–36.

The ESA system with the O_2 feedback carburetor, or electronic fuel injection (EFI), is used on many front-wheel drive Chrysler cars. (These systems are explained in Chapters 3 and 4.) An ignition resistor is not used in the primary circuit of these ESA systems. The Hall Effect switch in the distributor provides a signal to the computer each time the shutter blade on the rotor moves through the switch. A sync pickup is located in the distributor with the Hall Effect switch if this system is used on the multiport EFI system. The sync pickup signal is used for fuel-system control. A distributor with a Hall Effect switch is pictured in Figure 5–37.

A metal tab on the rotor grounds the shutter blade to the distributor shaft. If the ground tab does not contact the shaft, the engine will not start.

Diagnosis of Chrysler Electronic Ignition Systems

Diagnosis of Electronic Ignition Systems and ELB or ESA Systems with Conventional Pickup Coils

The test spark plug and 12V test-lamp procedure that was explained earlier in this chapter may be used to diagnose these systems if the engine fails to start. When a 12V test lamp is connected from the negative primary coil terminal to ground, and the lamp does not flutter while the engine is being cranked, the pickup coil or computer is defective. If the test lamp flutters while the engine is being cranked but the test spark plug will not fire when it is connected from the

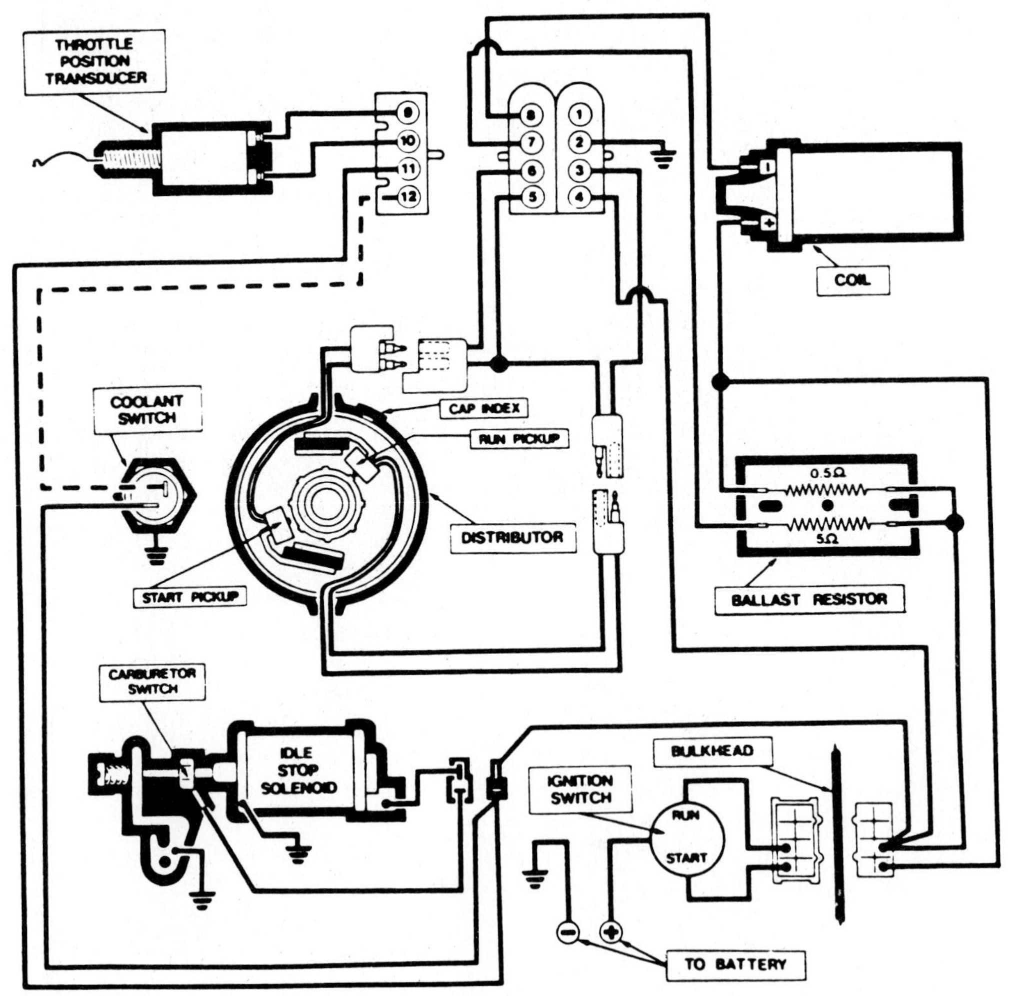

Figure 5–33 Chrysler Electronic Lean Burn System. (Courtesy of Chrysler Canada Ltd.)

coil high tension to ground, the coil is defective. The test spark plug being used must have the correct voltage requirement for Chrysler ignition systems.

Diagnosis of ESA System with Hall Effect Switch

If the engine fails to start, disconnect the distributor connector and momentarily connect a jumper wire between terminals 2 and 3 in the distributor connector that is connected to the computer. While the jumper wire is being connected, remove the coil secondary wire from the cap and hold it ¼ inch away from a good ground. If there is a good spark at the coil secondary wire each time the jumper wire is connected, but no spark occurs at the coil secondary wire while the engine is being cranked, the Hall Effect switch is defective. The jumper wire connection is illustrated in Figure 5–38.

If no spark occurs at the coil secondary wire when the jumper wire is connected, the coil, computer, or connecting wires are defective.

When the basic timing is being checked on an ESA system with an O_2 feedback carburetor, the carburetor switch must be grounded and the vacuum hose to the computer must be disconnected and plugged. (The carburetor switch is shown in Figure 5–36.)

Basic timing adjustment on ESA systems with EFI must be done with the computer in the limp-in mode. This is accomplished by disconnecting the coolant sensor wire at the thermostat housing until the power-loss lamp on the instrument panel is illuminated. Once this light is on, the sensor wire may be reconnected. After the timing is adjusted, the defective coolant sensor code should be cleared from the computer memory by disconnecting the red wire at the positive battery post with the engine not running.

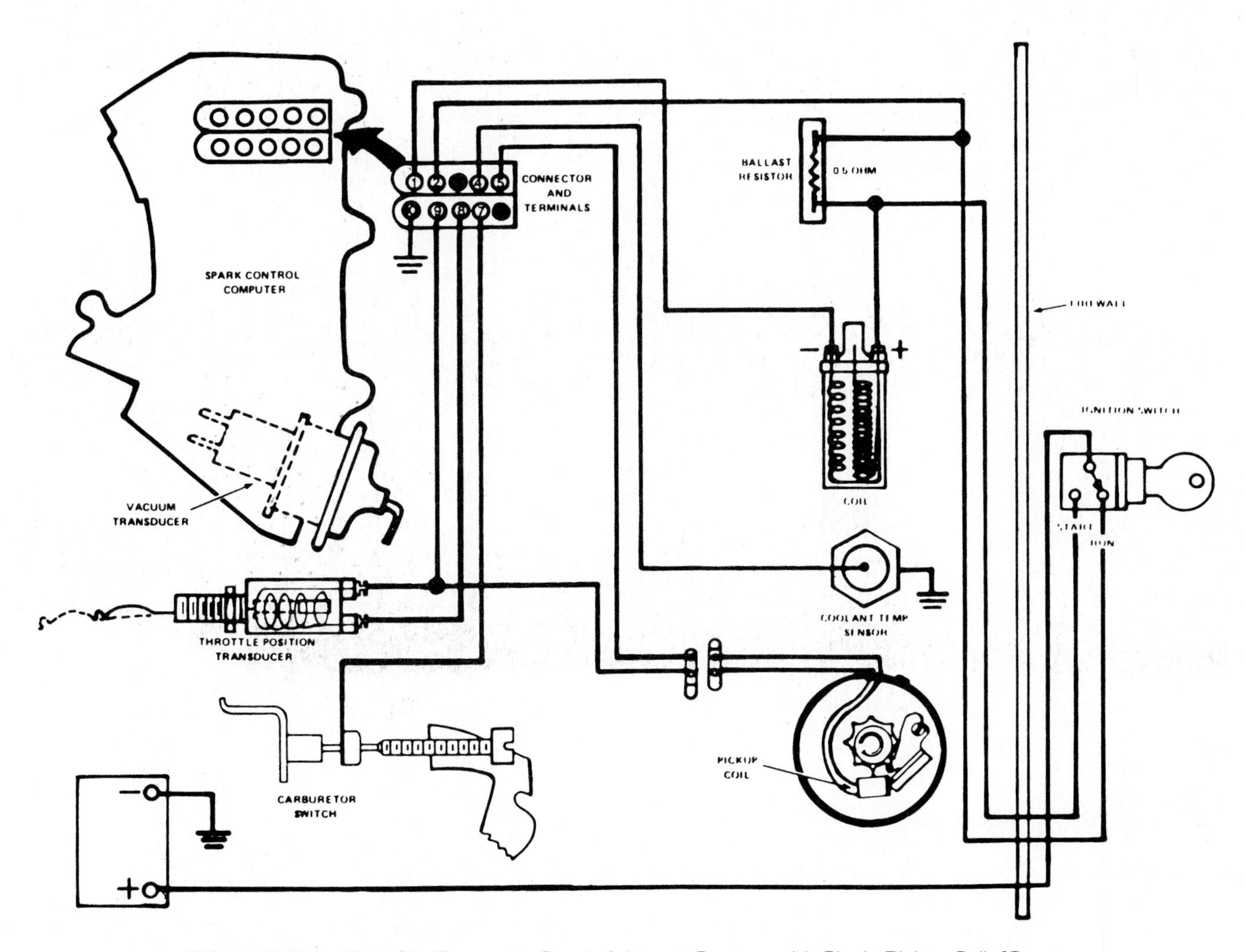

Figure 5–34 Chrysler Electronic Spark Advance System with Single Pickup Coil. (Courtesy of Sun Electric Corporation)

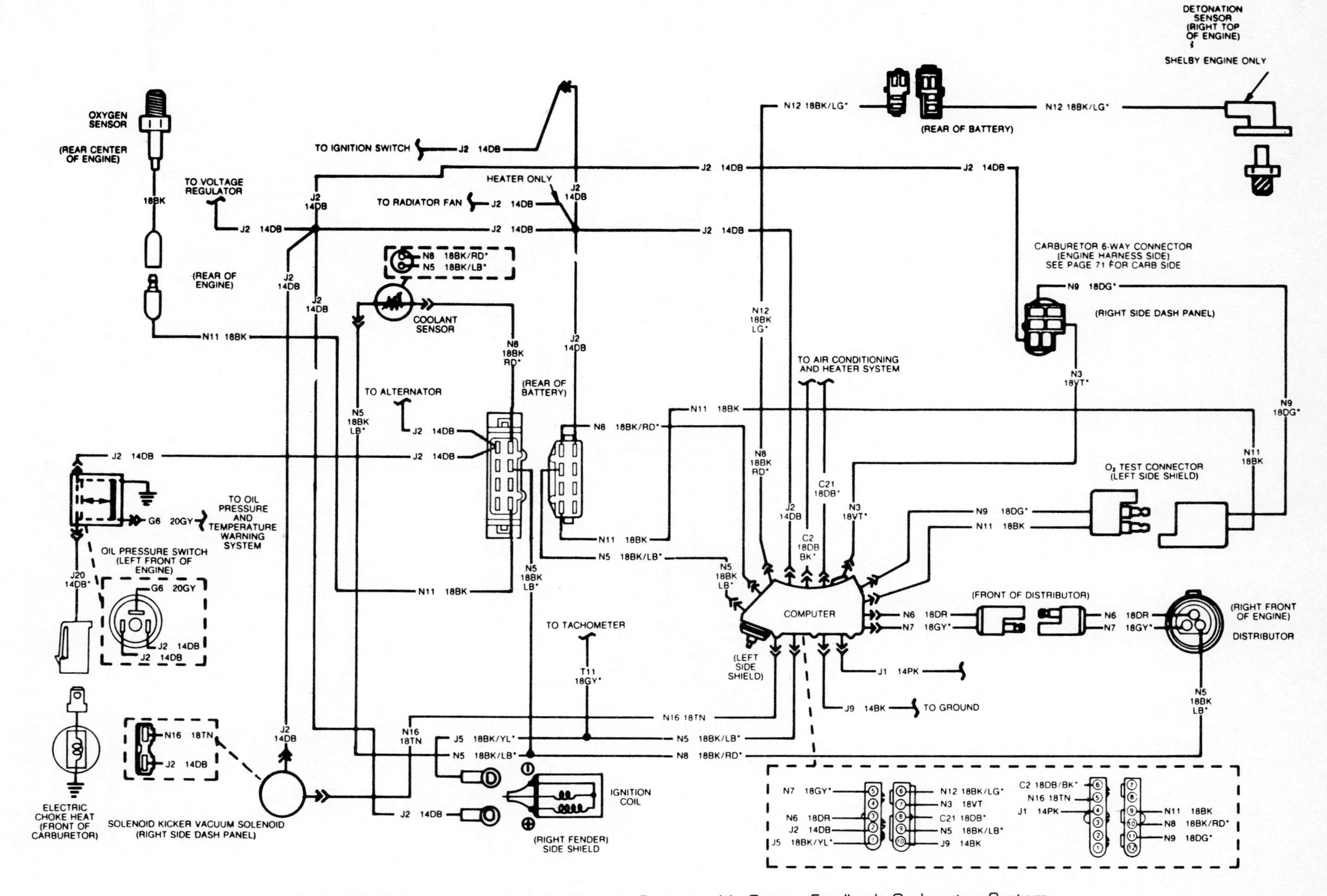

Figure 5–35 Electronic Spark Advance System with Oxygen-Feedback Carburetor System. [Courtesy of Chrysler Canada Ltd.]

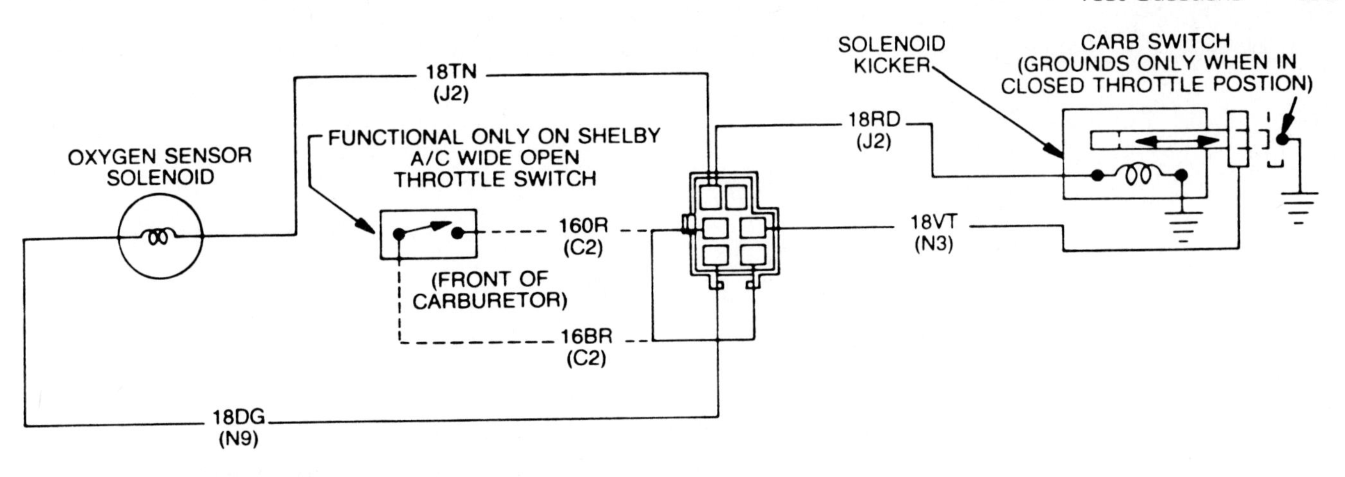

Figure 5-36 Carburetor Wiring Diagram for Oxygen-Feedback Carburetor System. [Courtesy of Chrysler Canada Ltd.]

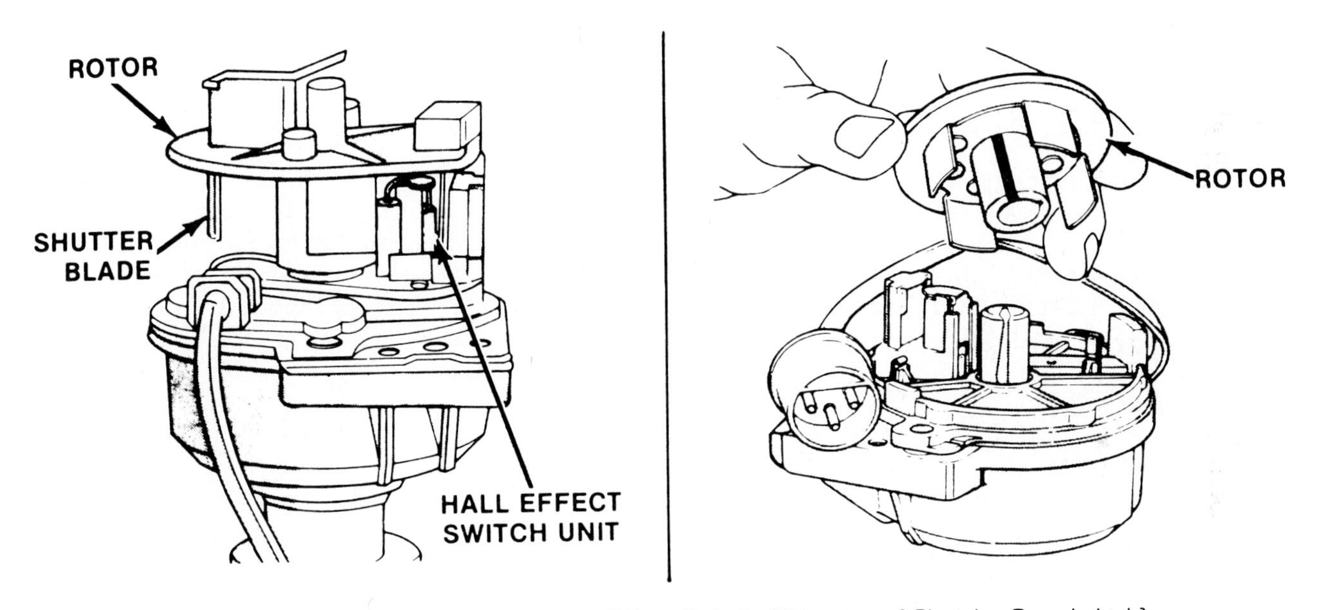

Figure 5-37 Distributor with Hall Effect Switch. [Courtesy of Chrysler Canada Ltd.]

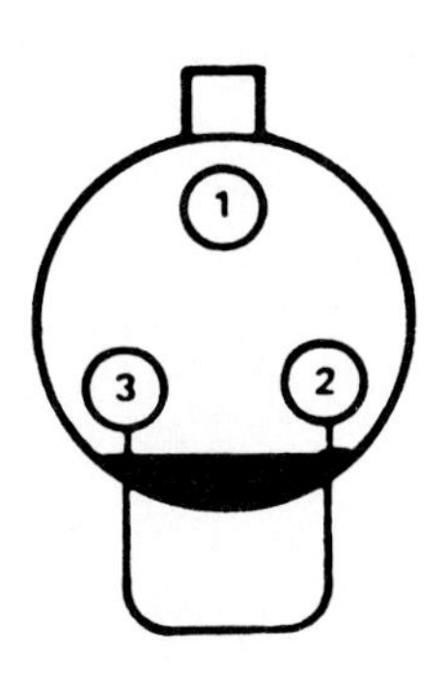

Figure 5-38 Testing ESA System with Hall Effect Switch. [Courtesy of Chrysler Canada Ltd.]

Test Questions

1. In an HEI system the dwell remains constant regardless of engine rpm. T F
2. In a 3C HEI system the pickup coil signal is sent through the ECM to the ignition module when the engine is running. T F
3. When an HEI distributor has a Hall Effect switch and a conventional pickup coil, the signal from the Hall Effect switch is used:
   (a) when the engine is being cranked.
   (b) when the engine is running.
   (c) only when the engine is running above 1200 rpm.

4. When a vacuum switch is connected to the additional three-wire connector on a Ford universal ignition module (UIM), the vacuum switch signal causes the module to:
   (a) retard the timing if the vacuum is below 6 in hg.
   (b) retard the timing if the vacuum is above 15 in hg.
   (c) advance the timing if the vacuum is below 6 in hg.
5. The basic timing is not adjustable on a Ford Duraspark III ignition system with a crankshaft position (CP) sensor. T F
6. If the shutter blade is not grounded to the distributor shaft in a Chrysler distributor with a Hall Effect switch, the engine will fail to start. T F
7. When an ohmmeter is connected across the pickup leads in an HEI distributor if an infinite reading is obtained, the pickup coil is:
   (a) grounded.
   (b) shorted.
   (c) open.
8. When a 12V test lamp is connected from the tach terminal to ground, if the lamp flutters while the engine is being cranked, the module is ______.
9. While the engine is being cranked, if a test spark fires when it is connected to the coil secondary wire, but fails to fire when connected to a spark plug wire, the defect could be in the:
   (a) ignition module.
   (b) distributor cap and rotor.
   (c) ignition coil.
10. When the basic timing is being checked on a Ford EEC IV TFI ignition system, the ______ wire in the distributor connector must be disconnected.
11. If a spark occurs at the coil secondary wire when terminals 2 and 3 are connected with a jumper wire in the distributor connector attached to the computer, but no spark occurs at the coil secondary wire while the engine is being cranked, the trouble could be a defective:
   (a) Hall Effect switch.
   (b) computer.
   (c) ignition coil.
12. When the basic timing is being checked on a Chrysler 2.2L engine with electronic fuel injection (EFI), the computer system must be placed in the ______ mode by disconnecting and reconnecting the ______ sensor.

6

Computer System with Body Computer Module

Practical Completion Objectives

1. Demonstrate an understanding of the body computer module (BCM) functions.
2. Perform the entry, exit, and erase procedures for BCM and electronic control module (ECM) system diagnosis.
3. Use the self-diagnostics to perform a complete diagnosis of ECM and BCM systems.

General Motors Computer Systems with Body Computer Module and Serial Data Line

System Design

Many 1986 and later model General Motors cars are equipped with a body computer module (BCM) and a serial data line that interconnects the ECM, BCM, instrument panel cluster (IPC), electronic climate control (ECC), voice/chime module, and heater ventilation and air conditioning (HVAC) programmer. The serial data line connecting these components is referred to as an 800 circuit, as indicated in Figure 6–1.

A permanent connector in the assembly line diagnostic link (ALDL) is part of the serial data line. The ALDL is similar to the ALCL in other systems. Since the serial data line is bidirectional, the system will still function if the permanent connector is removed from the ALDL. The serial data line may be compared to a party, or conference, telephone connection. When data is being transmitted between two components, the other components can listen to the data. For example, as the ECM sends data such as A/C clutch status, coolant temperature, and engine speed to the BCM, the other components have access to this information. The IPC uses the engine speed information to display engine revolutions per minute (rpm) for the driver.

Electronic Control Module Functions

The electronic control module (ECM) is similar to the ECM in other port fuel injection (PFI) systems. The input and output control functions are listed in Table 6-1, and the location of the ECM and the related inputs and outputs are shown in Figure 6–2.

The inputs and output control functions are similar to the ones described in Chapter 4, and the fuel system and fuel pump relay circuits are basically the same as those used in earlier PFI systems. Sequential energizing of the injectors is provided by the ECM. Since the electric cooling fan circuit is somewhat different from the systems described previously, this circuit is illustrated in Figure 6–3.

When a heavy-duty cooling system is ordered with the vehicle, an optional high-speed pusher fan is located in front of the radiator in addition to the conventional two-speed fan behind the radiator. The operation of the cooling fan circuit may be summarized as follows:

1. At 208°F (98°C) coolant temperature the ECM grounds the low fan relay winding. This action closes the relay contacts and supplies voltage through the relay contacts and the resistor to the low-speed fan motor. The resistor lowers the voltage at the motor, which results in low-speed fan operation.

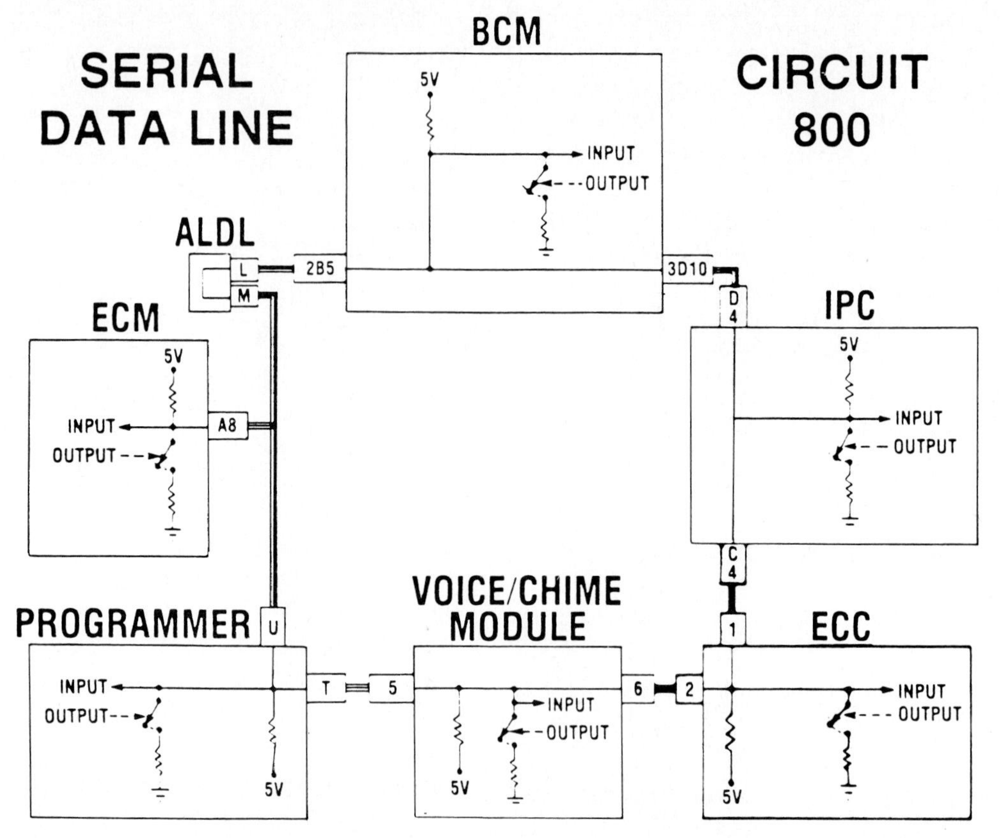

Figure 6–1 Serial Data Line and Interconnected Components. (Courtesy of General Motors Corporation)

TABLE 6–1 ECM Input and Output Control Functions.

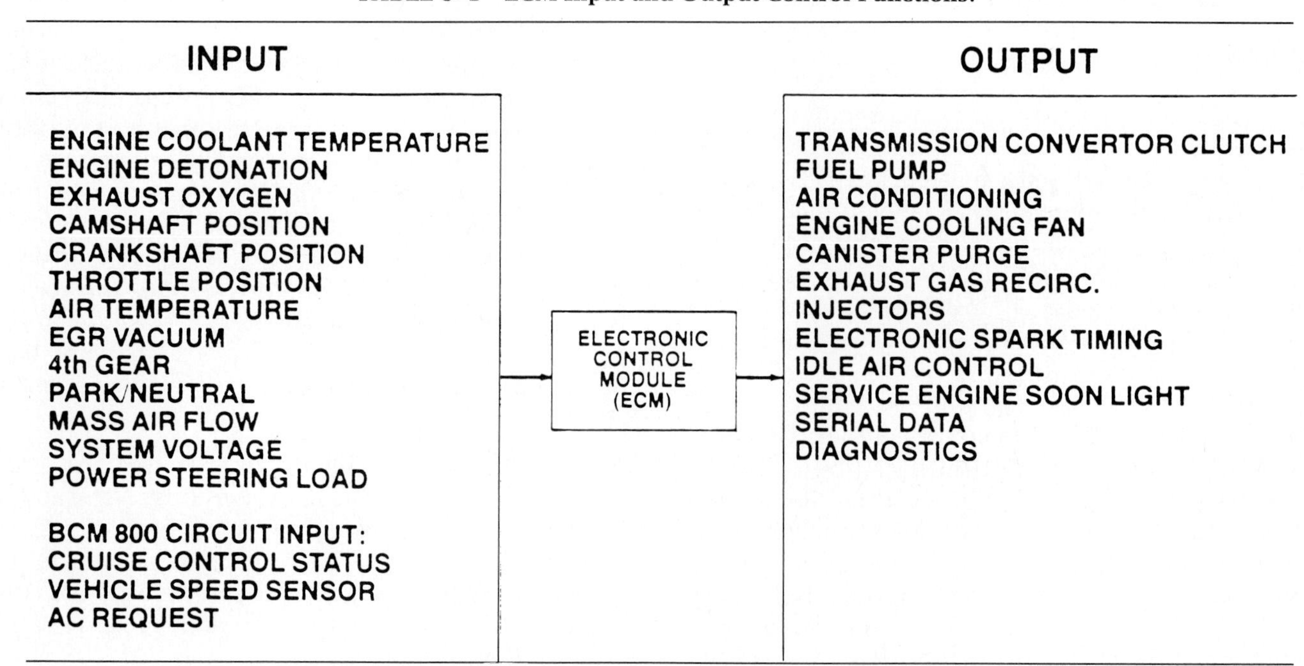

INPUT		OUTPUT
ENGINE COOLANT TEMPERATURE		TRANSMISSION CONVERTOR CLUTCH
ENGINE DETONATION		FUEL PUMP
EXHAUST OXYGEN		AIR CONDITIONING
CAMSHAFT POSITION		ENGINE COOLING FAN
CRANKSHAFT POSITION		CANISTER PURGE
THROTTLE POSITION		EXHAUST GAS RECIRC.
AIR TEMPERATURE		INJECTORS
EGR VACUUM	ELECTRONIC CONTROL MODULE (ECM)	ELECTRONIC SPARK TIMING
4th GEAR	→	IDLE AIR CONTROL
PARK/NEUTRAL		SERVICE ENGINE SOON LIGHT
MASS AIR FLOW		SERIAL DATA
SYSTEM VOLTAGE		DIAGNOSTICS
POWER STEERING LOAD		
BCM 800 CIRCUIT INPUT:		
CRUISE CONTROL STATUS		
VEHICLE SPEED SENSOR		
AC REQUEST		

(Courtesy of General Motors Corporation)

2. Low-speed fan operation also occurs when the low pressure contacts close in the A/C head pressure switch, before the ECM grounds the low-speed fan relay winding. These low-pressure contacts close when the A/C refrigerant pressure reaches 260 psi (1793 kPa).
3. The coolant temperature override switch closes and grounds both high-speed fan relays when the coolant temperature reaches 226°F (180°C). When this occurs, both relay contacts close, and the standard fan relay supplies voltage directly to the low-speed fan while the pusher fan relay supplies voltage to the pusher fan. This action results in high-speed operation of both fans.
4. When the A/C head pressure switch high-pressure contacts close at 300 psi (2069 kPa) refrigerant pressure, before the coolant temperature override switch closes, both high-speed fan relay windings are grounded through the high-pressure contacts, and high-speed operation of both fans occurs.

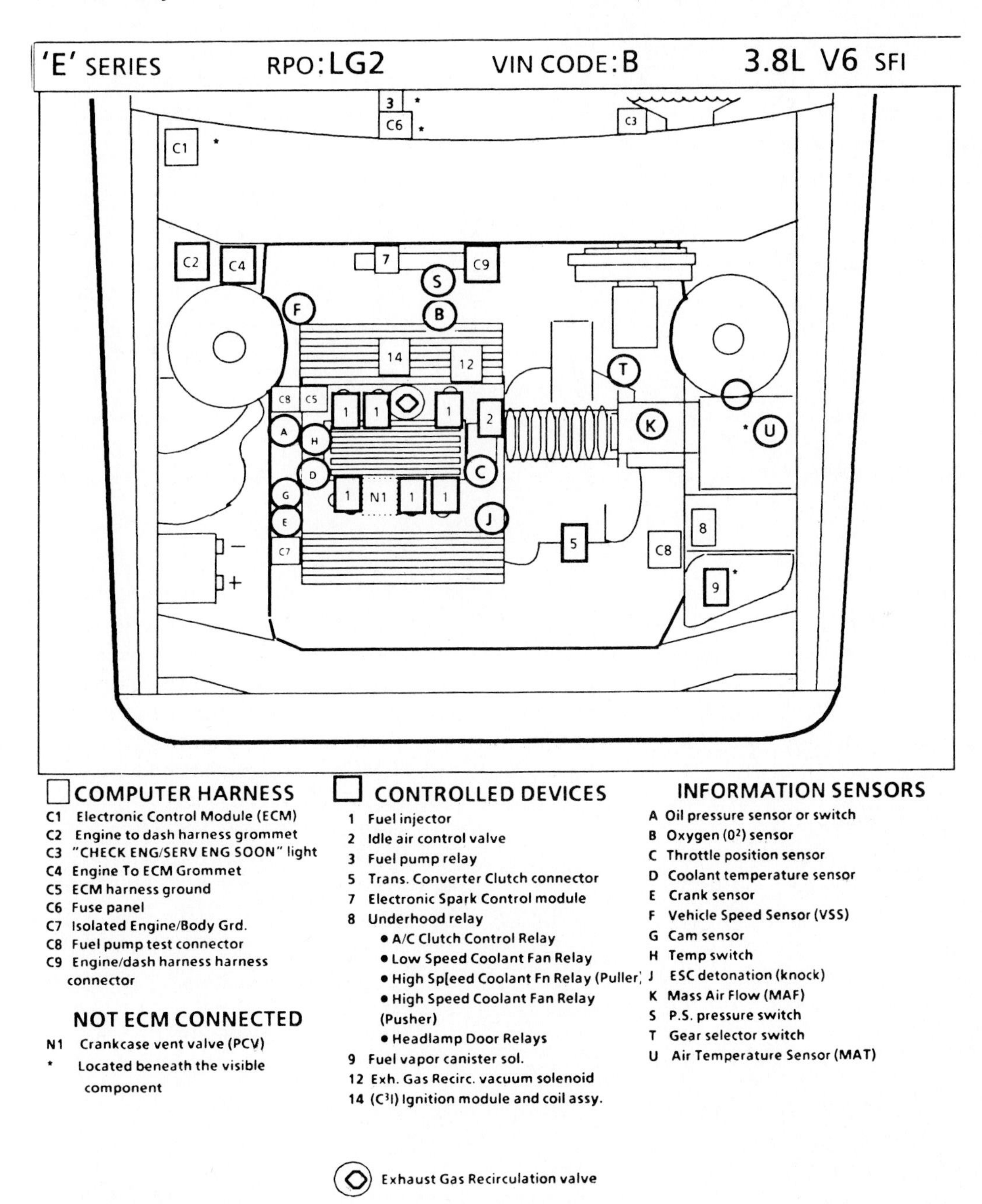

Figure 6–2 Location of ECM and Related Inputs and Outputs. (Courtesy of General Motors Corporation)

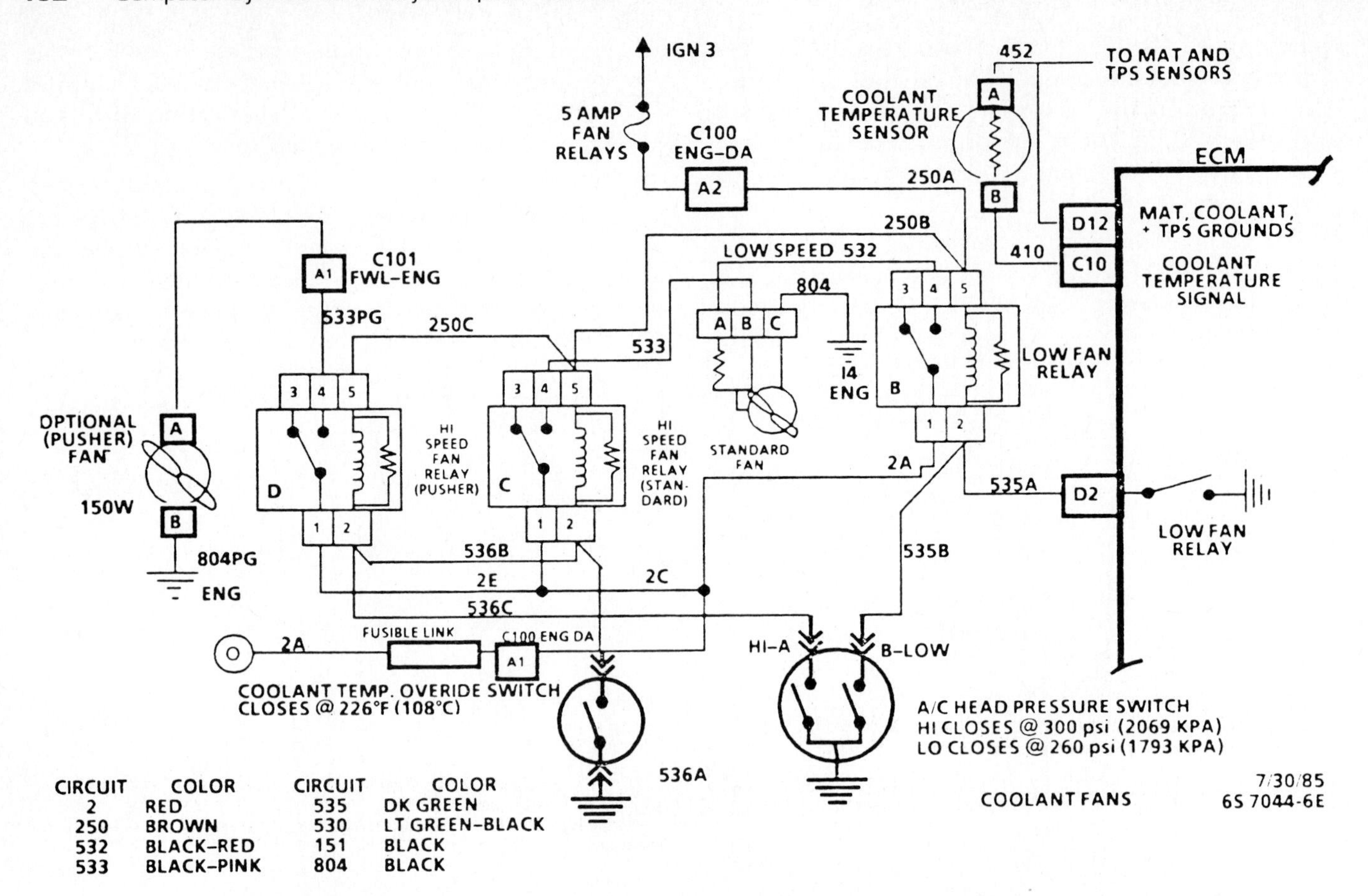

CIRCUIT	COLOR	CIRCUIT	COLOR
2	RED	535	DK GREEN
250	BROWN	530	LT GREEN-BLACK
532	BLACK-RED	151	BLACK
533	BLACK-PINK	804	BLACK

Figure 6–3 Electric Cooling Fan Circuit. (Courtesy of General Motors Corporation)

The distributor-less direct ignition system (DIS) used in these systems has a different camshaft position sensor and coil assembly than the computer-controlled coil ignition C³I systems described in Chapter 5. The ECM terminals and the voltage readings at each terminal are shown in Figure 6–4.

Central Power Supply

The central power supply (CPS) supplies 12V and 7V to the BCM, IPC, voice/chime module, and other electronic components regardless of input voltage variations. This 12V supply is available continuously from the battery through the CPS to the BCM, IPC, and voice module to maintain the computer memories and to operate the IPC. When the ignition switch is off, the 12V supply operates the voice/chime module. The 7V circuits are used for the BCM and IPC software. This 7V power supply is reduced to 5V by the BCM, and is then sent out as a reference voltage to the BCM input sensors. These 7V circuits function only when the CPS turns them on, which happens when the CPS receives a "wake-up" signal from the BCM. The ground circuits from the BCM, IPC, HVAC programmer, voice/chime module, and EEC panel are completed through the CPS. This ground circuit is connected directly to the negative battery terminal. Therefore, these ground circuits are not shared with any other components and electromagnetic interference (EMI) is reduced. The CPS output voltages and ground circuits are illustrated in Figure 6–5.

Normally the BCM will trigger the CP "wake-up" circuit when it receives an input signal from a door handle switch when a door is opened. On cars without illuminated entry, this input signal is supplied from the door jam switch. Alternate input signals that could also trigger the wake-up circuit are shown in Figure 6–6.

When the BCM receives a wake-up signal from a door handle switch, the BCM supplies 12V to the CPS. This action causes the CPS to activate the various 7V circuits to the BCM, IPC, and voice/chime module. The BCM and the CPS are usually located under the dash. The BCM and the CPS are pictured in Figure 6–7, and the location of the BCM, ECM, HVAC module, and voice module are shown in Figure 6–8.

This ECM voltage chart is for use with a digital voltmeter to further aid in diagnosis. The voltages you get may vary due to low battery charge or other reasons, but they should be very close.

THE FOLLOWING CONDITIONS MUST BE MET BEFORE TESTING:

- Engine at operating temperature
- Engine idling in closed loop (for "Engine Run" column)
- Test terminal not grounded
- ALCL tool not installed
- A/C Off

Note	KEY "ON" (VOLTAGE)	ENG. RUN (VOLTAGE)	OPEN CKT. (VOLTAGE)	CIRCUIT	PIN
④	.01	B+	0	FUEL PUMP RELAY	A1
⑤	.01	0	0	A/C CLUTCH CONTROL	A2
	B+	B+	0	CANISTER PURGE CONTROL	A3
	B+	B+	0	EGR SOLENOID	A4
	.79	B+	0	SERVICE ENGINE SOON LIGHT	A5
	B+	B+	B+	IGN #1 (ISO)	A6
	B+	B+	B+	TCC CONTROL	A7
	2.3 3.6	2.3 3.6	3.7	SERIAL DATA	A8
	5.0	4.9	5.0	DIAG. TERM	A9
①	B+	B+	B+	VSS SIGNAL	A10
	0	6.8	0	CAM HIGH	A11
	0	.02	0	GROUND	A12
				NOT USED	C1
				NOT USED	C2
	.86 11.8	12.3	3.8 10.5	IAC-B-LO	C3
	.86 11.8	.86	B+ .5	IAC-B-HI	C4
	.86 11.8	.86	B+ .5	IAC-A-HI	C5
	.86 11.8	12.3	.5 B+	IAC-A-LOW	C6
				NOT USED	C7
	0	0	B+	4TH GEAR SIGNAL	C8
				NOT USED	C9
②	2.24	1.75	5.0	COOLANT TEMP SIGNAL	C10
	2.2	1.75	5.0	AIR TEMP	C11
	B+	B+	0	INJECTOR 6	C12
	.45	.45	0	TPS SIGNAL	C13
	5.0	4.9	5.0	TPS 5V REF	C14
	B+	B+	0	INJECTOR 2	C15
	B+	B+	.5	B+	C16

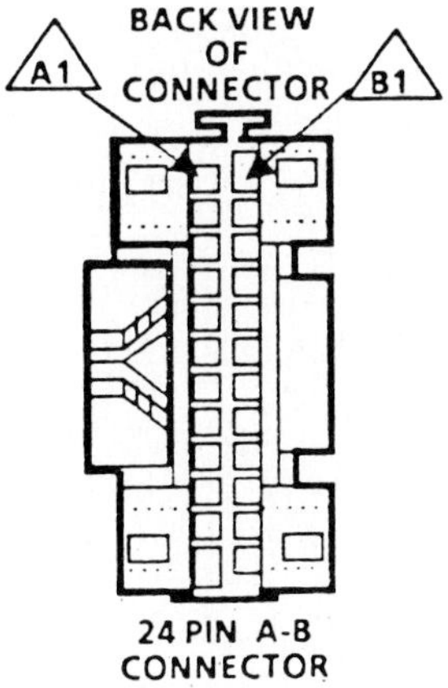

BACK VIEW OF CONNECTOR

A1 B1

24 PIN A-B CONNECTOR

WHEN TWO VALUES ARE GIVEN, THE VOLTAGE SIGNAL WILL CYCLE BETWEEN THE TWO VALUES

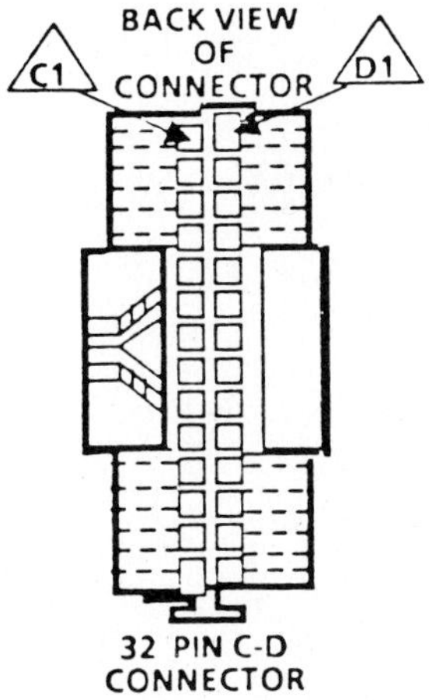

BACK VIEW OF CONNECTOR

C1 D1

32 PIN C-D CONNECTOR

PIN	CIRCUIT	KEY "ON" (VOLTAGE)	ENG RUN (VOLTAGE)	OPEN CKT. (VOLTAGE)	Note
B1	NOT USED				
B2	NOT USED				
B3	CRANK REF LOW	0	.0	3.0	
B4	EST CONTROL	.04	1.2	0	
B5	CRANK REF HI	0	4.0	0	
B6	MASS AIR FLOW SENSOR SIGNAL	2.0 3.5	2.6	5.0	③
B7	ESC SIGNAL	9.2	9.2	0	
B8	NOT USED				
B9	NOT USED				
B10	PARK/NEUTRAL SW. SIGNAL	.0	.0	B+	
B11	NOT USED				
B12	INJECTOR 5	B+	B+	0	
D1	GROUND	.0	.0	.0	
D2	COOLING FAN CONTROL	B+	1.57	.45	
D3	NOT USED				
D4	NOT USED				
D5	EST BYPASS SIGNAL	.0	3.75 4.85	.0	
D6	GRND (O_2) LOW	0	0	1.8	
D7	O_2 SENSOR SIGNAL	.42	.25 1.0	.42	③
D8	NOT USED				
D9	EVRV FBK SIGNAL	B+	B+	B+	
D10	GROUND	.0	.0	.0	
D11	POWER STEERING SW SIGNAL	B+	B+	0	
D12	MAT (AIR TEMP) & COOLANT TPS GROUND	.0	0	.0	
D13	NOT USED				
D14	INJECTOR 1	B+	B+	0	
D15	INJECTOR 3	B+	B+		
D16	INJECTOR 4	B+	B+	0	

① Varies from .60 to battery voltage depending on position of drive wheels.
② Normal operating temperature.
③ Varies
④ 12 V only for first 2 seconds unless engine is cranking or running.
⑤ 6.62 with A/C On (System will not reenergize if fuel rail pressure is high)

Figure 6–4 Electronic Control Module (ECM) Terminal Identification. (Courtesy of General Motors Corporation)

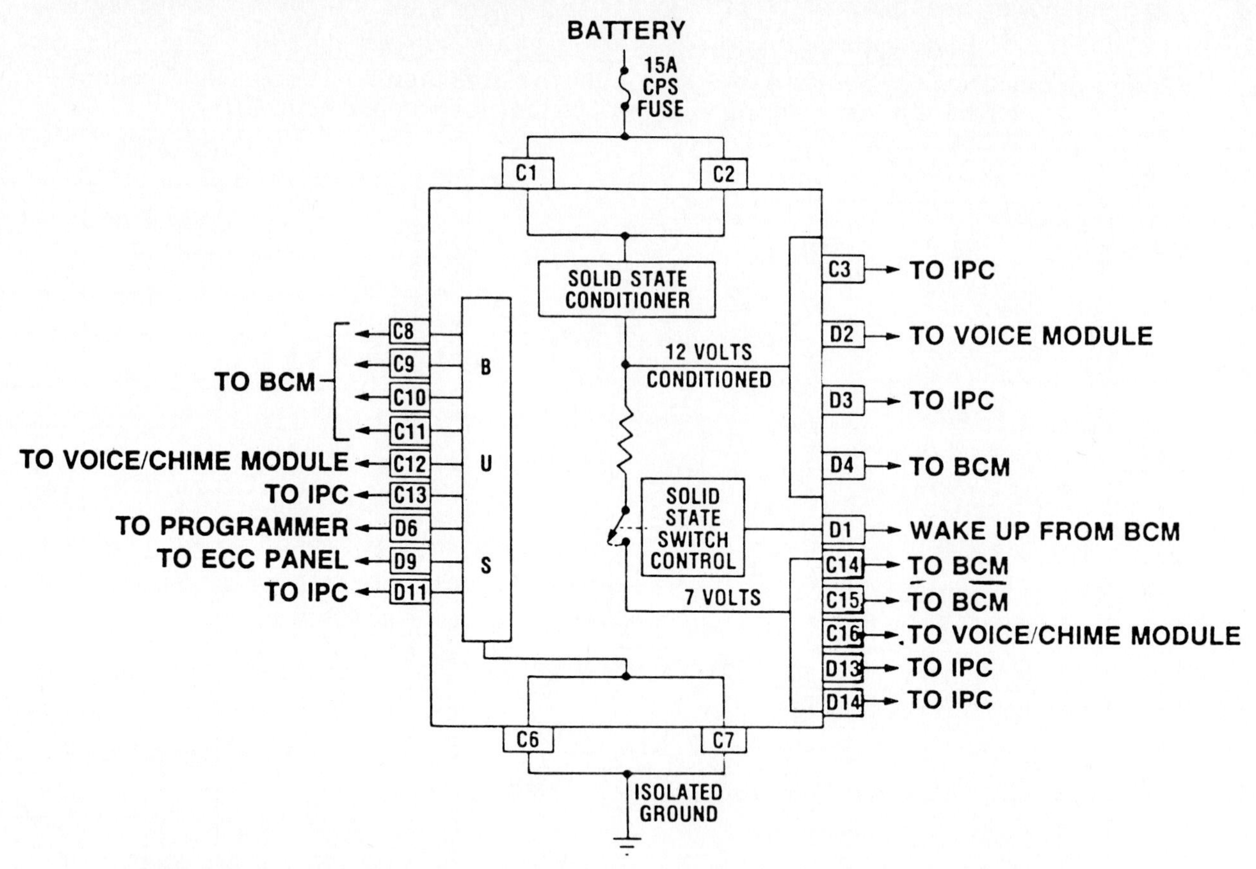

Figure 6-5 Central Power Supply Output Voltages and Ground Circuits. [Courtesy of General Motors Corporation]

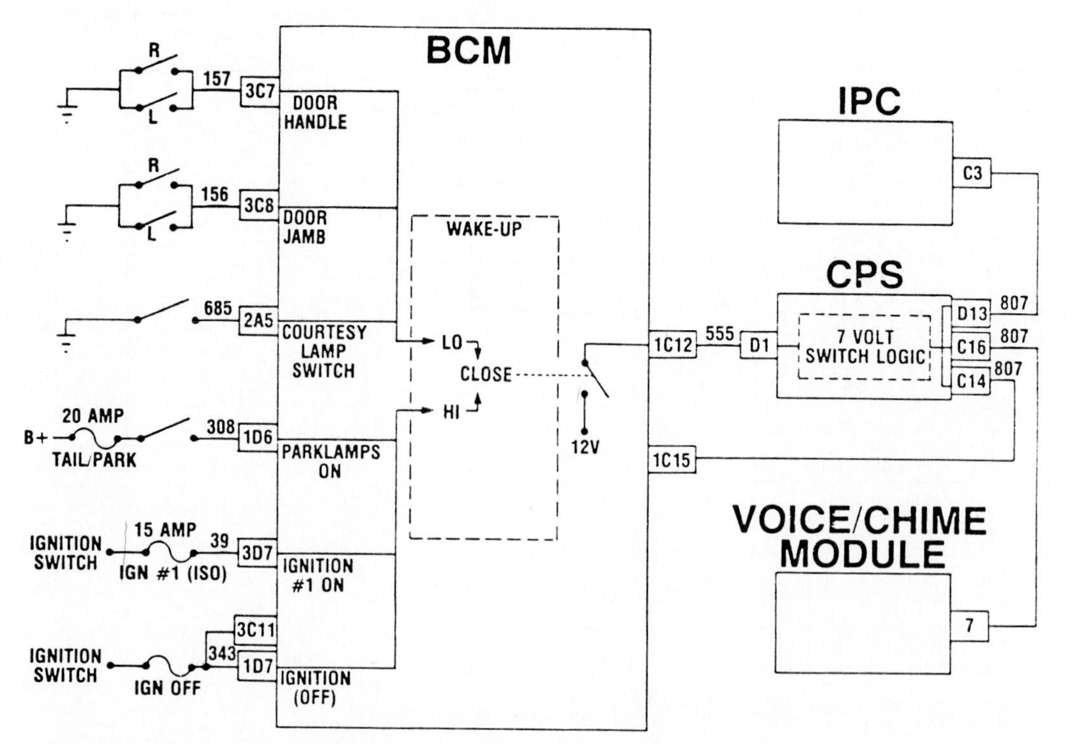

Figure 6-6 Wake-up Circuit to Body Computer Module and Central Power Supply. [Courtesy of General Motors Corporation]

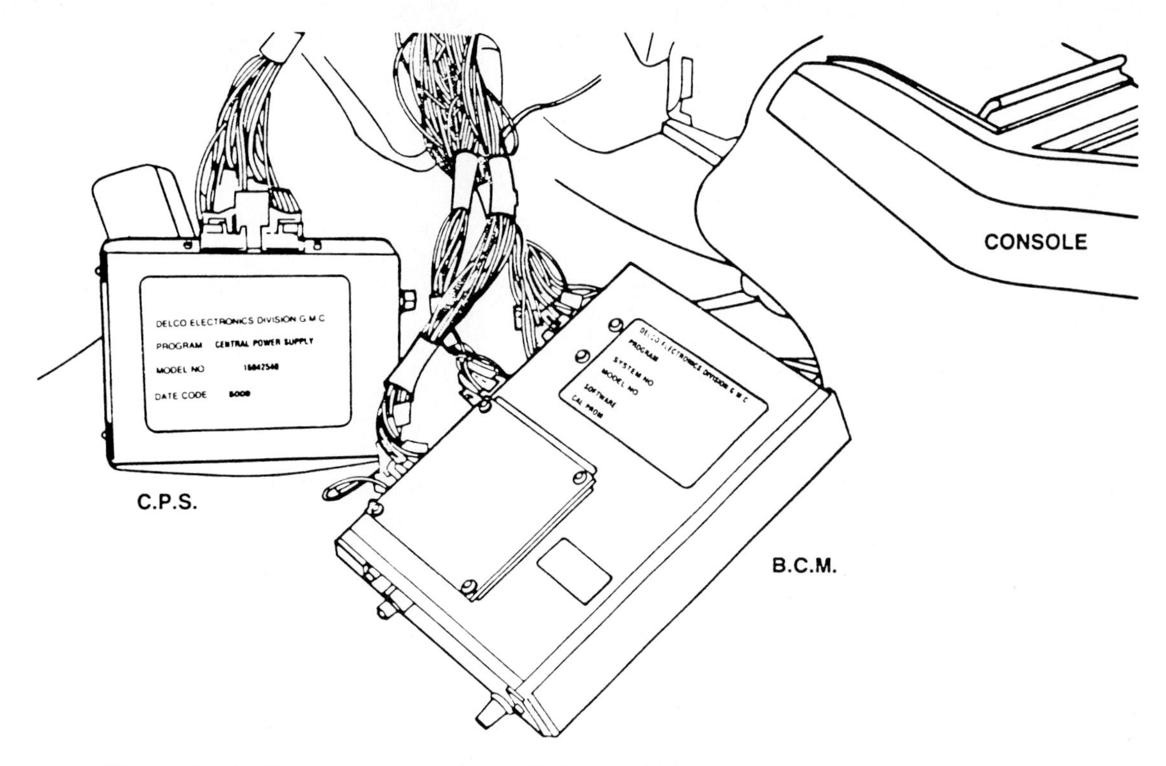

Figure 6–7 Body Computer Module (BCM) and Central Power Supply (CPS) Location. (Courtesy of General Motors Corporation)

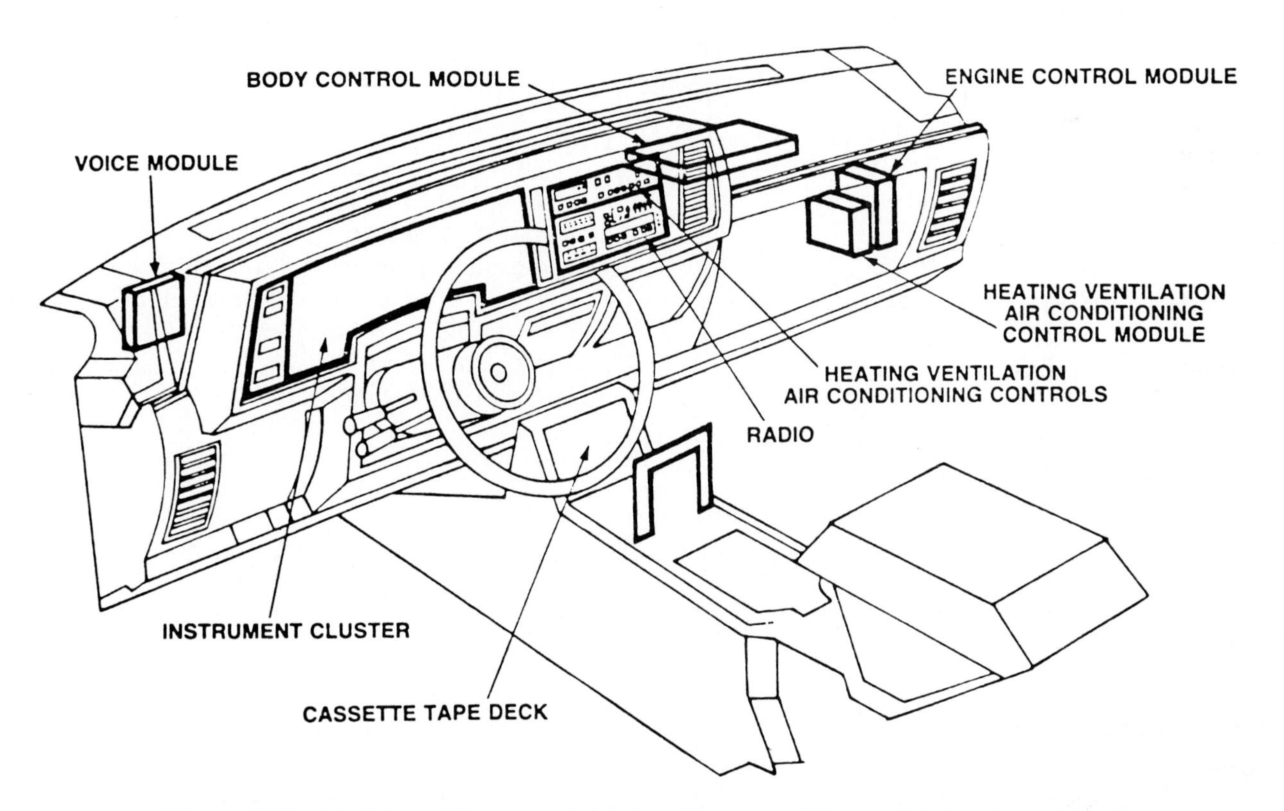

Figure 6–8 Module and Computer Location. (Courtesy of General Motors Corporation)

Body Computer Module

General Functions

The body computer module (BCM) is the center of communications for the multiple computer system, and it performs the following functions:

1. Controls electronic climate control system
2. Controls cruise control system
3. Controls the fuel level display and performs fuel data calculations
4. Controls vehicle status message systems such as "Door Ajar" warning.
5. Provides vehicle speed data
6. Stores and updates odometer information
7. Provides transaxle shifter position data for display (This feature was not available on 1986 early production models.)
8. Calculates English/metric conversions for display
9. Controls courtesy lights including the optional illuminated entry
10. Controls optional twilight sentinel
11. Controls instrument panel dimming
12. Recognizes and compensates for BCM system failures and stores diagnostic codes to be displayed for diagnostic purposes.

The BCM functions listed above are those of a 1986 Toronado. BCM functions vary depending on the application and year of vehicle.

Electronic Climate Control Operation

Body Computer Module

The body computer module (BCM) is the control center for the electronic climate control (ECC) system. A microprocessor is located in the BCM and is supported by random access memories (RAMs), read only memories (ROMs), and programmable read only memories (PROMs). The microprocessor also contains stored programs and input/output interfaces.

Sources of Input and Output Control Functions

The BCM receives input signals from these sources:

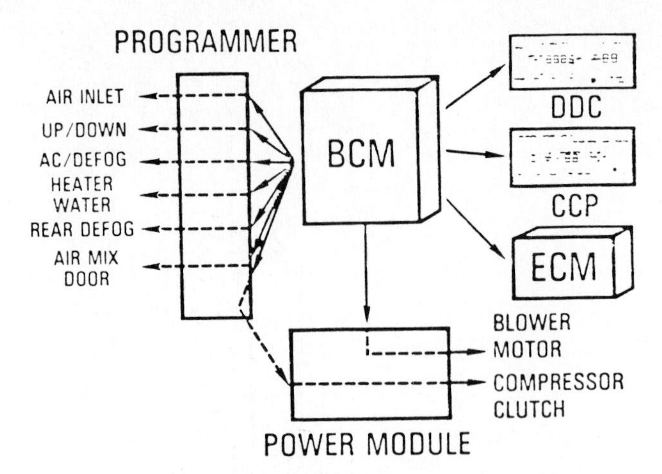

Figure 6–9 BCM Output Control Functions. (Courtesy of General Motors Corporation)

1. ECM data
2. ECC setting
3. Outside temperature
4. In-car temperature
5. Sunload temperature
6. Vehicle speed
7. Low refrigerant pressure switch
8. A/C high side temperature
9. A/C low side temperature

The BCM controls the output functions illustrated in Figure 6–9.

The BCM displays information for the driver on the climate control panel (CCP), or on a diesel data center (DDC) if the car has a diesel engine. When the driver desires a specific ECC mode, he or she depresses the appropriate button on the CCP, which signals the BCM to perform the necessary output function.

Data links are connected between the BCM and the electronic control module (ECM). The ECM is a computer that controls such functions as air–fuel ratio and spark advance. (Refer to Chapter 2 for a description of data links.) Data links transfer data from the BCM to the ECM, and data is also transmitted from the ECM to the BCM. The bidirectional data links between the ECM and the BCM are identified as the 800 circuit.

Compressor Clutch Control

The BCM and the ECC power module operate the air conditioning compressor clutch. If any of the following conditions exist, the compressor clutch remains off:

1. Outside air temperature below 45°F (10°C).
2. Engine coolant or compressor output temperatures too high.
3. Refrigerant pressure, or charge, too low.
4. Throttle wide open.

When all the BCM inputs are within calibrated values, the BCM signals the ECM to engage the air conditioning compressor clutch. This action causes the ECM to transmit a voltage signal to the power module, and this module grounds the compressor clutch relay winding, which closes the relay points and supplies voltage to the compressor clutch. If the air conditioning compressor clutch is on, any of the following conditions will signal the BCM to disengage the clutch:

1. A/C low pressure switch open.
2. A/C low side temperature below 30°F (−1°C).
3. A/C high side temperature above 199°F (93°C).
4. Coolant temperature above 259°F (125°C).
5. Open or shorted A/C low side sensor circuit.

These conditions will force the ECM to disengage the compressor clutch.

1. Wide open throttle signal from throttle position sensor (TPS).
2. System over-voltage.
3. System under-voltage.
4. Power steering cutout switch open.

The compressor clutch circuit is shown in Figure 6–10.

Blower Speed Control

For every electronic climate control (ECC) setting and program number, the BCM sends a calculated blower voltage signal through the A/C programmer to the power module. This module amplifies the signal and sends it to the blower motor to supply the desired blower speed. Feedback information is sent from the blower motor back to the programmer. The blower motor circuit is shown in Figure 6–11, and the ECC control panel is illustrated in Figure 6–12.

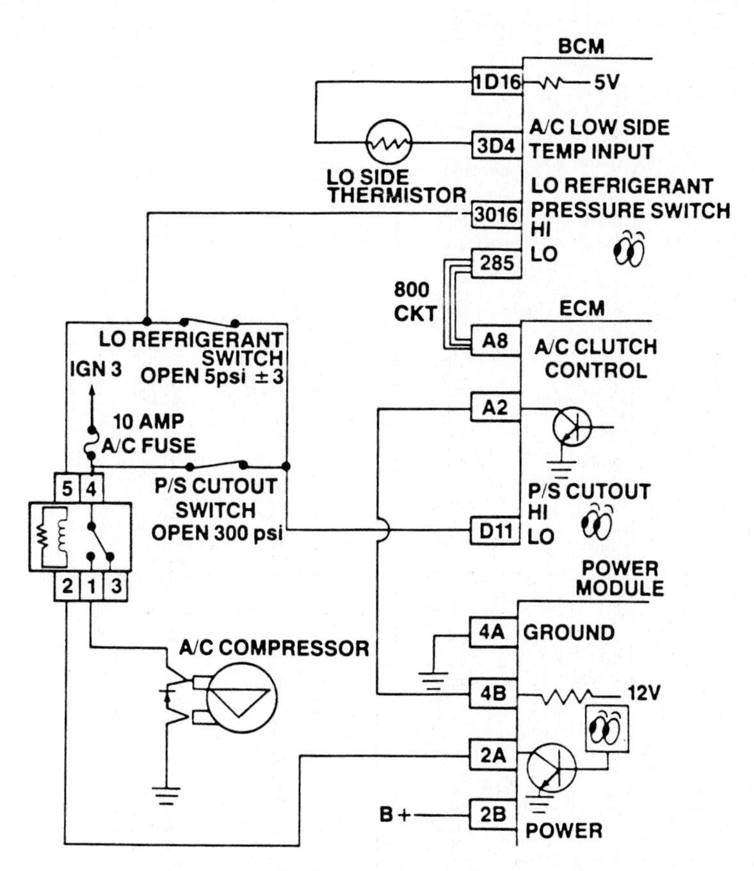

Figure 6–10 Air Conditioning Compressor Clutch Circuit. (Courtesy of General Motors Corporation)

Mode Door Control

The BCM operates a group of solenoids in the programmer, and these solenoids turn the vacuum on and off at the mode door actuator diaphragms. When the BCM energizes a solenoid, vacuum is supplied to an actuator diaphragm which moves the appropriate door. The programmer vacuum system, actuator diaphragm, and mode doors are illustrated in Figure 6–13.

If vacuum to the air inlet door actuator is shut off, this door provides 100 percent outside air to the ECC system. When vacuum is supplied to the air inlet door actuator diaphragm, the door is moved to provide 80 percent inside air and 20 percent outside air to the ECC system.

A vacuum supply to the up–down door actuator diaphragm results in door movement that provides air flow to the A/C-defog door with an air bleed to the heater outlets. When the programmer shuts off the vacuum to the up-down door actuator, this door directs air flow to the heater outlets with an air bleed to the A/C-defog door.

If vacuum is supplied to the A/C-defog door actuator diaphragm, that door directs air to the A/C outlets. The A/C-defog door supplies air flow to the defog outlets if the vacuum supply to the actuator diaphragm is shut off.

When vacuum is supplied to the heater water valve, that valve blocks coolant flow through the heater core, whereas a zero vacuum at the valve actuator causes the valve to open.

Air-Mix Door Operation

An air-mix door motor in the programmer is linked to the air-mix door. The BCM sends a signal to the programmer informing the programmer to drive the motor and provide the exact air-mix door position to obtain the condition requested by the driver. A po-

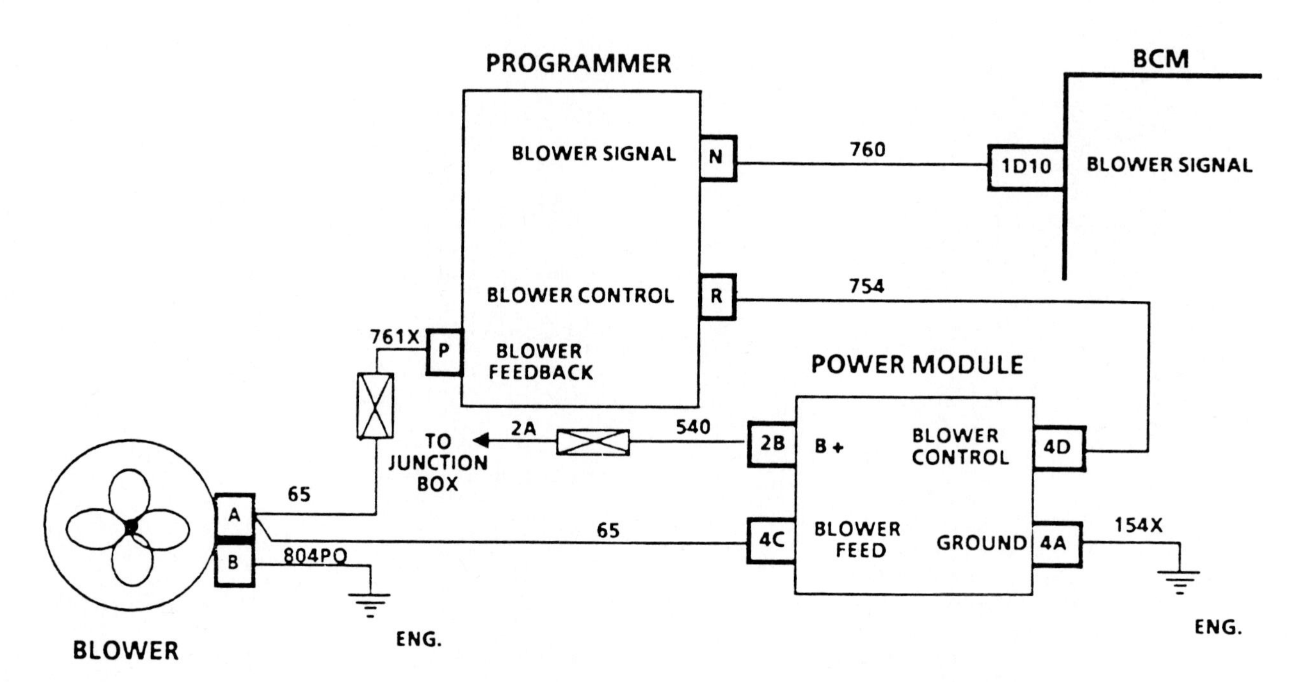

Figure 6–11 Blower Motor Circuit. [Courtesy of General Motors Corporation]

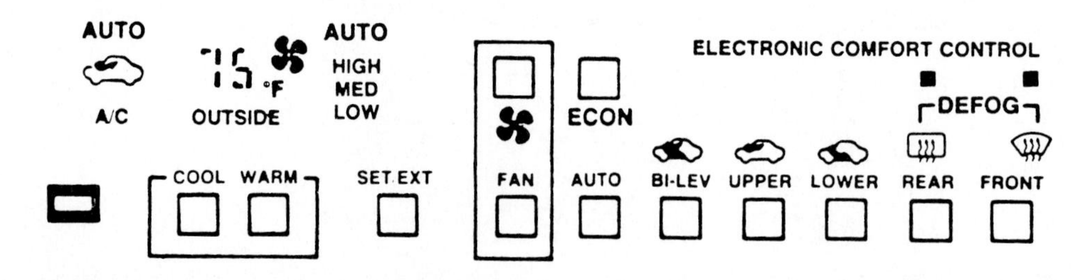

Figure 6–12 Electronic Climate Control Panel. [Courtesy of General Motors Corporation]

tentiometer connected to the air-mix door sends a feedback signal to the BCM which provides precise information regarding air-mix door position. The BCM checks this feedback signal before it commands the air-mix door to move. A diagram of the air-mix door control system is provided in Figure 6–14, and the temperature sensor circuits connected to the BCM are illustrated in Figure 6–15.

Rear Defog Control

When the driver presses the rear defog button on the CCP, a signal is sent from the CCP to the BCM. When this signal is received, the BCM informs the ECC programmer to ground the rear defog relay winding, which closes the relay contacts and energizes the rear defog. The rear defog circuit is pictured in Figure 6–16.

The BCM is capable of diagnosing the air conditioning system and many other functions explained later in this chapter.

Cruise Control Operation

The BCM has taken over the function of the external electronic cruise control system controller used in the past. A cruise set speed display appears on the IPC when the cruise is engaged or resumed, or when the

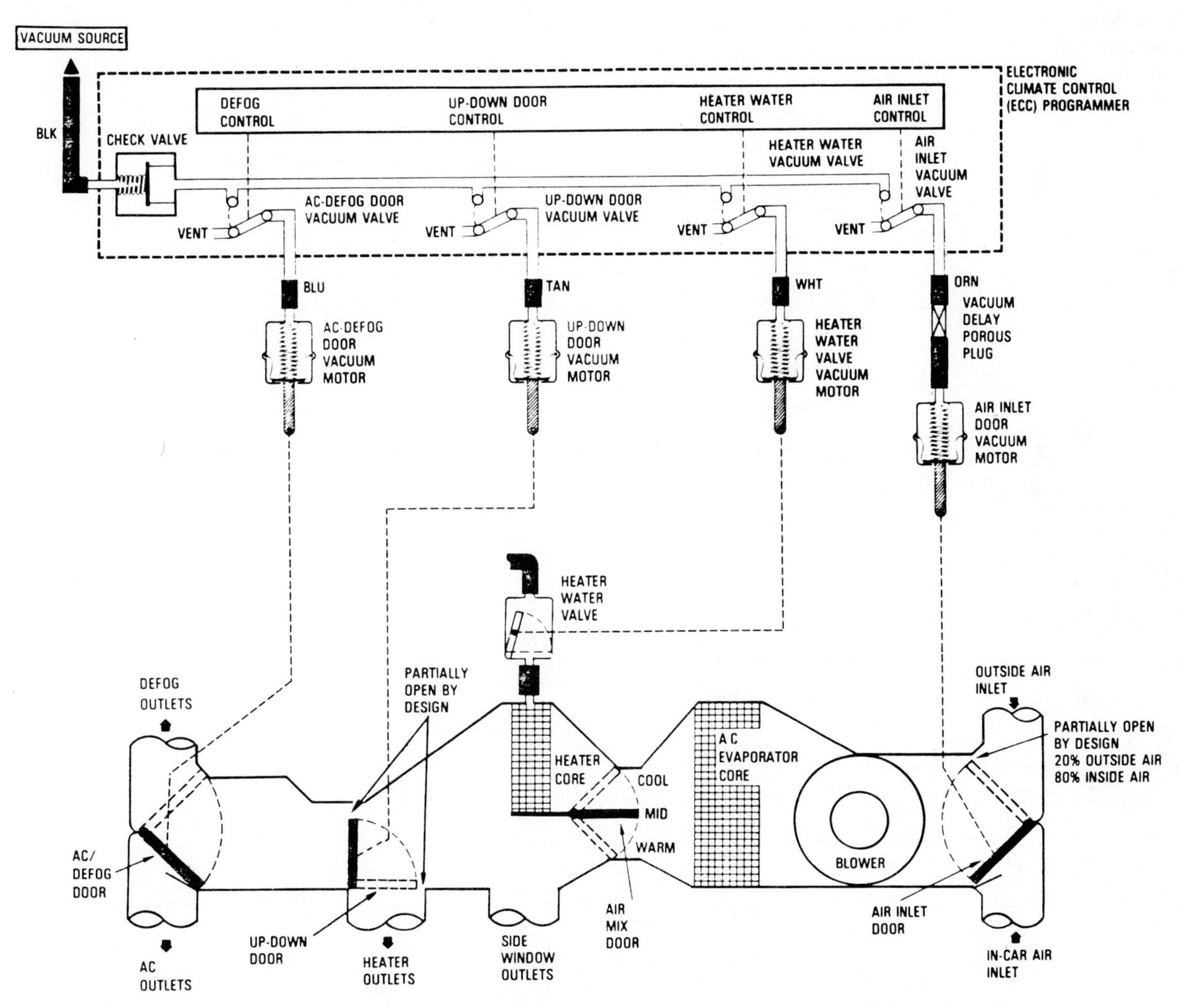

Figure 6–13 Programmer Vacuum System, Actuator Diaphragms, and Mode Doors. [Courtesy of General Motors Corporation]

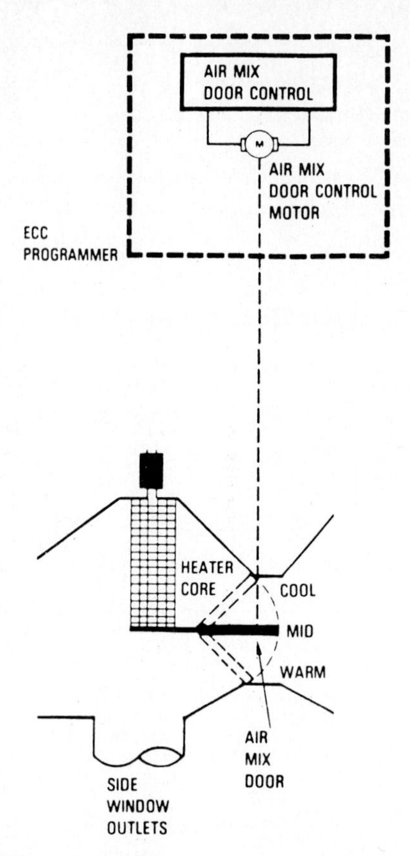

Figure 6–14 Air-Mix Door Control System. (Courtesy of General Motors Corporation)

speed is changed. The BCM receives input signals from the cruise on–off switch, set-coast switch, and resume–accelerate switch on the turn signal lever. These normally-open switches provide a voltage input signal to the BCM when they are closed by the driver. If the brake pedal is depressed, the brake switch opens, and in response to this signal the BCM deactivates the cruise control. Other inputs are the vehicle speed sensor (VSS) and ECM-to-BCM communications on the serial data circuit which the BCM uses for vehicle speed and gear position information. The BCM operates the vent and vacuum solenoids in the cruise servo to control the servo vacuum and the cruise set speed. Information is also supplied from the servo to the BCM to monitor cruise control operation. The cruise control circuit is shown in Figure 6–17.

Instrument Panel Cluster

General Information

The instrument panel cluster (IPC) contains a microprocessor that communicates with the rest of the system through the serial data circuit. Vacuum fluorescent displays as well as incandescent telltale bulbs are used in the IPC. (Refer to Chapter 11 for an explanation of vacuum fluorescent displays.) The vac-

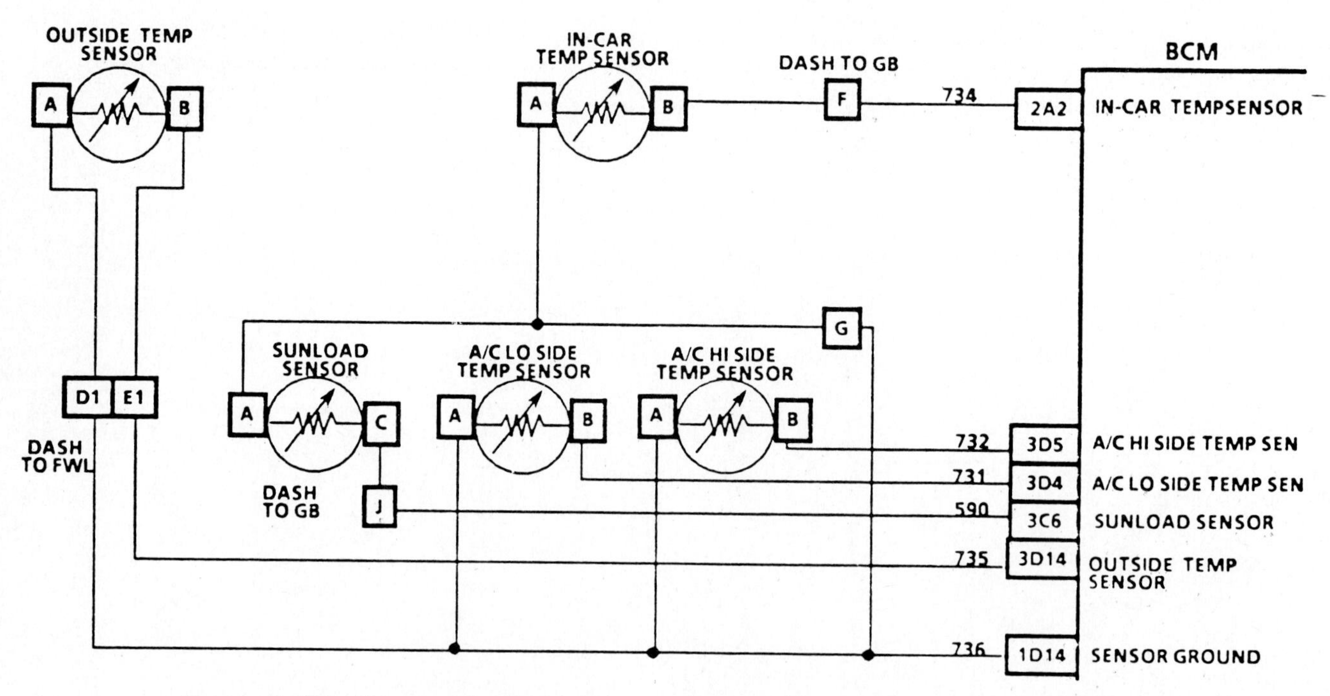

Figure 6–15 Temperature Sensor Circuits. (Courtesy of General Motors Corporation)

uum fluorescent displays perform these functions:

1. Display speed on digital readout
2. Provide odometer information
3. Provide bar-graph display of system voltage
4. Provide bar-graph display of engine oil pressure
5. Provide bar-graph display of coolant temperature
6. Provide bar-graph display of fuel level
7. Indicate PRNDL position for transaxle gear range selection. (This was not available on early 1986 production.)

Vacuum fluorescent indicators are used for these items:

1. Turn signal indicators
2. High beam indicator
3. Lights-on indicator
4. Cruise-on indicator

Driver-requested functions are:

1. Bar-graph tachometer display
2. Fuel and trip information

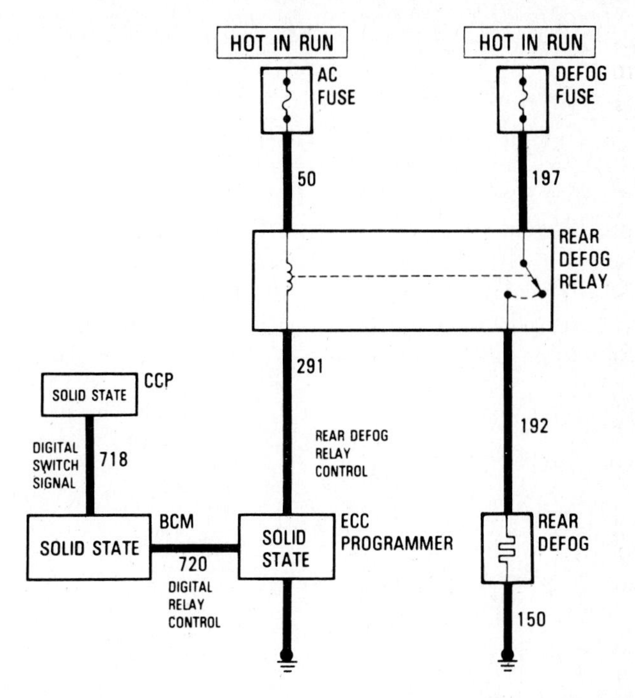

Figure 6–16 Rear Defog Control. (Courtesy of General Motors Corporation)

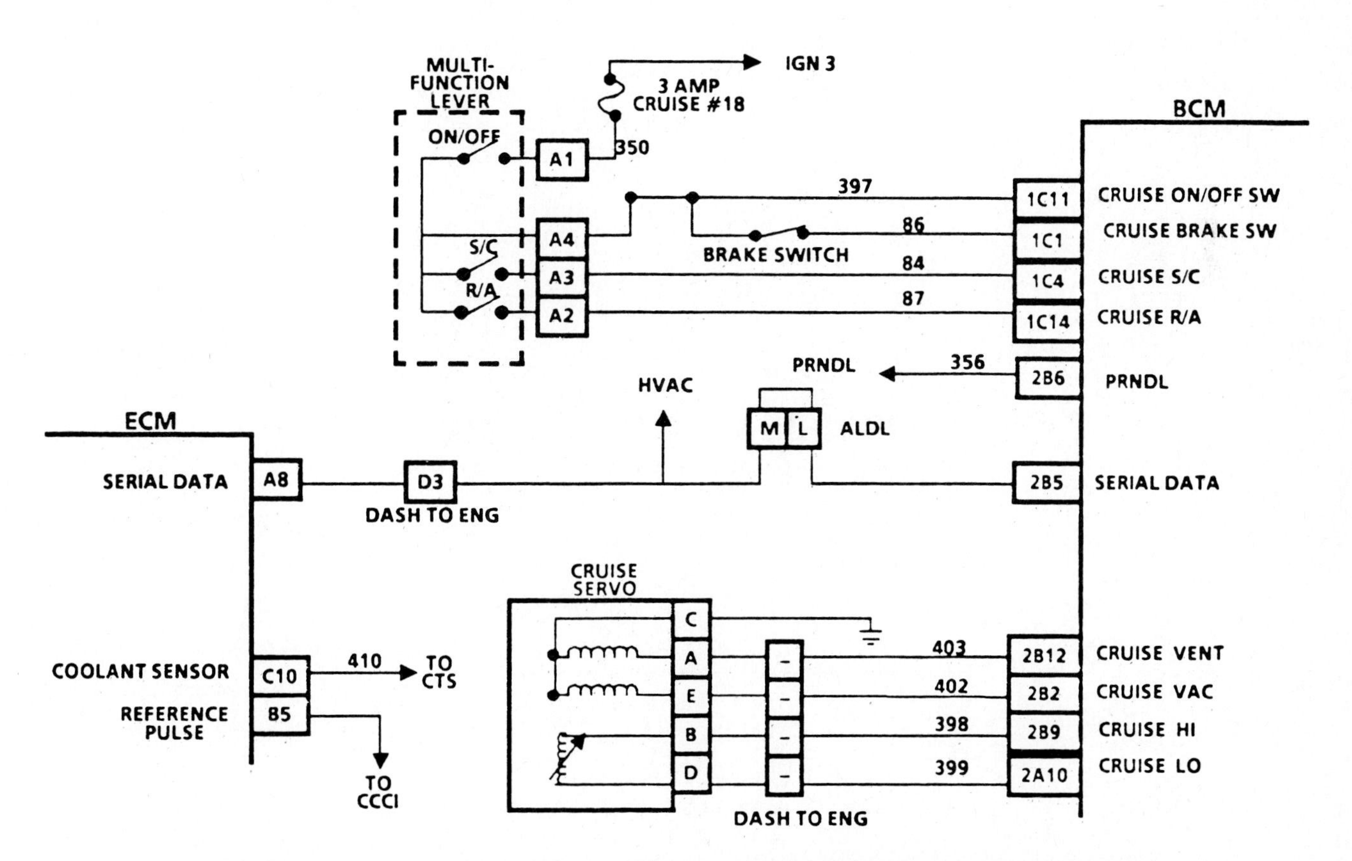

Figure 6–17 Cruise Control Circuit. (Courtesy of General Motors Corporation)

When the ignition switch is turned on, all segments of the IPC are illuminated for a few seconds, as indicated in the top diagram in Figure 6–18.

After the initial IPC display, each bar-graph will have brightly illuminated bars to provide specific indications. For example, if the fuel tank is half full, the bars on the fuel gauge will be brightly illuminated to the center position on the fuel gauge. The IPC information center display has a 20-character vacuum fluorescent dot matrix display with these messages available:

1. Fuel level
2. Engine hot
3. Low coolant
4. Low refrigerant
5. Electrical problem
6. Generator problem
7. Low brake fluid
8. Park brake
9. Low washer fluid
10. Lighting problem
11. A/C system problem
12. Low oil pressure
13. Right door open
14. Left door open
15. Both doors open
16. Headlamp out
17. Tail lamp out
18. Park lamp out
19. Stop lamp out
20. Lamp fuse out
21. Cruise set speed

The following messages are provided by incandescent telltale lamps:

1. Security
2. Fasten belts
3. Service engine soon
4. Brake
5. Lights on

Switch panels on each side of the IPC supply driver-initiated inputs to the IPC. The left switch panel is shown in Figure 6–19, and the right switch panel is pictured in Figure 6–20.

Left Switch Panel Operation

The odo/trip button switches the mileage display from season odometer to trip odometer. If this button is pressed a second time, the mileage display changes

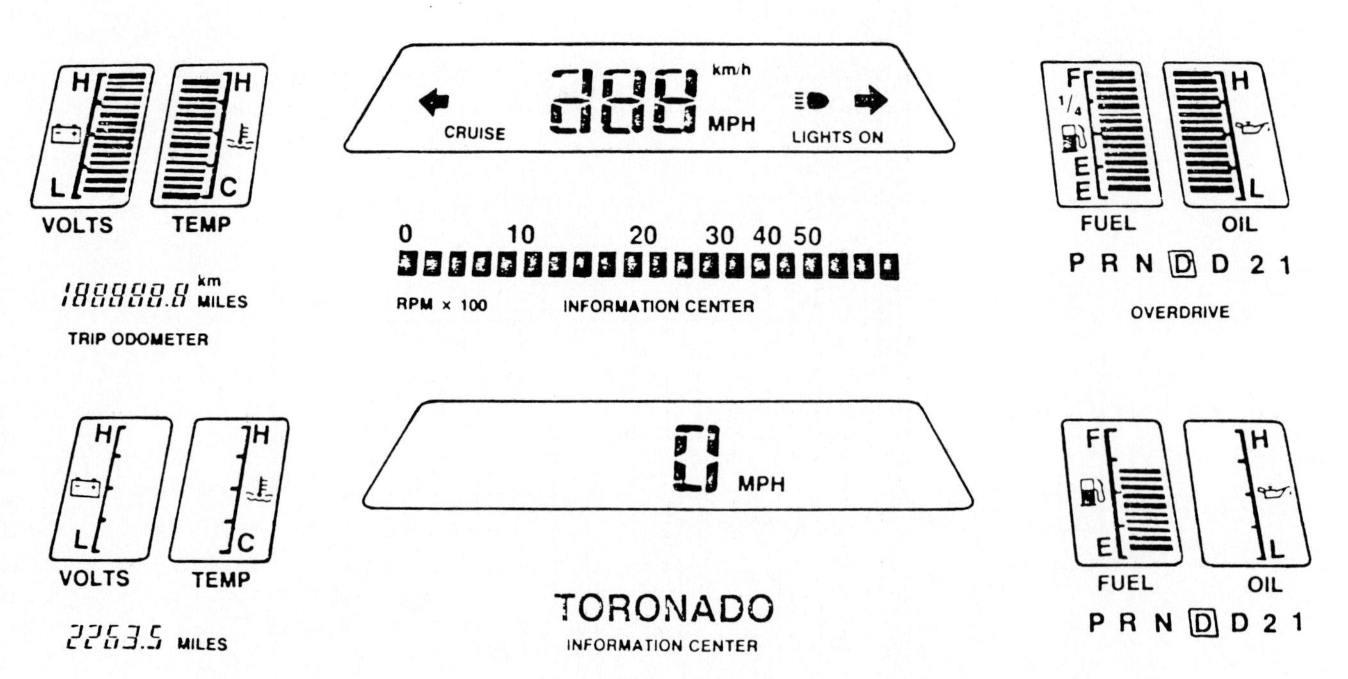

Figure 6–18 Instrument Panel Cluster. [Courtesy of General Motors Corporation]

back to season odometer. When the trip reset button is pressed, the trip odometer is reset to zero.

The English/metric button is used to switch the IPC and electronic climate control (ECC) displays from English to metric units for speed, distance, fuel consumption, and temperature.

A vacuum fluorescent display check is initiated when the driver presses the system monitor button. This button is also used to acknowledge messages in the information center and to redisplay these messages.

The left switch panel also contains the headlight and park light switches, the panel dimming switch, and the twilight sentinel control switch.

Right Switch Panel Operation

When the driver presses the tach button, the tachometer display appears in the information center.

The gauge scale button switches the fuel level gauge from full scale to an expanded quarter-tank scale. After a specific length of time, the display changes back to full scale.

When the fuel-used button is pressed, the information center displays the amount of fuel used since the last driver-requested reset.

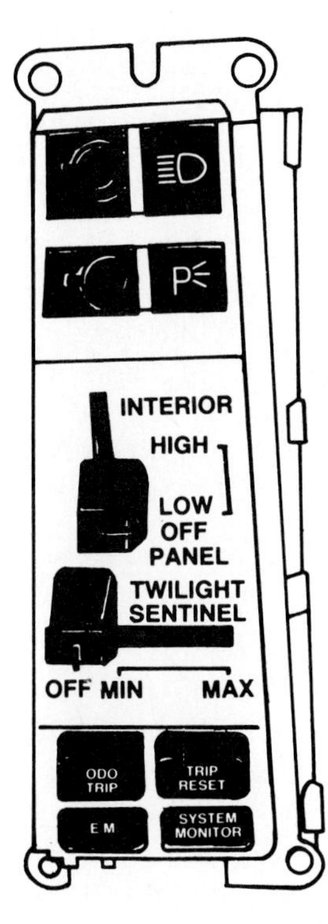

Figure 6–19 Left Switch Panel. [Courtesy of General Motors Corporation]

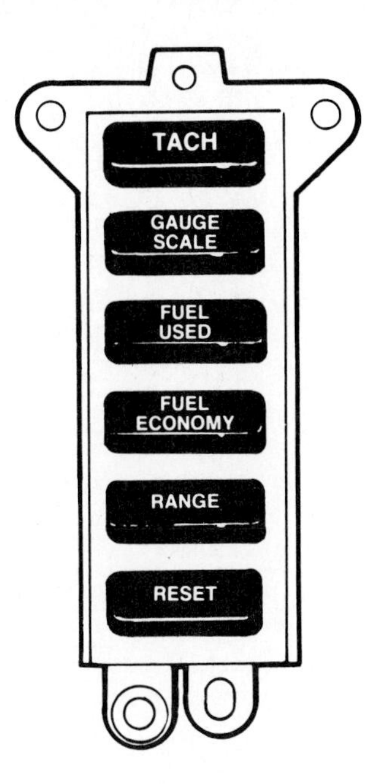

Figure 6–20 Right Switch Panel. [Courtesy of General Motors Corporation]

Pressing the fuel economy button causes fuel consumption information to be displayed in the information center. Either instant fuel economy or average fuel economy since the last reset may be displayed.

The range button allows the driver to request fuel range information. This value is based on the amount of fuel in the tank and the average fuel economy during the last 25 miles (40 km) of driving.

The reset button allows the driver to reset the fuel-used and average fuel economy readings to zero, depending on which reading is displayed in the information center.

Specific Instrument Panel Cluster Functions

Data Communication

Serial data from the BCM is the most important input to the IPC. The IPC controls its displays on the basis of the information received from the BCM on the serial data link. If this input information from the BCM is not available, the IPC will not function. Under this condition, ''Electrical Problem'' will be displayed in the information center. An IPC wiring diagram that includes the serial data link to the BCM is shown in Figure 6–21.

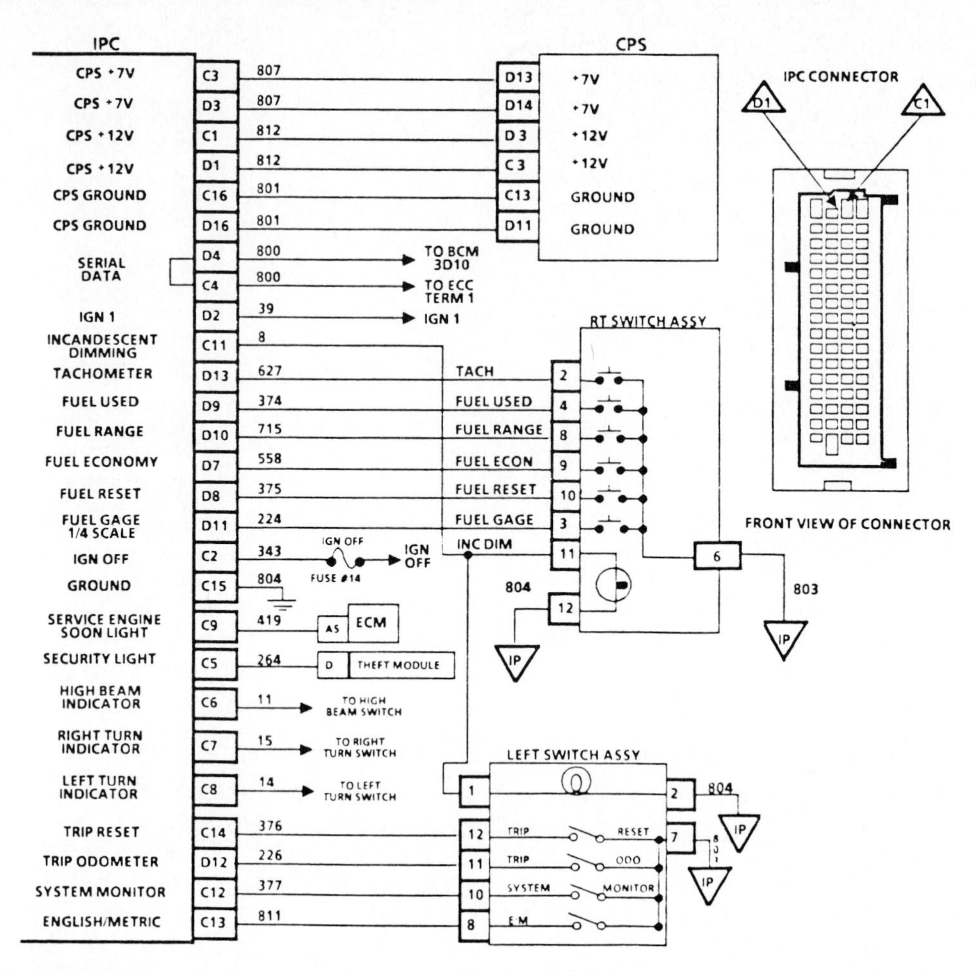

Figure 6–21 Instrument Panel Cluster Wiring Diagram. (Courtesy of General Motors Corporation)

Speedometer and Odometer Operation

The permanent magnet-type vehicle speed sensor (VSS) is located in the transaxle. This sensor generates an AC voltage pulse that is proportional to vehicle speed. A buffer circuit in the BCM changes this signal into a DC square-wave digital signal. The BCM communicates the vehicle speed information to the IPC on the serial data circuit, and a decoder in the IPC decodes this information before the IPC microprocessor illuminates the speedometer display.

The BCM counts the VSS signal pulses to calculate mileage for the season odometer display. This total accumulated mileage is continuously written into the BCM's electronically erasable programmable read-only memory (EPROM). If battery voltage is removed from the BCM, this information is retained in the EPROM. The BCM also uses VSS input to operate the trip odometer, but this information is stored in the BCM's random-access memory (RAM). Removal of battery voltage from the BCM results in the loss of trip odometer information. The IPC will display only season odometer information or trip odometer information at any one time. As mentioned previously, this selection is made by pressing the odo/trip button.

The IPC and BCM circuits related to speedometer and odometer operation are illustrated in Figure 6–22.

Park-Reverse-Neutral-Drive-Low Display

The park-reverse-neutral-drive-low (PRNDL) display in the IPC is activated from the ignition-off circuit to the IPC. This circuit supplies voltage to the IPC when the ignition switch is in the off or run position. Service and towing considerations require this display with the ignition off. In this ignition switch position, no other IPC display is illuminated. The gear selector connections to the BCM and ECM are shown in Figure 6–23.

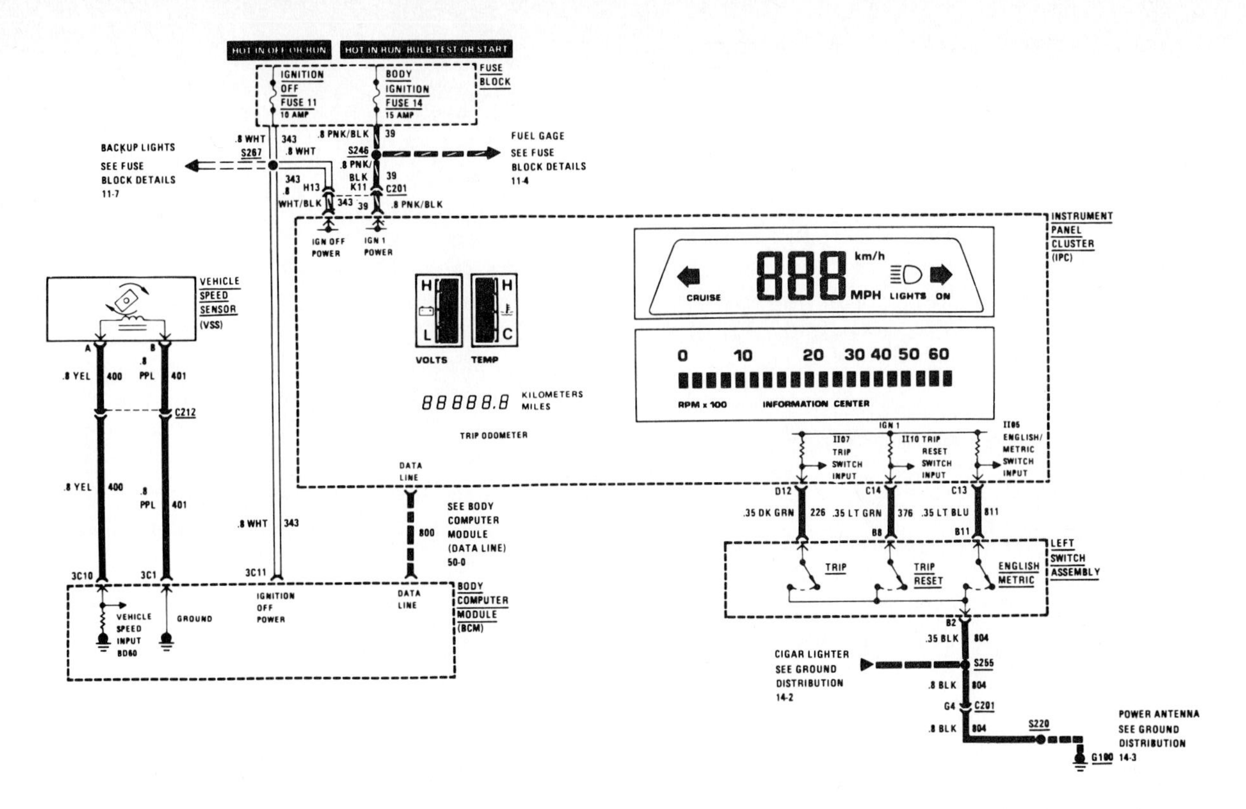

Figure 6–22 Instrument Panel Cluster and Body Computer Module Circuits Related to Speedometer and Odometer Operation. (Courtesy of General Motors Corporation)

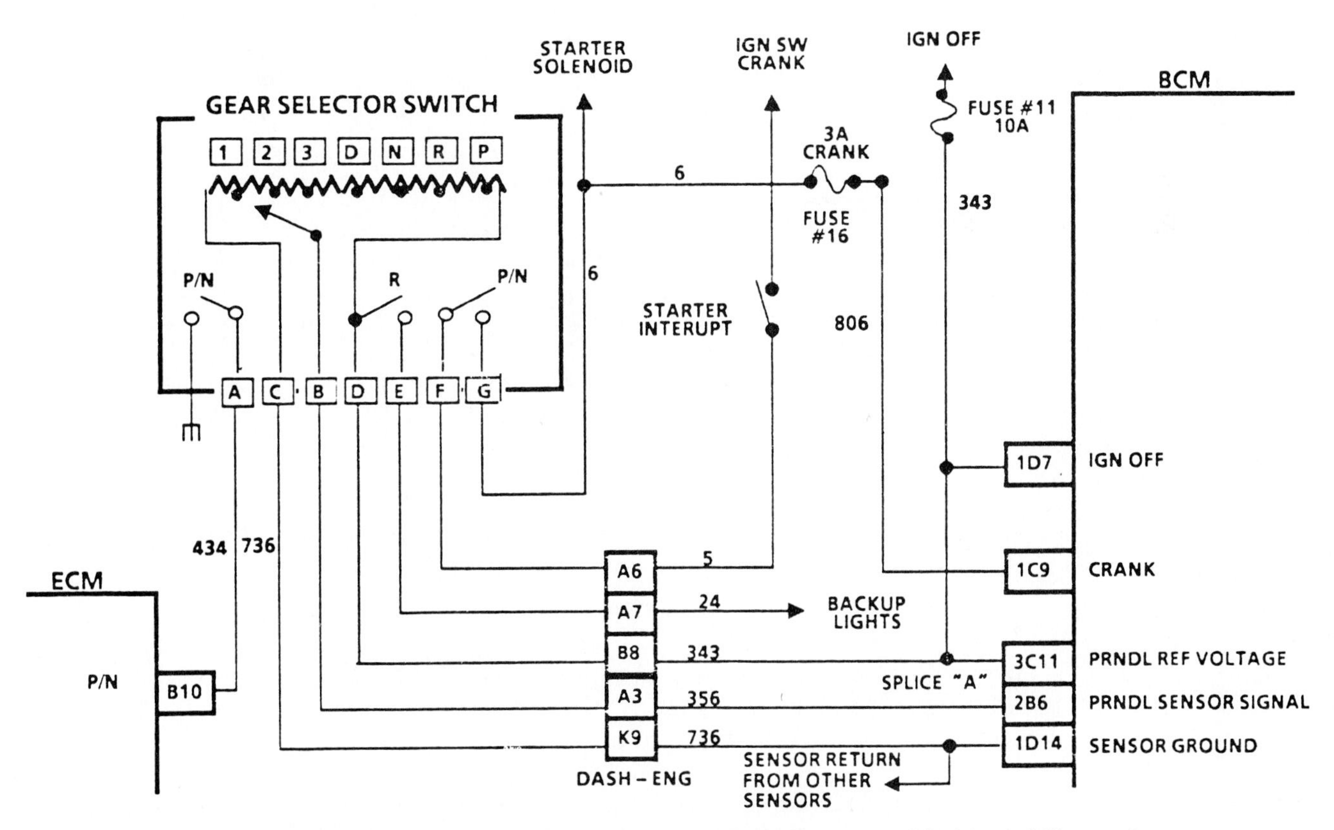

Figure 6–23 Gear Selector Switch Circuit to Body Computer Module and Electronic Control Module. (Courtesy of General Motors Corporation)

Voltage Display

The BCM monitors voltage directly from the Ignition 1 circuit, and compares this voltage to stored value limits. A digital signal from the BCM communicates this voltage information to the IPC, where the signal is decoded. On the basis of this information, the IPC microprocessor decides how many segments to illuminate on the voltage bar-graph.

Coolant Temperature and Oil Pressure Displays

The coolant temperature sensor sends an input signal to the ECM. This input affects many ECM outputs. The ECM transmits this information the the BCM on the serial data line, and the BCM repeats this signal and sends it to the IPC. The BCM also uses coolant temperature information for A/C system control. When the IPC microprocessor receives this coolant temperature information, it compares the value received to its stored program to decide how many segments of the coolant temperature bar-graph should be illuminated.

An oil pressure sensor in the oil gallery near the oil filter sends an input signal to the BCM in relation to oil pressure. This information is converted to a digital signal by the BCM and then sent to the IPC. When the information is received, the IPC compares it to the stored program before deciding how many bars to illuminate in the oil pressure bar-graph. Some early production models had an off–on type oil pressure switch. On these applications, the IPC illuminated half of the oil pressure bar-graph segments as long as minimum oil pressure was available.

Tachometer Display

An engine speed signal is sent from the distributorless direct ignition system (DIS) module to the ECM. This signal is used by the ECM to control some of its output functions. The ECM sends this signal to the

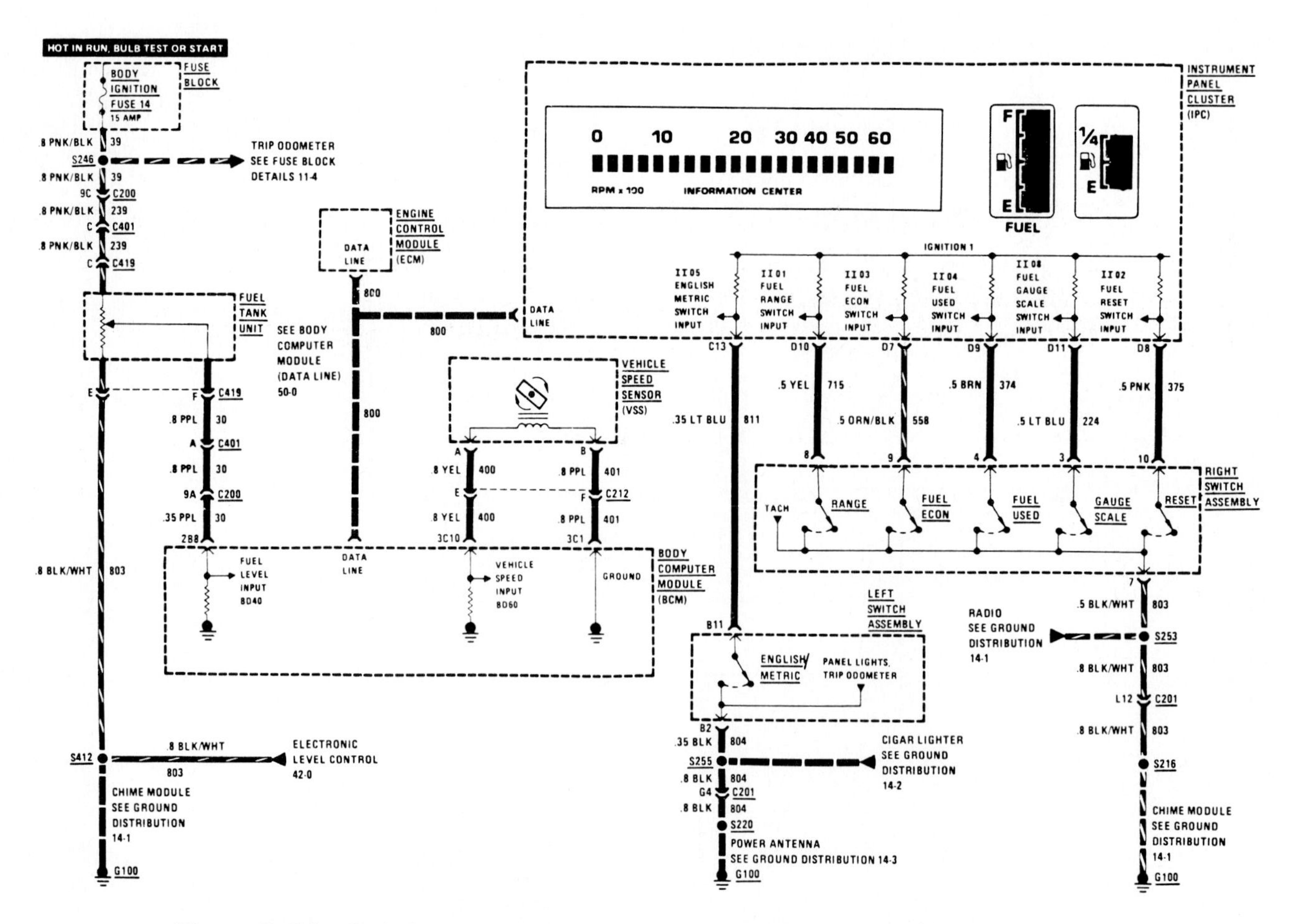

Figure 6–24 Body Computer Module and Instrument Panel Cluster Circuits Related to Fuel Data Display. (Courtesy of General Motors Corporation)

BCM on the serial data line, and the IPC eavesdrops on this ECM-to-BCM information. If the driver has pressed the tach button, the IPC uses this engine speed signal to accurately illuminate the tach display.

Fuel Data Calculations

The fuel tank sending unit sends an input signal to the BCM in relation to the fuel level in the tank. Sending unit resistance varies with fuel level, and the BCM measures the voltage drop across the sending unit. A distance-travelled calculation is completed by the BCM from the vehicle speed sensor (VSS) signal. The ECM sends injector pulse width and flow rate information to the BCM. On the basis of these inputs, the BCM determines fuel consumption and supplies this data to the IPC on the serial data line. When the driver requests any fuel data function, the IPC uses the information from the BCM to provide the correct display. The fuel tank sending unit signal to the BCM is also used by the IPC to illuminate the fuel gauge bar-graph. The BCM and IPC circuits related to fuel data displays are illustrated in Figure 6–24.

Lamp Monitor Operation

The lamp monitor circuits provide "lamp out" messages on the information center. Four circuits are connected between the BCM and the lamp monitor module: headlamp, out, tail lamp out, stop lamp out, and park lamp out. When all the lamps are working normally, the lamp monitor module connects these circuits to ground and causes a low circuit voltage. The input circuit from each lamp switch is connected through two equal-resistance wires to the lamp monitor module. Output wires from these same lamp monitor module terminals are connected to the appropriate lamps. The lamp monitor circuit for the front and rear exterior lamps is shown in Figure 6–25.

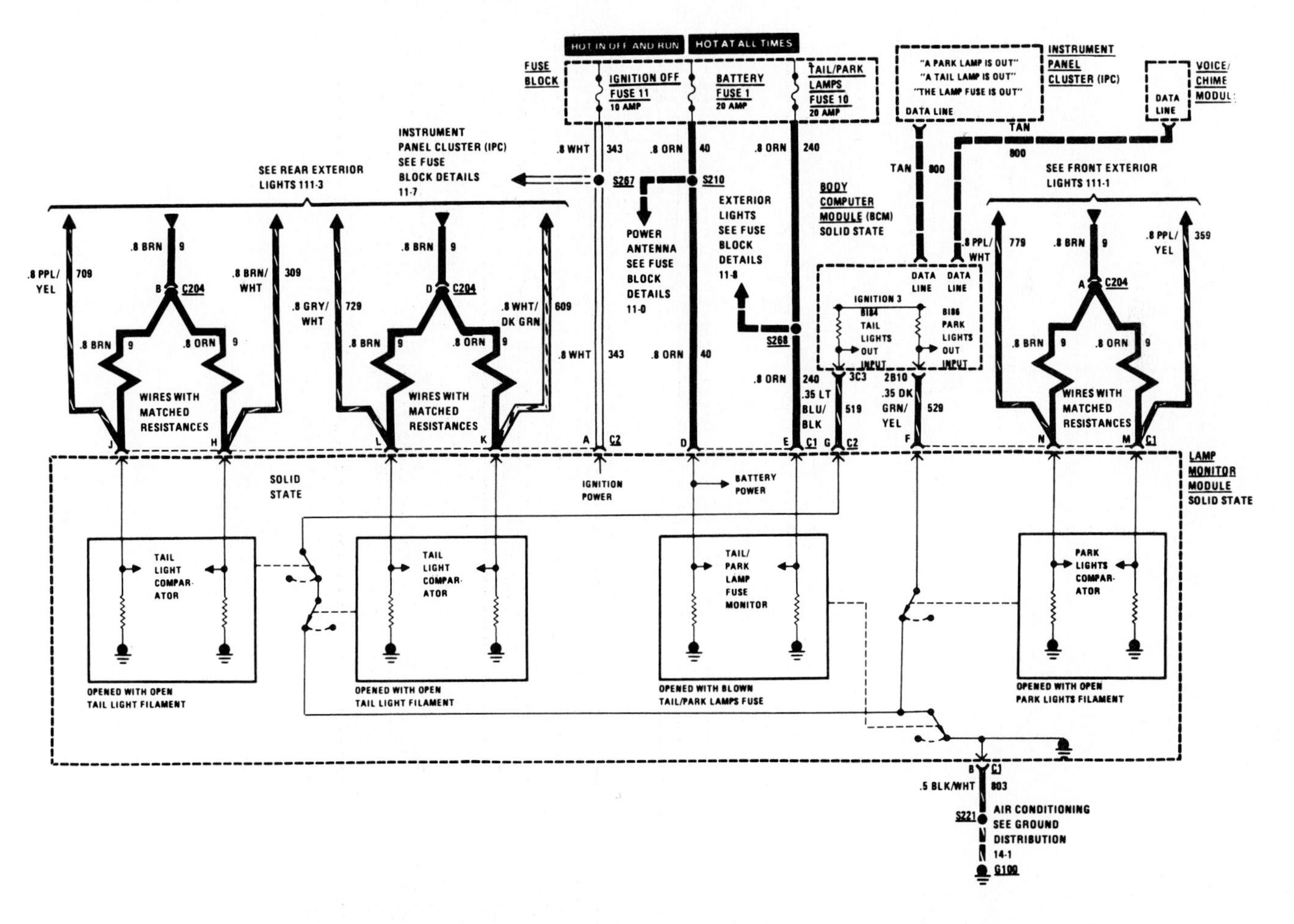

Figure 6–25 Lamp Monitor Circuit for Front and Rear Exterior Lamps. (Courtesy of General Motors Corporation)

If a lamp burns out, an open circuit is created and the voltage increases at the lamp monitor module terminal to which the burned out lamp is connected. This voltage increase causes the lamp monitor module to open the appropriate lamp-out circuit from the BCM. This action results in the appropriate lamp-out communication from the BCM to the IPC, and the IPC microprocessor displays the message in the information center.

Twilight Sentinel Operation

The twilight sentinel system keeps the exterior lights on for an adjustable length of time after the ignition switch is turned off. This system is operated by the twilight delay control on the left switch panel. When this switch is off, the lights and the head lamp doors operate manually through the head lamp switch. If the twilight sentinel delay control is moved away from the off position, a variable resistance in the control is connected in series with the twilight photocell and both of these components are connected to the BCM. The twilight photocell is a photoresistor that senses ambient light, and its resistance increases in dark conditions. The BCM senses the voltage drop across the photocell and the delay control. If the delay control is on the photocell resistance is above a specific value because of dark conditions, the BCM turns the lights on. The BCM performs this function by grounding the twilight head lamp relay and twilight park lamp relay windings. When this occurs the head lamps, park lamps, and head lamp doors are turned on just as though the driver-controlled head lamp switch had been turned on. When the ignition switch is turned off, the BCM keeps the lights on for a specific length of time, adjustable with the twilight delay control. The twilight sentinel circuit is illustrated in Figure 6–26.

Illuminated Entry

Battery voltage is applied directly through a fuse to the courtesy light bulbs, and the circuit is completed from each bulb through the courtesy light relay con-

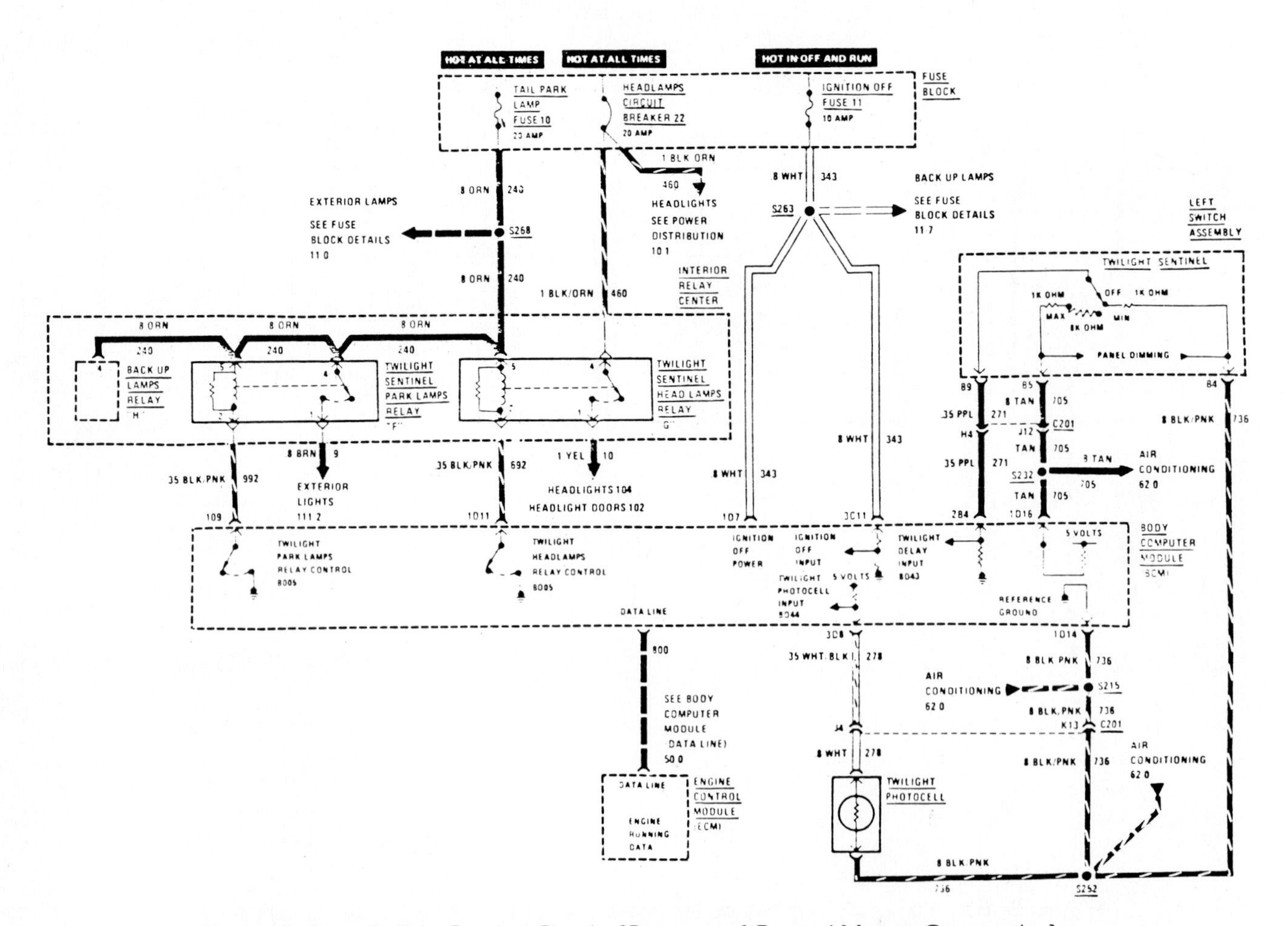

Figure 6–26 Twilight Sentinel Circuit. (Courtesy of General Motors Corporation)

tacts to ground. These contacts are normally open. Voltage is also applied through the same fuse to the courtesy light relay winding, and the circuit from this winding is completed to ground through the BCM. Normally-open switches are located in the door locks and door jams. If an exterior door handle is pushed, the door-lock switch closes and signals the BCM to ground the courtesy light relay winding. This BCM action turns on the interior courtesy lights, and the lights remain on for 20 seconds. When the door is opened, the door jam switch signals the BCM to keep the courtesy lights on. When the door is closed, the courtesy lights go out immediately if the ignition switch is turned to the run position. The twilight photocell input signal to the BCM affects the operation of the courtesy lights. On a bright sunny day this signal will inform the BCM that the courtesy lights are not required. Under this condition the BCM will not ground the courtesy light relay winding. The courtesy light circuit is illustrated in Figure 6–27 and 6–28.

Panel Dimming

The BCM controls the illumination of the left and right switch panels as well as the radio, electronic climate control (ECC), and ash tray. Dimming of these lights is also controlled by the BCM in response to the input signal from the panel dimming control in the left switch panel. When the ignition switch is turned on, a 5V signal is supplied from the BCM to the panel dimming control. The potentiometer in the panel dimming switch is grounded through the BCM. As the panel dimming control is moved toward the dim position, the voltage signal decreases from the switch to the BCM. When this signal is received the BCM dims the switch panel and display panel illumination. A pulse width modulated (PWM) signal is sent from the BCM to the ECC panel and the radio on circuit 724 to control the intensity of the vacuum fluorescent displays (VFDs) in these components. When the BCM increases the on time of the PWM signal, VFD intensity increases. The 724 circuit from the

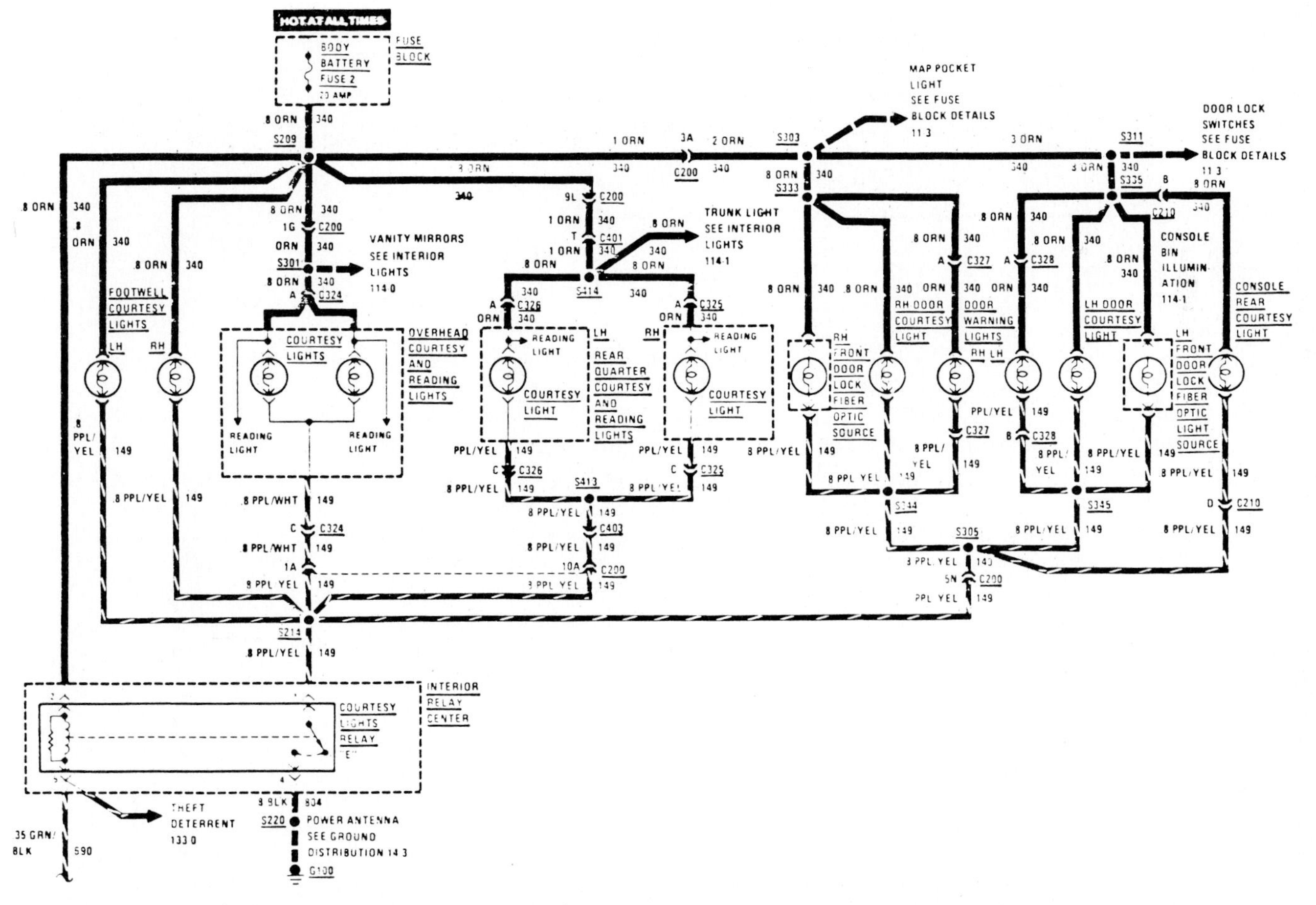

Figure 6–27 Illuminated Entry Courtesy Light Circuit. (Courtesy of General Motors Corporation)

BCM to the EEC panel and the radio is shown in Figure 6–29.

The BCM also sends a panel dimming signal on the serial data line to the IPC. The IPC microprocessor uses this signal to control the intensity of the VFD in the IPC. A PWM signal is also sent from the IPC to the incandescent illumination bulbs in the left and right switch panels, ash tray, and face plates in the EEC panel, radio, and tape deck to control the intensity of these bulbs, as illustrated in Figure 6–30.

Charging Circuit Monitor

Many 1986 and later model General Motors vehicles have a new Delco Remy integral alternator. The integral regulator in this alternator has four terminals. A phase (P) terminal on the regulator may be used as a speed, or tachometer, signal. The L terminal is connected to the charge indicator bulb and parallel resistor. This terminal may also be connected to the BCM. The field (F) monitor terminal is connected to the BCM, and the sense (S) terminal is connected to the positive battery terminal. When the ignition switch is turned on, voltage is supplied through the charge indicator bulb and the BCM to the L terminal and the lamp driver in the regulator. This action causes the transistor to turn on, which allows current to flow from the battery terminal through the transistor and the alternator field coil to ground. When the engine is started, the regulator controls the field current and limits the alternator voltage. While the alternator is charging, the regulator cycles the field current on and off, and this cycling signal is applied from the F terminal to the BCM. This signal informs the BCM when a defect occurs in the charging circuit, and trouble codes are set in the BCM memory. The integral alternator circuit is shown in Figure 6–31, and the connections between the BCM and the regulator are illustrated in Figure 6–32.

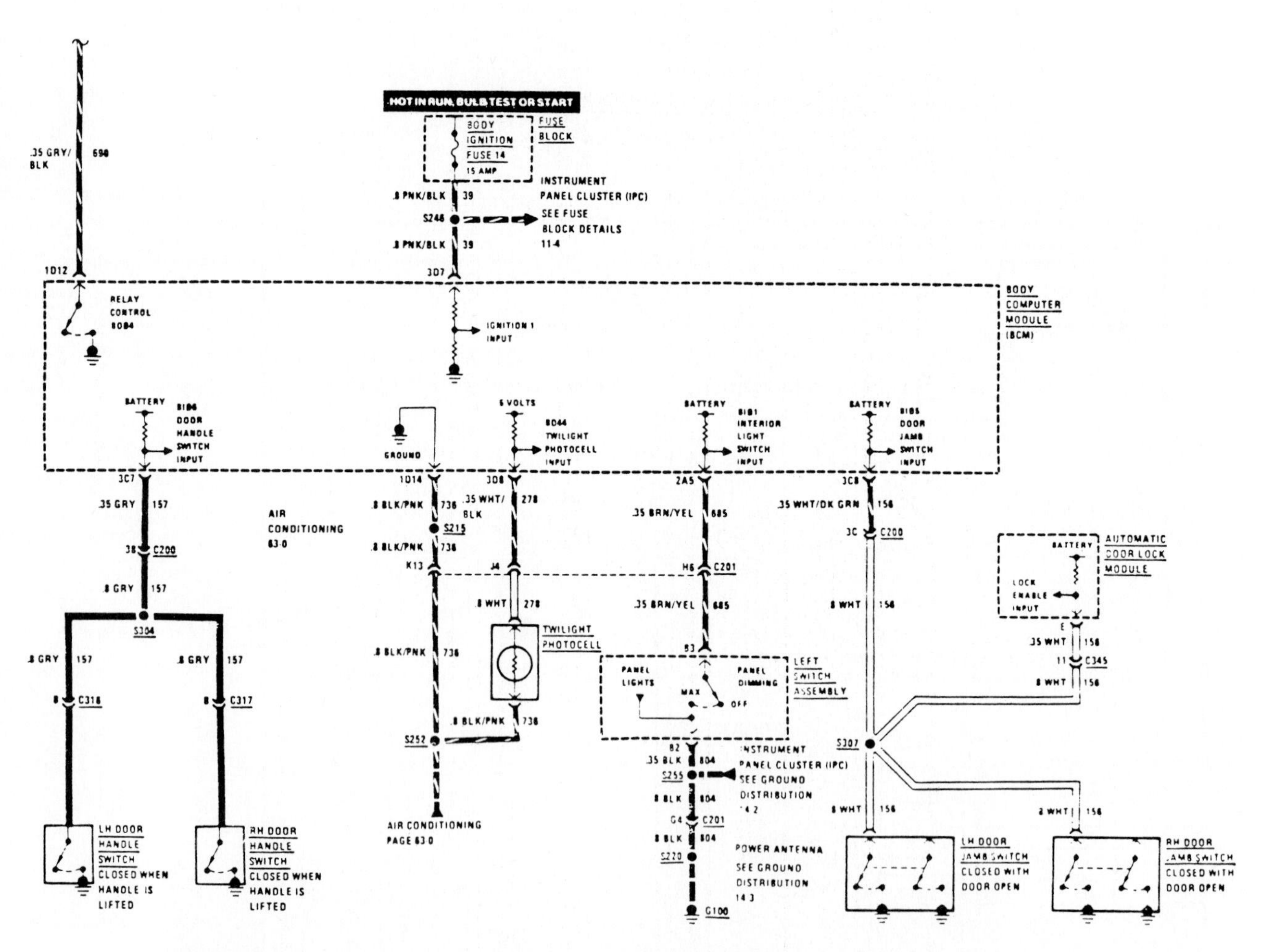

Figure 6–28 Illuminated Entry Courtesy Light Circuit. (Courtesy of General Motors Corporation)

Voice-Chime Warning System

Some General Motors vehicles are equipped with a chime module, while others have an optional voice-chime module. The chime module provides chime warnings and the voice-chime module gives voice warnings through the left front radio speaker with the chime warnings. When the BCM receives an input signal that indicates a driver warning is necessary, it sends a signal to the IPC on the serial data circuit and the IPC displays a visual warning on the information center. At the same time the BCM will signal the chime or voice and chime warning. These chime warnings and information center displays are available:

1. Left door open — continuous slow chime
2. Right door open — continuous slow chime
3. Both doors open — continuous slow chime
4. Headlamp out — five-second medium chime
5. Tail lamp out — five-second medium chime
6. Park lamp out — five-second medium chime
7. Stop lamp out — five-second medium chime
8. Lamp fuse out — five-second medium chime
9. Low washer — five-second slow chime
10. A/C system problem — five-second fast chime
11. Generator problem — five-second fast chime
12. Electrical problem — five-second fast chime
13. Fasten seat belt — five-second slow chime

The chime module circuit is illustrated in Figure 6–33.

On vehicles equipped with a voice-chime module, the BCM inputs and voice-chime warnings are:

1. Lights on — voice warning with a continuous fast chime
2. Key in ignition — voice warning with a continuous fast chime

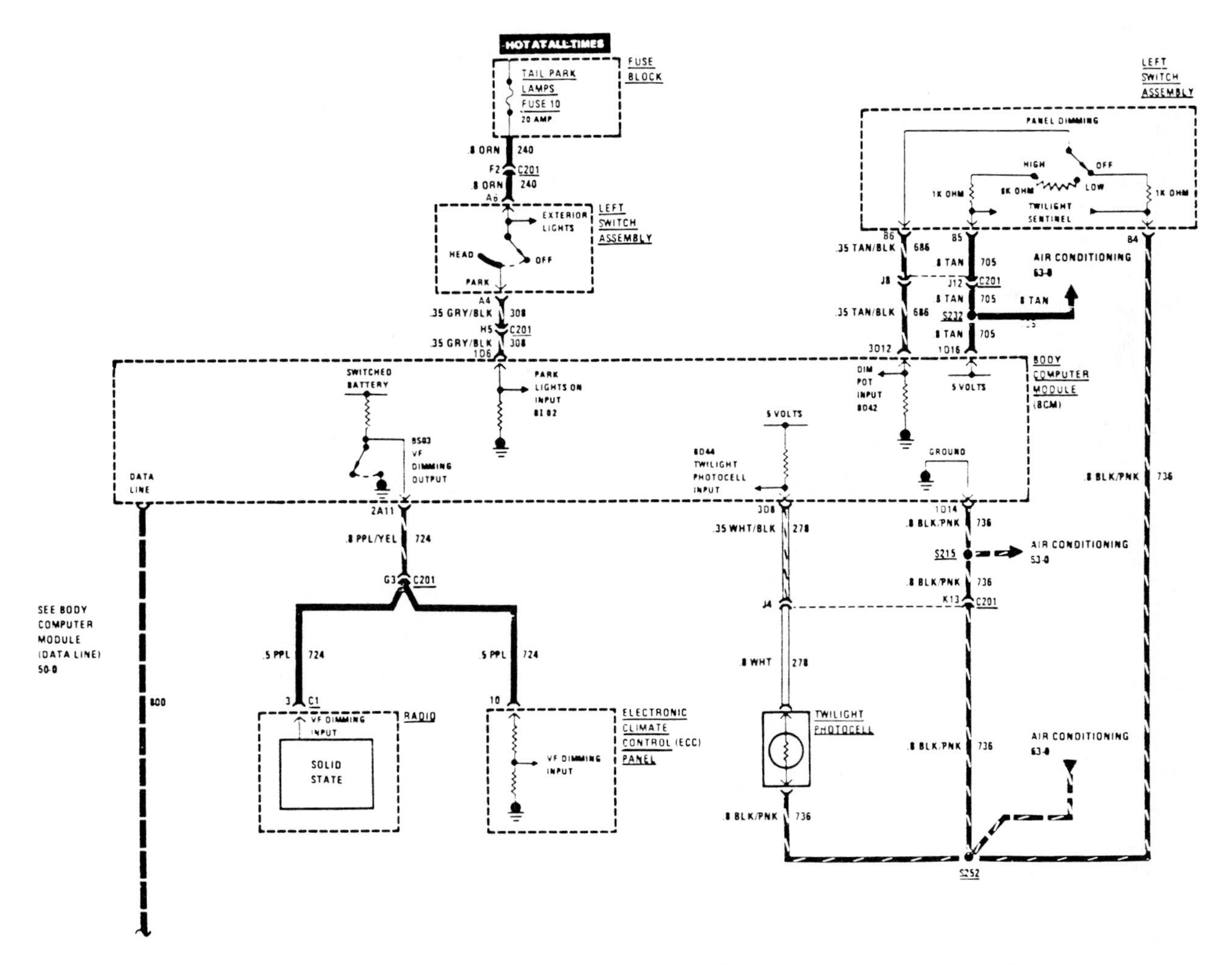

Figure 6–29 Panel Dimming Circuit. [Courtesy of General Motors Corporation]

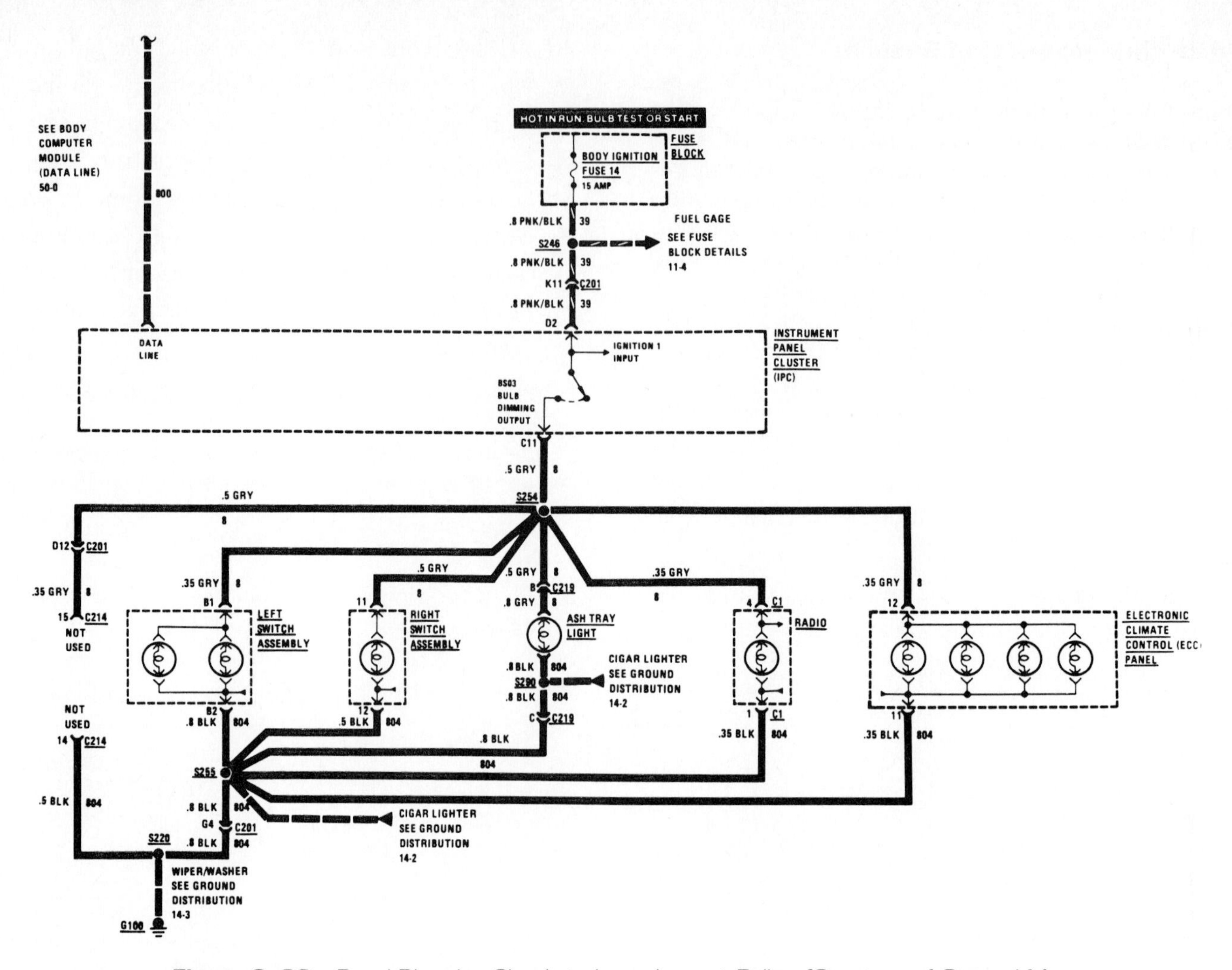

Figure 6-30 Panel Dimming Circuit to Incandescent Bulbs. (Courtesy of General Motors Corporation)

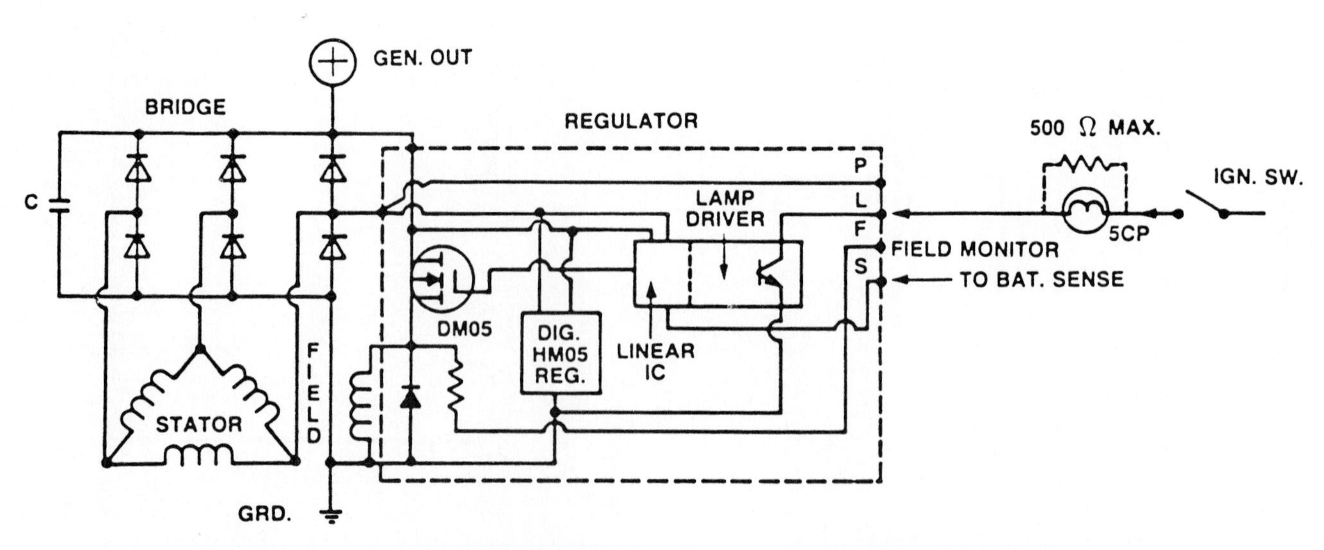

Figure 6-31 Integral Voltage Regulator to Body Computer Module Connections. (Courtesy of General Motors Corporation)

3. (a) Engine hot
   First warning — voice warning to turn off A/C and continuous fast chime
   (b) Second warning — voice warning to idle engine in park with continuous fast chime and information center display
   (c) Third warning — voice warning to turn off ignition switch with continuous fast chime and information center display
   (d) Fourth message — voice message that engine temperature has returned to normal
4. Parking brake — voice warning with a continuous medium chime and an information center display
5. Brake fluid level — voice warning with a continuous fast chime and an information center display
6. Engine oil pressure — voice warning with a continuous medium chime and an information center display

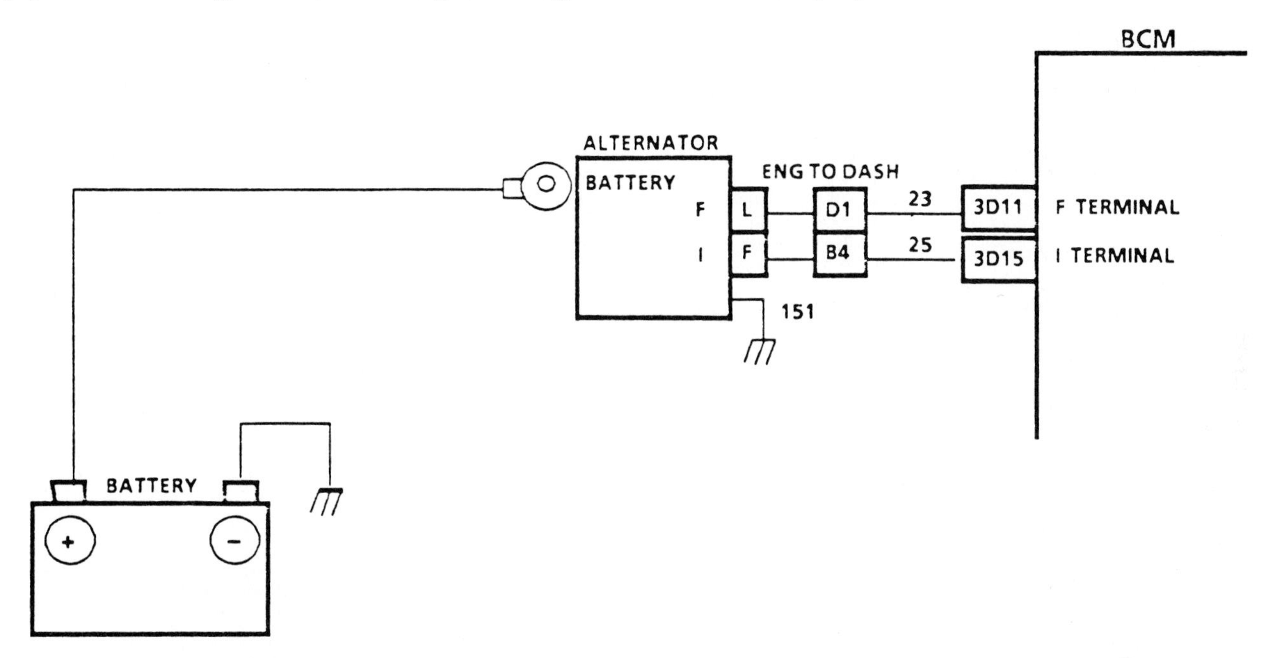

Figure 6–32 Integral Voltage Regulator to Body Computer Module Connections. (Courtesy of General Motors Corporation)

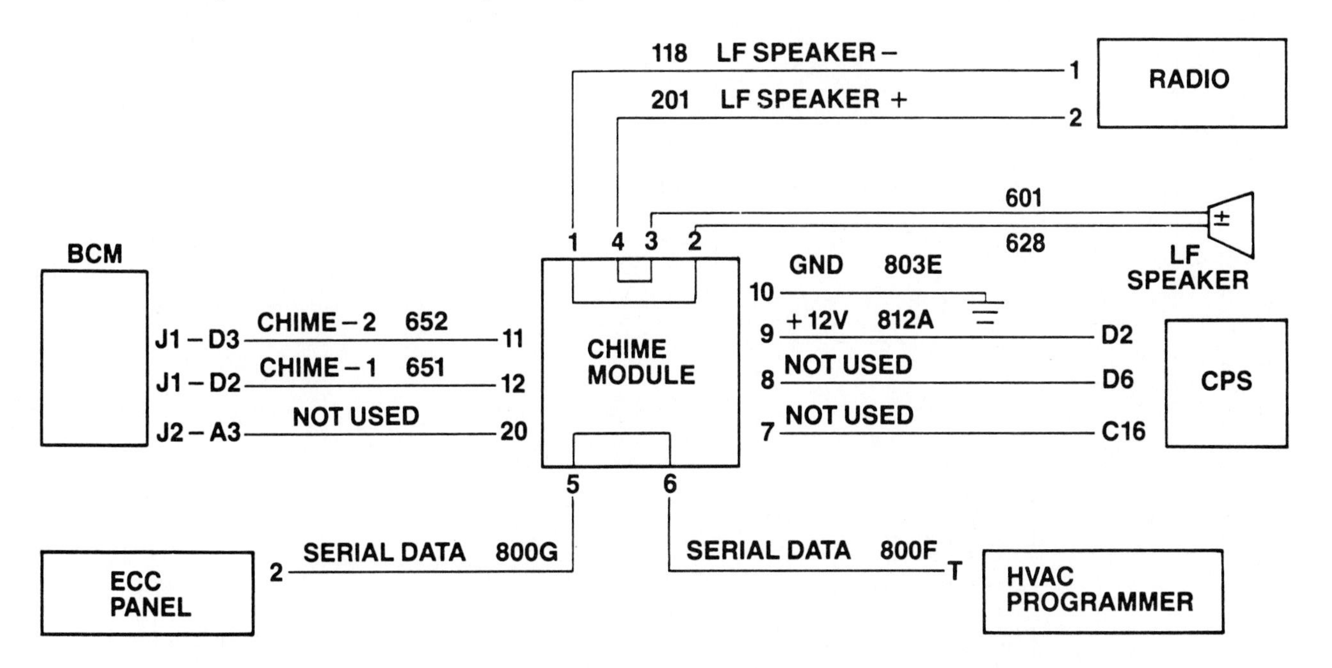

Figure 6–33 Chime Module Circuit. (Courtesy of General Motors Corporation)

The same wiring harness connector will fit the chime module or voice-chime module. A voice-chime module circuit is shown in Figure 6–34.

Relay Centers

Many General Motors vehicles have two relay centers. One of these centers is usually located under the hood and the other is positioned under the dash or in the console.

Each relay is identified on a decal in the relay cover. A fuse panel may be located with the relay center, as pictured in Figure 6–35.

The BCM wiring varies depending on the make of car. A typical BCM wiring diagram is illustrated in Figure 6–36.

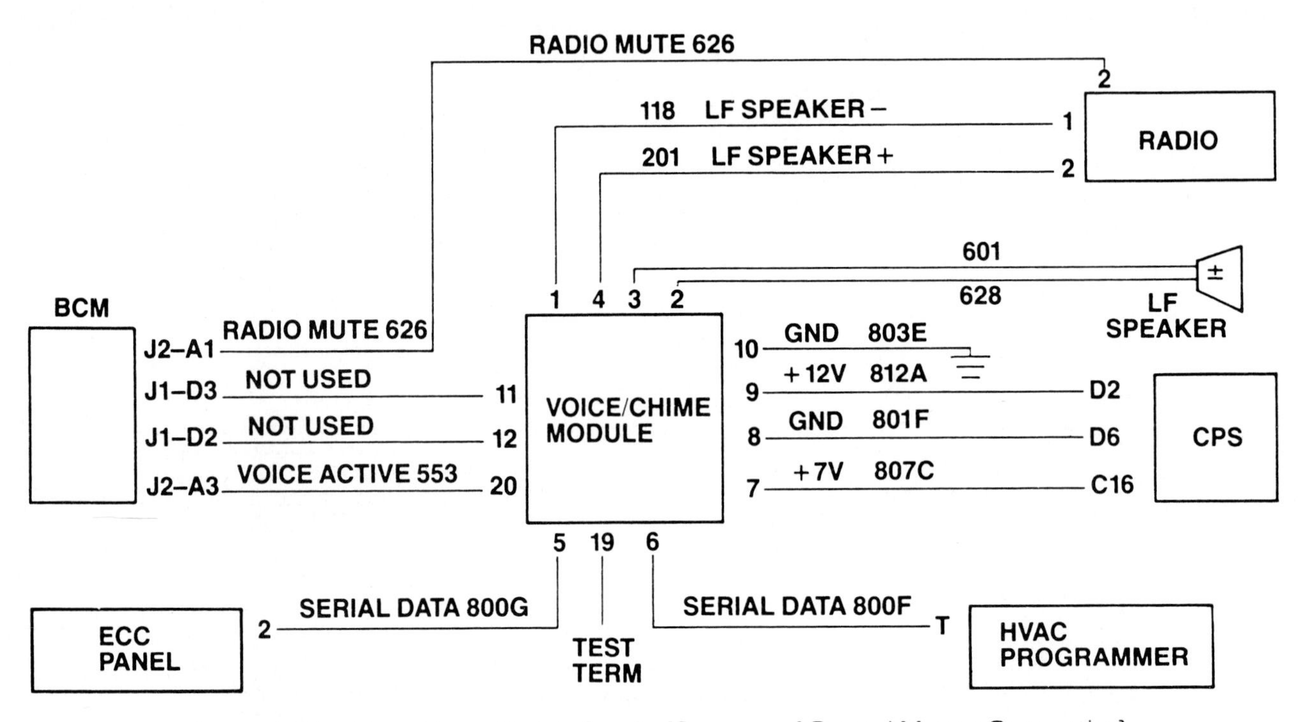

Figure 6–34 Voice-Chime Module Circuit. (Courtesy of General Motors Corporation)

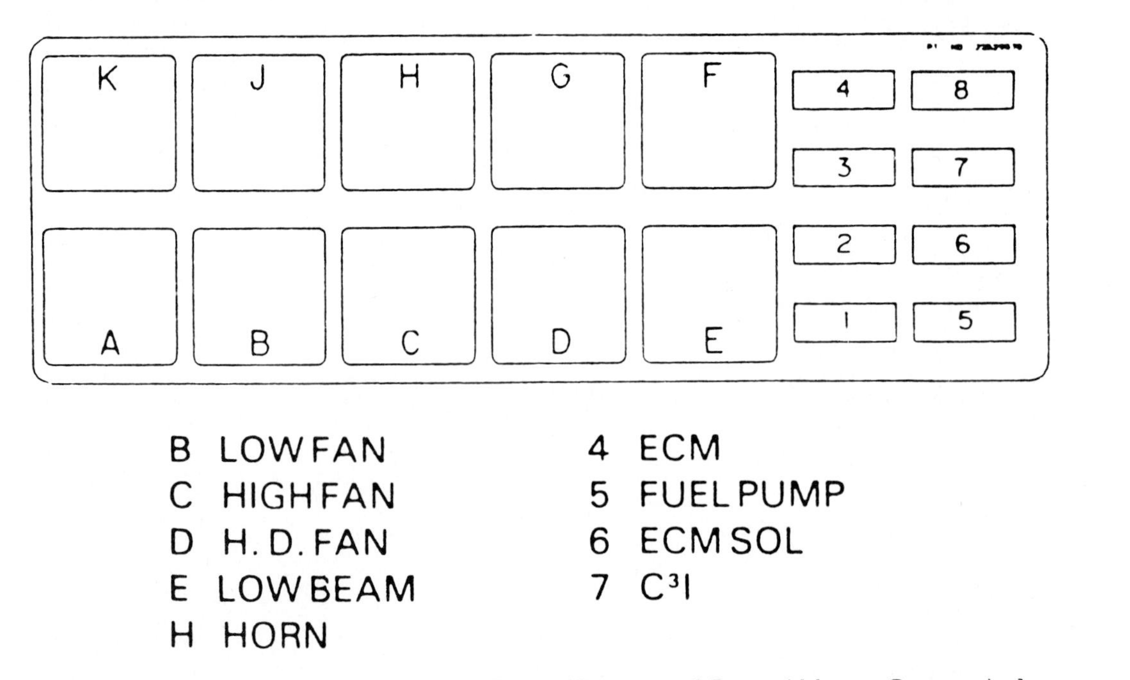

Figure 6–35 Relay Center and Fuse Panel. (Courtesy of General Motors Corporation)

Diagnosis of General Motors Computer Systems with Body Computer Module

General Diagnostic Procedure

The diagnostic procedure is started by pressing the off and warm buttons in the electronic climate control panel simultaneously for 3 seconds with the ignition switch on. An exit from the diagnostic mode occurs when the bi-level button is pressed. If the off button is pressed, the diagnostic procedure will return to the next selection in the previous test mode. The hi and lo buttons are used to select level, test type, or device in the diagnosis. The hi button could be considered to be a "yes" button, whereas the lo button is a "no" command. For example, if a specific test mode appears in the instrument panel cluster (IPC) information center, the technician may select this test mode by pressing the hi button. When the

#1 — BLACK (C1–C16 / D1–D16)

CIR DESC	WIRE COL	CKT NO	Pin	Pin	CKT NO	WIRE COL	CKT DESC
CRUISE BRAKE SW	BRN	86	1	1	801	BLK/RED	CPS GROUND
BRAKE FLD LEV SW	TAN/WHT	33	2	2	651	PPL/YEL	CHIME 1
STOP LAMP OUT	PPL/WHT	549	3	3	652	PPL/WHT	CHIME 2
C.C. SET/COAST SW	DK BLU	84	4	4	733	LT BLU	MIX DR. WIPER
PARK BRAKE SW	TAN/WHT	233	5	5	801	BLK/RED	CPS GROUND
	SPARE		6	6	308	GRA/BLK	PARKLAMPS ON
	SPARE		7	7	343	WHT	IGNITION OFF
HEADLAMPS ON	YEL	10	8	8		SPARE	
CRANK INPUT	LT GRN	80	9	9	992	BLK/PNK	TWILIGHT PK REL
CPS 12 VOLT	RED/WHT	812A	10	10	760	PPL/WHT	BLOWER CONTROL
CRUISE ON/OFF	GRA/WHT	397	11	11	692	BLK/PNK	TWILIGHT H/L REL
CPS WAKE-UP	DK BLU/WHT	555	12	12	690	GRA/BLK	COURTESY LT REL
	SPARE		13	13		SPARE	
C.C. RESUME/ACCEL	GRA/BLK	87	14	14	736	BLK/PNK	5V RETURN (GND)
CPS 7 VOLT	PPL/WHT	807	15	15	801	BLK/RED	CPS GROUND
CPS 7 VOLT	PPL/WHT	807	16	16	705	TAN	5V REFERENCE

#2 — BLACK (A1–A12 / B1–B12)

CIR DESC	WIRE COL	CKT NO	Pin	Pin	CKT NO	WIRE COL	CKT DESC
RADIO MUTE	DK GRN/ORN	626	1	1		SPARE	
IN CAR TEMP	DK GRN/WHT	734	2	2	402	LT GRN	CRUISE VAC ON
VOICE ACTIVE	TAN/WHT	553	3	3		SPARE	
WASHER FLUID LEV	BLK/WHT	99	4	4	271	PPL	TWILIGHT DELAY
COURTESY LT SW	BRN/YEL	685	5	5	800A	TAN	SERIAL DATA
	SPARE		6	6	356	GRA/BLK	PRNDL
HEADLAMP OUT	DK BLU/YEL	539	7	7		SPARE	
CPS GROUND	BLK/RED	801	8	8	30	PPL	FUEL LEVEL WIPER
	SPARE		9	9	398	TAN	CRUISE SERVO HI
CRUISE SERVO LO	LT BLU/BLK	399	10	10	529	DK GRN/YEL	PARK LAMP OUT
VF DIMMING	PPL/YEL	724	11	11		SPARE	
	SPARE		12	12	403	DK BLU/WHT	CRUISE VENT ON

#3 — RED (C1–C16 / D1–D16)

CIR DESC	WIRE COL	CKT NO	Pin	Pin	CKT NO	WIRE COL	CKT DESC
VSS RETURN	PPL	401	1	1		SPARE	
	SPARE		2	2		SPARE	
TAILLAMP OUT	LT BLU/BLK	519	3	3		SPARE	
VSS TO ECM	BRN	437	4	4	731	GRA/RED	A/C LO SIDE TEMP
	SPARE		5	5	732	DK BLU	A/C HI SIDE TEMP
SUNLOAD SENSOR	LT BLU/YEL	590	6	6		SPARE	
DOOR HANDLE SW	GRA	157	7	7	39	PNK/BLK	IGNITION REF
DOOR JAMB SW	WHT/DK GRN	156	8	8	278	WHT/BLK	TWI PHOTOCELL
	SPARE		9	9	313	LT GRN	OIL PRESSURE
VSS FEED	YEL	400	10	10	800	TAN	SERIAL DATA
IGNITION OFF	WHT	343	11	11	23	GRA/WHT	GEN F TERMINAL
DRIVER DOOR AJAR	GRA/BLK	147	12	12	686	TAN/BLK	DIM CONTROL
KEY IN IGNITION	LT GRN	80	13	13	69	GRA	COOLANT LEVEL
PASS.DOOR AJAR	BLK/ORN	158	14	14	735	LT GRN/BLK	OUTSIDE TEMP
SEAT BELT SW	BLK/PNK	238	15	15	25	BRN	GEN I TERMINAL
PARK SIGNAL	LT GRN/BLK	275	16	16	721	WHT	LO REF PRESSURE

6E OLDS

Figure 6-36 Body Computer Module Wiring. (Courtesy of General Motors Corporation)

technician does not want a certain test mode indicated in the information center, the lo button is pressed. The hi button is used to move on to the next parameter in a specific test mode, while the lo button returns the diagnostic procedure to the previous parameter. The diagnostic codes or parameters appear in the IPC information center, and certain override values appear in the electronic climate control (ECC) display. For example, an override value of 0 to 99 is given for blower speed. The warm button increases the override value and blower speed, whereas the cool button is used to decrease these values. All of the diagnostic functions performed by the EEC buttons are illustrated in Figure 6–37.

ECM and BCM Diagnostic Codes

When the diagnostic mode is entered, all the segments in the IPC and ECC panel will be illuminated for a short time, after which ECM and BCM codes will be displayed in the IPC information center. The ECM codes begin with the letter E, and the BCM codes begin with B. The ECM codes are listed in Figure 6–38, followed by the BCM codes in Figure 6–39 and diagnostic code comments in Figure 6–40.

ECM Diagnostic Functions

When the ECM and BCM codes have been displayed, specific ECM diagnostic functions are displayed in the information center. These functions include data, inputs, and outputs. If the technician wants to enter one of these functions, the hi button must be pressed. When the lo button is pressed, the next function choice appears. Once a specific function is selected, the hi button will advance the diagnosis to the next parameter in that function. The diagnosis of ECM data, inputs, and outputs is shown in Figure 6–41.

After the "ECM Outputs" display, "Clear ECM Codes" will appear in the information center. If ECM codes are in the computer memory, they may be erased by pressing the hi button for three seconds.

BCM Diagnostic Functions

The BCM diagnostic functions are available following the ECM diagnostic functions. BCM diagnostic functions include data, inputs, outputs, and overrides, and these functions are selected using the same procedure as the ECM diagnostic functions. The BCM diagnostic functions are illustrated in Figure 6–42.

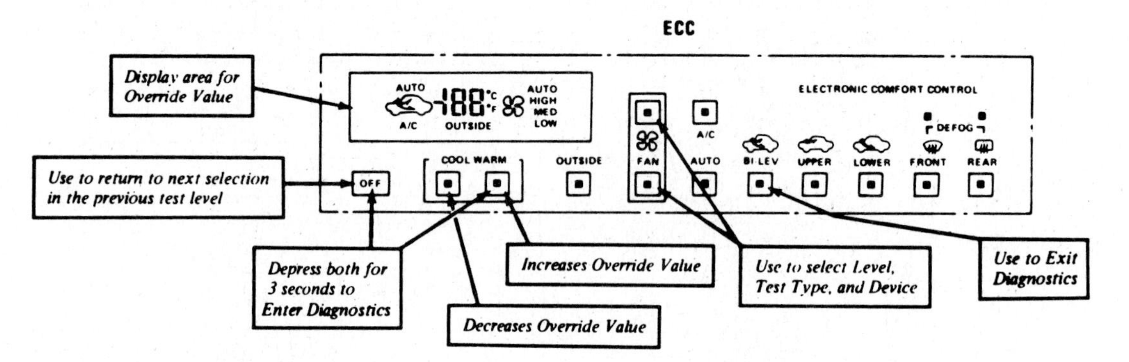

Figure 6–37 Diagnostic Functions of Electronic Climate Control Buttons. [Courtesy of General Motors Corporation]

CODE	DESCRIPTION	COMMENTS
E013	Open Oxygen Sensor Circuit [Canister Purge]	Ⓐ
E014	Coolant Sensor Temperature Too High	Ⓐ/Ⓕ/Ⓜ
E015	Coolant Sensor Temperature Too Low	Ⓐ/Ⓕ/Ⓜ
E016	System Voltage Out of Range [All Solenoids]	Ⓐ
E021	Throttle Position Sensor Voltage Too High [TCC]	Ⓐ
E022	Throttle Position Sensor Voltage Too Low [TCC]	Ⓐ
E024	Speed Sensor Circuit [TCC]	Ⓐ
E029	Open Fourth Gear Circuit	Ⓐ
E032	EGR System Fault	Ⓐ
E033	MAF Sensor Voltage High	Ⓐ
E034	MAF Sensor Voltage Low	Ⓐ

CODE	DESCRIPTION	COMMENTS
E037	MAT (ATS) Sensor Temperature Too High	Ⓐ
E038	MAT (ATS) Sensor Temperature Too Low	Ⓐ
E040	Open Power Steering Pressure Switch Circuit [A/C Clutch]	Ⓐ
E041	Cam Sensor Circuit -(C^3I)	Ⓐ
E042	Ignition System -(C^3I)	Ⓐ/Ⓙ
E043	ESC System	Ⓐ
E044	Lean Exhaust Signal	Ⓐ/Ⓘ
E045	Rich Exhaust Signal	Ⓐ/Ⓘ
E047	BCM-ECM Communication [A/C Clutch & Cruise]	Ⓐ
E051	ECM PROM Error	Ⓐ/Ⓙ/Ⓚ
E052	Calpak Error	Ⓐ
E055	ECM Error	Ⓐ

Figure 6–38 Electronic Control Module [ECM] Codes. [Courtesy of General Motors Corporation]

At the conclusion of the BCM diagnostic functions, "Clear BCM Codes" will appear in the information center. These codes are erased in the same way as ECM codes. The IPC inputs shown in Figure 6–43 are available after the BCM diagnostic functions.

The BCM and ECM codes and diagnostic functions will vary depending on the model and year of vehicle. The codes and functions shown in this chapter are from a 1986 Toronado. (Refer to Chapter 4 for operation and diagnosis of General Motors Computer systems with conventional ECMs and electronic fuel injection.)

Test Questions

1. When the diagnostic mode is entered, the:
   (a) off and hi buttons must be pressed simultaneously.
   (b) off and lo buttons must be pressed simultaneously.
   (c) off and warm buttons must be pressed simultaneously.
2. When an exit is made from the diagnostic mode the:
   (a) off button must be pressed.
   (b) bilevel button must be pressed.
   (c) rear defog button must be pressed.
3. The ECM and BCM diagnostic codes are displayed in the:
   (a) electronic climate control panel.
   (b) instrument panel cluster (IPC) information center.
   (c) radio-stereo panel.
4. The buttons used to increase or decrease the BCM override values are the:
   (a) hi and lo buttons.
   (b) off and auto buttons.
   (c) cool and warm buttons.
5. The cruise control module is contained in the:
   (a) BCM.
   (b) ECM.
   (c) IPC.

CODE	DESCRIPTION	COMMENTS
B110	Outside Air Temperature Circuit	Ⓑ / Ⓗ
B111	A/C High Side Temperature Circuit	Ⓑ
B112	A/C Low Side Temperature Circuit [A/C Clutch]	Ⓑ / Ⓔ
B113	In-Car Temperature Circuit	Ⓑ
B115	Sunload Temperature Circuit	Ⓑ
B118	Door Jam/Ajar Circuit	Ⓒ
B119	Twilight Sentinel Photosensor Circuit	Ⓑ
B120	Twilight Sentinel Delay Pot Circuit	Ⓑ
B122	Panel Lamp Dimming Pot Circuit	Ⓑ
B123	Courtesy Lamps On Circuit	Ⓑ
B124	Speed Sensor Circuit [Cruise]	Ⓗ
B127	PRNDL Sensor Circuit [Cruise]	Ⓜ
B131	Oil Pressure Sensor Circuit	Ⓜ
B132	Oil Pressure Sensor Circuit	Ⓜ
B334	Loss of ECM Serial Data [Cruise and A/C Clutch]	Ⓐ / Ⓑ / Ⓔ / Ⓖ
B335	Loss of ECC Serial	Ⓑ / Ⓖ / Ⓛ
B336	Loss of IPC Serial Data	Ⓑ / Ⓖ
B337	Loss of Programmer Serial Data [A/C Clutch]	Ⓑ / Ⓔ / Ⓖ
B338	Loss of Voice Serial Data	Ⓑ / Ⓖ
B409	Generator Detected Condition	Ⓑ
B411	Battery Volts Too Low [Cruise]	Ⓑ
B412	Battery Volts Too High [Cruise]	Ⓑ
B440	Air Mix Door	Ⓑ
B445	Compressor Clutch Engagement [A/C Clutch]	Ⓑ / Ⓔ
B446	Low A/C Refrigerant Condition Warning	Ⓑ
B447	Very Low A/C Refrigerant [A/C Clutch]	Ⓑ / Ⓔ
B448	Very Low A/C Refrigerant Pressure Condition [A/C Clutch]	Ⓑ / Ⓔ
B449	A/C High Side Temperature Too High [A/C Clutch]	Ⓒ / Ⓔ
B450	Coolant Temperature Too High [A/C Clutch]	Ⓒ / Ⓔ
B552	BCM Memory Reset Indicator	
B556	BCM EEPROM Error	Ⓓ
B660	Cruise - Transmission Not In Drive [Cruise]	Ⓒ
B663	Cruise - Car Speed and Set Speed Difference Too High [Cruise]	Ⓒ
B664	Cruise - Car Acceleration Too High [Cruise]	Ⓒ
B667	Cruise - Cruise Switch Shorted [Cruise]	Ⓑ
B671	Cruise - Servo Position Sensor Circuit [Cruise]	Ⓑ
B672	Cruise - Vent Solenoid Circuit [Cruise]	Ⓑ
B673	Cruise - Vacuum Solenoid Circuit [Cruise]	Ⓑ

Figure 6–39 Body Computer Module (BCM) Codes. (Courtesy of General Motors Corporation)

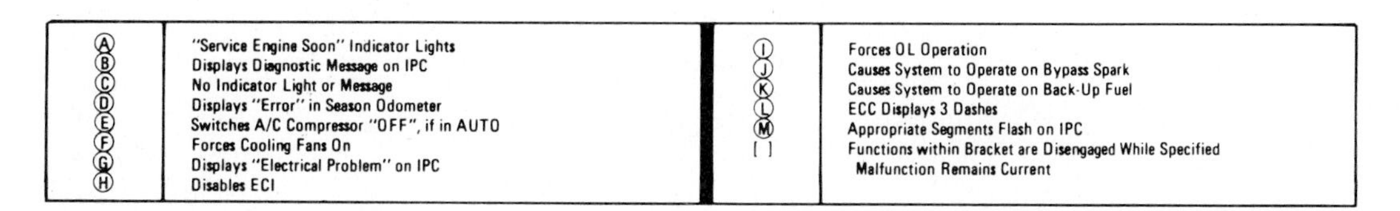

Ⓐ	"Service Engine Soon" Indicator Lights
Ⓑ	Displays Diagnostic Message on IPC
Ⓒ	No Indicator Light or Message
Ⓓ	Displays "Error" in Season Odometer
Ⓔ	Switches A/C Compressor "OFF", if in AUTO
Ⓕ	Forces Cooling Fans On
Ⓖ	Displays "Electrical Problem" on IPC
Ⓗ	Disables ECI
Ⓘ	Forces OL Operation
Ⓙ	Causes System to Operate on Bypass Spark
Ⓚ	Causes System to Operate on Back-Up Fuel
Ⓛ	ECC Displays 3 Dashes
Ⓜ	Appropriate Segments Flash on IPC
[ ]	Functions within Bracket are Disengaged While Specified Malfunction Remains Current

Figure 6–40 Diagnostic Code Comments. (Courtesy of General Motors Corporation)

6. The central power supply provides a constant 7V and 12V to the:
   (a) ECM.
   (b) BCM only.
   (c) BCM, IPC, and voice module.

7. The 7V outputs from the central power supply are available:
   (a) at all times.
   (b) when the ignition switch is turned on.
   (c) when a vehicle door is opened.

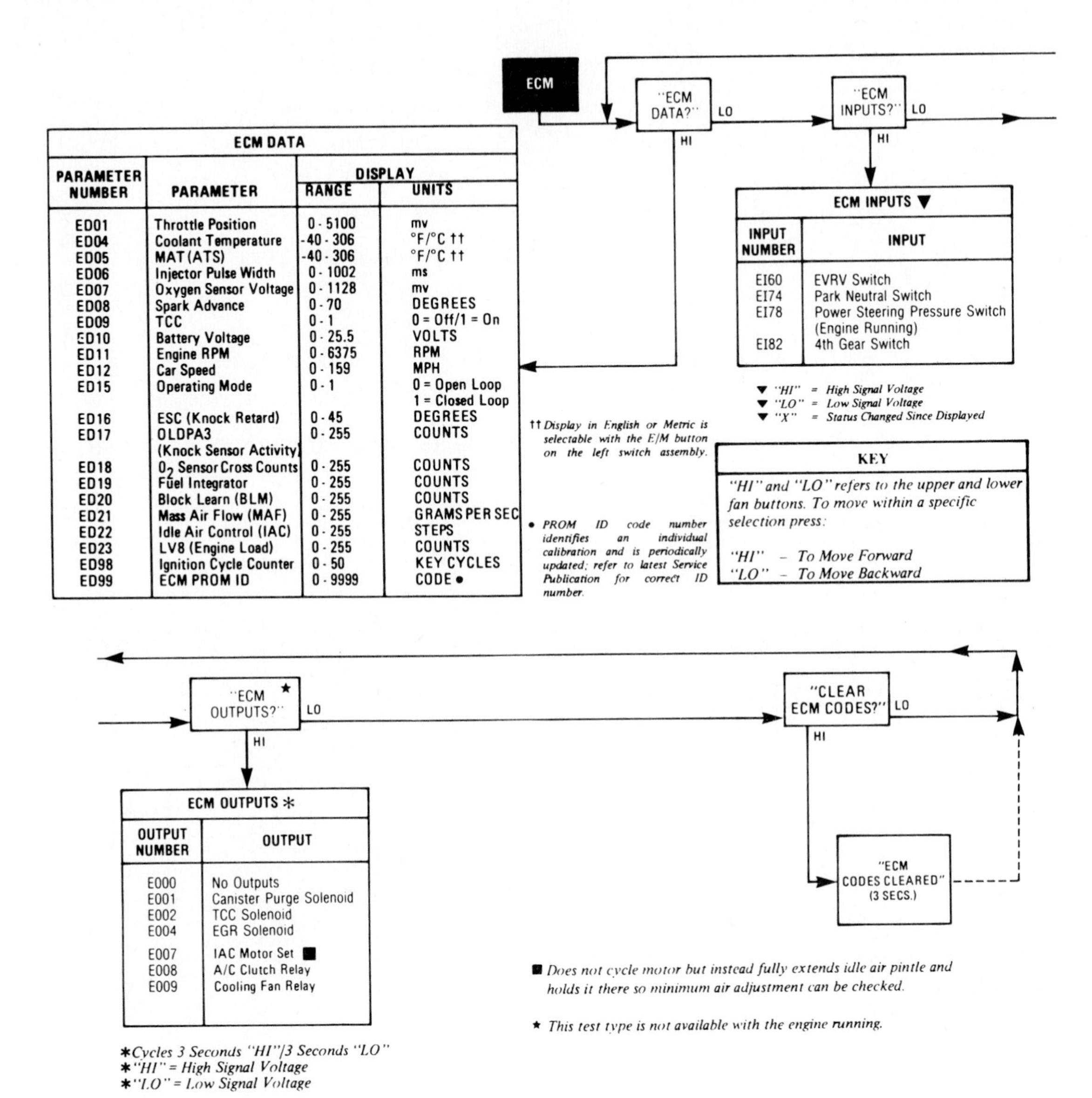

Figure 6–41 Electronic Control Module (ECM) Diagnostic Functions. (Courtesy of General Motors Corporation)

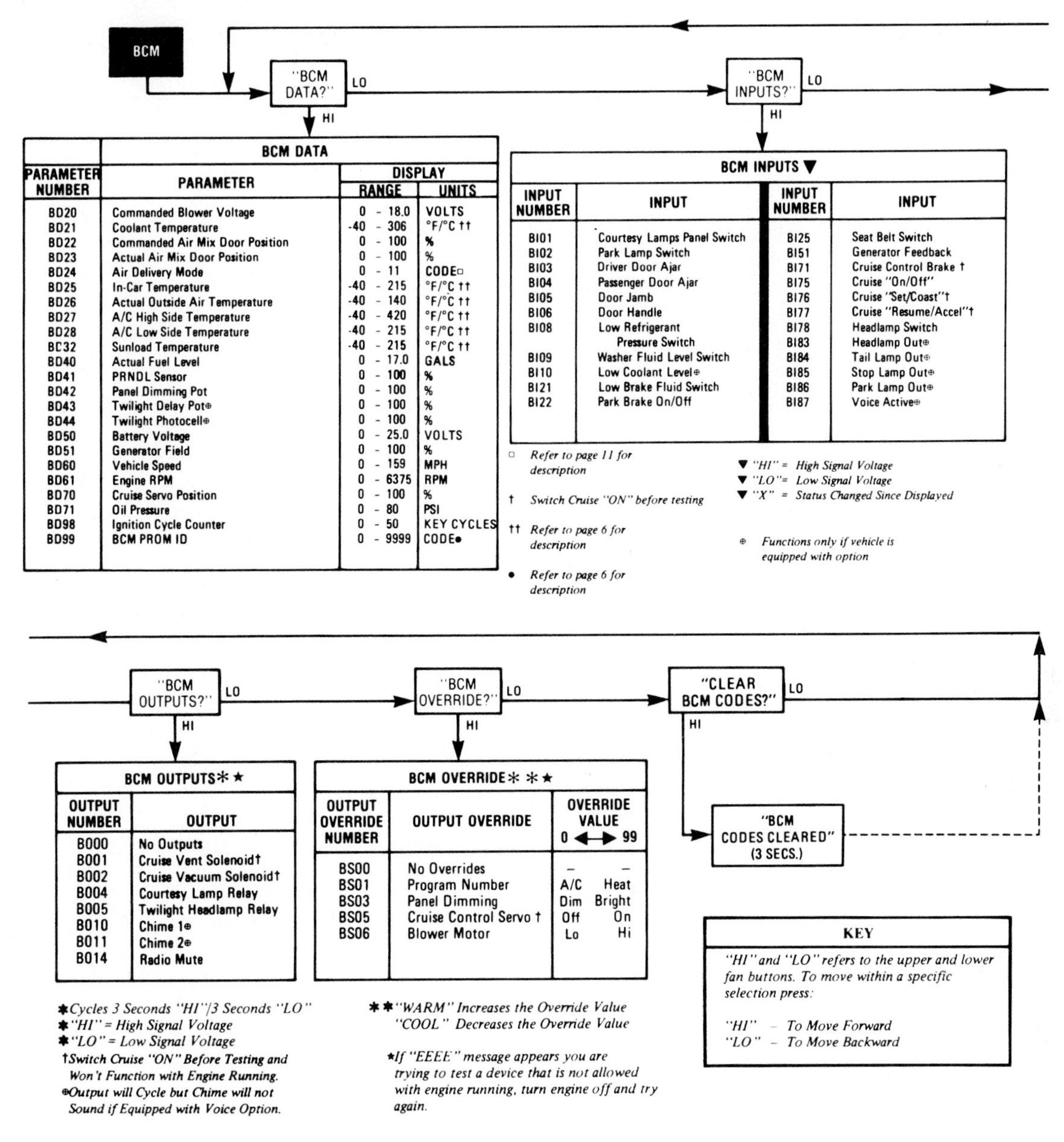

BCM
"BCM DATA?"
LO
HI
"BCM INPUTS?"
LO
HI
"BCM OUTPUTS?"
LO
HI
"BCM OVERRIDE?"
LO
HI
"CLEAR BCM CODES?"
LO
HI
"BCM CODES CLEARED" (3 SECS.)
□ Refer to page 11 for description
† Switch Cruise "ON" before testing
†† Refer to page 6 for description
• Refer to page 6 for description
▼ "HI" = High Signal Voltage
▼ "LO" = Low Signal Voltage
▼ "X" = Status Changed Since Displayed
⊕ Functions only if vehicle is equipped with option
✱Cycles 3 Seconds "HI"/3 Seconds "LO"
✱"HI" = High Signal Voltage
✱"LO" = Low Signal Voltage
†Switch Cruise "ON" Before Testing and Won't Function with Engine Running.
⊕Output will Cycle but Chime will not Sound if Equipped with Voice Option.
✱✱"WARM" Increases the Override Value
"COOL" Decreases the Override Value
★If "EEEE" message appears you are trying to test a device that is not allowed with engine running, turn engine off and try again.
KEY
"HI" and "LO" refers to the upper and lower fan buttons. To move within a specific selection press:
"HI" – To Move Forward
"LO" – To Move Backward

BCM DATA

PARAMETER NUMBER	PARAMETER	DISPLAY RANGE	DISPLAY UNITS
BD20	Commanded Blower Voltage	0 - 18.0	VOLTS
BD21	Coolant Temperature	-40 - 306	°F/°C ††
BD22	Commanded Air Mix Door Position	0 - 100	%
BD23	Actual Air Mix Door Position	0 - 100	%
BD24	Air Delivery Mode	0 - 11	CODE□
BD25	In-Car Temperature	-40 - 215	°F/°C ††
BD26	Actual Outside Air Temperature	-40 - 140	°F/°C ††
BD27	A/C High Side Temperature	-40 - 420	°F/°C ††
BD28	A/C Low Side Temperature	-40 - 215	°F/°C ††
BC32	Sunload Temperature	-40 - 215	°F/°C ††
BD40	Actual Fuel Level	0 - 17.0	GALS
BD41	PRNDL Sensor	0 - 100	%
BD42	Panel Dimming Pot	0 - 100	%
BD43	Twilight Delay Pot⊕	0 - 100	%
BD44	Twilight Photocell⊕	0 - 100	%
BD50	Battery Voltage	0 - 25.0	VOLTS
BD51	Generator Field	0 - 100	%
BD60	Vehicle Speed	0 - 159	MPH
BD61	Engine RPM	0 - 6375	RPM
BD70	Cruise Servo Position	0 - 100	%
BD71	Oil Pressure	0 - 80	PSI
BD98	Ignition Cycle Counter	0 - 50	KEY CYCLES
BD99	BCM PROM ID	0 - 9999	CODE•

BCM INPUTS ▼

INPUT NUMBER	INPUT	INPUT NUMBER	INPUT
BI01	Courtesy Lamps Panel Switch	BI25	Seat Belt Switch
BI02	Park Lamp Switch	BI51	Generator Feedback
BI03	Driver Door Ajar	BI71	Cruise Control Brake †
BI04	Passenger Door Ajar	BI75	Cruise "On/Off"
BI05	Door Jamb	BI76	Cruise "Set/Coast"†
BI06	Door Handle	BI77	Cruise "Resume/Accel"†
BI08	Low Refrigerant Pressure Switch	BI78	Headlamp Switch
		BI83	Headlamp Out⊕
BI09	Washer Fluid Level Switch	BI84	Tail Lamp Out⊕
BI10	Low Coolant Level⊕	BI85	Stop Lamp Out⊕
BI21	Low Brake Fluid Switch	BI86	Park Lamp Out⊕
BI22	Park Brake On/Off	BI87	Voice Active⊕

BCM OUTPUTS✱ ★

OUTPUT NUMBER	OUTPUT
B000	No Outputs
B001	Cruise Vent Solenoid†
B002	Cruise Vacuum Solenoid†
B004	Courtesy Lamp Relay
B005	Twilight Headlamp Relay
B010	Chime 1⊕
B011	Chime 2⊕
B014	Radio Mute

BCM OVERRIDE✱ ✱★

OUTPUT OVERRIDE NUMBER	OUTPUT OVERRIDE	OVERRIDE VALUE 0	◀▶ 99
BS00	No Overrides	–	–
BS01	Program Number	A/C	Heat
BS03	Panel Dimming	Dim	Bright
BS05	Cruise Control Servo †	Off	On
BS06	Blower Motor	Lo	Hi

Figure 6–42 Body Computer Module [BCM] Diagnostic Functions. [Courtesy of General Motors Corporation]

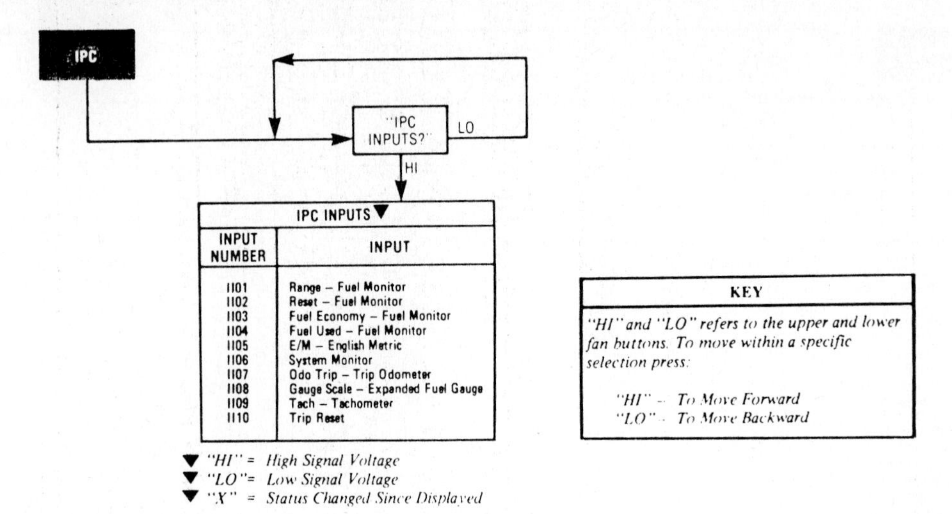

Figure 6-43 Instrument Panel Cluster (IPC) Input Diagnosis. (Courtesy of General Motors Corporation)

7

Computer Systems with Transmission Control

Practical Completion Objectives

1. Demonstrate an understanding of the electronically controlled transmission (ECT) system.
2. Diagnose the ECT system with the self-diagnostics in the electronic control unit (ECU).
3. Use a voltmeter to diagnose the throttle-position sensor, brake switch, and shift timing.

Toyota Electronically Controlled Transmission

Purpose

The electronically controlled transmission (ECT) is designed to improve fuel economy and performance. Since shift points and converter clutch lockup are controlled by the electronic control unit (ECU), these functions are optimized for economy and performance. The ECT system is available on four-speed automatic overdrive transmissions in rear wheel drive cars and also on four-speed automatic transaxles in front wheel drive cars.

Electronic Control System

A separate ECU is used for the ECT system. The inputs to this ECU are the following:

1. Driving pattern selector
2. Neutral/start switch
3. Throttle-position sensor
4. Brake-light switch
5. Vehicle-speed sensors
6. Overdrive switch

The ECU operates three solenoids in the transmission to control the converter clutch lockup and transmission shift points. A block diagram of the ECU inputs and outputs is shown in Figure 7–1, and the location of the ECT components is pictured in Figure 7–2.

Throttle Position Sensor A direct or an indirect type of throttle-position sensor may be used in the ECT system. The direct type of throttle-position sensor contains an idle (IDL) contact that indicates a fully closed throttle, while three other contacts, L1, L2, and L3, provide signals at intermediate throttle positions. An E2 contact is a ground terminal. The direct type of sensor is illustrated in Figure 7–3, and Figure 7–4 indicates the percentage of throttle opening at which the various contacts provide signals to the ECU.

The indirect type of throttle-position sensor contains a variable resistor. A 5V reference signal is sent from the Toyota Computer Controlled System (TCCS) ECU to the VC terminal in the sensor. The variable signal from the sensor TA terminal is sent to the TCCS ECU. A signal converter in the TCCS ECU relays this signal to the ECT ECU. The IDL terminal in the sensor sends a fully closed throttle signal to the ECT ECU, and the E terminal provides a ground connection between the sensor and the TCCS ECU. Outputs controlled by the TCCS ECU include electronic fuel injection (EFI) and spark advance. (These output control functions are similar to the ones explained in Chapters 3 and 4.) An indirect type of throttle-position sensor is shown in Figure 7–5.

Speed Sensors The main, number 1, speed sensor is located in the transmission extension housing. A magnet attached to the output shaft rotates past the sensor. This rotating magnet operates a reed switch in the sensor that sends a vehicle-speed signal to the

ECT ECU. The signals from the throttle-position sensor and the main speed sensor are used by the ECU to determine the transmission shift points and the converter clutch lockup schedule. The main, number 1, speed sensor is illustrated in Figure 7–6.

The backup, number 2, speed sensor is mounted in the speedometer. This sensor also contains a rotating magnet and a reed switch that generates four pulses for each speedometer cable revolution. When both speed sensors are operating normally, the ECU

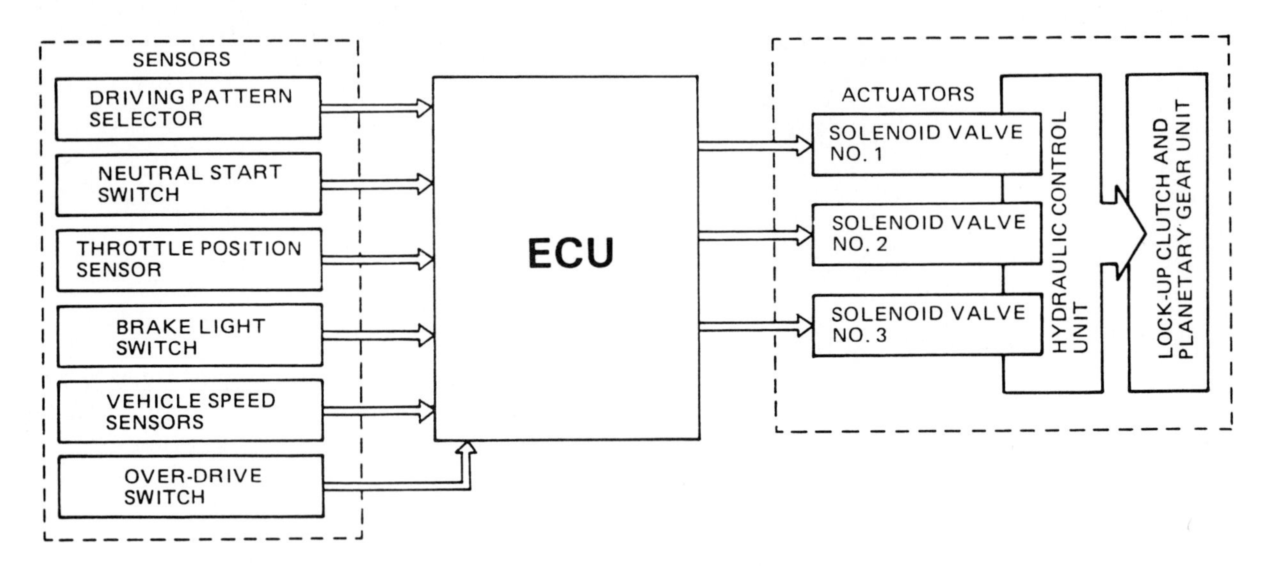

Figure 7–1 Electronically Controlled Transmission (ECT) System Inputs and Outputs. (Courtesy of Toyota Motor Corporation, Japan)

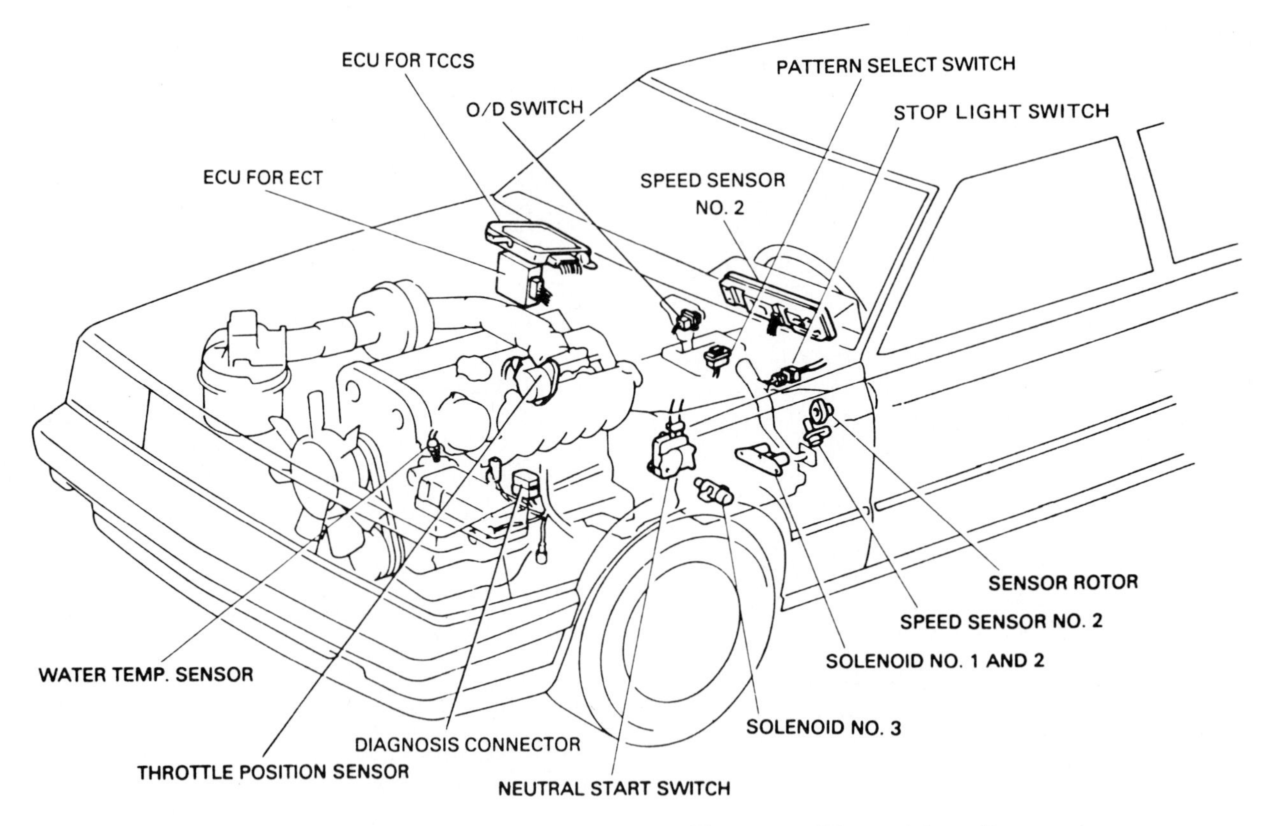

Figure 7–2 ECT System Component Location. (Courtesy of Toyota Motor Corporation, Japan)

uses the signal from the main speed sensor. If the main speed sensor is defective, the ECU is programmed to use the signal from the backup speed sensor. The speed sensor signal replaces governor pressure in a conventional hydraulically operated automatic transmission. A backup, number 2, speed sensor is pictured in Figure 7–7.

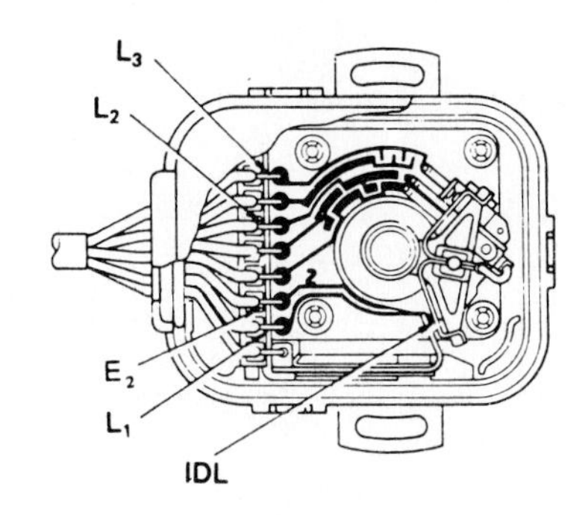

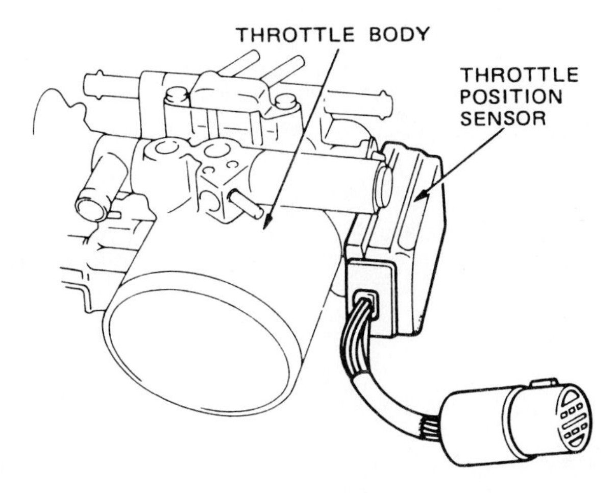

Figure 7–3 Direct Type of Throttle-Position Sensor. [Courtesy of Toyota Motor Corporation, Japan]

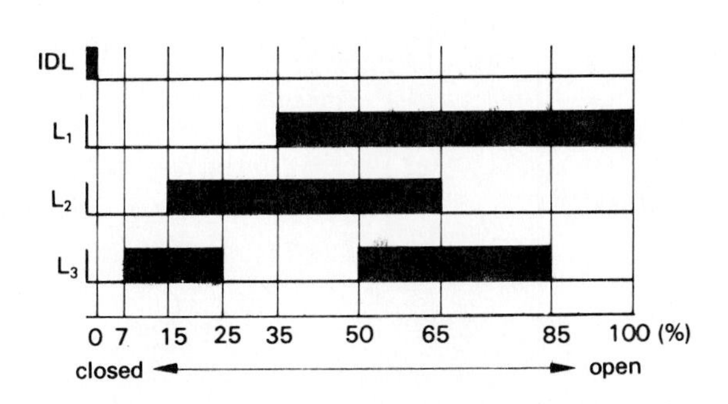

Figure 7–4 Direct Type of Throttle-Position Sensor Signals in Relation to Throttle Opening. [Courtesy of Toyota Motor Corporation, Japan]

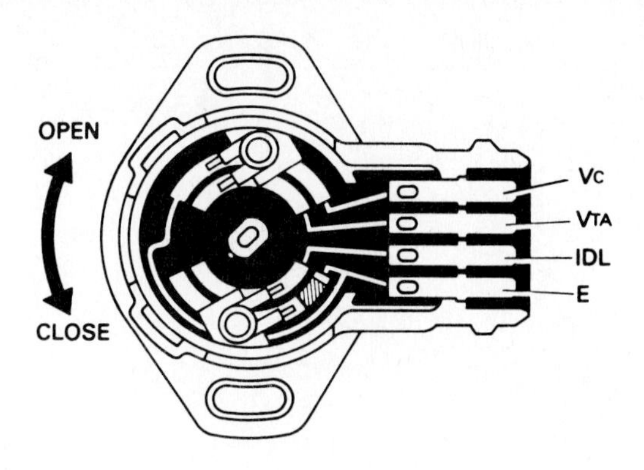

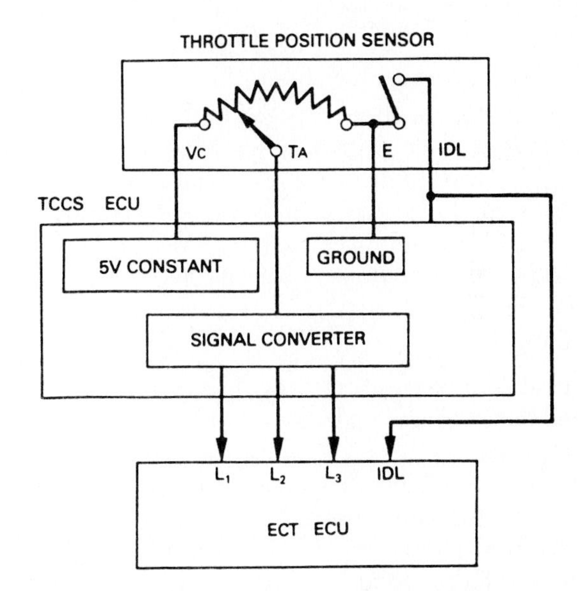

Figure 7–5 Indirect Type of Throttle-Position Sensor. [Courtesy of Toyota Motor Corporation, Japan]

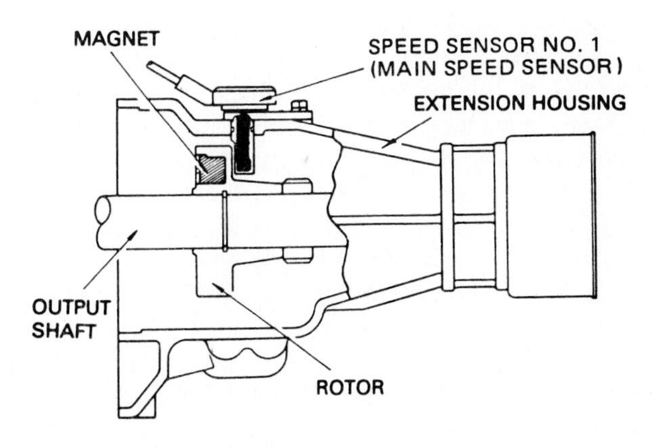

Figure 7–6 Main Speed Sensor. [Courtesy of Toyota Motor Corporation, Japan]

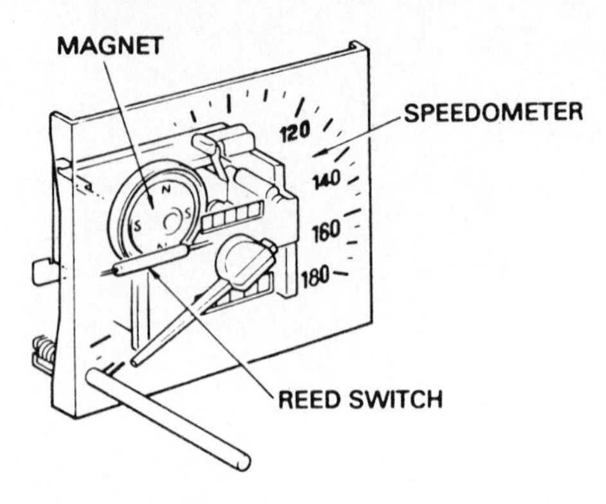

Figure 7–7 Backup Speed Sensor. [Courtesy of Toyota Motor Corporation, Japan]

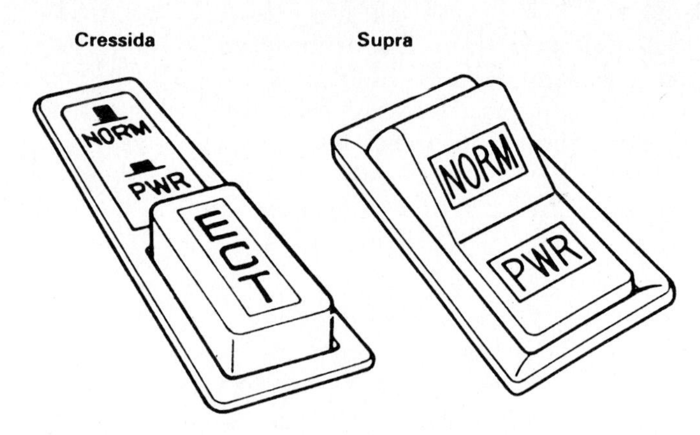

Figure 7–8 Pattern Select Switches. [Courtesy of Toyota Motor Corporation, Japan]

Pattern Select Switch The pattern select switch is mounted in the transmission shift console. This switch allows the driver to select various ECU modes that provide different transmission shift points. Some pattern select switches have power or normal modes, whereas other switches provide economy, power, and normal modes. Pattern select switches are shown in Figure 7–8.

Neutral/Start Switch The neutral/start switch performs the conventional starter solenoid control function. However, this switch also informs the ECT ECU regarding the transmission selector lever position. When the transmission selector is moved from neutral (N) to drive (D), the ECU uses the signal from the neutral safety switch N contact to initiate a shift to third gear and then to first gear, which prevents sudden transmission shocks. This action will only be initiated if the vehicle is stopped, the brake pedal is depressed, and the throttle is in the idle position. The neutral/start switch is illustrated in Figure 7–9.

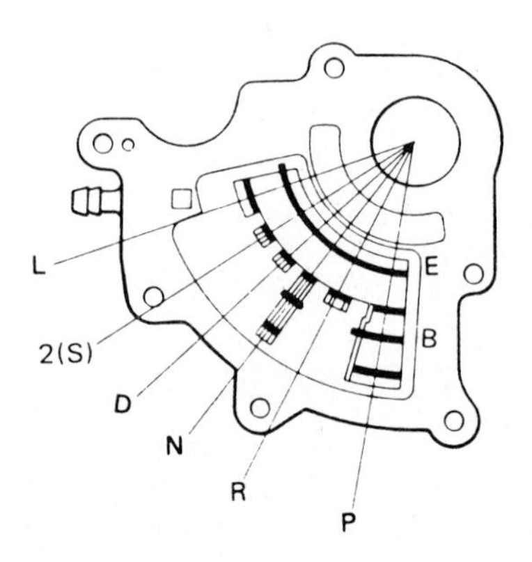

Figure 7–9 Neutral/Start Switch. [Courtesy of Toyota Motor Corporation, Japan]

Overdrive Switch The overdrive (O/D) switch is located in the instrument panel. When this switch is in the on position, the switch contacts are open. Under this condition a 12V signal is supplied through the O/D off lamp to the ECU. This signal informs the ECU to allow the transmission to shift into O/D when the correct input signals are available.

If the O/D switch is moved to the off position, the switch contacts will close. When this occurs, current will flow through the O/D off lamp and the switch contacts to ground. Under this condition, the O/D off lamp will be illuminated, and an OV signal is applied to the ECU, which informs the ECU to cancel the transmission O/D operation. The operation of the O/D switch and off lamp is outlined in Figure 7–10.

The Toyota Computer Controlled System ECU and the cruise control computer will signal the ECT

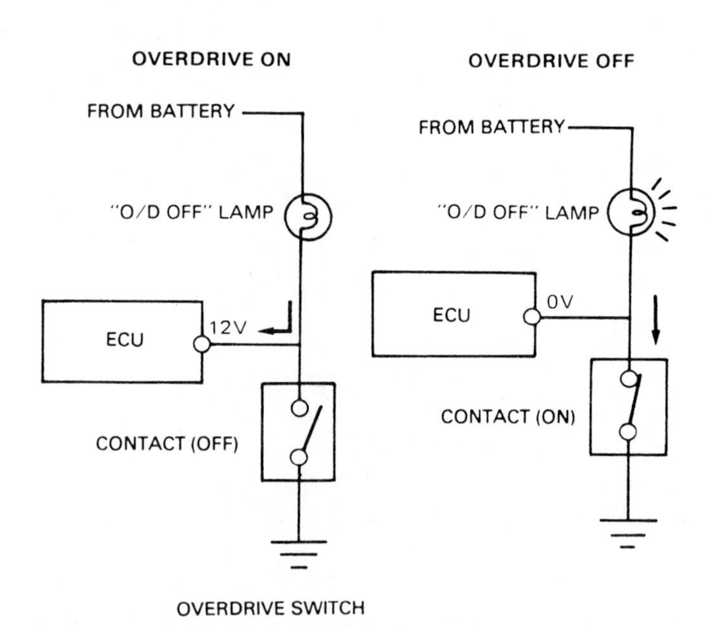

Figure 7–10 Overdrive Switch and Off-Lamp Operation. [Courtesy of Toyota Motor Corporation, Japan]

ECU to prevent overdrive operation if the coolant temperature is below 122°F (50°C). This action will also occur if the cruise control is operating and the preset vehicle speed exceeds the actual vehicle speed by 4 to 6 mph (6 to 10 kph).

Brake-Light Switch The brake-light switch is mounted on the brake-pedal bracket. When the brake pedal is depressed, this switch signals the ECT ECU that the brakes have been applied, and the ECU disengages the converter lockup clutch. This action prevents engine stalling if the rear wheels lock up during a brake application.

Solenoid Valves The number 1 and number 2 solenoid valves are mounted in the transmission valve body. These solenoids contain an electromagnet and a movable plunger. Each plunger tip closes a bleed port in the line pressure hydraulic system. When the solenoids are not energized, the plunger tips keep the bleed ports closed. If either solenoid winding is energized by the ECT ECU, the plunger is lifted and the bleed port is opened. This action reduces the line pressure applied to one of the shift valves, which results in shift valve movement and a transmission shift. The number 1 and number 2 solenoid valves are pictured in Figure 7–11, and the application of these valves in each transmission gear is outlined in Table 7–1.

The solenoid valves are shown with the transmission shift valves in Figure 7–12.

The number 3 solenoid valve is located in the transmission valve body on some models, or mounted in the transmission case on other models. This solenoid also contains an electromagnet and a movable plunger, and the plunger tip opens and closes a bleed port in the line pressure hydraulic system. However, the action of the number 3 solenoid valve controls a lock-up control valve to operate the converter lock-up clutch. (The operation of this lock-up torque converter is similar to the lock-up converter explained in Chapter 3.) Solenoid number 3 is shown in Figure 7–13.

TABLE 7–1 Number 1 and Number 2 Solenoid Application.

Solenoid valve \ Gear	1st	2nd	3rd	OD
No. 1	ON	ON	OFF	OFF
No. 2	OFF	ON	ON	OFF

(Courtesy of Toyota Motor Corporation, Japan)

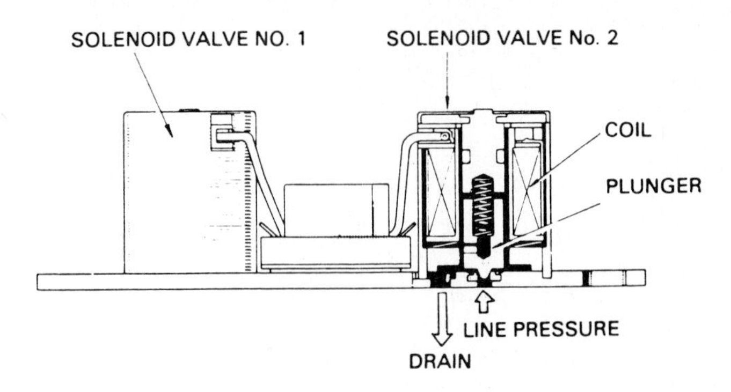

Figure 7–11 Number 1 and Number 2 Solenoid Valves. (Courtesy of Toyota Motor Corporation, Japan)

When the number 3 solenoid valve is not energized, the lock-up control valve is positioned so it directs fluid through the hollow input shaft into the converter. This action moves the lock-up clutch away from the front of the converter, and the clutch remains disengaged. When the ECT ECU energizes the number 3 solenoid valve, the plunger opens the bleed port, which reduces line pressure on one end of the lock-up control valve. When this occurs, the lock-up control valve moves upward, and fluid is then directed through the converter hub into the converter. This fluid forces the lock-up clutch against the front of the converter, which provides a locking action between the front of the converter and the transmission input shaft. The lock-up clutch is applied in second, third, and O/D gears with the transmission selector in drive (D). The ECU controls lock-up timing to prevent shock during transmission shifts.

ECT Electronic Control Unit (ECU)

The electronic control unit (ECU) is a microprocessor with a memory that contains all the shift schedules and the lock-up converter schedule. The normal, power, or economy shift schedules that may be selected by the driver are also contained permanently in the ECU memory. The normal shift schedule is a well-balanced shift and lock-up schedule for general driving requirements, whereas the economy schedule provides increased fuel efficiency especially for city driving. When mountainous driving is encountered, or faster acceleration is desired, the power schedule changes the transmission shift points and converter lock-up timing to take full advantage of torque multiplication in the converter, which improves engine performance and power. The normal and power shift schedules are provided in Figure 7–14 and Figure 7–15.

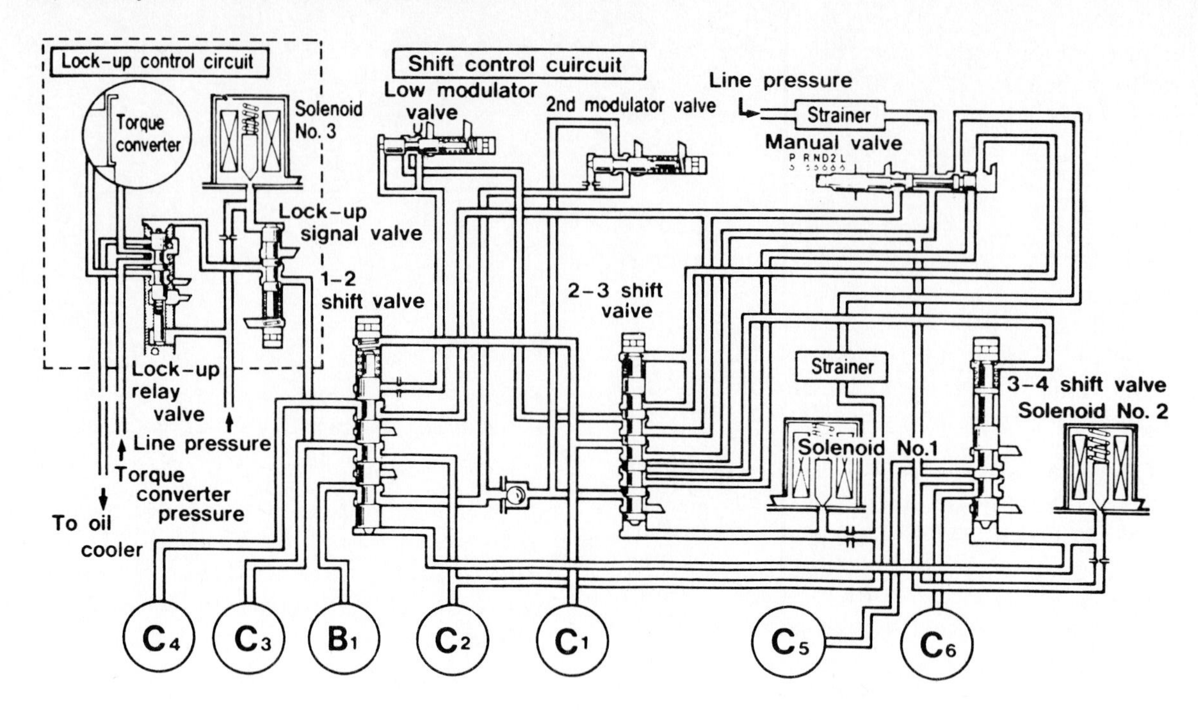

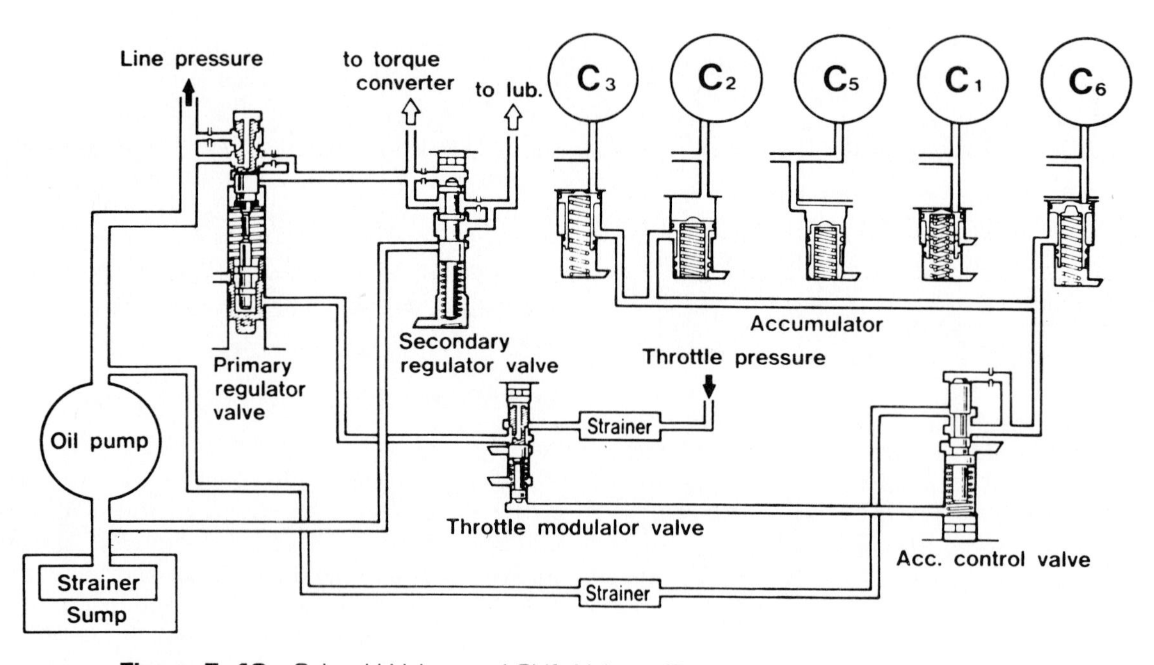

Figure 7-12 Solenoid Valves and Shift Valves. [Reprinted with Permission of Society of Automotive Engineers © 1984]

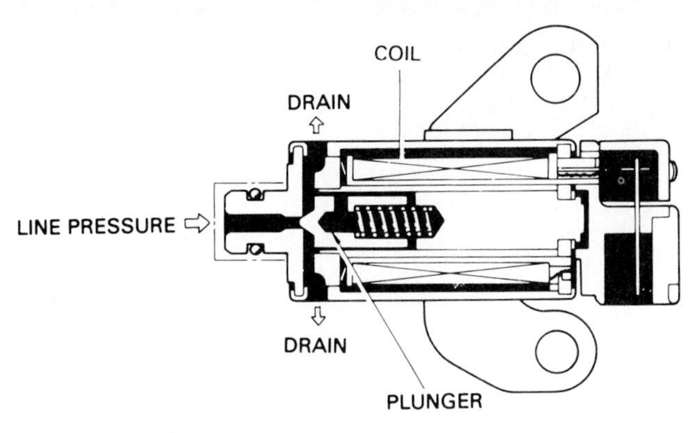

Figure 7–13 Number 3 Solenoid Valve. [Courtesy of Toyota Motor Corporation, Japan]

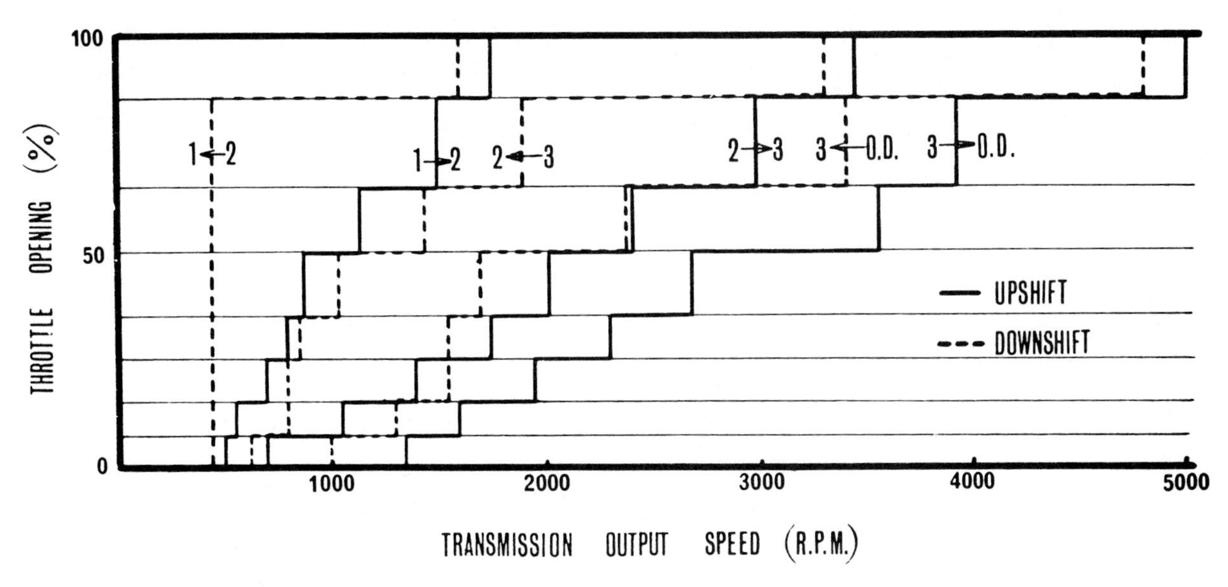

Figure 7–14 Normal Shift Schedule. [Reprinted with Permission of Society of Automotive Engineers © 1982]

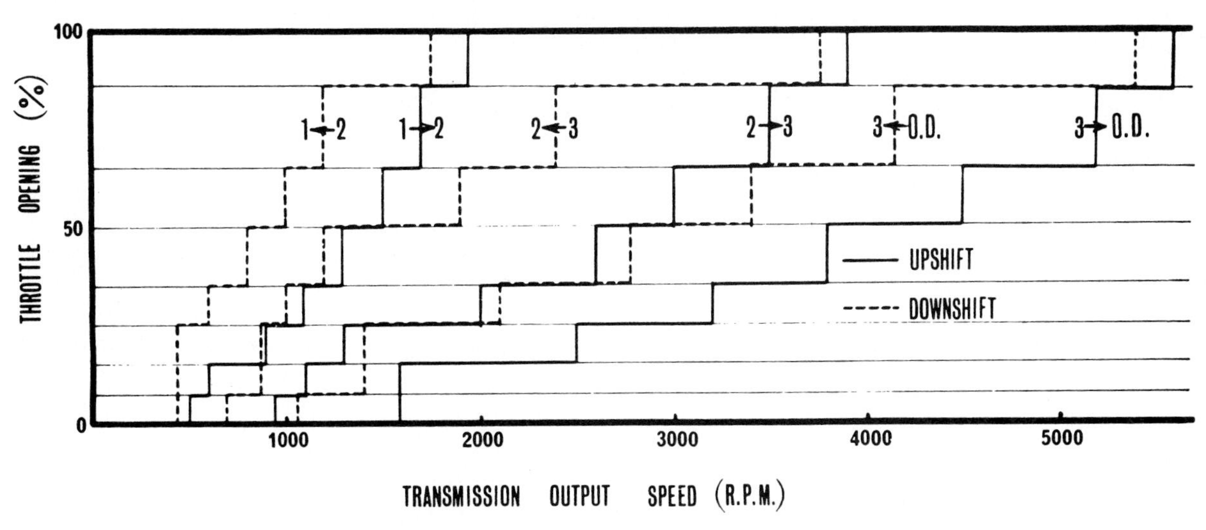

Figure 7–15 Power Shift Schedule. [Reprinted with Permission of Society of Automotive Engineers © 1982]

The shift schedules vary depending on the application. A lock-up converter economy shift schedule is shown in Figure 7–16.

The transmission is controlled electronically while the vehicle is moving ahead. If the transmission selector is in neutral, park, or reverse, the transmission is controlled mechanically.

Failsafe Functions

If solenoid number 1 or number 2 becomes defective, the ECT ECU will continue to operate the solenoid that is still working normally. Therefore, some transmission shifts will occur, and the vehicle may be driven in the forward gears. When a major electronic malfunction occurs and none of the solenoid valves will operate, the transmission may be shifted with the selector level. When the transmission selector is moved to the low (L), second (S), and drive (D) positions, the transmission will operate in first gear, third gear, and overdrive (O/D).

Electrical Circuit

The electrical circuit varies considerably depending on the model and year of vehicle. A complete ECT wiring diagram for a 1985 Cressida is shown in Figures 7–17 and 7–18.

Diagnosis of Electronically Controlled Transmission

Self-Diagnosis

The ECT ECU will cause the O/D lamp to flash on and off if a defect occurs in the speed sensors or solenoid valves. When the ignition switch is turned off, the defect will be stored in the ECU memory. Diagnostic codes representing system defects may be obtained when the ignition switch is turned on and the ECT terminal is connected to the E1 terminal in the diagnostic connector. On Supra models, a single diagnostic terminal must be grounded to obtain the diagnostic codes. Both diagnostic connections are indicated in Figure 7–19.

When the diagnostic connection is completed with the ignition switch on, the O/D off lamp will indicate various codes. If the O/D off lamp flashes every 0.25 seconds, the ECT system is in normal condition. When the O/D off lamp flashes six times followed by a 1.5-second pause and two more flashes, a code 62 is indicated. A normal code is compared to a code 62 in Figure 7-20, and the five defective codes that apply to the ECI system are illustrated in Figure 7–21.

If a defect occurs in the number 3 converter clutch lock-up solenoid, the ECU will not flash the O/D off

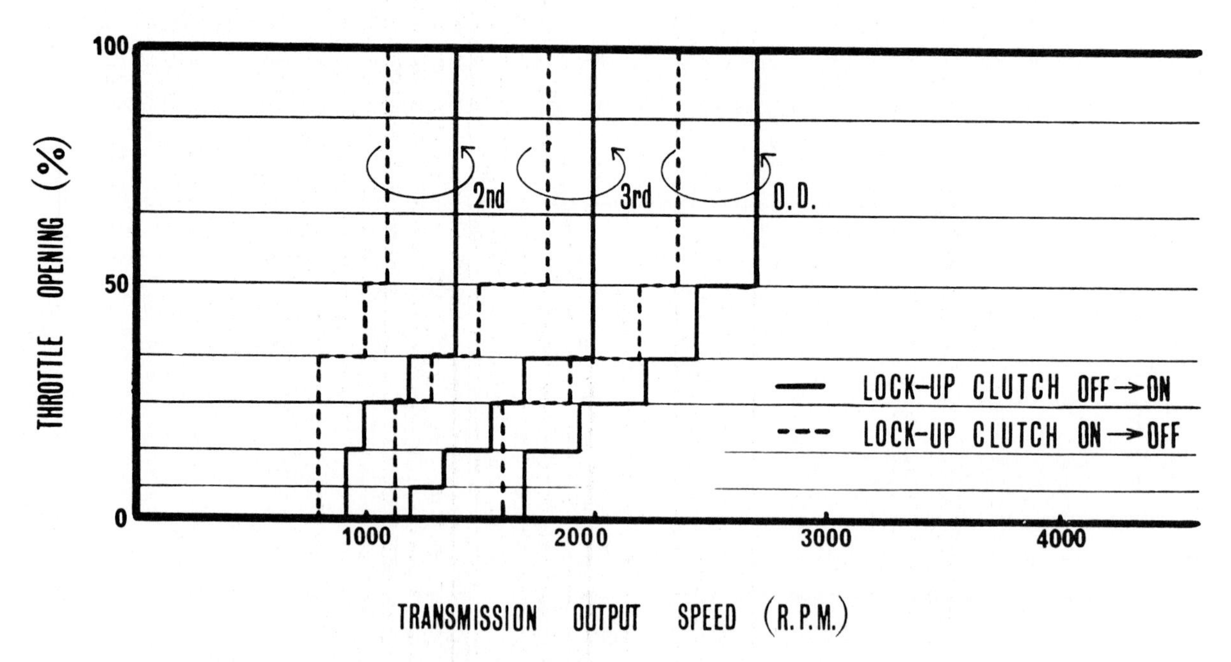

Figure 7–16 Lock-up Converter Economy Shift Schedule. [Reprinted with Permission of Society of Automotive Engineers © 1982]

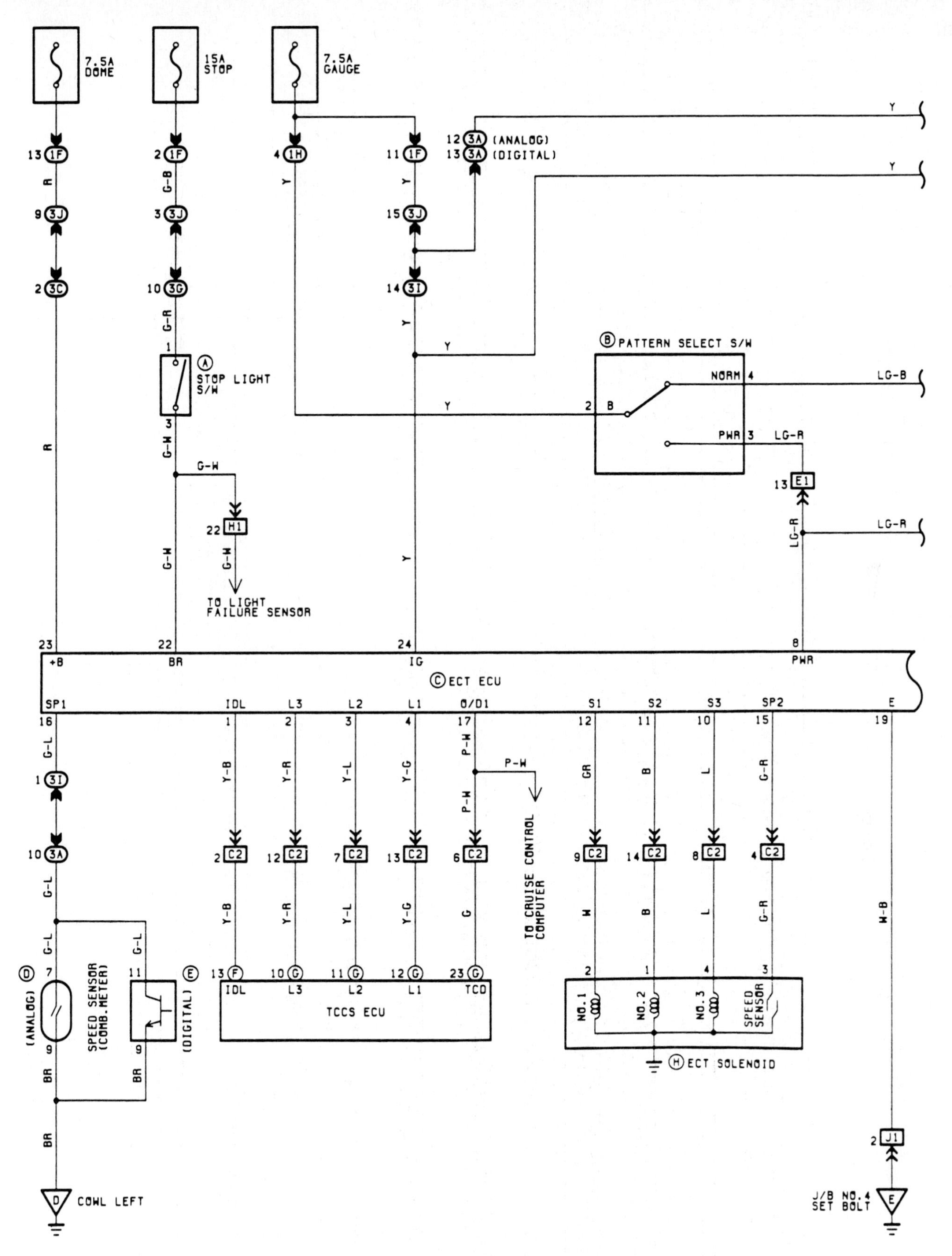

Figure 7-17 ECT Wiring Diagram Part One. [Courtesy of Toyota Motor Corporation, Japan]

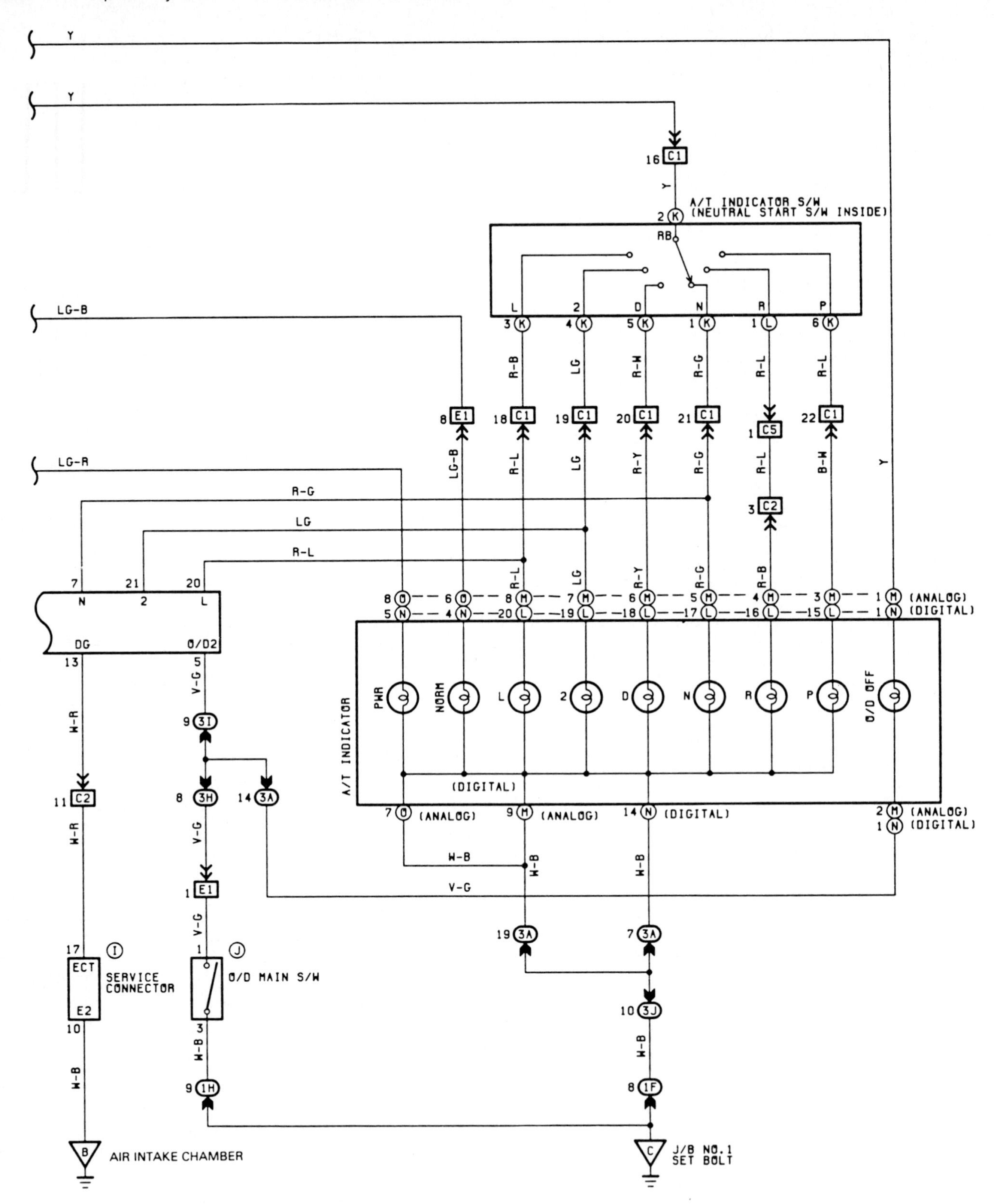

Figure 7–18 ECT Wiring Diagram Part Two. (Courtesy of Toyota Motor Corporation, Japan)

lamp. However, under this condition a code 64 will be stored in the ECU memory. The diagnostic codes are obtained from the flashing O/D lamp on Cressida and Supra models.

On other models, a voltmeter must be connected from the ECT or DG terminal to ground with the ignition switch on. The voltmeter reading will indicate normal or defective conditions as shown in Table 7–2.

Operation Check Function

Throttle-Position Sensor Check When the ignition switch is turned on and a voltmeter is connected from the ECT or DG terminal to ground, the throttle-position sensor, brake switch signal, and shift timing can be checked. If the throttle is opened from the idle position the voltmeter readings shown in Figure 7–22 should be obtained if the throttle-position sensor is

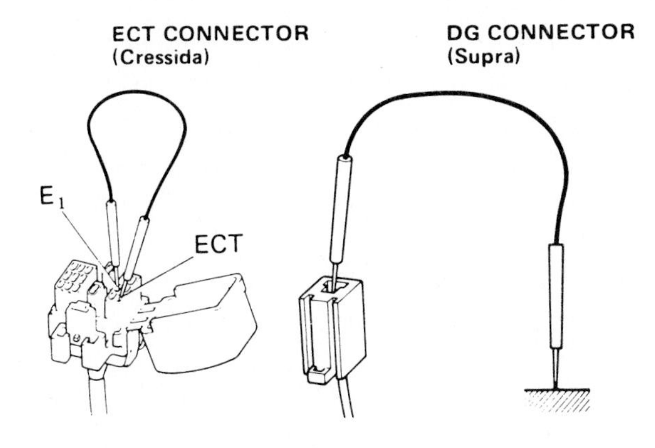

Figure 7–19 Diagnostic Test Connections. [Courtesy of Toyota Motor Corporation, Japan]

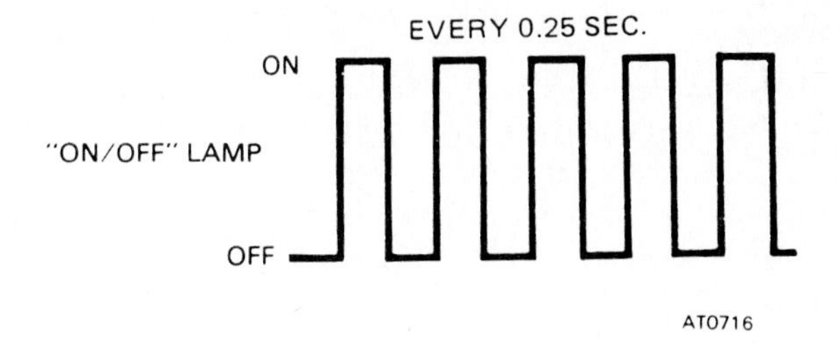

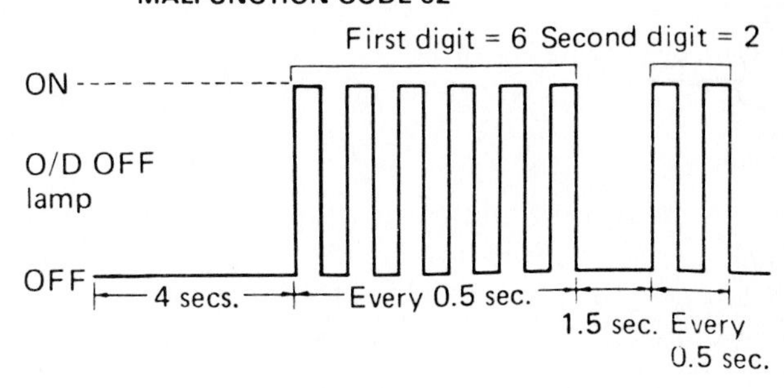

Figure 7–20 Normal Code and Code 62. [Courtesy of Toyota Motor Corporation, Japan]

TABLE 7–2 Voltmeter Diagnosis of ECT.

Voltage reading	Implication
0 V	Normal
4 V	Speed sensor defective
8 V	Solenoid valves defective

(Courtesy of Toyota Motor Corporation, Japan)

Code No.	Light Pattern	Diagnosis System
42		Defective No. 1 speed sensor (in combintion meter) Severed wire harness or short circuit
61		Defective No. 2 speed sensor (in ATM) Severed wire harness or short circuit
62		Severed No. 1 solenoid or short circuit Severed wire harness or short circuit
63		Severed No. 2 solenoid or short circuit Severed wire harness or short circuit
64		Severed No. 3 solenoid or short circuit Severed wire harness or short circuit

Figure 7–21 ECT Defective Codes. [Courtesy of Toyota Motor Corporation, Japan]

working normally. Do not touch the brake pedal during the throttle-position sensor tests.

Brake Switch Tests With the ignition switch on and the throttle wide open, the voltmeter connected from the ECT or DG terminal to ground should read 8V. When the brake pedal is depressed, the voltmeter reading should be 0V if the brake switch is working normally.

Shift Timing Tests When the vehicle is driven on the road with the voltmeter connected from the ECT or DG terminal to ground, the voltages provided in Table 7–3 should be obtained in each transmission gear.

The converter clutch lock-up may be cycled on and off in second and third gear by depressing and releasing the accelerator pedal. If the correct transmission gear is not available at the specified voltage, one of the solenoid valves may be sticking or a defect may exist in the transmission valve body. The ECU cannot diagnose a sticking solenoid valve.

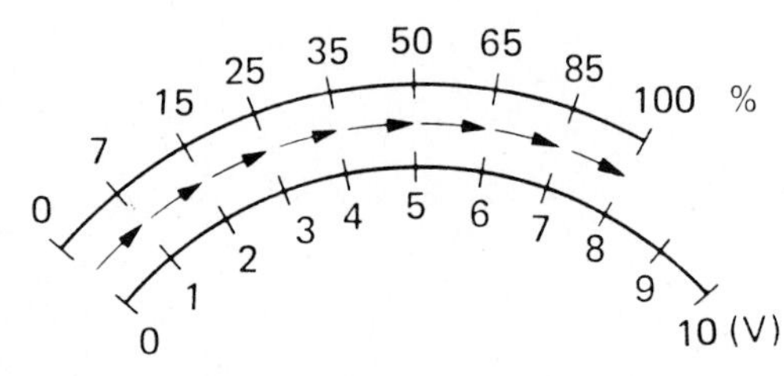

Figure 7–22 Voltage Readings from Throttle-Position Sensor. (Courtesy of Toyota Motor Corporation, Japan)

TABLE 7–3 Voltage Readings in each Transmission Gear.

Gear	1st	2nd		3rd		OD	
lock-up	—	OFF	ON	OFF	ON	OFF	ON
DG term. voltage	0	2	3	4	5	6	7

(Courtesy of Toyota Motor Corporation, Japan)

Test Questions

1. The two main input signals used by the electronic control unit (ECU) in the electronically controlled transmission (ECT) system to determine the transmission shift points are the:
   (a) overdrive switch and neutral/start switch signals.
   (b) throttle-position sensor and vehicle-speed sensor signals.
   (c) neutral/start switch and driving pattern selector signals.
2. If the main speed sensor in the transmission is defective, the ECU is programmed to use a vehicle-speed signal from the:
   (a) overdrive switch.
   (b) throttle-position sensor.
   (c) back-up speed sensor in the speedometer.
3. If a major electronic defect occurs such as a defective ECU in the ECT system, the car cannot be driven in the forward gears. T F
4. When the power mode is selected, the ECU will cause the transmission upshifts to occur at a lower engine speed. T F
5. On a Toyota Cressida with the ignition switch on and the ECT terminal connected to the E1 terminal in the diagnostic connector, if the overdrive (O/D) off lamp flashes steadily every 0.25 seconds the:
   (a) number 3 solenoid valve is defective.
   (b) number 1 solenoid valve is defective.
   (c) ECT system is in normal condition.
6. A flashing overdrive (O/D) off lamp when the vehicle is driven at 50 mph (80 kph) indicates:
   (a) that the transmission is cycling from overdrive to third gear.
   (b) a defect in the ECT system.
   (c) a pulsating overdrive switch.

8

Exhaust Turbochargers

Practical Completion Objectives

1. Observe service precautions when turbochargers and turbocharged engines are being serviced.
2. Test boost pressure on a turbocharged engine.
3. Diagnose electronic spark control (ESC) systems on turbocharged engines.

Exhaust Turbochargers

Operation

In a turbocharged engine, the normally wasted energy of the exhaust gases is converted to mechanical energy by pressurizing the intake manifold with denser air. The exhaust gases flow past a vaned turbine wheel, which causes the turbine to rotate at high speed. A shaft is connected from the turbine wheel to a vaned compressor wheel mounted in the air intake system. The compressor wheel must rotate with the turbine wheel, and the high speed rotation of the compressor wheel forces air into the intake manifold. This forces more air–fuel mixture into the cylinders on the intake strokes, which creates higher cylinder pressures on the compression and power strokes and provides increased engine power. The Chrysler 2.2L turbocharged engine provides 160 foot pounds (ft/lb), 217.6 newton meters (Nm) of torque, and 142 horsepower. This turbocharger provides a 35 percent increase in torque and a 50 percent improvement in horsepower compared to the same engine without a turbocharger. The basic turbocharger principle is illustrated in Figure 8–1.

On a normally aspirated nonturbocharged engine, a vacuum is created in the intake manifold by the downward piston movement on the intake strokes. This manifold vacuum is used to move the air–fuel mixture through the carburetor into the cylinders. When the turbocharger pressurizes the intake manifold, more air–fuel mixture can be forced into the cylinders compared to a normally aspirated engine.

Design

The main components in the turbocharger are shown in Figure 8–2.

The turbine and compressor wheels turn at speeds above 100,000 rpm. An oil line supplies a continuous flow of oil from the main oil gallery to the turbocharger housing and bearings. The oil is drained from the turbocharger housing to the crank-

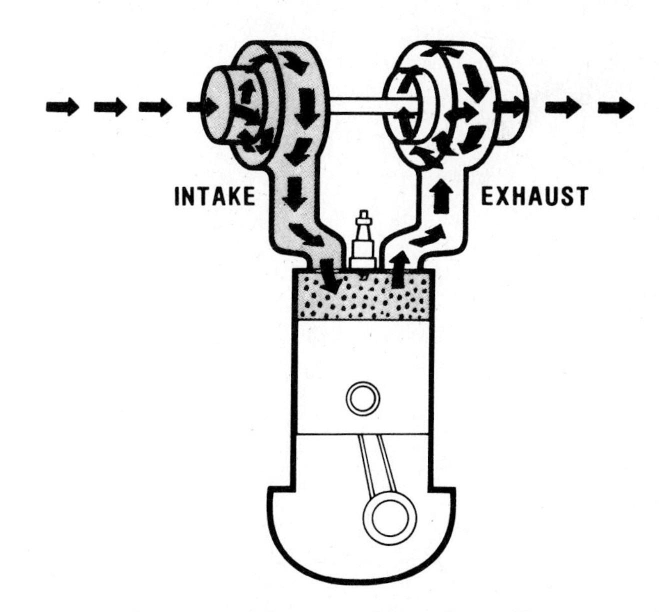

Figure 8–1 Turbocharger Operation. (Courtesy of Chrysler Canada Ltd.)

case through an oil drainback tube. On the Chrysler 2.2L turbocharged engine, engine coolant is also circulated through the turbocharger housing. This coolant circulation maintains the turbocharger bearing temperature below 300°F (149°C). When turbochargers do not have coolant circulation, excessive heat buildup after a hot engine is shut off may cause the engine oil to boil and harden in the bearings, which shortens bearing life. Seals on each side of the turbocharger shaft prevent oil from leaking into the intake or exhaust system. In most automotive turbochargers, the center housing assembly, with the shaft, bearings, seals, compressor wheel, and turbine wheel, is replaced as a unit. The coolant lines connected to the turbocharger are shown in Figure 8–3.

The turbocharger housing is bolted to the exhaust manifold so that all the exhaust must flow past the turbine wheel. An exhaust pipe is connected from the turbocharger exhaust outlet to the catalytic converter. The air intake hose is connected from the throttle body assembly to the air inlet on the compressor wheel housing. Air is forced from the compressor wheel through the air discharge hose into the intake manifold. The turbocharger mounting with the oil lines and coolant lines is illustrated in Figure 8–4.

Many turbocharged engines are equipped with electronic fuel injection (EFI). These engines usually have fuel injectors located in the intake manifold near the intake ports. On this type of engine the turbocharger forces air into the intake manifold, and the fuel is injected into the intake ports by the fuel injectors. The throttle in the throttle body assembly controls engine speed by regulating the amount of air entering the turbocharger compressor housing. A flexible hose is connected from the throttle body assembly to the air cleaner. (Refer to Chapter 4 for a description of the EFI system on the 2.2L engine.) The turbocharger and throttle body assembly are illustrated in Figure 8–5.

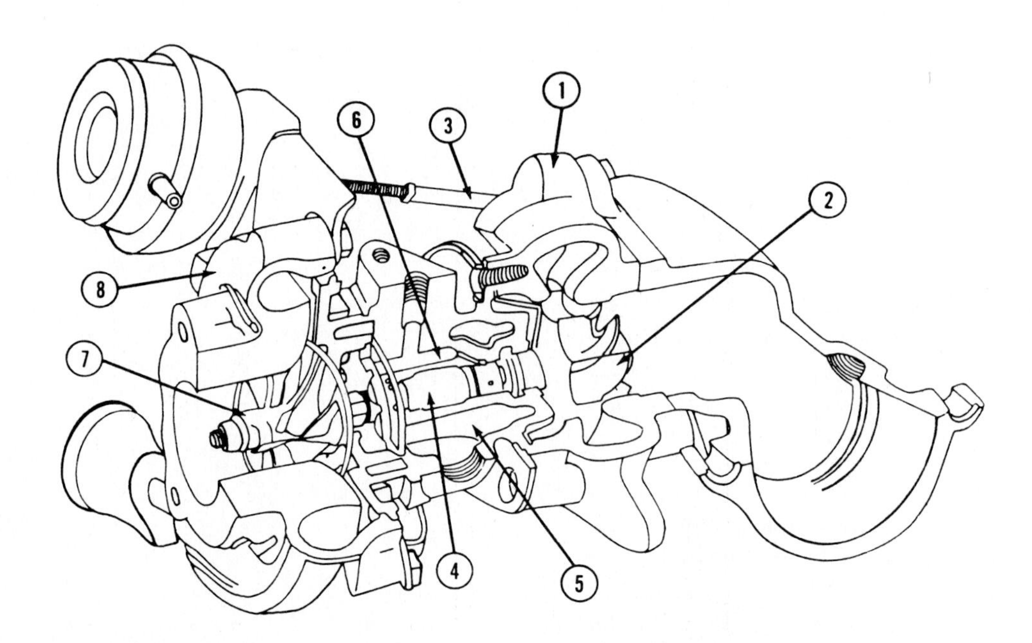

COMPONENTS OF A TURBOCHARGER

Turbine Housing: (Hot Side) (1)
Directs exhaust gages to turbine wheel.

Turbine Wheel: (Hot side) (2)
Driven by exhaust gages which drives the compressor through a shaft.

Waste Gate: (3)
Limits amount of boost pressure.

Shaft Wheel Assembly: (4)
Supported by bearings that float in oil.

Water Passage: (5)
Water circulation helps cool hot side of turbocharger.

Oil Passage: (6)
Lubricates and helps cool turbocharger.

Compressor Wheel: (7)
Compresses air to be forced into intake manifold.

Compressor Housing: (8)
Collects and directs pressurized air to intake manifold.

Figure 8–2 Turbocharger Components. (Courtesy of Chrysler Canada Ltd.)

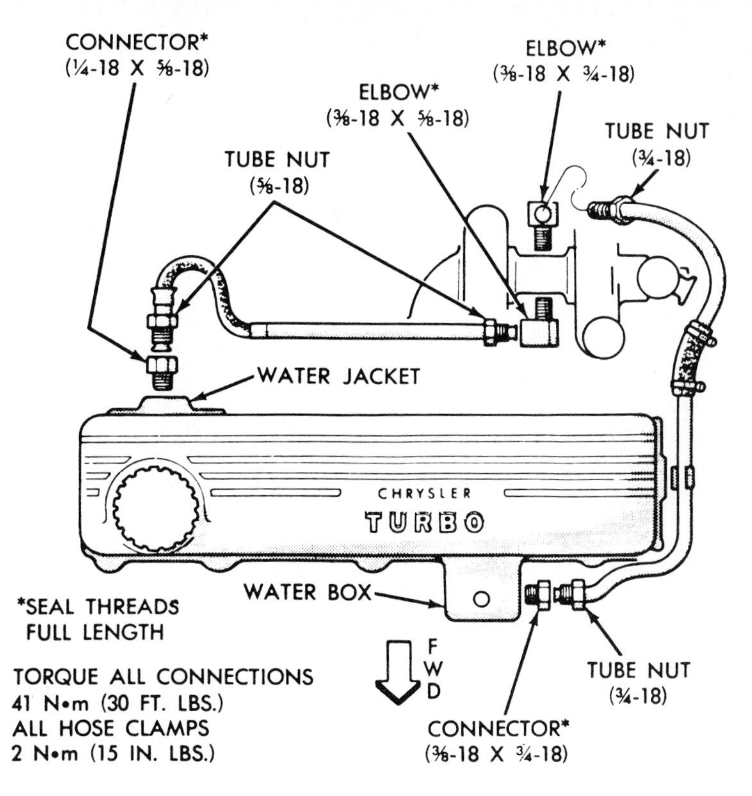

Figure 8–3 Turbocharger Coolant Lines. (Courtesy of Chrysler Canada Ltd.)

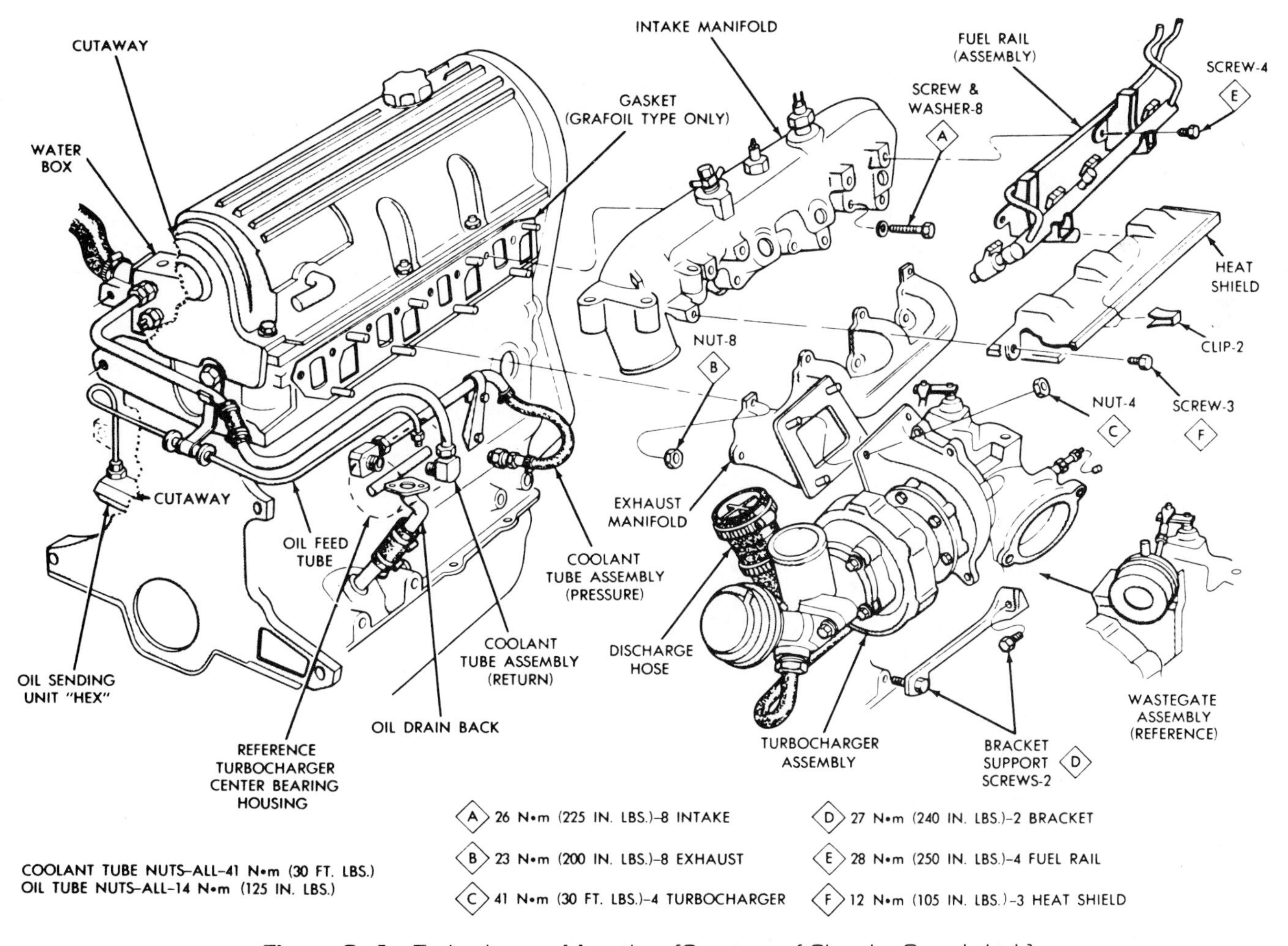

Figure 8–4 Turbocharger Mounting. (Courtesy of Chrysler Canada Ltd.)

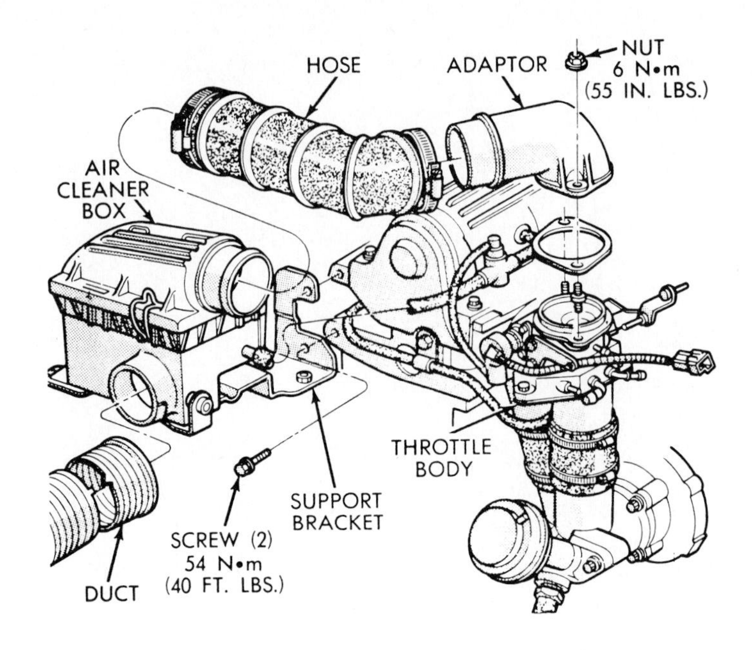

Figure 8–5 Turbocharger and Throttle Body Assembly. [Courtesy of Chrysler Canada Ltd.]

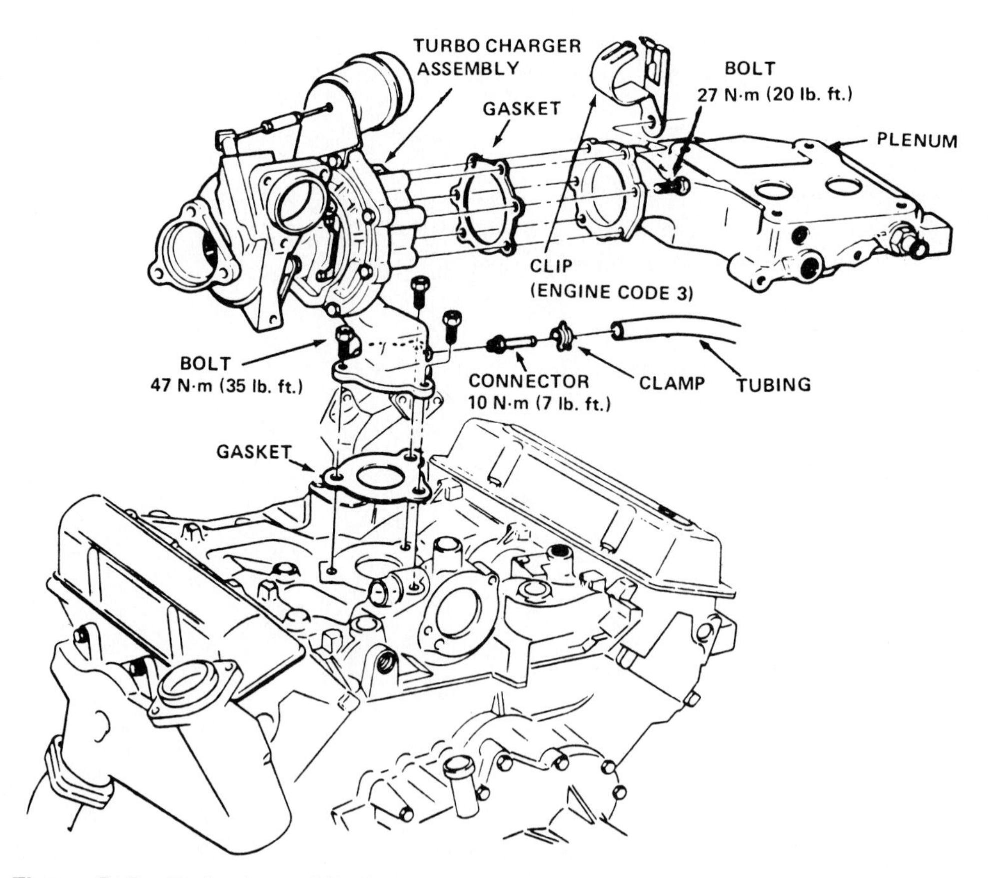

Figure 8–6 Turbocharged Engine with Carburetor. [Courtesy of General Motors Corporation.]

Some turbocharged engines are equipped with a carburetor rather than an EFI system. On these engines the compressor housing is connected to a plenum plate under the carburetor. The compressor wheel forces the air–fuel mixture into the intake manifold. A turbocharged engine with a carburetor is shown in Figure 8–6.

Boost Pressure Control

A wastegate poppet valve is located in the turbine housing. This valve is connected to the wastegate diaphragm by an operating rod. Boost pressure from the compressor housing is applied to the wastegate diaphragm cover. On many applications, vacuum from the throttle body assembly is applied to the spring chamber under the wastegate diaphragm. When the boost pressure reaches a preset value, the boost pressure moves the wastegate diaphragm against the spring pressure and opens the poppet valve. This allows some of the exhaust to bypass the turbine wheel, which limits the turbine wheel and boost pressure. On the Chrysler 2.2L turbocharged engine, the boost pressure is limited to 7.2 psi (49.6 kPa). This turbocharger begins to pressurize the intake manifold at 1200 engine rpm, and it supplies full boost pressure at 2050 rpm. The wastegate diaphragm and poppet valve are pictured in Figure 8–7.

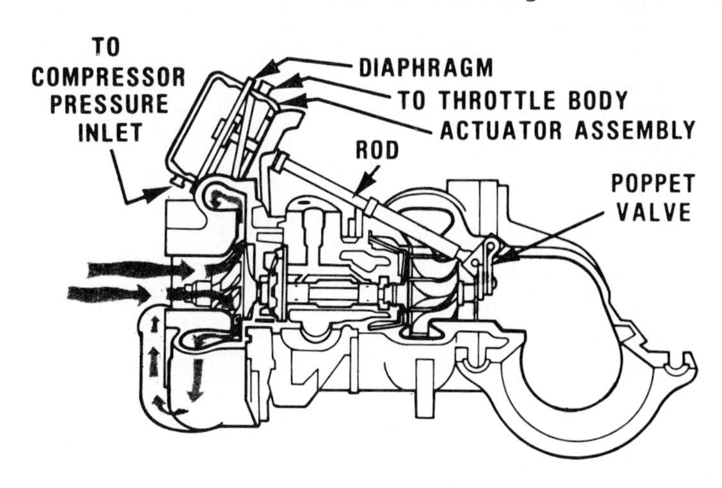

Figure 8–7 Wastegate Diaphragm and Poppet Valve. (Courtesy of Chrysler Canada Ltd.)

Computer Control of Boost Pressure On many engines equipped with electronic fuel injection, the turbocharger boost pressure is controlled by the electronic control module (ECM). In this type of boost control system, an ECM-operated solenoid is connected in the wastegate diaphragm boost pressure hose as illustrated in Figure 8–8.

The ECM senses boost pressure from the manifold absolute pressure (MAP) sensor signal. When the boost pressure reaches the maximum safe limit, the ECM will energize the wastegate solenoid. This action opens the solenoid and vents some of the boost

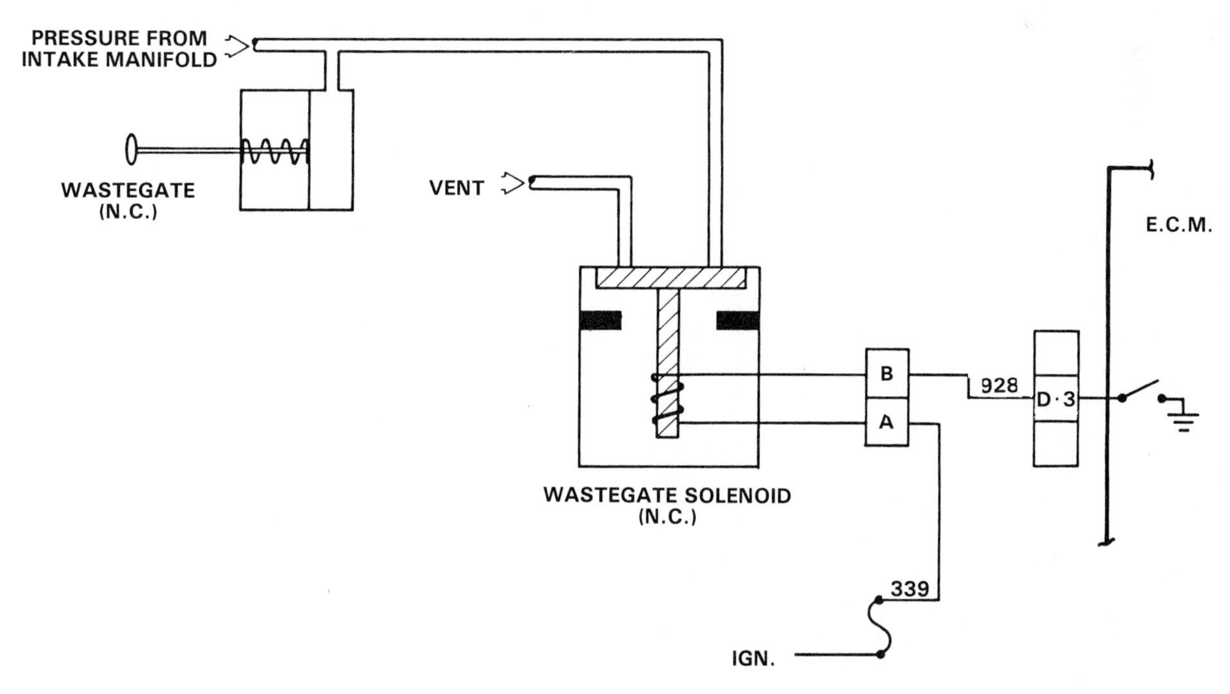

Figure 8–8 Computer-Controlled Boost Pressure. (Courtesy of General Motors Corporation.)

pressure from the wastegate diaphragm to control wastegate movement and boost pressure. The ECM pulses the wastegate solenoid on and off to control boost pressure. During periods of hard acceleration on some models, the ECM will allow a higher boost pressure for a few seconds to improve engine performance. For example, the ECM may allow a boost pressure of 10 psi (68.9 kPa) for 10 seconds when the engine is accelerated at wide-open throttle. After 10 seconds the ECM will return the boost pressure to the normal maximum limit if the engine is still operating at high speed. (Refer to Chapter 4 for a complete description of computer systems used with electronic fuel injection.)

Boost Pressure Indicators Most turbocharged engines have a boost indicator on the instrument panel. Some vehicles have a boost indicator light in the instrument panel that is illuminated when the turbocharger develops full boost pressure. Other applications have a second boost indicator light, which is illuminated when the turbocharger begins to develop boost pressure. These indicator lights are operated by a pressure switch in the intake manifold. Some turbocharged vehicles have a boost pressure gauge in the instrument panel as indicated in Figure 8–9.

Boost Pressure Diagnosis If the vehicle is not equipped with a boost indicator gauge, a pressure gauge may be connected to the intake manifold for test purposes. When the vehicle is accelerated at wide-open throttle from 0–60 mph (96 kph) the boost pressure specified by the manufacturer should be indicated on the gauge. A boost pressure that is higher than specified indicates a defective or sticking wastegate. The wastegate diaphragm may be checked for leaks with a hand-operated vacuum pump. After vacuum has been applied to the diaphragm, if the vacuum gauge reading on the pump slowly decreases, the diaphragm is leaking. Many wastegate operating rods on automotive turbochargers are not adjustable. If the boost pressure is less than specified, the wastegate, or the turbocharger, is defective.

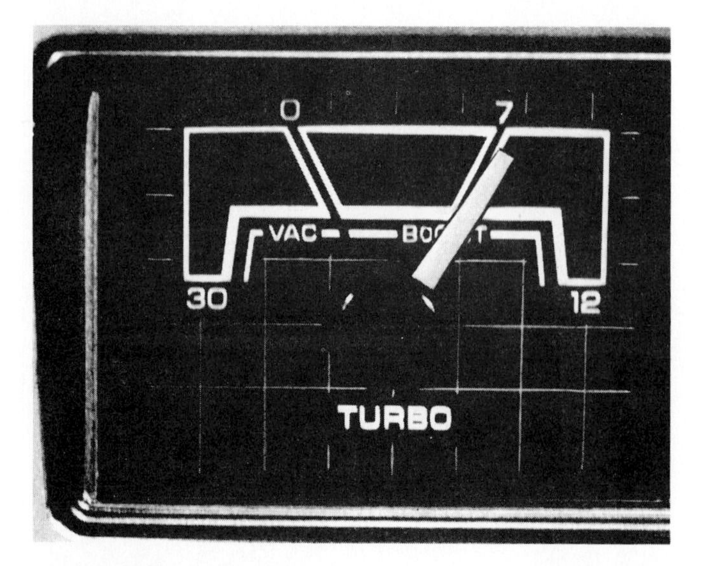

Figure 8–9 Boost Pressure Gauge. (Courtesy of Chrysler Canada Ltd.)

Many of the components in a turbocharged engine are designed with improved durability because of the higher pressures and temperatures encountered in turbocharged engines. In the Chrysler 2.2L turbocharged engine, the following components have improved durability, or performance characteristics, compared to the same components in the normally aspirated 2.2L engine.

1. The camshaft has reduced overlap and duration, which lowers exhaust temperature.
2. Valve springs have a higher closing rate.
3. The head gasket has improved sealing qualities.
4. Nickel alloy exhaust valves and silicon alloy intake valves are used in the turbocharged engine.
5. Dished pistons provide a compression ratio of 8.2:1 compared to 9:1 compression ratio on the normally aspirated engine.
6. Piston rings are the high performance type.
7. Select-fit connecting rod and main bearings provide improved lubrication.
8. The oil pump has higher output, and the tension of the bypass valve is increased.
9. High silicon alloy modular iron is used in the exhaust manifold.
10. A higher capacity radiator is used.
11. The larger 2¼ in (6.3 cm) diameter exhaust system has increased flow capacity.
12. The heat-shielded heavy duty starting motor has silver-soldered electrical connections.

Service Precautions on 2.2L Turbocharged Engine

Severe turbocharger or engine damage can be caused by incorrect service procedures. Therefore, the following service precautions must be followed.

1. The turbocharged engine must not be operated with the air cleaner removed because small dirt particles can destroy the compressor wheel.
2. Never disconnect the air hose between the throttle body and the turbocharger, or the hose connected from the turbocharger to the intake manifold, with the engine running. If either of those hoses are

disconnected, the throttle will not control the amount of air entering the engine and excessive engine speed may damage the engine.

3. To provide adequate turbocharger lubrication, the engine oil and filter must be changed at the manufacturer's recommended intervals. The engine oil level must be maintained at the full mark on the dipstick.
4. The coolant level must be maintained between the "maximum" and "minimum" marks on the coolant recovery bottle. When any of the coolant is drained from the system, the air must be bled from the system as it is refilled. This is done by removing the temperature switch from the top of the thermostat housing. As the cooling system is refilled, air will be bled from the temperature switch opening.

Turbocharger Spark Retard Systems

Electronic Spark Control (ESC) System

The ESC system is connected to the General Motors high-energy ignition (HEI) system. Conventional vacuum and centrifugal advance mechanisms are used in the HEI distributor. The distributor pickup coil leads are connected to the ignition module, but these leads are also connected to the ESC controller, which is mounted under the dash on most cars. A fifth terminal on the ignition module is also connected to the ESC controller, and a ground wire is connected from the controller to the distributor housing as shown in Figure 8–10.

A detonation sensor is mounted at the rear of the engine block or in the intake manifold. This sensor

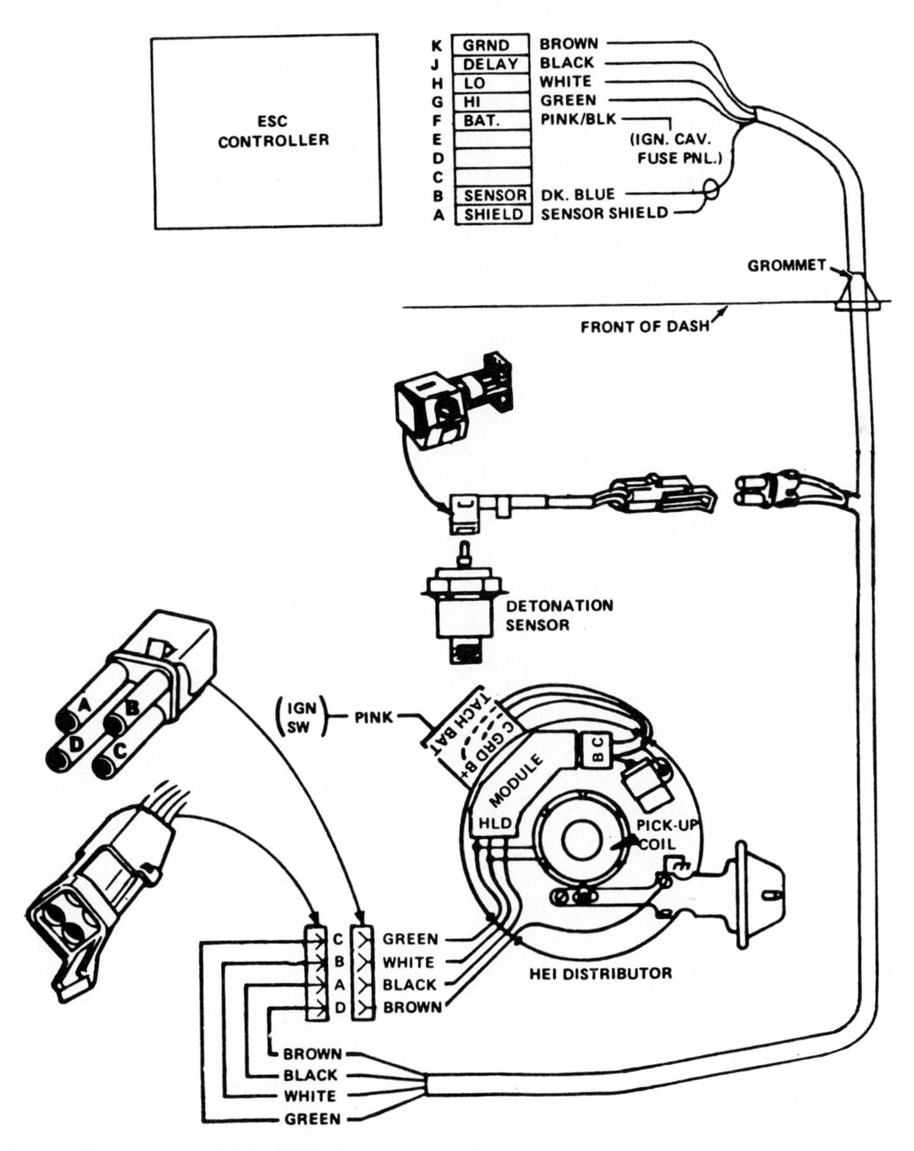

Figure 8–10 Electronic Spark Control (ESC) System. (Courtesy of General Motors Corporation.)

is also connected to the ESC controller. Detonation may occur in a turbocharged engine because of the higher cylinder pressures and temperature. When the engine detonates, the detonation sensor signals the controller to interfere with the pickup coil and retard the timing 4°. The ESC system can provide a total retard of 20°, if the engine continues to detonate. Once the detonation stops, the controller will restore the spark advance in a few seconds. On later model vehicles the ESC system is combined with the computer command control (3C) system. (This system is described in Chapter 3.) The electronic fuel injection (EFI) system used in the Chrysler 2.2L turbocharged engine has a detonation sensor to retard the spark advance if the engine detonates. (Refer to Chapter 4 for a description of this system.)

Diagnosis of ESC System A defective controller may prevent the engine from starting. The controller may be bypassed by disconnecting the distributor four-wire connector and connecting a jumper wire from the green to black wires in the distributor connector. If this connection causes the engine to start, the controller is defective, or there is no power being supplied to the pink wire on the controller with the ignition switch on.

If the engine detonates, the ESC system may be checked by tapping the engine near the detonation sensor with a small hammer while the engine is running at fast idle. When the engine is tapped at 1-second intervals, the timing marks should retard 4° each time the engine is tapped until a maximum retard of 20° is reached. The timing marks may be observed with an advance type of timing light to determine the exact amount of spark retard. If the timing does not retard when the engine is tapped, the detonation sensor or controller is defective. The detonation sensor may be checked by connecting an ohmmeter from the sensor terminal to ground with the ignition switch off. The resistance of the sensor should be 175 to 375 ohms.

Test Questions

1. A worn seal in the turbocharger could allow oil to leak into the intake system. T F
2. When engine coolant is circulated through the turbocharger housing, the bearing temperature is kept below ____________ degrees F.
3. Excessive boost pressure could be caused by a defective wastegate diaphragm. T F
4. Turbocharged engines have a ____________ compression ratio than normally aspirated engines.
5. The electronic spark control (ESC) system provides an instant spark control of 20° if the engine detonates. T F
6. When the Chrysler 2.2L turbocharged engine is operated with the hose removed between the turbocharger and the intake manifold, the engine will:
   (a) stall.
   (b) overspeed.
   (c) detonate.
7. When the Chrysler 2.2L turbocharged engine is being refilled with coolant, ____________ must be bled from the system.
8. When the engine is operating at 1500 rpm, if the engine is tapped near the detonation sensor with a hammer and the timing does not retard, the trouble could be a defective:
   (a) detonation sensor.
   (b) distributor pickup coil.
   (c) ignition coil.

9

Voice Alert Systems

Practical Completion Objectives

1. Diagnose an 11-function voice alert system.
2. Diagnose a 24-function monitor and voice alert system.

Eleven-Function Voice Alert System

Operation

The messages that may be provided by this system are the following:

1. Your headlights are on.
2. Don't forget your keys.
3. Your washer fluid is low.
4. Your fuel is low.
5. Your electrical system is malfunctioning; prompt service is required.
6. Your parking brake is on.
7. Your door is ajar.
8. Please fasten your seat belts.
9. Your engine is overheating; prompt service is required.
10. Your engine oil pressure is low; prompt service is required.
11. All monitored systems are functioning.
12. Thank you.

Message 1 will be heard after the following sequence of events:

1. The driver's door is closed.
2. Headlights are on.
3. Ignition switch is turned on and off.
4. Key is removed from ignition switch.
5. Driver's door is opened.

Message 5 will occur if the charging voltage is below 11.75V and the engine is running above idle speed for several minutes. If the engine temperature is above 270°F (132°C) for 1 minute and engine speed is above idle for 1 minute, message 9 will be heard. Message 10 will be provided if the oil pressure is low for a minimum of 2 seconds and the engine is running. Forward vehicle motion is necessary, plus the other applicable conditions before messages 6, 8, or 11 will be heard. Message 7 is provided if a door is ajar and the vehicle is in forward or reverse motion.

The heart of the electronic voice alert system is a microprocessor, or module, mounted above the glove compartment. A switch or sensor located in each monitored component sends the necessary input signal to the module. If an unsatisfactory signal is received, the control module will provide the appropriate message. Forward and reverse vehicle motion signals are provided by a speed sensor and the backup light switch.

A volume control is located on the underside of the module. The on/off switch on the module will cancel the voice signal if the switch is moved toward the rear of the vehicle. This switch is accessible through an opening in the top right inside the glove box. A pulsating beep is provided before the audible message, and a tone following the message. The on/off module switch only cancels the audible message.

Diagnosis of 11-Function Electronic Voice Alert System

A module test may be performed with the following procedure:

1. Remove the keys from the ignition switch.
2. Open and close the driver's door.
3. Wait one minute.
4. Open the driver's door.
5. Press and hold the door ajar switch located on the left B post.

When these steps are completed, the module should deliver all 12 audible messages. If the module misses some messages, module replacement is necessary. If there are no audible messages, be sure the on/off switch is moved toward the front of the vehicle. If audible messages are still not available, check battery feed to module, ground circuit to module, left door ajar switch, and left door courtesy lamp switch. The left door courtesy lamp switch is grounded and the door ajar switch is not grounded in the test mode. If these circuits and components are satisfactory, replace the module. This test sequence determines the ability of the module to supply audible messages, but it does not test individual switches or circuits.

Individual circuits may be tested by creating the necessary conditions to provide the audible message with a known good chimes module inserted in place of the electronic voice alert module. If the ignition switch is turned on and off and the keys are left in the switch, a *Don't forget your keys* message should be heard when the driver's door is opened. If no message is heard, connect a known good chimes module in place of the electronic voice alert module. When no chimes are available, test the circuit from the module to the ignition switch. Each circuit that is monitored by the module may be tested in the same way. The number of functions and the conditions required to provide audible messages will vary depending on the year of the vehicle.

Twenty-Four Function Monitor and Voice Alert System

Operation

The 24-function message center is referred to as an "electronic monitor." This system provides the driver with visual and audible warnings. Visual messages are presented on a two-line, ten-character, blue-green vacuum fluorescent display. Orange- and lemon-colored warning symbols are activated simultaneously. Audible messages are provided by a voice synthesis microprocessor. The monitor system contains an electronic monitor module with two vacuum fluorescent displays and an electronic voice alert microprocessor, or module. The monitor is located in the center dash area, and the voice alert module is located above the glove box. This system is capable of displaying and verbalizing 24 warning conditions on the vehicle. The actual number of warning conditions varies depending on the vehicle options. Messages displayed and verbalized include three categories as follows:

1. Safety—Passenger door ajar, driver door ajar, hatch ajar, fasten seat belts.
2. Operational—Oil pressure low, engine temperature high, fuel level low, transmission pressure low, voltage low.
3. Convenience—Coolant level low, brake fluid low, disc brake pads worn, washer fluid low, rear washer fluid low, engine oil level low, headlight out, brake lamp out, tail lamp out, parking brake engaged, keys in ignition, and exterior lights on.

The electronic monitor senses various defective conditions from the inputs. When a defective condition exists, this module will display a warning message and send a tone and talk signal to the voice alert module. This will cause the voice alert module to generate a short tone and provide the appropriate audible message. When more than one defect exists, the same sequence is followed for each fault, and each message is displayed for 4 seconds. If the defect is driver correctable, the monitor module will signal the voice alert module to provide a "thank you" tone. The voice alert module does not generate tones through the radio speaker, as in the 11-function system. Instead, an external sound transducer in the voice alert module provides the tone signals. If the radio is on, the audible messages will interrupt the radio, and these messages are delivered through the radio speaker as in the 11-function system. Two visual messages and audible messages are shown in Figure 9–1.

Requirements to Obtain Messages A *keys in ignition*, or *exterior lamps on* message is provided if these conditions are present and the driver's door is opened with the ignition switch in the off, lock, or acc positions. The audible message is followed by a pulsating tone for keys in ignition, or a continuous

Visual Messages		Audible Messages
	WASHER FLUID LOW	"Your washer fluid is low"
	RR WASHER FLUID LOW	"Your rear washer fluid is low"

Figure 9–1 Visual Messages and Corresponding Audible Messages. (Courtesy of Chrysler Canada Ltd.)

tone for exterior lamps on, and these tones will continue until the condition is corrected.

A *fasten seat belts* message is provided for 6 seconds when the ignition switch is turned on. It will continue to be displayed if the driver's seat belt is not buckled, or until the car has been moved 16 to 24 in (40.6 to 60.9 cm), at which time the audible message is given.

The *monitored systems OK* message is provided when the ignition switch has been on for 6 seconds and no defects have been found.

Some visual messages such as *park brake engaged*, all *door ajar* messages, and *trunk* or *hatch ajar*, will be displayed when the ignition switch is on and the corresponding fault is sensed. The audible message will be provided with the visual message after the vehicle has been moved 16 to 24 in (40.6 to 60.9 cm). When the fault is corrected, a short tone will be heard.

Faults in the following systems will provide visual and audible messages if the ignition switch is on and the defective condition has been sensed for 15 seconds: *washer fluid low*, *rear washer fluid low*, *fuel level low*, *coolant level low*, *brake fluid low*, and *disc brake pads worn*. After the defective condition has been corrected, the ignition switch must be turned off to clear these messages from the monitor module.

Any lamp message such as *headlamp*, *tail lamp*, or *brake lamp out* will be displayed and heard if the ignition switch is on and the fault has been sensed for 3.5 seconds. Correction of the fault will clear the failure message.

A *low oil pressure* message will be provided if low oil pressure is sensed and the engine rpm is between 300 and 1500 rpm. The failure message will be cleared when the fault is corrected.

The *engine temperature high* message is displayed and heard when the engine temperature is sensed high and engine rpm is above 300 for 30 seconds. A second-level audible message will be heard after the same fault conditions are present for 60 seconds. The failure message is cleared when the fault is corrected or the engine speed drops below 300 rpm.

A *voltage low* audible and visual message is given if the battery voltage is below 12.35V and the engine speed is above 1500 rpm for 15 seconds. The ignition switch must be turned off after the fault is corrected to clear this message.

When the transmission pressure is sensed low for 15 seconds, and the engine speed is above 300 rpm with the vehicle not in reverse, a *low transmission pressure* audible and visual message is provided. The ignition switch must be turned off after the correction procedure to clear this message.

An *engine oil level low* audible and visual message is provided if the monitor module is powered by the time delay relay (TDR) only, and the engine oil level is sensed low. The TDR supplies power to the monitor for a specific time period after the ignition is turned off. This message is also given with the ignition on and the TDR still on. The ignition switch must be turned off after the condition is corrected to clear this message.

A system check button is located on front of the monitor. When this button is pushed twice within a 5-second interval, the system will be muted and a *mute engaged* message will be displayed for 4 seconds. This message is only visual. The system may also be muted with the switch on the voice alert module, as in the 11-function system.

The complete 24-function system is illustrated in Figure 9–2.

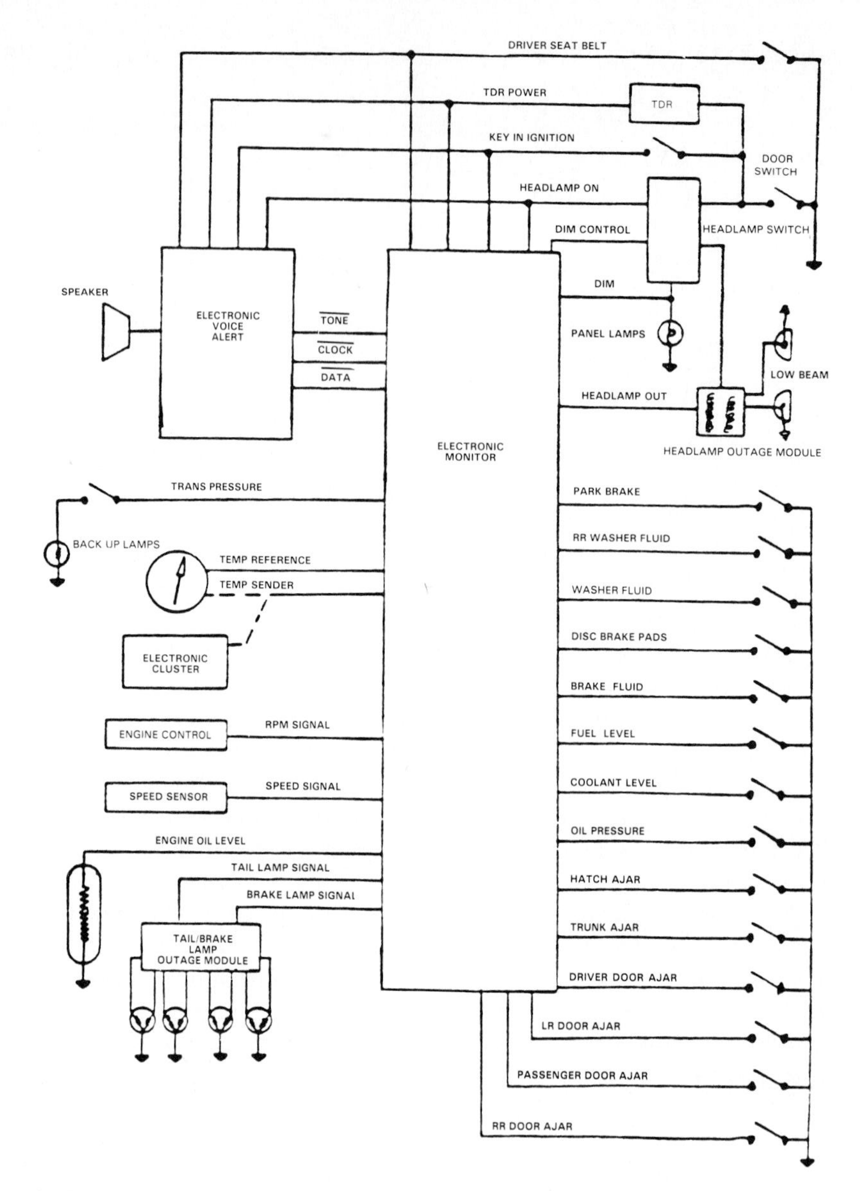

Figure 9–2 Twenty-Four Function Monitor and Voice Alert System. (Courtesy of Chrysler Canada Ltd.)

Diagnosis of 24-Function Monitor and Voice Alert System

When the system check button on front of the monitor is pressed once, a tone will sound and many of the audible and visible messages will be presented in a demo sequence. When this sequence is completed, the system automatically returns to normal operation. The demo sequence will be aborted if the system check button is pressed during the sequence. This sequence proves that the monitor module and the voice alert module are capable of delivering visual and audible messages.

All the system fault signals to the monitor module are provided by a ground switch except the voltage low signal, lamp out signals, and oil level low signal. If a fault signal is not given, and that signal is controlled by a ground switch, proceed as follows:

1. Check the operating conditions required to give the fault message.
2. Check the wiring from the monitor module to the specific switch that is not providing a message.
3. Check engine rpm if it applies to the missing fault message.
4. Check the distance input if it applies to the missing fault message.

5. Connect a jumper wire from the input switch terminal to ground, and meet all other requirements for the fault message being diagnosed. If a fault message is given, the switch is defective. When the fault message is not provided, the monitor module is defective.

If a fault message is still given after the fault condition has been corrected, proceed with the following diagnosis:

1. Check the conditions that are required to provide the fault message.
2. Check the wiring to the switch that provides the fault message.
3. Disconnect the switch; if the fault message stops, replace the switch. Monitor module replacement is necessary if the fault message continues.

The lamp out signals are provided by front and rear lamp modules, which are shown in Figure 9–2. These modules send an alternating current (AC) signal to the monitor module if a light burns out. If a *lamp out* message is not provided, proceed as follows:

1. Check the conditions that are required to provide the fault message.
2. Check the wiring.
3. Remove a tail/stop light bulb and disconnect a headlight. If a *lamp out* message is not displayed, replace the monitor module. When only some failures occur, replace the lamp module that corresponds to the missing *lamp out* message.

The engine oil level input sensor is a variable resistor that is built into the dipstick. If an *oil level low* message is not given, use the following diagnosis:

1. Check the conditions that are required to provide the fault message.
2. Check wiring to the dipstick.
3. Remove the dipstick and connect a jumper wire across the wiring harness connector. If a fault message is given, replace the dipstick. When the fault message is not given, replace the monitor module.

Electronic Navigator

An electronic navigator is available on many vehicles, which reads these nine functions:

1. Miles until empty.
2. Estimated time to arrival and time of arrival.

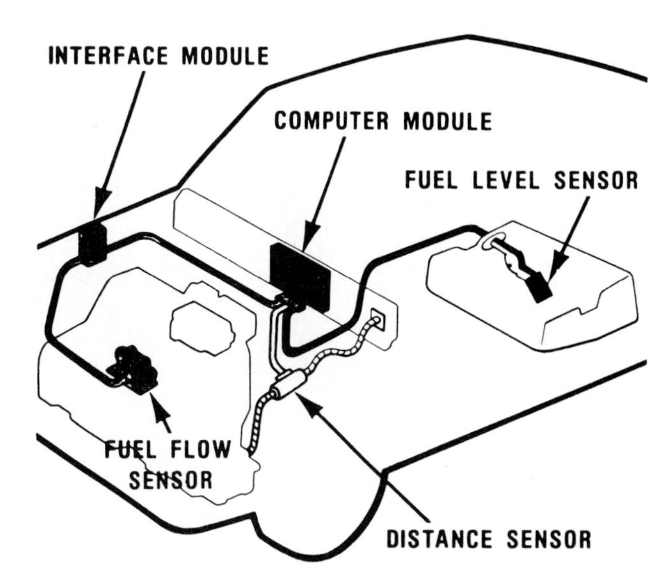

Figure 9–3 Electronic Navigator System Components. (Courtesy of Chrysler Canada Ltd.)

3. Distance to destination.
4. Time, day, month, and date.
5. Present and average miles per gallon.
6. Fuel consumed.
7. Average speed.
8. Miles traveled.
9. Elapsed driving time.

The main components in the electronic navigator system are illustrated in Figure 9–3. (A detailed description of electronic navigators and travel computers is provided in Chapter 10.)

Figure 9–4 illustrates an instrument panel which contains an electronic instrument cluster, electronic navigator, and the message center for the 24-function monitor and voice alert system.

Test Questions

1. An 11-function or 24-function voice alert system may be muted when the voice alert module switch is moved toward the ____________ of the vehicle.
2. In a 24-function monitor and voice alert system the diagnostic mode is entered by:
   (a) depressing the system check button once with the ignition on.
   (b) depressing the system check button twice in a 5-second interval with the ignition on.

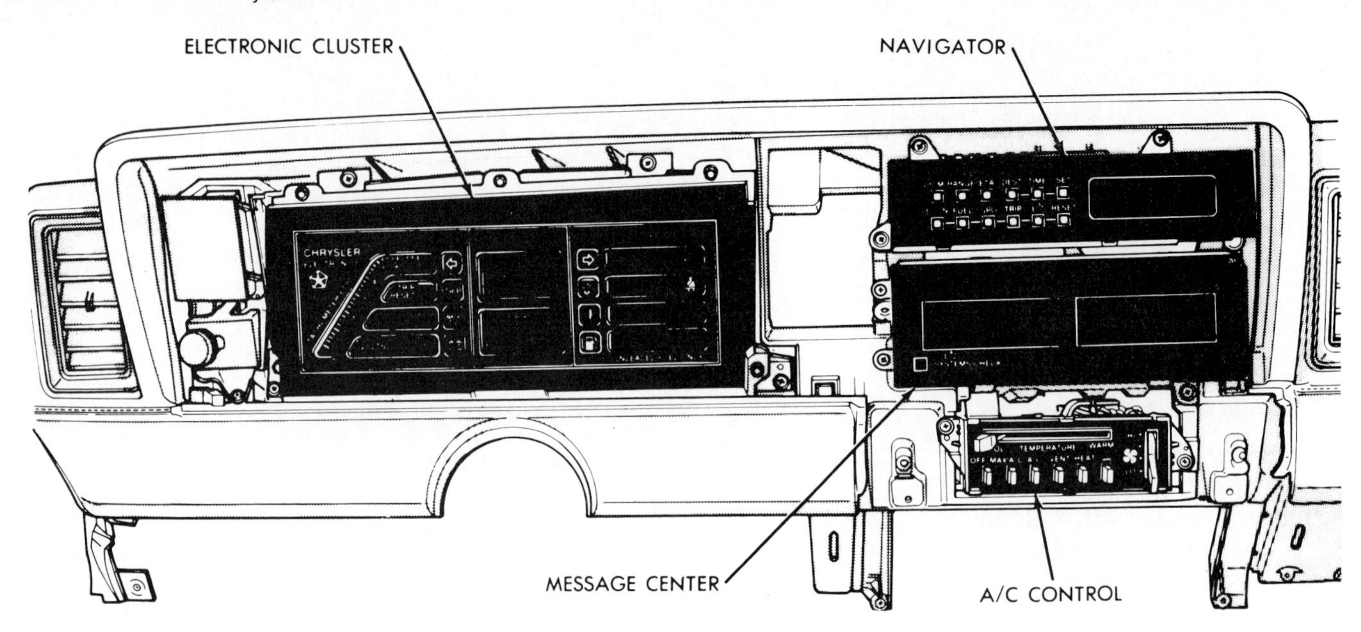

Figure 9–4 Instrument Panel with Electronic Cluster, Navigator, and Message Center for 24-Function Monitor and Voice Alert System. (Courtesy of Chrysler Canada Ltd.)

(c) depressing the system check button three times in a 5-second interval with the ignition on.

3. In a 24-function monitor and voice alert system, a *voltage low* message is given if the engine speed is above 1500 rpm for 15 seconds and the battery voltage is below:
(a) 11.75V.
(b) 12.0V.
(c) 12.35V.

10

Trip Computers

Practical Completion Objectives

1. Demonstrate the use of the various navigator functions.
2. Perform a general diagnosis on the navigator system.
3. Use the speed, fuel flow, and fuel level diagnostic function to diagnose the navigator system.

Electronic Navigator Operation

Inputs

The electronic navigator supplies the driver with trip information that is not supplied by standard instrumentation. (A complete electronic navigator is shown in Chapter 9, Figure 9–4.)

A vehicle speed and distance traveled signal is supplied from the speed sensor to the module in the navigator. This speed sensor signal is also used in the electronic fuel-injection (EFI) system, and the electronic voice alert system. (The EFI system is explained in Chapter 4, and the electronic voice alert system is described in Chapter 9.) An input signal regarding the amount of fuel consumed is sent from the logic module in the EFI system to the navigator module. On some trip computer systems this signal is generated by a fuel sensor in the fuel line. The fuel gauge sending unit in the fuel tank sends an input signal to the navigator module in relation to the amount of fuel in the tank. This signal is transmitted on the G4 circuit. The navigator control buttons and wiring diagram are illustrated in Figure 10–1.

Range Function When the RANGE button is pressed, the navigator digital display will indicate the number of miles that can be driven on the fuel remaining in the tank. The navigator module multiplies the amount of fuel in the tank by the projected fuel mileage to provide this calculation. An update of the range reading is provided every few seconds by the navigator module as fuel is used out of the tank.

Distance to Destination Function The distance to destination (DEST) must be set before this funtion will operate. When the DEST is entered, press the DEST button and then press the SET button within 5 seconds. This action will cause the navigator display to indicate 0 miles. At this time enter the DEST with the appropriate numbered navigator buttons. When the SET button is pressed, the navigator module will begin the countdown of distance traveled. The maximum distance setting is 9999 miles (16,091 km). After the DEST is entered, the navigator will display the remaining DEST when the DEST button is pressed. When the destination is reached, the navigator will display TRIP COMPLETED followed by several audible tones. The electronic navigator system is connected to the voice alert system. The navigator will display English or metric readings if the US/M button is pressed.

Time Function When the engine is started, or the TIME button is pressed, the navigator will continuously display the time of day, day of week, month of year, and day of month. The display may be reset if the time button is pressed followed by the set button within 5 seconds. When this action is taken, an arrow will appear on the display that indicates that the hours are to be set. The hour display is advanced if the RESET button is pressed, or the US/M button will back up the display. After the hour display has been set, press the SET button again and the arrow will

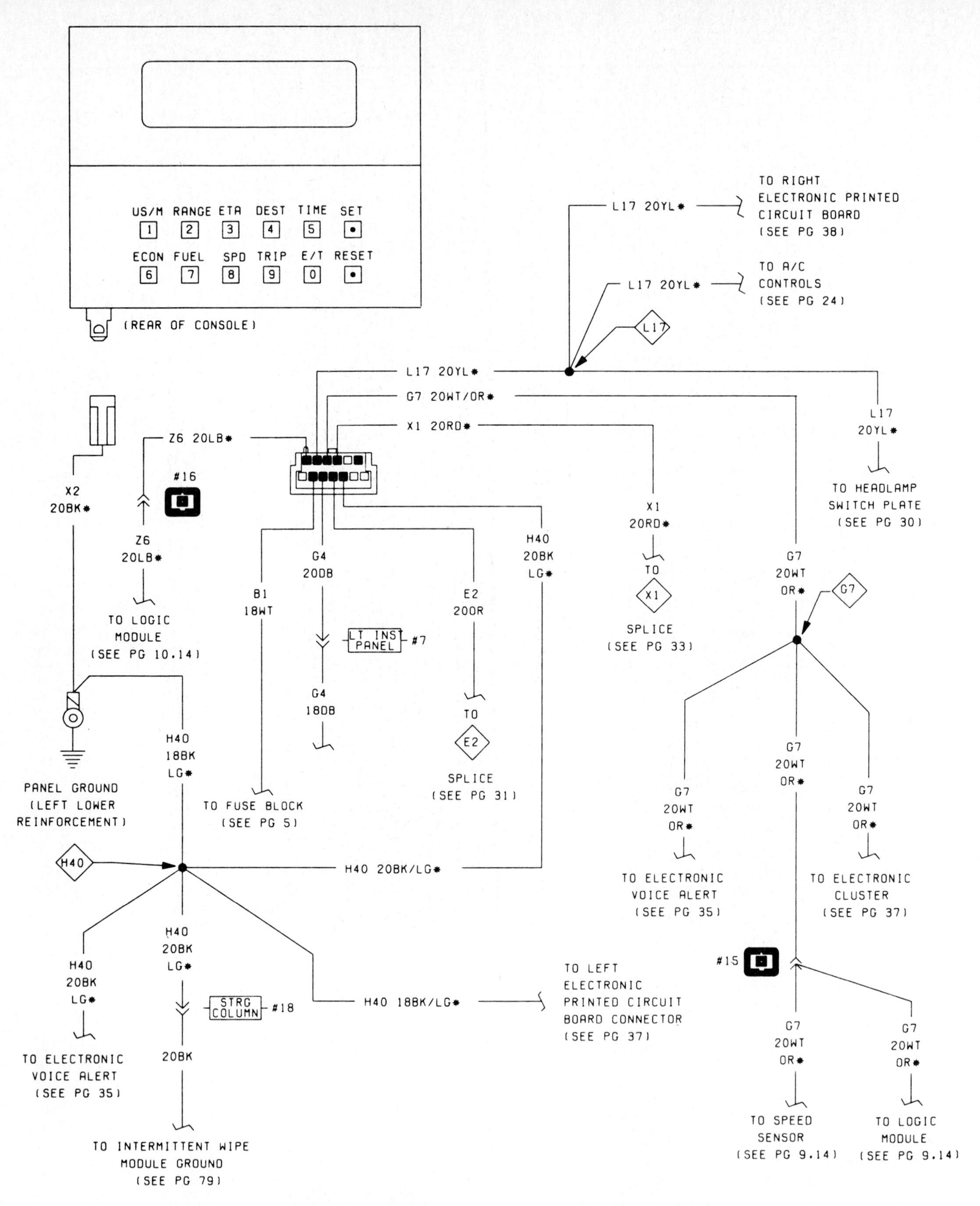

Figure 10–1 Electronic Navigator Control Buttons and Wiring Diagram. [Courtesy of Chrysler Canada Ltd.]

point to the minute display. The same procedure may be used to advance or back up each display. When the minutes are set, repeat the procedure for day, month, and date. Any portion of the time display may be bypassed if the SET button is pressed until the arrow indicates the time function that requires setting. When the entire time display is correct, press the SET button a final time to establish the readings. If the battery has been disconnected, the time display will require setting.

Estimated Time of Arrival Function When the estimated time of arrival (ETA) button is pressed, the navigator will display the estimated driving time to destination for 5 seconds. After this time, the display will switch to a continuous reading of the time and date that you will arrive at a previously entered destination. If the vehicle is operating at normal or high speeds, the navigator module calculates the ETA from the current vehicle speed. At low vehicle speeds the module uses the trip average speed to make this calculation. If the ETA exceeds 100 hours, TRIP OVER 100 HOURS will be displayed. When the destination has been passed, TRIP COMPLETED will appear on the display.

Economy Function If the economy (ECON) button is pressed, the average miles per gallon (mpg) since the last reset will be displayed for 5 seconds. After this time the display will indicate the present mpg, and this reading is updated continuously. The navigator module will begin a new average mpg calculation if the RESET button is pressed while the average mpg is displayed. Updating of the average mpg reading occurs every 16 seconds, whereas the present mpg display is updated every 2 seconds.

Fuel Function When the FUEL button is pressed, the navigator display will indicate the number of gallons consumed since the last reset. The highest possible display reading is 999.9 gallons, and the display is updated every few seconds. A new fuel consumed calculation is started if the RESET button is pressed within 5 seconds after the FUEL button is pressed.

Speed Function When the speed (SPD) button is pressed, the average miles per hour (mph) since the last reset will be displayed. This reading will be updated every 8 seconds, and the highest display is 85 mph. If the RESET button is pressed within 5 seconds after the SPD button is pressed, a new average mph calculation will be initiated. Since low speed ETA calculations are based on average speed, the driver may wish to reset the average speed after the DEST is entered.

Trip Function If the TRIP button is pressed, the navigator display will indicate the accumulated trip miles since the RESET button was pressed. The maximum displayed mileage is 999.9 miles, and this reading is updated every .5 seconds. When this mileage is reached, the display will return to zero. A new trip mileage calculation is initiated if the RESET button is pressed within 5 seconds after the TRIP button is pressed.

Elapsed Time Function When the elapsed time (E/T) button is pressed, the amount of driving time since the RESET button was pressed will be indicated on the navigator display. The highest reading is 99 hours and 59 minutes. After this time is reached, the reading returns to zero. During the first hour after the RESET button has been pressed the E/T will be displayed in minutes and seconds. After this time hours and minutes will be shown on the E/T display. If the RESET button is pressed within 5 seconds after the E/T button is pressed a new E/T calculation will be initiated.

Reset Function The reset button is used as previously described to clear various functions and begin new calculations. All trip information may be cleared simultaneously if the RESET button is pressed twice within 5 seconds after any of the following buttons are pressed: ECON, FUEL, SPD, TRIP, ET. When this action is taken the navigator display will indicate TRIP RESET for 5 seconds.

Electronic Navigator Diagnosis

General Diagnosis

When the navigator display is inoperative, fails to respond to function buttons, or goes to minimum brightness with the headlamps on, proceed with the diagnosis in Figure 10-2.

Speed Diagnostic Function

If all the navigator functions illustrated in Figure 10-3 fail to operate, the speed diagnostic function may be used to diagnose the system.

During the speed diagnostic function use the following procedure:

1. Press the SPD button and within 5 seconds simultaneously press the US/M and RESET buttons followed by the SET button.
2. Enter 122 on the navigator.

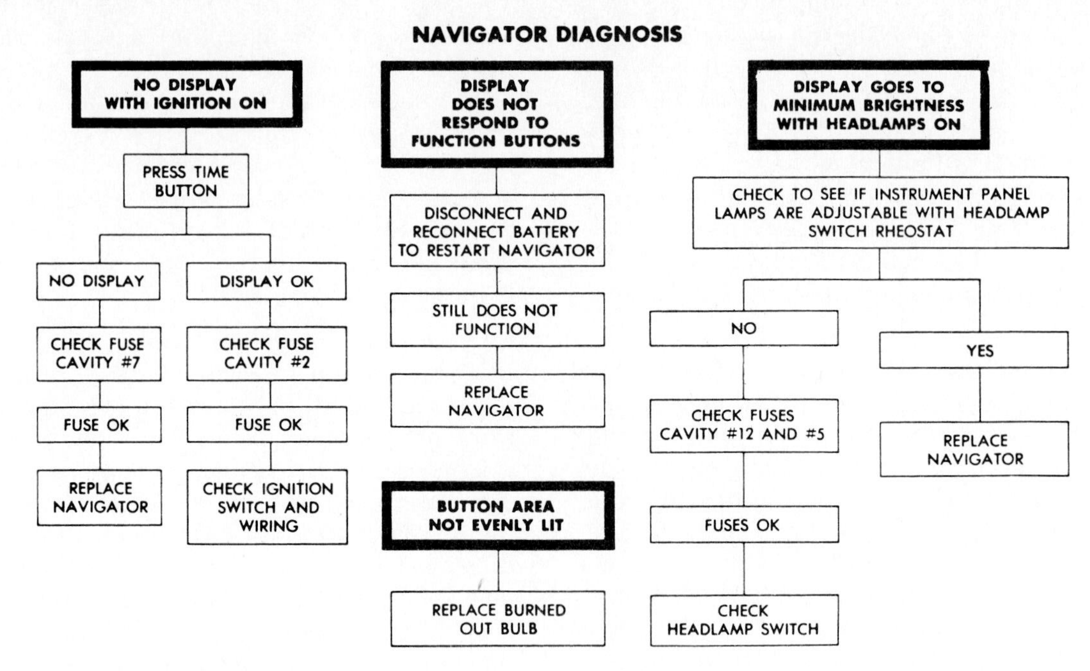

Figure 10-2 General Navigator Diagnosis. [Courtesy of Chrysler Canada Ltd.]

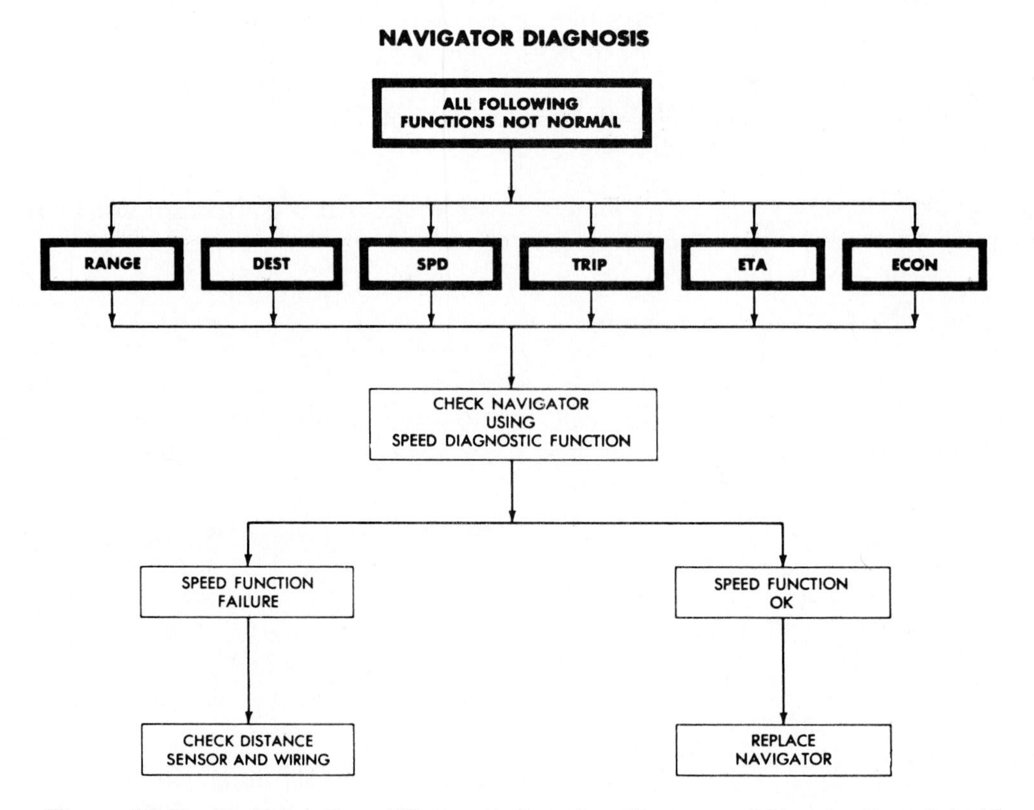

Figure 10-3 Navigator Speed Diagnostic Function. [Courtesy of Chrysler Canada Ltd.]

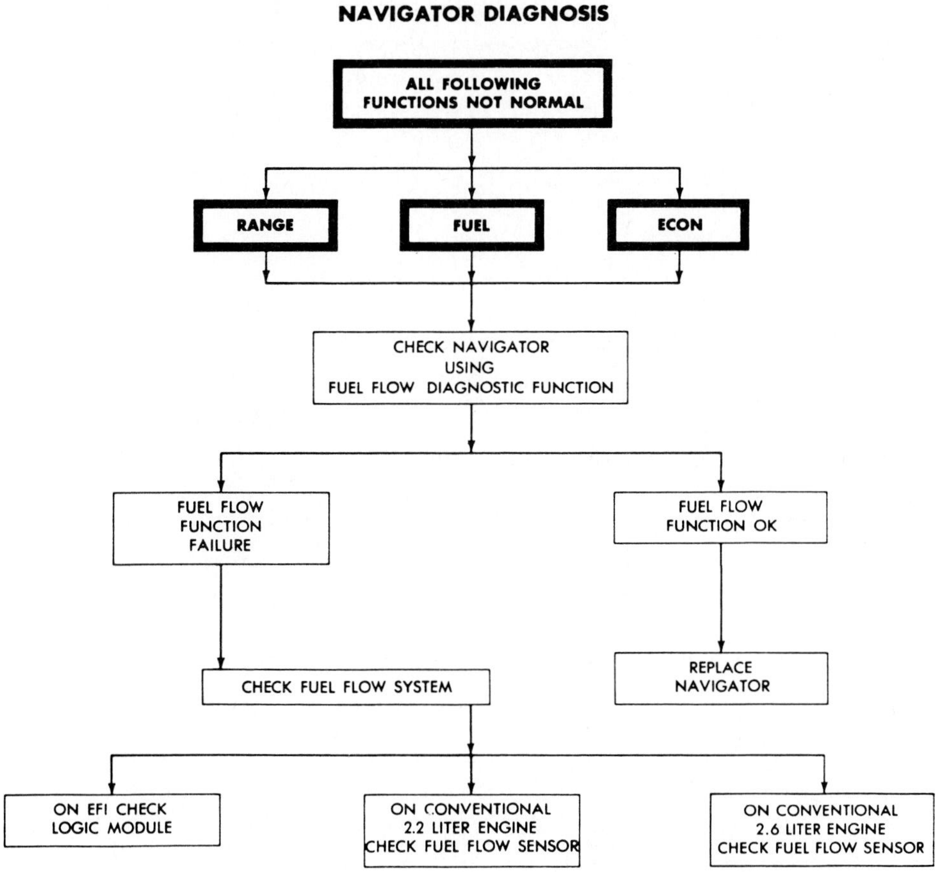

NAVIGATOR DIAGNOSIS

ALL FOLLOWING FUNCTIONS NOT NORMAL

RANGE

FUEL

ECON

CHECK NAVIGATOR USING FUEL FLOW DIAGNOSTIC FUNCTION

FUEL FLOW FUNCTION FAILURE

FUEL FLOW FUNCTION OK

CHECK FUEL FLOW SYSTEM

REPLACE NAVIGATOR

ON EFI CHECK LOGIC MODULE

ON CONVENTIONAL 2.2 LITER ENGINE CHECK FUEL FLOW SENSOR

ON CONVENTIONAL 2.6 LITER ENGINE CHECK FUEL FLOW SENSOR

Figure 10-4 Fuel Flow Diagnostic Function. (Courtesy of Chrysler Canada Ltd.)

3. The display should indicate 1.11 times the vehicle speed. For example, at 10 mph, the display should read 11.
4. Exit from the speed diagnostic function by pressing the SET button followed by another function button.

Fuel Flow Diagnostic Function

If the navigator functions shown in Figure 10-4 do not operate normally, use the fuel flow diagnostic function to diagnose the system as outlined in this table.

Proceed as follows for the fuel flow diagnostic function:

1. Press the SPD button and within 5 seconds simultaneously press the US/M and RESET buttons followed by the SET button.
2. Enter 123 on the navigator.
3. The display should indicate 2 to 4 with the engine idling.
4. The SET button is pressed followed by any other function button to exit from the fuel flow diagnostic function.

Fuel Level Diagnostic Function

If the range function does not operate normally, the fuel level diagnostic function shown in Figure 10-5 may be used to diagnose the system.

The following procedures must be used during the fuel level diagnostic function:

1. Press the SPD button and within 5 seconds press the US/M and RESET buttons at the same time, and then press the SET button.
2. Enter 124 on the navigator.
3. The display reading should be near 0 if the fuel

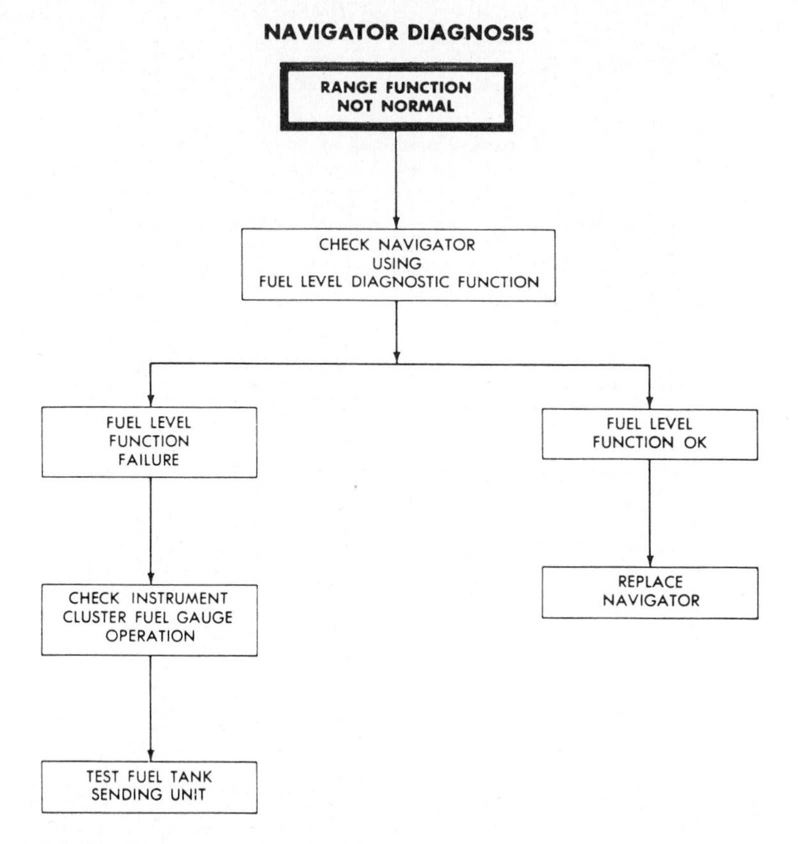

Figure 10-5 Fuel Level Diagnostic Function. (Courtesy of Chrysler Canada Ltd.)

tank is full, or above 229 if the tank is empty. The readings between these figures are proportional to the amount of fuel in the tank. For example, with 5 gallons of fuel in the tank, the display reading should be approximately 195.

4. Exit from the fuel level diagnostic function by pressing the SET button followed by any other function button.

Test Questions

1. An input signal regarding the amount of fuel consumed is sent to the navigator module from the:
   (a) electric fuel pump.
   (b) logic module or fuel flow sensor.
   (c) fuel injectors.
2. The navigator will display the time function when the engine is started. T F
3. If the battery is disconnected, the navigator will maintain the correct time. T F
4. All trip information may be cleared simultaneously if the ECON button is pressed and then followed within 5 seconds by pressing the:
   (a) RESET button twice.
   (b) SET button once.
   (c) E/T and TIME buttons simultaneously.
5. During the speed diagnostic function, the number that must be entered with the navigator button is
   (a) 111.
   (b) 122.
   (c) 229.

11

Electronic Instrumentation

Practical Completion Objectives

1. Diagnose vacuum fluorescent display (VFD) instrument clusters.
2. Observe safety precautions when vehicles with VFDS are being serviced.

Types of Electronic Instrumentation Displays

Light Emitting Diodes

A light emitting diode (LED) is a diode that emits light as electric current flows through it. Small dotted segments are arranged in the LED display so that numbers and letters can be formed when selected segments are turned on. LEDs usually emit red light and they can be seen easily in the dark; however, they are less visible in direct sunlight. The tachometer on the electronic instrument cluster in Figure 11–1 has 36 LEDs.

Liquid Crystal Displays

The liquid crystal display (LCD) uses a polarized light principle on a nematic liquid crystal to display numbers and characters. An LCD display uses extremely low electrical power, but it requires back-lighting to be viewed in the dark. A complete LCD instrument display is illustrated in Figure 11–2.

The retainer is made of heat-resistant mineral and glass-filled polyester, and it provides a closely dimensioned support for the LCD cells. A dark background is used on the nematic LCDs, and active areas change from dark to clear when they are energized. Polarizers in the LCDs provide the proper balance of contrast, transmission, and hue. The protective matte film on the front surface of the LCDs reduces reflections and provides resistance to scratches and chemicals.

A thin polycarbonate transflector is mounted behind the LCD to provide color. The front surface of the transflector is silk-screened with translucent fluorescent inks, which make use of front-incident ambient light and rear-incident back-lighting to obtain proper day and night intensity levels.

Backlighting is achieved with the use of clear acrylic light pipes. Light entrance areas and special optical patterns on the rear surfaces of the light pipes provide optimum backlighting balance and intensity. White polycarbonate reflectors are placed behind the light pipes to use the light escaping from their rear surfaces.

Hundreds of individual electrical connections are made through the elastometric connectors, which make contact between the indium tin oxide conductor pads on the LCDs to corresponding contacts on the driver board.

The driver board contains 8 static LCD driver integrated circuits (ICs), which are individually responsible for driving 32 LCD segments.

The microprocessor on the logic board receives input data from various sensors and switches and then determines whether each LCD segment should be on or off. A 12-pin connector is used to connect the logic board to the driver board, whereas 24-pin and 36-pin connectors complete the electrical connections between the microprocessor and the vehicle electrical system. The inputs to the microprocessor in the LCD instrument cluster and the output control features are outlined in block form in Figure 11–3.

Figure 11-1 Electronic Tachometer with LEDs. [Courtesy of Chrysler Canada Ltd.]

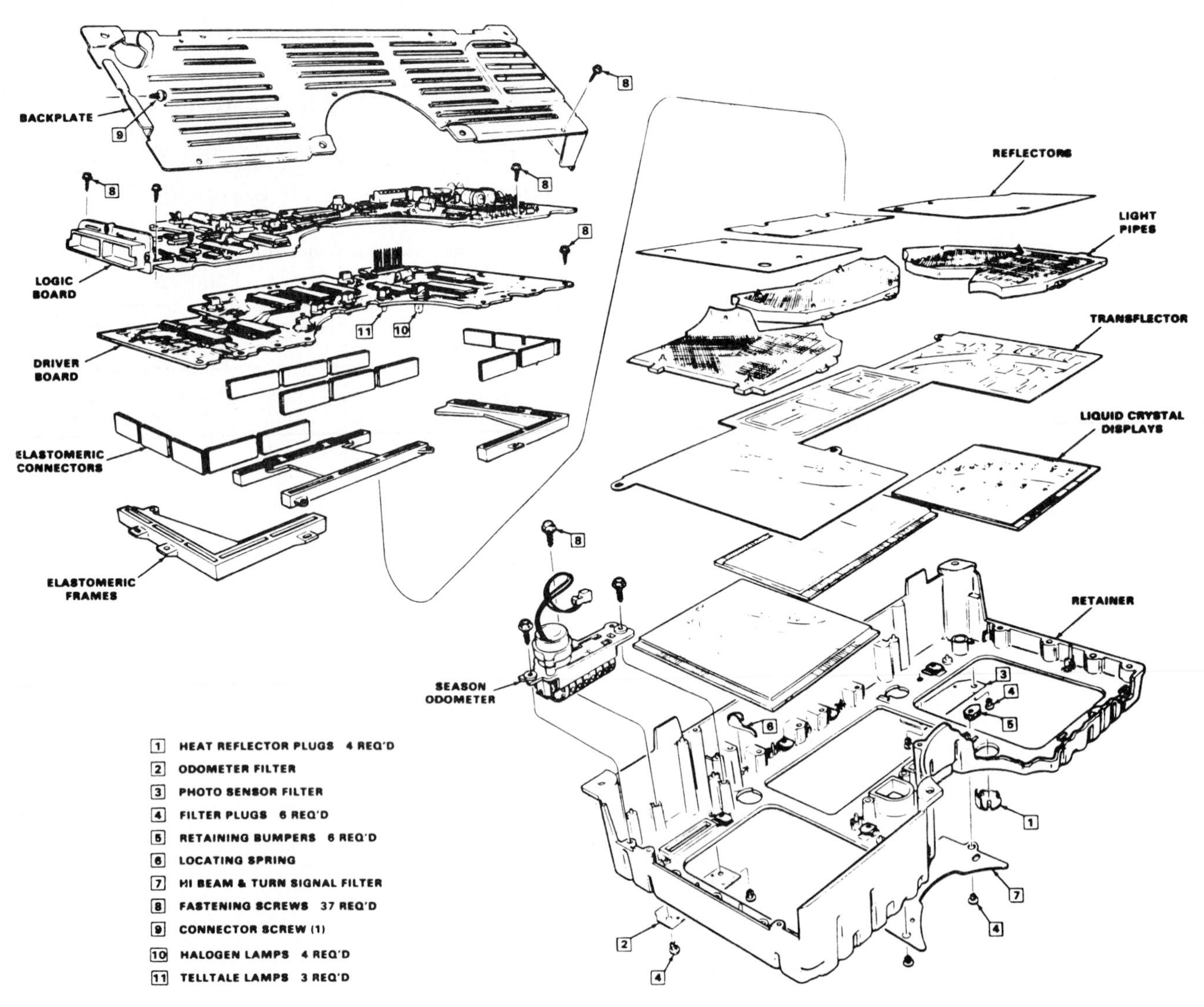

Figure 11-2 Liquid Crystal Display Components. [Reprinted with Permission of Society of Automotive Engineers © 1983].

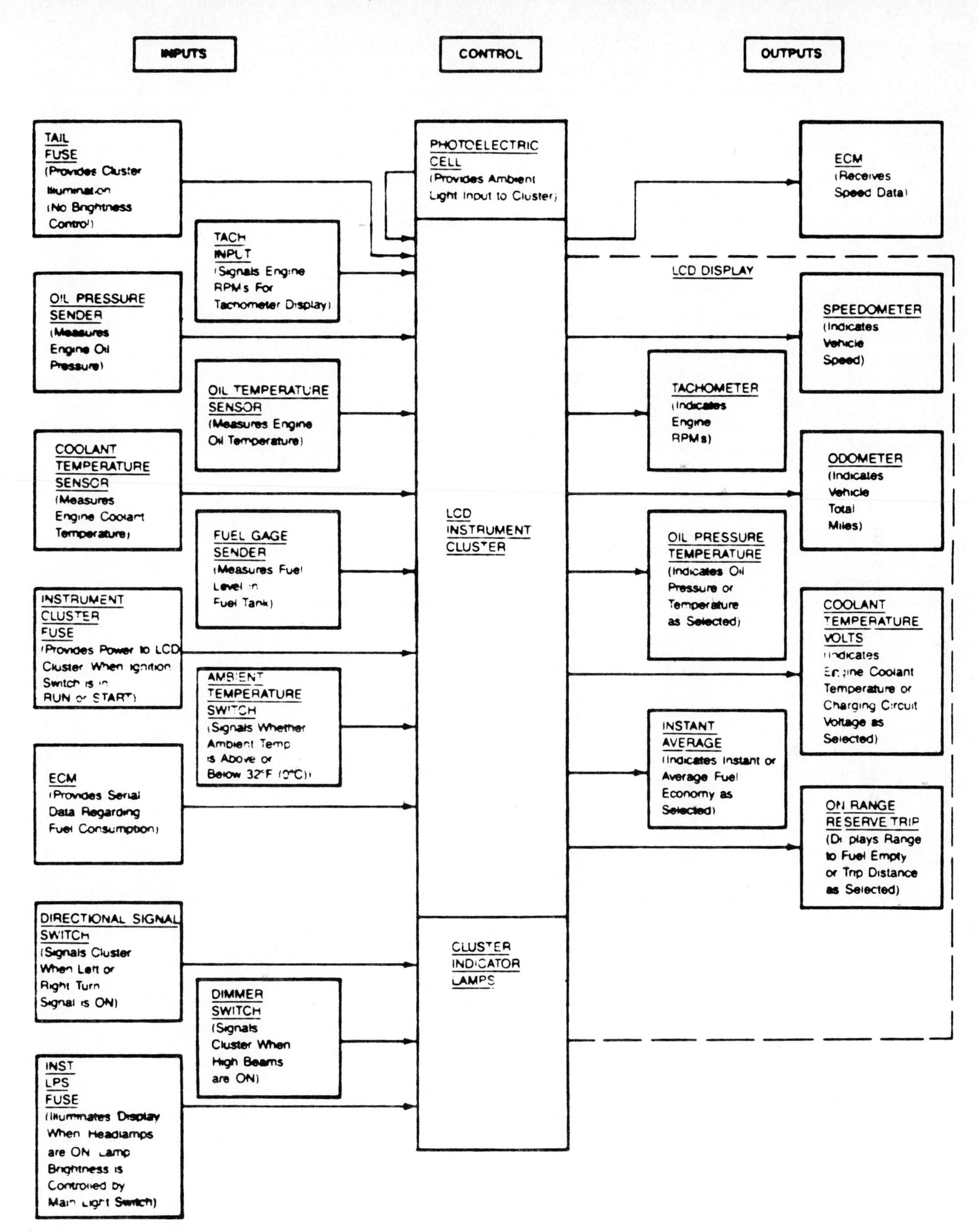

Figure 11-3 Inputs and Outputs, LCD Instrument Cluster Microprocessor [Courtesy of General Motors Corporation]

The front of the LCD instrument cluster is pictured in Figure 11–4.

The switch and telltale console, shown in Figure 11–5, is mounted near the LCD instrument cluster.

The LCD instrument cluster is continuously backlit by four halogen lamps. Their intensity is automatically controlled by pulse-width modulation to provide proper display intensity under varied natural lighting conditions. A photocell in the upper left corner senses natural light conditions, and the microprocessor then determines the correct pulse-width cycle for the halogen lamps, which provides the proper LCD display intensity.

When the ignition switch is turned on, the LCD cluster is lit at full intensity for 2 seconds, and all the LCD segments are activated. During this time, the microprocessor measures all inputs so it can display them immediately after the initialization sequence. The driver may adjust the brilliance of the LCD display by rotating the rheostat on the headlight switch.

LCD Instrument Cluster Functions

Vehicle Speed

The speedometer displays vehicle speed in both digital and analog bar graph form. Metric or English values may be selected on the LCD displays with the Metric/English switch on the switch and telltale console. The yellow digital speedometer displays speed from 0 to 157 mph or 0 to 255 kph. A green 41-segment bar graph displays speed from 5 to 85 mph (8 to 137 kph). A multipole permanent magnet speed sensor in the transaxle provides a speed signal to the LCD microprocessor. The odometer is driven with a

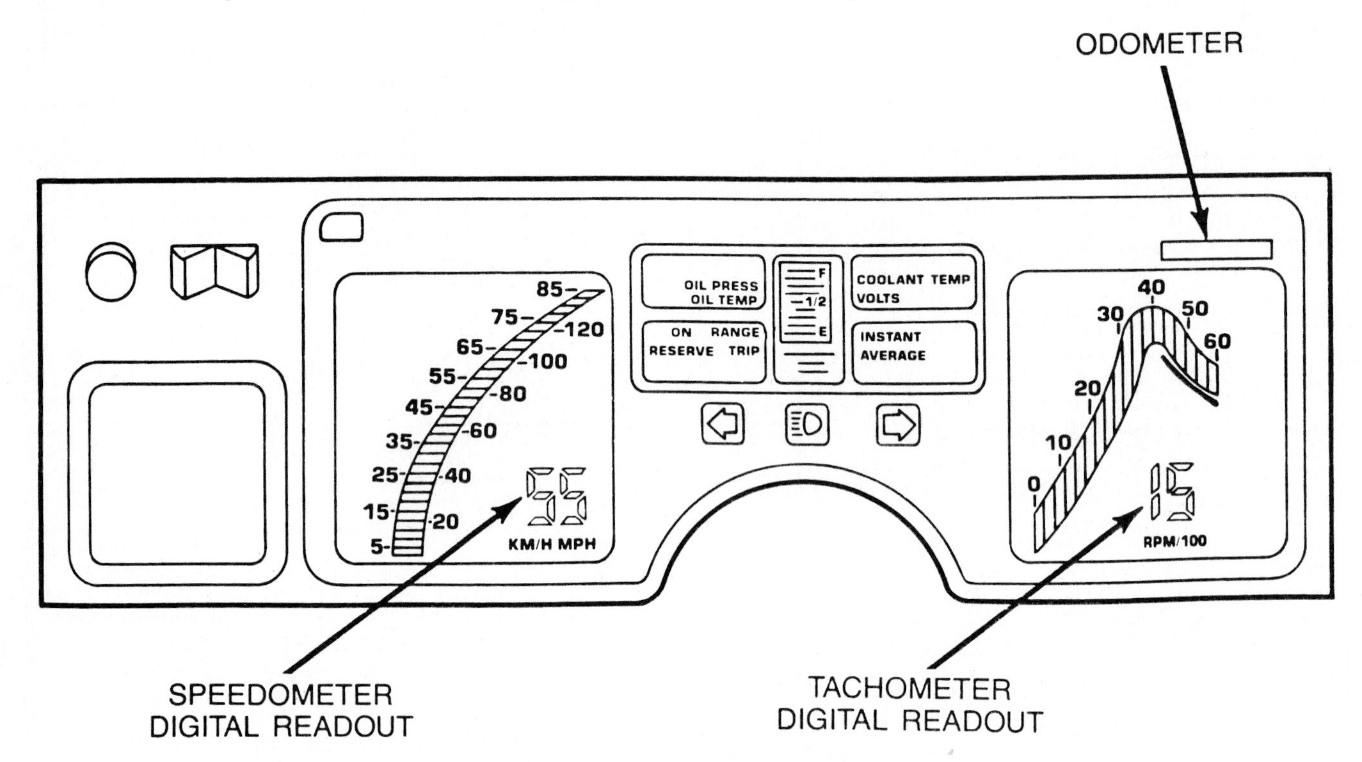

Figure 11–4 LCD Instrument Cluster. [Courtesy of General Motors Corporation]

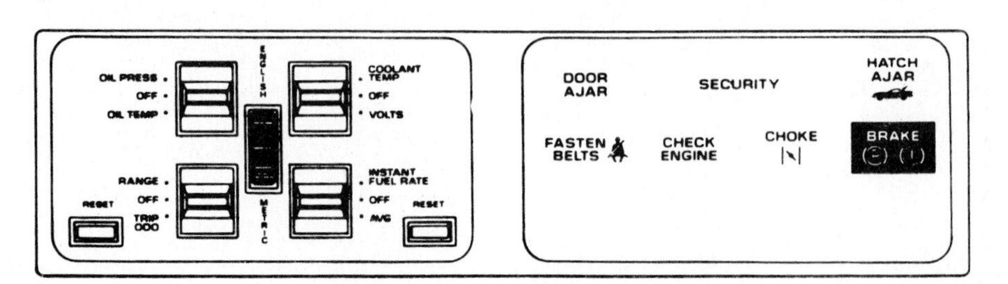

Figure 11–5 Switch and Telltale Console Used with LCD Instrument Cluster. [Courtesy of General Motors Corporation]

stepper motor. A trip odometer is updated every 0.1 mile (mi) or 0.1 kilometer (km). This odometer counts from 0 to 999.9 mi (or 999.9 km), and then it will count from 1000 to 4000 mi (1000 to 6436 km). The trip odometer is set on zero when the reset button is depressed.

Engine Speed

The tachometer displays engine rpm in analog and digital forms. Yellow digits display engine speed from 0 to 7000 rpm. A 31-segment bar graph is green from 0 to 4300 rpm, yellow to 5100 rpm, and red to 6000 rpm. Vehicle speed and engine speed displays are updated 16 times per second. An input signal from the high-energy ignition (HEI) module is used by the microprocessor to control the tachometer display.

Oil Pressure and Temperature

The driver may select an oil pressure or oil temperature reading on the LCD cluster with the selector switch in the switch and telltale console. Oil pressure displays range from 0 to 80 psi (0 to 560 kPa), while oil temperature displays read from 149° to 320°F (65° to 160°C). When the oil temperature exceeds 300°F (149°C), an out-of-normal-limits warning is activated in the oil display.

Coolant Temperature and Volts

Coolant temperature, or a voltage reading, may be selected on the LCD display when the appropriate selector button is depressed on the switch and telltale console. Coolant temperature will be displayed from 104°F to 302°F (40° to 150°C) and electrical system voltage is displayed in the 11.5 to 16.5V range. When coolant temperature exceeds 255°F (124°C), the out-of-normal-limits warning is shown on the temperature display.

Fuel Economy

Instant or average fuel economy is displayed in the lower right LCD quadrant. This display will indicate miles per gallon (mpg) or liters per 100 kilometers (L/100 km). Speed sensor pulses are counted by the LCD microprocessor to determine the distance traveled. The electronic control module (ECM) on the vehicle controls the fuel system. (Refer to Chapter 4 for an explanation of the ECM.) A serial data link from the ECM to the LCD microprocessor provides the necessary information regarding the amount of fuel consumed. This signal is updated each 0.675 second, and the instant fuel economy display is updated each 0.75 second. Average fuel economy is computed from the distance traveled and the fuel used since the reset button was depressed.

Fuel Level

An illuminated bar graph fuel guage is located in the center of the LCD cluster. When the fuel tank is full, all the bars are brightly illuminated. As the fuel level is lowered in the tank, the bars gradually fade out from the top down. An amber low fuel warning light is activated when only two bars remain illuminated. When this occurs the lower left reading in the LCD display switches from trip distance, or range, to a display of the distance traveled with the low fuel warning activated. Miles or kilometers may be displayed in this reserve fuel mode. When two bars are left illuminated on the fuel gauge, the range display will read zero. At higher levels in the fuel tank, the range display indicates the distance that may be traveled on the fuel remaining in the tank. If the vehicle is operating in the reserve fuel mode, the driver may display the trip odometer reading for 5 seconds by selecting "off" or "range" and then switching to "trip odometer." After 5 seconds this display will change back to display the distance traveled since the low fuel warning was activated.

Vacuum Fluorescent Displays

A vacuum fluorescent display (VFD) generates its light based on the same basic principles as a television picture tube. In the VFD a heated filament emits electrons that strike a phosphor material and emit a blue-green light. The filament is a resistance wire that is heated by electric current. A coating on the heated filament emits electrons that are accelerated by the electric field of the accelerating grid. The anode is charged with a high voltage that attracts the electrons from the grid. A VFD computer supplies high voltage to the specific anode segments that are needed to emit light for any given message. The operating principles of a VFD are shown in Figure 11–6.

The brightness of the VFD may be intensified by increasing the voltage on the accelerating grid. Another method of controlling VFD brilliance is pulse-width timing. When this method is used, the VFD is turned on and off very rapidly. A shorter on-time

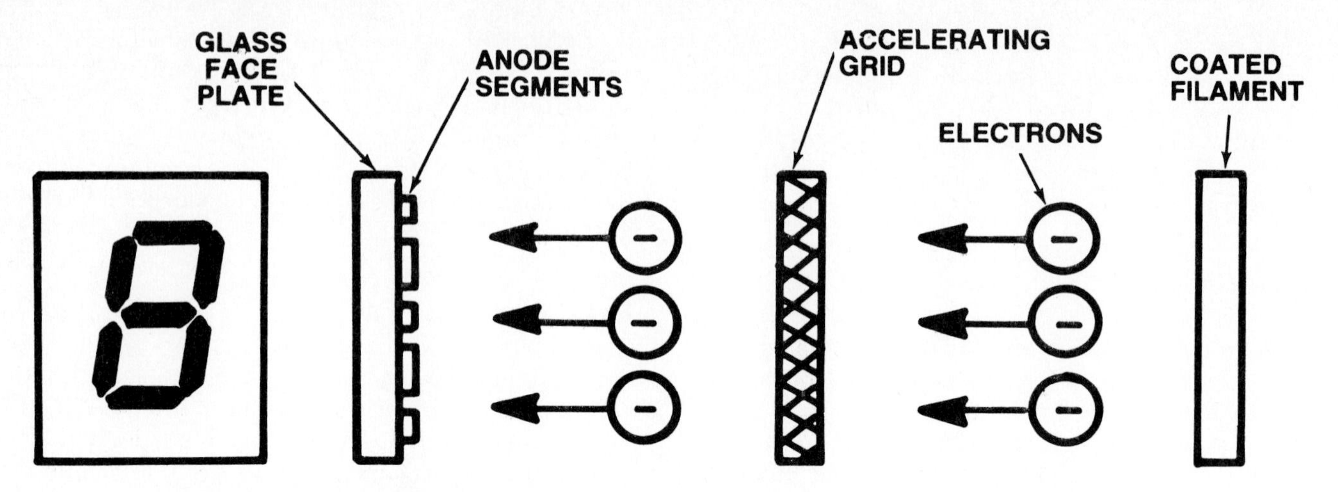

Figure 11–6 Vacuum Fluorescent Display Principles. (Courtesy of Chrysler Canada Ltd.)

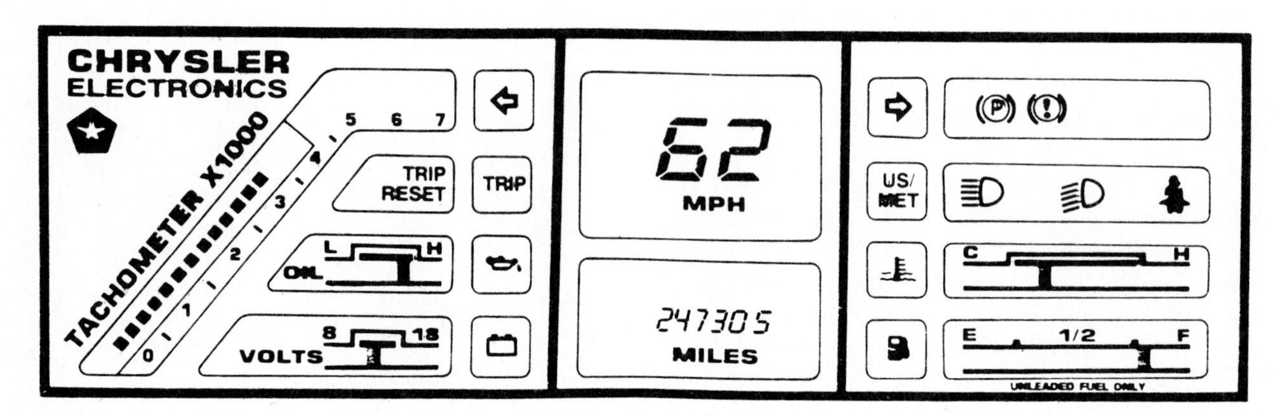

Figure 11–7 Vacuum Fluorescent Display. (Courtesy of Chrysler Canada Ltd.)

dims the display whereas a longer on-time increases the brilliance. Shock can damage a VFD display; therefore, it must be handled with care. A typical VFD is pictured in Figure 11–7.

Diagnosis of VFD

If no display appears when the ignition switch is turned on, check the fuses that supply power to the VFD. On the 21-pin connector, pin 20 is ground, pin 21 is ignition, and pin 10 is battery feed. The 21-pin VFD connector and terminal identification is shown in Figure 11–8.

If the VFD cluster illuminates, use the following sequence for cluster self-diagnosis:

1. Press the trip and trip-reset buttons with the ignition switch off. While these buttons are depressed, turn on the ignition switch. The letters EIC appear in the odometer, and the low oil lamp will illuminate. If EIC does not appear, replace the cluster.
2. When EIC is present, press the US/M button and all the VFD displays should illuminate. If the word "fail" and a code appears, and some displays do not illuminate, a fault exists in the VFD. For example, a fail code 4 indicates a defective, or incorrectly installed, odometer chip.
3. When 5 appears in the speedometer window, the first 5 tests are completed. Test 6 is initiated by waiting 5 seconds and depressing the US/M button, which should cause all the individual segments to appear in sequence. If any two segments in the same digit appear simultaneously, such as a horizontal line and a vertical line, replace the cluster. If more than one image appears at once in the gauge sequence, cluster replacement is also necessary.
4. After a 5-second pause, press the US/M button to initiate test 7. This test will illuminate the gauge scales and sequence the speedometer and odometer. If this test fails to occur, replace the cluster.
5. Wait 5 seconds and press the US/M button to be-

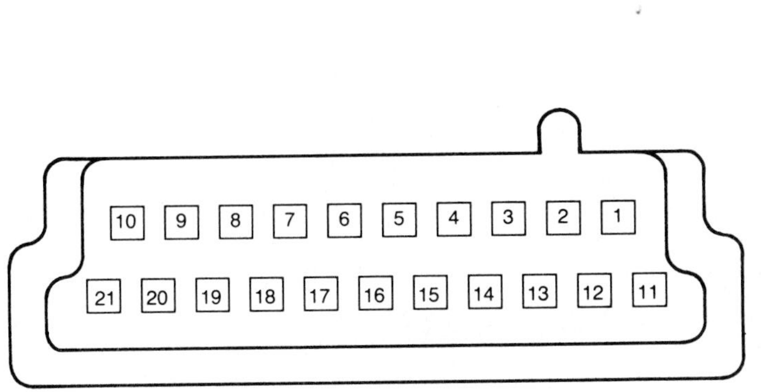

RIGHT PRINTED CIRCUIT BOARD CONNECTOR

CAV	CIRCUIT	GAUGE	COLOR	FUNCTION
1	P51	20	GY *	LAMP CHECK SWITCH
2	P2	20	BK *	PARK BRAKE SWITCH
3	E2	20	OR *	ILLUMINATION LAMPS
4	L7	20	BK/YL*	PARKING LAMPS
5	K3	20	BK/RD*	POWER LOSS LAMP
6	H4	20	BK	LAMP GROUND
7	L4	18	VT *	LOW BEAM LAMP
8	D5	20	TN	RIGHT TURN SIGNAL LAMP
9	L5	18	RD	HIGH BEAM LAMP
10	X1	20	RD *	BATTERY FEED
11				
12				
13	G7	20	WT/OR*	SPEED SENSOR
14	T11	20	GY/RD*	TACHOMETER
15	L17	20	YL *	DISPLAY DIMMING
16	G4	20	DB	FUEL SENDER
17	G60	20	GY/YL*	OIL SENDER
18				
19	G20	20	VT/YL*	TEMPERATURE SENDER
20	H40	18	BK/LG*	CHASIS GROUND
21	G5	20	DB *	IGNITION RUN/START

Figure 11–8 Vacuum Fluorescent Display 21-Pin Connector and Terminal Identification. [Courtesy of Chrysler Canada Ltd.]

gin test 8. On systems with warning lamps, the battery, temperature, and fuel lamps should light in sequence followed by the appearance of an 8 in the speedometer. If these lamps do not light, test the individual bulbs. Replace the cluster if bulb replacement does not correct the problem. On systems without warning lamps, the 8 in the speedometer will appear immediately when the US/M button is depressed. Once the 8 appears, the VFD test sequence is completed, and the cluster is confirmed good if all the tests were satisfactory. Press the US/M button or the TRIP RESET button to return the VFD to normal operation.

VFD Cluster or Odometer Chip Replacement

If the VFD cluster is replaced, the odometer memory chip should be removed from the old cluster and installed in the new cluster. This will retain the original miles or kilometers on the odometer. A special tool is available for odometer chip removal. The odometer chip may be ruined by static electric charges. Therefore, extreme care must be used when this chip is handled. Do not carry the odometer chip around unless the foam retainer is installed on the chip pins. New odometer chips are supplied with the foam retainer in place, and the retainer must not be removed until the chip is to be installed in the VFD. Be careful not to bend the chip pins during installation.

The odometer chip must be installed as indicated in Figure 11–9.

When the fuel, temperature, and oil gauges all provide a maximum reading, check the pull-up module connector at the back of the cluster as indicated in Figure 11–10.

VFD Service Precautions

The following precautions should be observed when VFD-equipped vehicles are serviced.

1. The ignition switch should be turned on before battery booster cables are connected to the battery. When the VFD is illuminated, a protection circuit prevents static charge damage to the odometer chip.
2. Nonresistor plugs or nonresistance-type plug wires must not be used or VFD damage may result.
3. When secondary ignition system spark is being tested, do not allow spark to jump more than 1/4 in (6.35 mm).

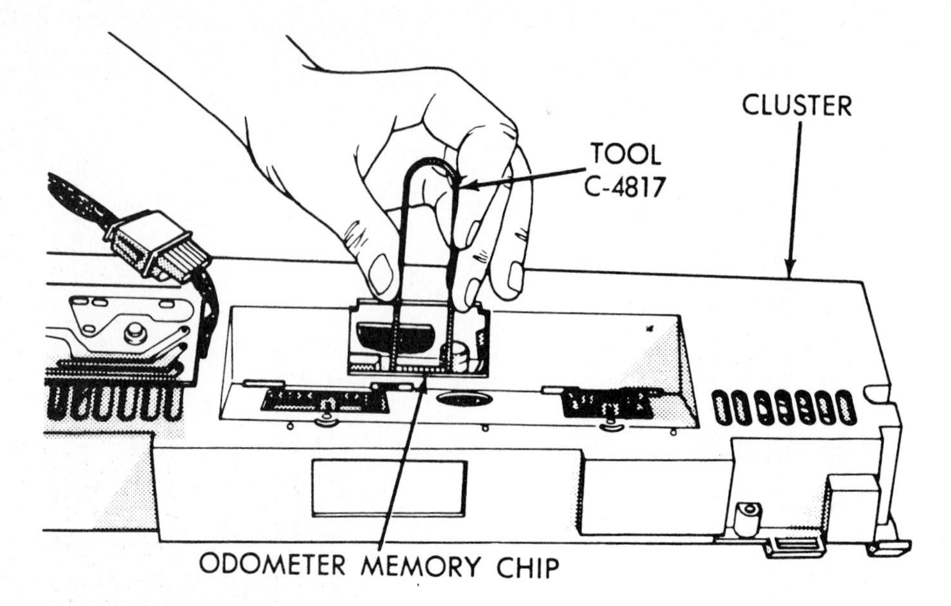

Figure 11–9 Odometer Chip Installation. (Courtesy of Chrysler Canada Ltd.)

Figure 11–10 Pull-Up Module Connector. (Courtesy of Chrysler Canada Ltd.)

Test Questions

1. A liquid crystal display (LCD) instrument cluster requires back-lighting. T F
2. In the LCD instrument cluster shown in Figure 11–4, the low fuel warning light is activated when:
   (a) two bars are lit on the fuel gauge.
   (b) three bars are lit on the fuel gauge.
   (c) four bars are lit on the fuel gauge.

The brightness of a vacuum fluorescent display may be intensified by increasing the voltage applied to the ____________.

On the VFD in Figure 11–7, the diagnostic mode is entered by:

(a) depressing the US/M button with the ignition switch on.
(b) holding the trip and trip reset buttons in the depressed position and turning the ignition on.
(c) depressing the trip and trip reset buttons with the ignition on.

5. When the VFD in Figure 11–7 is being diagnosed and the letters EIC do not appear, VFD cluster replacement is necessary. T F
6. An odometer chip in a VFD must not be carried unless the ____________ is in place on the chip.
7. Nonresistor-type spark plugs may be used with VFD instrument clusters. T F

12

Computer-Controlled Antilock Brake Systems

Practical Completion Objectives

1. Observe safety precautions when the antilock brake system is being serviced.
2. Use the correct procedure to fill the master cylinder and bleed the brakes.
3. Replace paper adjusting discs and adjust wheel sensors.
4. Replace front and rear sensor rings.

Antilock Brake System Components

Master Cylinder and Hydraulic Brake Booster

The master cylinder contains a hydraulic pump motor that supplies pressure to a booster piston and provides power assist during a brake application. A fluid reservoir on the master cylinder has two main chambers that are connected through hoses to the master cylinder and the pump motor. Integral fluid level switches are located in the reservoir cover. A solenoid valve body that is mounted on the master cylinder contains three pairs of electrically operated solenoid valves. One pair of solenoid valves is connected to each front wheel, and the other pair is connected to the rear wheels. A master cylinder used in the antilock brake system is shown in Figure 12–1.

The accumulator is a gas-filled pressure chamber that is mounted on the pump and motor assembly in the master cylinder as indicated in Figure 12–2.

Electronic Controller

The electronic controller is located in the luggage compartment underneath the package tray. This controller contains two microprocessors that monitor the brake system during normal driving conditions and also while the system is providing an antilock function. When the controller receives input signals from the wheel sensors that indicate the beginning of wheel lock-up conditions, the controller operates the master cylinder solenoid valves to control master cylinder pressure and prevent wheel lock-up. If a defect causes the electronic system to be inoperative, normal power-assist braking will be maintained. The electronic controller is illustrated in Figure 12–3.

Wheel Sensors

Wheel sensors at each wheel are mounted directly above a toothed sensor ring. These sensors provide input signals to the electronic controller. During a hard brake application the wheels may lock up and begin to skid. This condition may cause the vehicle to spin out of control. When the wheel sensor signals indicate a wheel lock-up condition, the controller immediately operates the solenoid valves to control master cylinder pressure. A rear wheel sensor is pictured in Figure 12–4, and a front wheel sensor is shown in Figure 12–5.

Antilock Brake System Operation

Master Cylinder Operation

The hydraulic pump maintains a pressure of 2030 to 2610 psi (14,000 to 18,000 kPa) in the accumulator. This pressure is also supplied to a control valve that is controlled by brake pedal movement. If the brakes are applied, the scissor lever mechanism activates the control valve, which allows pressure from the hydraulic pump to enter the boost chamber. This pres-

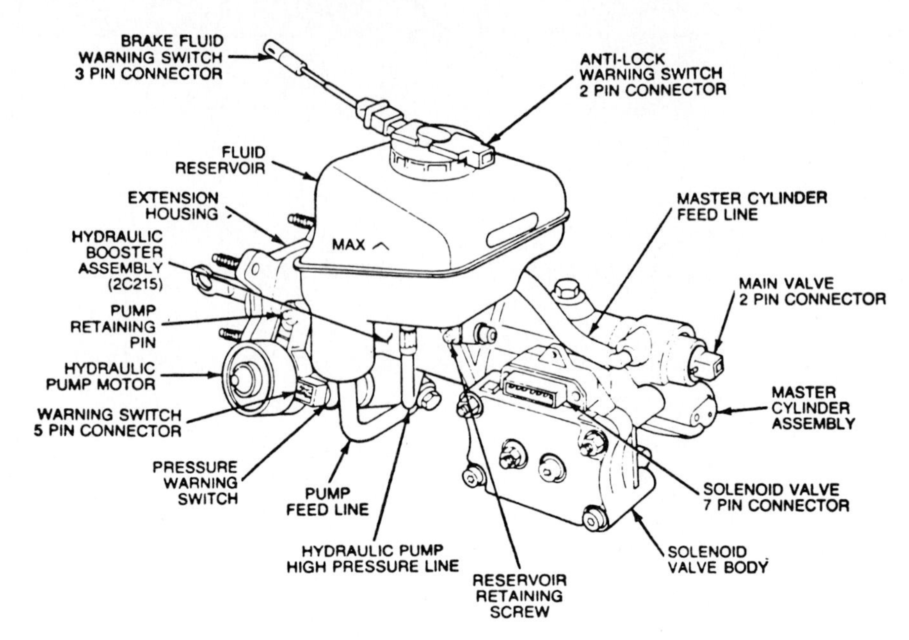

Figure 12–1 Master Cylinder, Antilock Brake System. (Courtesy of Ford Motor Co.)

sure will be proportional to brake pedal travel. Pressure to the boost chamber is also supplied through the normally open solenoid valve, and the proportioning valve, to the rear brakes. The boost chamber pressure and the brake pedal force moves the master cylinder pistons, which supplies fluid pressure through the normally open solenoid valves to the front brakes. Therefore, the hydraulic pump pressure assists the driver to apply the brakes, which reduces brake pedal effort. The electronic controller closes the normally open solenoid valves and opens the normally closed solenoid valves to reduce the master cylinder pressure and prevent wheel lock-up. When the normally closed solenoid valves are opened, fluid returns from the calipers, or wheel cylinders, to the reservoir.

During normal braking, the brake pedal application will feel identical to a normal power brake system. In the antilock mode, slight pulsations will be felt on the brake pedal and a small change in pedal height will be experienced. The antilock brake hydraulic system is shown in Figure 12–6.

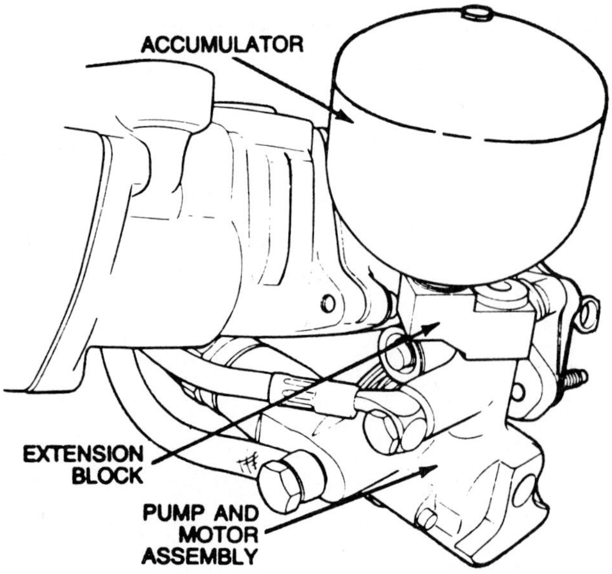

ACCUMULATOR
EXTENSION
BLOCK
PUMP AND
MOTOR
ASSEMBLY

Figure 12–2 Accumulator on Pump and Motor Assembly. (Courtesy of Ford Motor Co.)

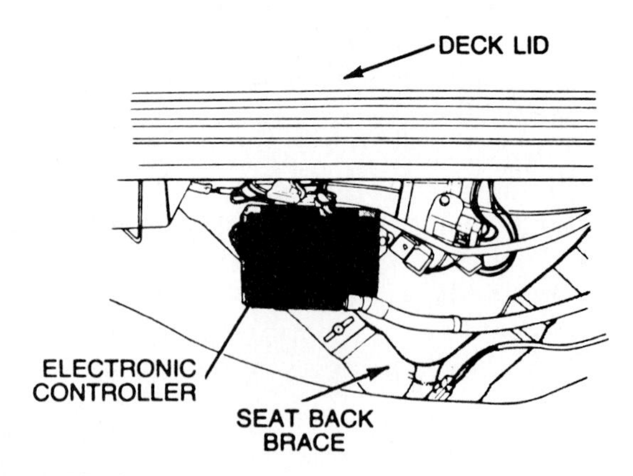

Figure 12–3 Electronic Controller for Antilock Brakes. (Courtesy of Ford Motor Co.)

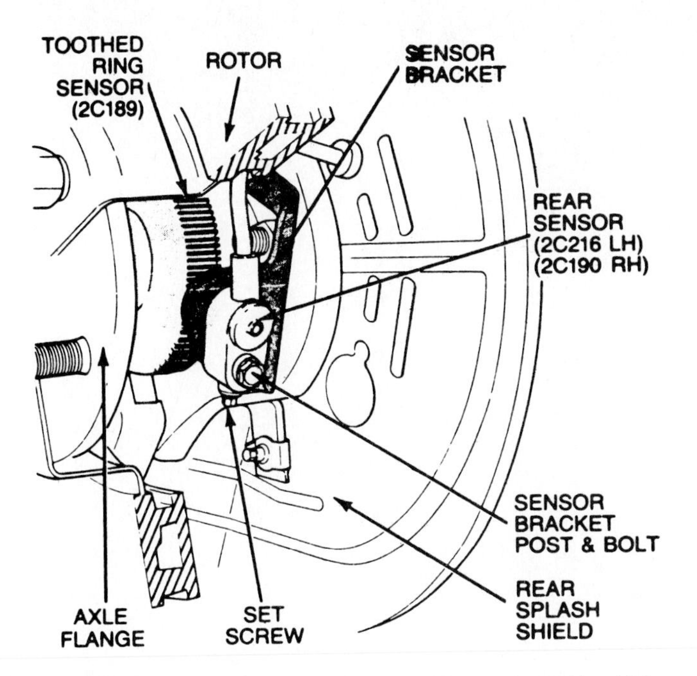

Figure 12-4 Rear Wheel Sensor. (Courtesy of Ford Motor Co.)

Figure 12-5 Front Wheel Sensor. (Courtesy of Ford Motor Co.)

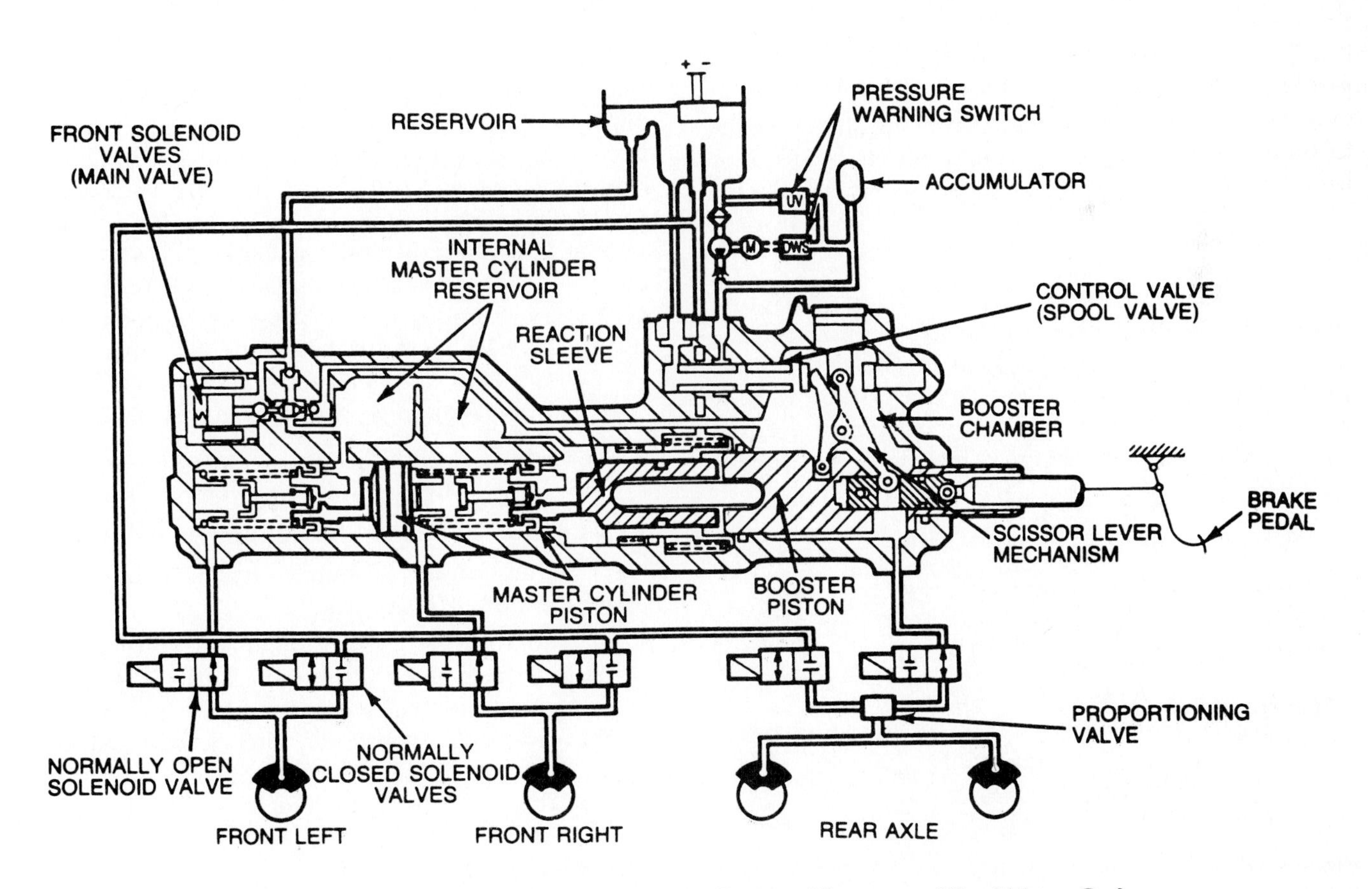

Figure 12-6 Hydraulic System, Antilock Brakes. (Courtesy of Ford Motor Co.)

Service and Diagnosis of Antilock Brake Systems

Antilock Brake Warning Light

A check antilock brake warning light is located in the roof console, and the conventional brake warning light is positioned in the instrument panel. When the ignition switch is turned on, the electronic controller performs a check of all the anitlock brake system components. This check requires 3 to 4 seconds, and during this time the check antilock brake light is illuminated. If either, or both, warning lights are illuminated after this 3- to 4-second interval, a defect exists in the brake system. The check antilock brake warning light in the roof console is illustrated in Figure 12–7.

Checking and Filling Brake Fluid Reservoir

When the brake fluid level is checked in the master cylinder reservoir, the following procedure must be used:

1. Turn the ignition switch on and pump the brake pedal until the hydraulic pump motor starts.
2. Wait until the hydraulic pump motor shuts off.
3. Check the brake fluid level in the master cylinder reservoir. If this level is below the "max" fill line on the reservoir, bring the level up to this line with the type of brake fluid specified by the manufacturer.

The fluid level may be above the "max" fill line depending on the accumulator state of charge. Always use the correct procedure to check the fluid level in the reservoir. If the reservoir is overfilled, it may overflow when the accumulator discharges during normal system operation. The "max" fill line on the master cylinder reservoir is shown in Figure 12–8.

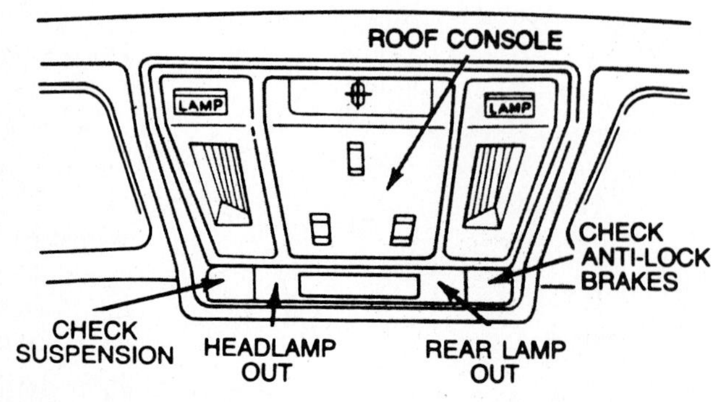

Figure 12–7 Check Antilock Brake Warning Light. (Courtesy of Ford Motor Co.)

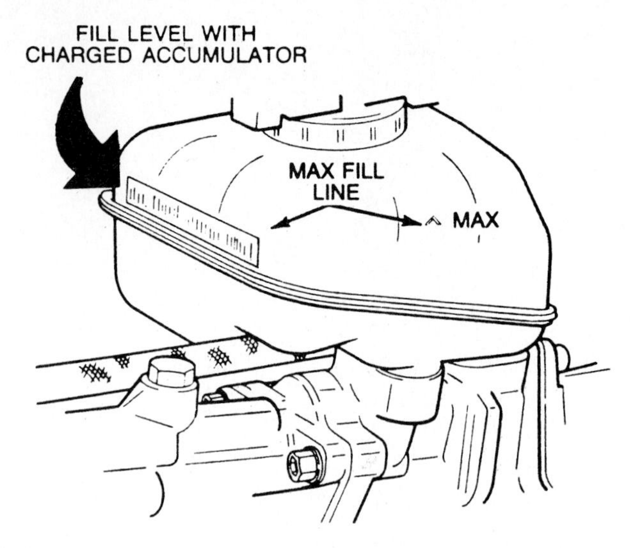

Figure 12–8 Checking Fluid Level in Master Cylinder Reservoir. (Courtesy of Ford Motor Co.)

Antilock Brake System Bleeding

The front brakes can be bled in the conventional manner with or without the accumulator being charged. [Refer to *Automechanics: Understanding the New Technology* (1987) by Don Knowles, published by Reston Publishing Company, Prentice-Hall, Inc. A Division of Simon & Schuster, for a description of conventional brake systems and service procedures.] The rear brakes must be bled with a fully charged accumulator, or a pressure bleeder may be used to bleed the rear brakes.

When a fully charged accumulator is used during the rear brake bleeding procedure, follow these steps:

1. Turn the ignition switch on and pump the brake pedal until the hydraulic pump motor starts.
2. Wait until the hydraulic pump motor shuts off.
3. With the ignition key switch on, have an assistant apply the brake pedal.
4. Loosen the rear caliper bleeder screws individually. Repeat steps 1 through 4 until a clear stream of brake fluid without air bubbles is discharged from each bleeder screw. Use caution when the bleeder screws are loosened, because of the high system pressures. A bleeder hose should be connected from the bleeder screws to a brake fluid container.
5. When the bleeding process is complete, refill the master cylinder reservoir as outlined previously.

If a pressure bleeder is used to bleed the rear brakes, connect the pressure bleeder to the master cylinder reservoir and maintain the pressure bleeder pressure at 35 psi (240 kPa). With the ignition switch off and the brake pedal released, open each rear caliper bleeder screw for 10 seconds at a time. Close the bleeder screws when a clear stream of brake fluid without air bubbles is discharged from each caliper. When the rear brake bleeding operation is completed, disconnect the pressure bleeder, turn on the ignition switch, and pump the brake pedal several times. Siphon the excess brake fluid from the master cylinder and adjust the fluid level as outlined previously. A pressure bleeder is pictured in Figure 12–9.

Wheel Sensor Adjustment

Mounting bolts hold each wheel sensor bracket in place, and a set screw retains each wheel sensor in its bracket. (The front and rear wheel sensors are pictured in Figures 12–4 and 12–5.) The caliper and rotor assemblies must be removed on front and rear wheels before the sensor brackets can be removed. It is possible to remove the front wheel sensors from their brackets without disturbing the calipers or rotors. A paper adjusting disc is located on each sensor pole. The disc controls the setting of the air gap between the sensor pole and the toothed sensor ring. The paper disc must be replaced whenever the sensor position is disturbed. A dull knife should be used to scrape the old disc from the sensor pole. When a new disc is attached to the sensor pole, use this sensor adjustment procedure:

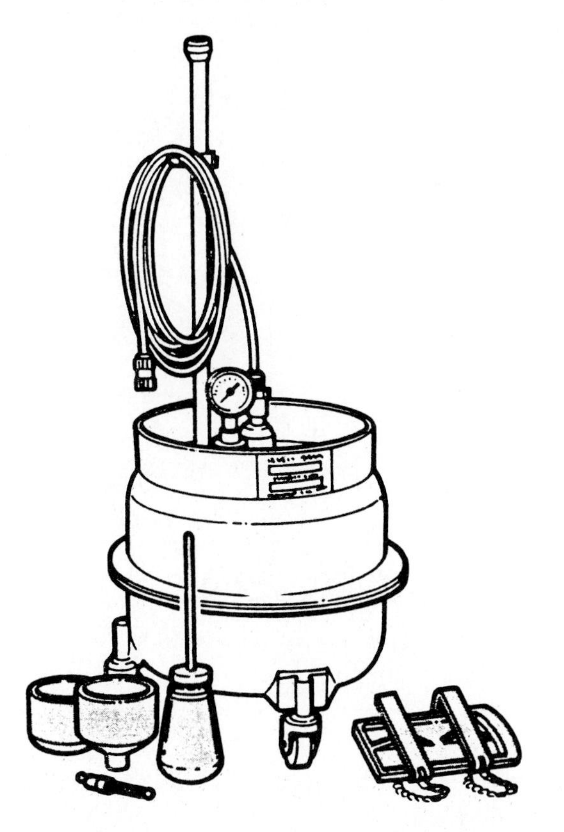

Figure 12–9 Pressure Bleeder. (Courtesy of Ford Motor Co.)

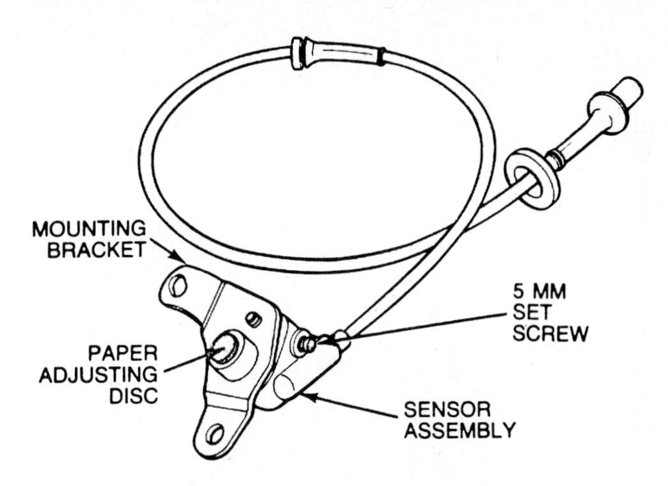

Figure 12–10 Sensor and Bracket Assembly. (Courtesy of Ford Motor Co.)

1. Loosen the set screw and slide the sensor into the bracket.
2. Install the sensor bracket mounting bolts if the bracket has been removed. Torque these to 40 to 60 lb in (4.5 to 6.8 Nm).
3. Slide the sensor through the bracket until the paper disc touches the toothed ring. Hold the sensor in this position and tighten the sensor screw to 21 to 26 lb in (2.4 to 3 Nm). A complete sensor and bracket assembly is shown in Figure 12–10.

Sensor Ring Replacement

Hydraulic pressure must be discharged from the system before any hydraulic component is serviced. This pressure discharge is accomplished by turning the ignition switch off and pumping the brake pedal at least 20 times until an increase in pedal pressure is clearly felt. When the front rotor and hub assembly has been removed, press each wheel stud out of the hub until the stud contacts the sensor ring, as illustrated in Figure 12–11.

When all the wheel studs have been pressed out until they contact the sensor ring, install the sensor ring removal tool on the studs and press the studs and sensor ring out of the hub, as pictured in Figure 12–12.

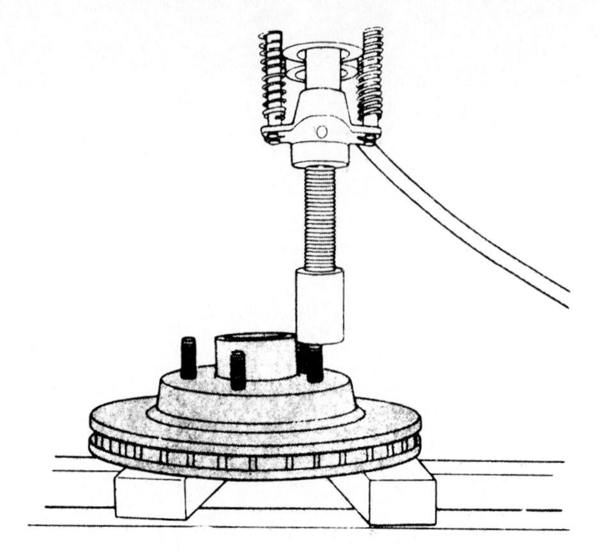

Figure 12–11 Pressing Out Front Wheel Studs. (Courtesy of Ford Motor Co.)

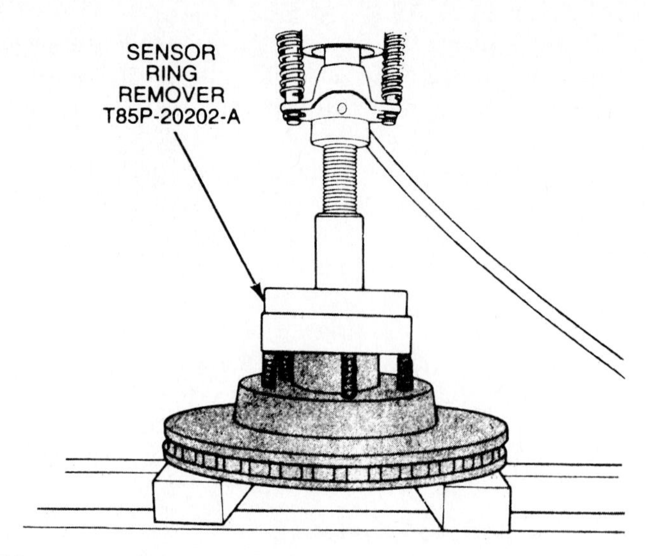

Figure 12–12 Front Wheel Stud and Sensor Ring Removal. (Courtesy of Ford Motor Co.)

After the new wheel studs have been pressed into the hub, the sensor ring should be pressed onto the hub until the sensor ring contacts the hub flange. A sensor ring replacer tool is used to press the ring into the hub, as indicated in Figure 12–13.

After the rear axles have been removed, the rear sensor ring may be pressed from the axle, as illustrated in Figure 12–14.

When the rear sensor rings are pressed into the axles, a 1.8 in (47 mm) gap must be left between the outer edge of the sensor ring and outer edge of the axle flange as shown in Figure 12–15.

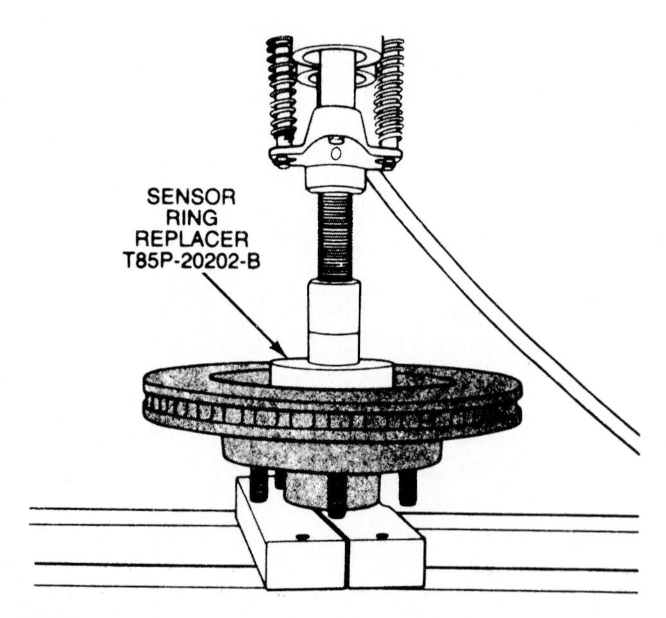

Figure 12–13 Front Sensor Ring Installation. (Courtesy of Ford Motor Co.)

Test Questions

1. The normal fluid pressure supplied from the hydraulic pump in the master cylinder to the accumulator would be:
   (a) 1000 to 1200 psi (6895 to 8274 kPa)
   (b) 1500 to 1600 psi (10,342 to 11,032 kPa)
   (c) 2000 to 2600 psi (13,790 to 17,927 kPa)

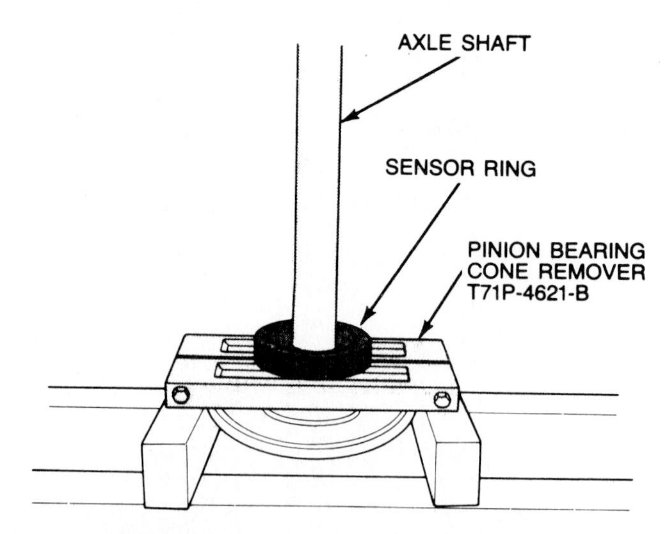

Figure 12–14 Rear Sensor Ring Removal. (Courtesy of Ford Motor Co.)

2. The rear brakes may be bled without a pressure bleeder if the accumulator pressure is discharged. T F
3. The wheel sensors must be pushed down in their brackets until the paper adjusting disc contacts the :
   (a) sensor ring.
   (b) rotor.
   (c) brake pad.
4. The accumulator pressure must be discharged before hydraulic components are removed. T F
5. In the antilock brake system, brake pedal assist is provided by:
   (a) vacuum booster diaphragm.
   (b) hydraulic pump pressure.
   (c) power steering pump pressure.

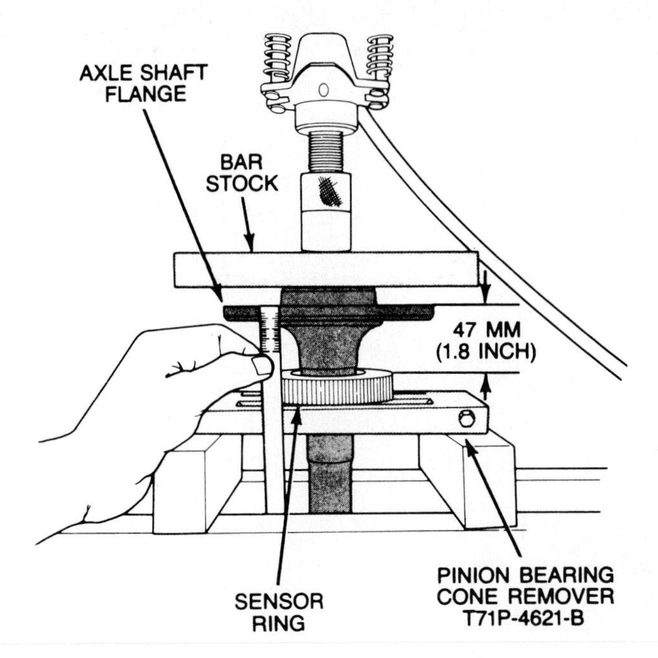

Figure 12–15 Rear Sensor Ring Installation. (Courtesy of Ford Motor Co.)

13

Computer-Controlled Suspension Systems

Practical Completion Objectives

1. Observe safety precautions when electronic air suspension systems are being serviced.
2. Perform diagnostic tests on an electronic air suspension system.
3. Complete the air spring fill procedure.
4. Adjust height sensors.

System Components

Air Springs

The air springs replace the coil springs in conventional suspension systems. These air springs have a composite rubber and plastic membrane that is clamped to a piston located in the lower end of the spring. An end cap is clamped to the top of the membrane, and an air spring valve is positioned in the end cap. The air springs are inflated or deflated to provide a constant vehicle height. Front air springs are mounted between the control arms and the cross-member. The lower ends of these air springs are retained in the control arm with a clip, and the upper ends are positioned in a cross-member spring seat. A front air spring is shown in Figure 13–1.

The rear air springs are the same as the front air springs except for their mounting. The lower ends of the rear springs are bolted to the rear suspension arms, and the upper ends of these springs are attached to the frame. Rear air spring mounting is illustrated in Figure 13–2.

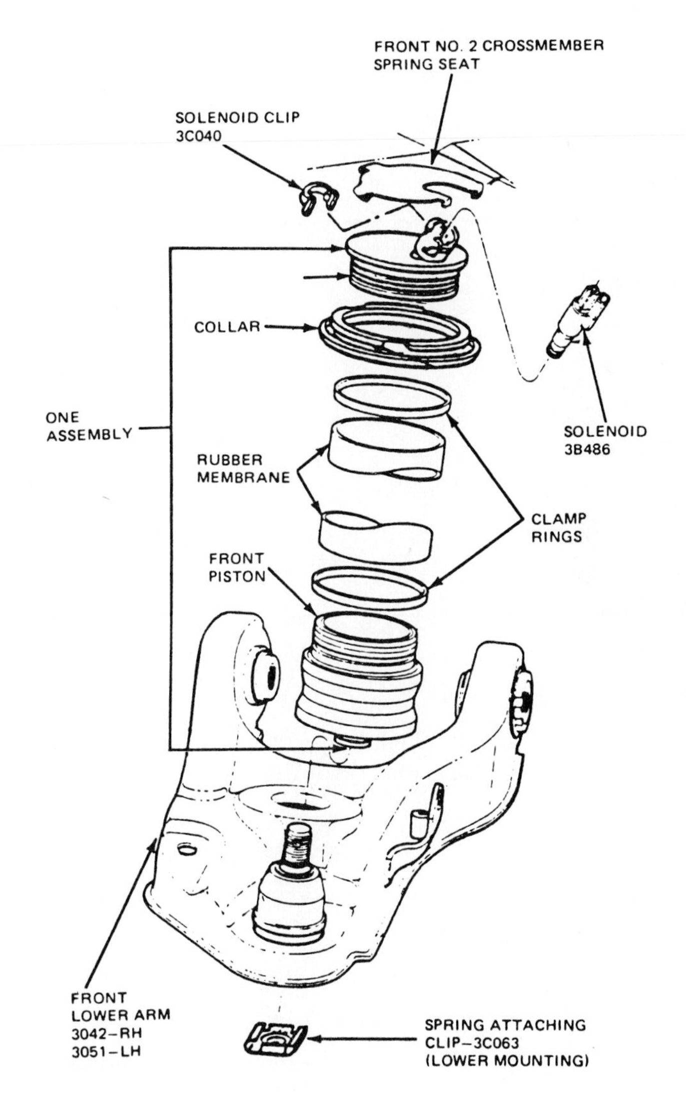

Figure 13–1 Front Air Spring. [Courtesy of Ford Motor Co.]

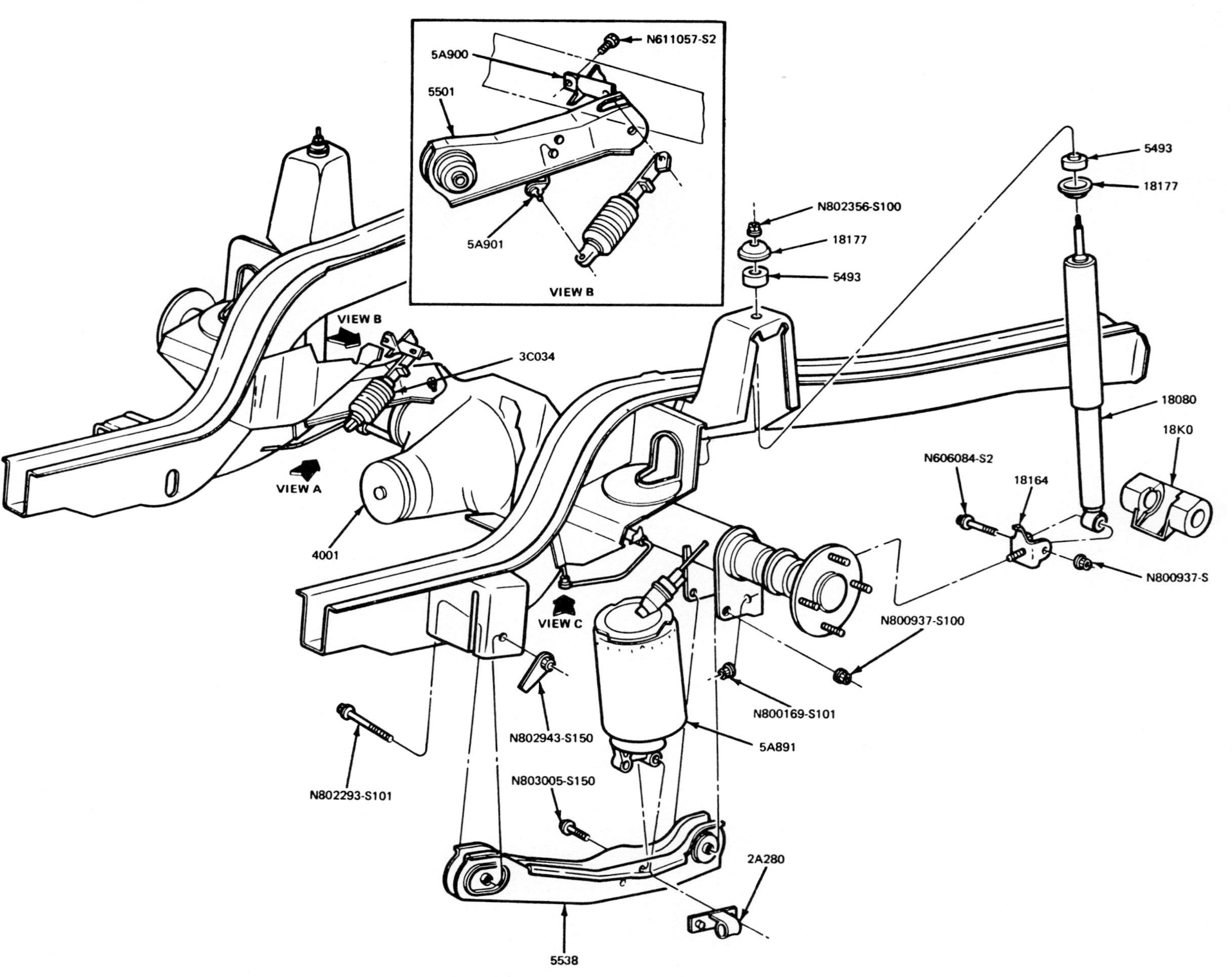

Figure 13–2 Rear Air Spring Mounting. [Courtesy of Ford Motor Co.]

Air Spring Valves

An air spring valve is mounted in the top of each air spring. These valves are an electric solenoid type of valve that is normally closed. When the valve winding is energized, plunger movement will open the air passage to the air spring. Under this condition air may enter, or be exhausted from, the air spring. Two "O" ring seals are located on the end of the valves to seal them into the air spring cap. The valves are installed in the air spring cap with a two-stage rotating action similar to a radiator pressure cap. An air spring valve is pictured in Figure 13–3.

Air Compressor

A single piston in the air compressor is moved up and down in the cylinder by a crankshaft and connecting rod. The armature is connected to the crankshaft, and therefore the rotating action of the armature moves the piston up and down. Armature rotation occurs when 12V is supplied to the compressor input terminal. Intake and discharge valves are located in the cylinder head. An air dryer that contains a silica gel is mounted on the compressor. This silica gel removes moisture from the air as it enters the system. The compressor is shown in Figure 13–4.

Nylon air lines are connected from the compressor outlets to the air spring valves. The compressor will only operate when it is necessary to force air into one or more air springs to restore the vehicle trim height.

An air vent valve is located in the compressor cylinder head. This normally closed electric solenoid valve allows air to be vented from the system. When it is necessary to exhaust air from an air spring, the air spring valve and the vent valve must be energized at the same time with the compressor shut off. Air

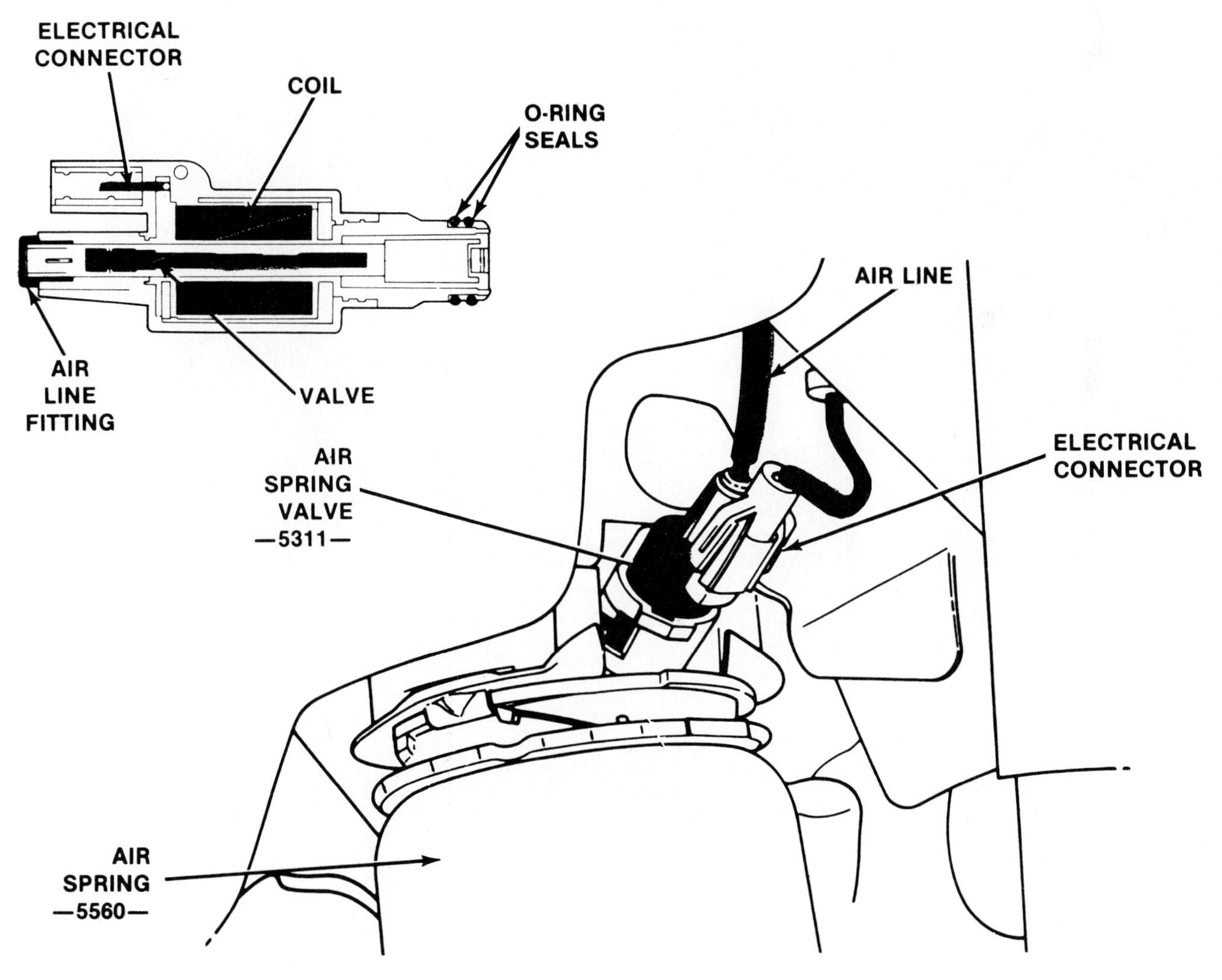

Figure 13–3 Air Spring Valve. (Courtesy of Ford Motor Co.)

exhausting is necessary if the vehicle is too high. The air vent valve is pictured in Figure 13–5.

Compressor Relay

When the compressor relay is energized it supplies 12V through the relay contacts to the compressor input terminal. The relay contacts open the circuit to the compressor if the relay is deenergized. Figure 13–6 illustrates the compressor relay.

Control Module

The control module is a microprocessor that operates the compressor, vent valve, and air spring valves to control the amount of air in the air springs and maintain a specific distance between the vehicle chassis and the road surface. This distance is referred to as the trim height. The control module is located in the trunk, as pictured in Figure 13–7.

The control module turns on the service indicator light in the roof panel to alert the driver that a system defect is present. Diagnostic capabilities are also designed into the control module.

On/Off Switch

The on/off switch opens the circuit to the control module. This switch is located in the trunk near the control module, as shown in Figure 13–7. When the

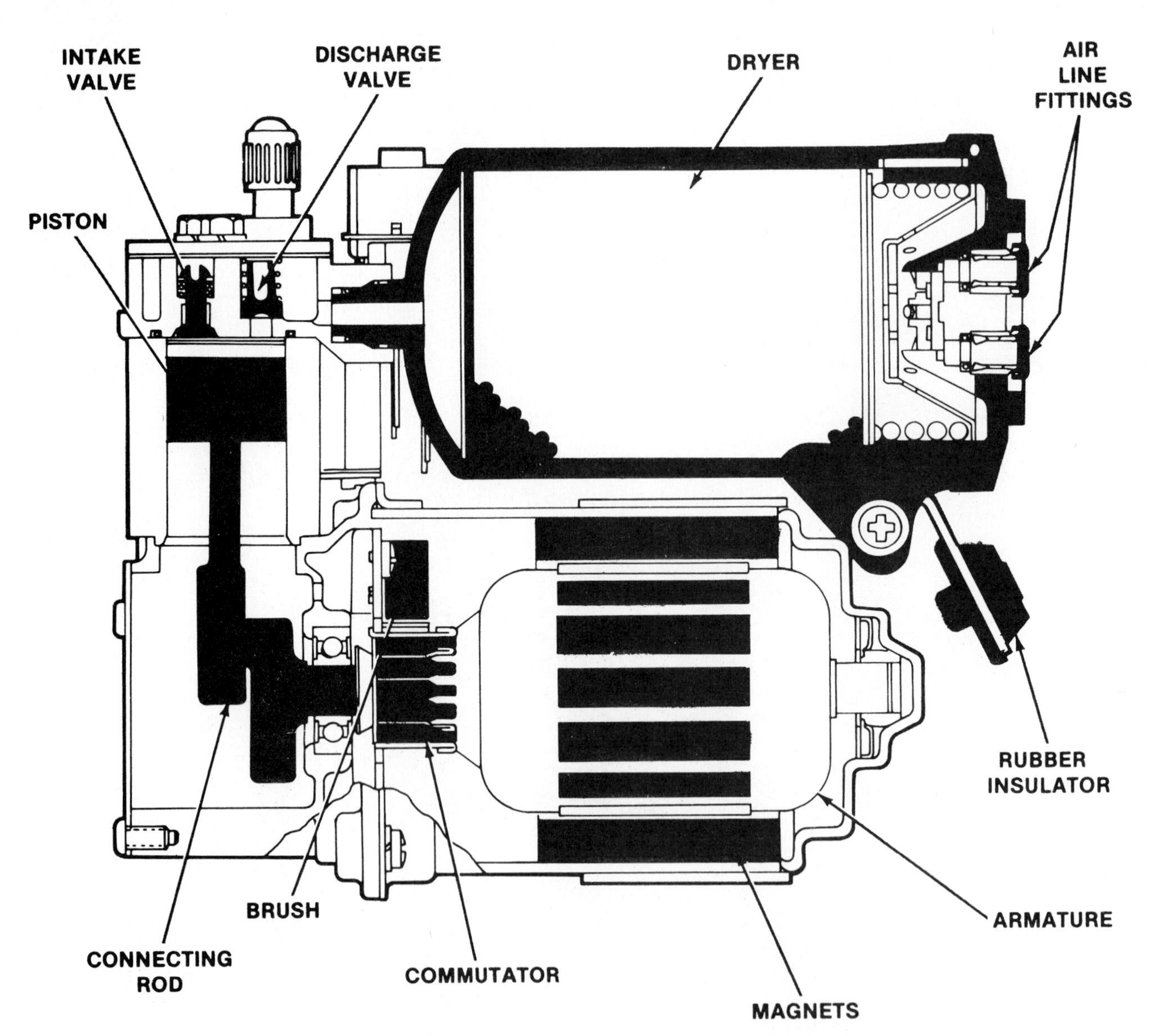

Figure 13–4 Air Compressor. (Courtesy of Ford Motor Co.)

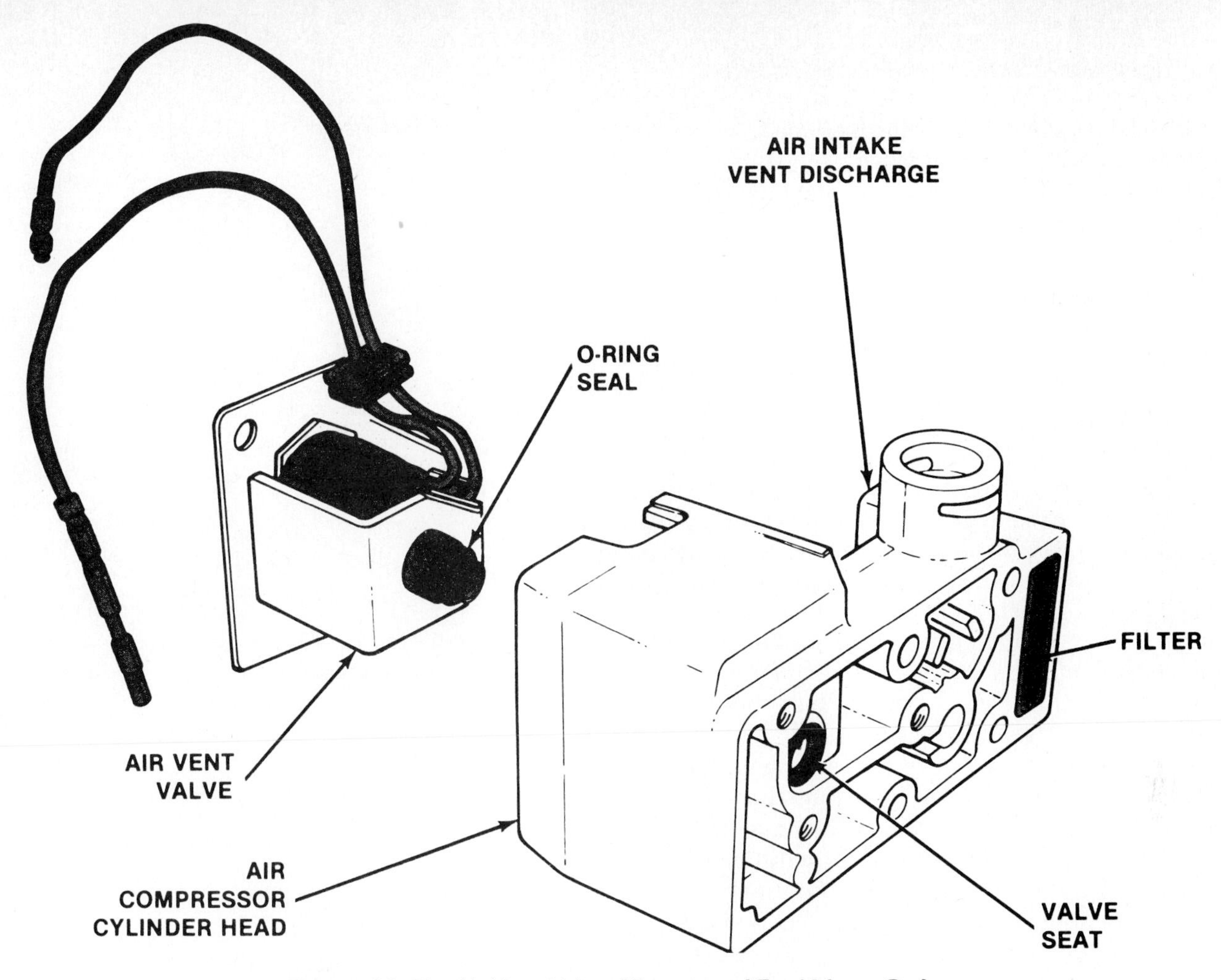

Figure 13-5 Air Vent Valve. (Courtesy of Ford Motor Co.)

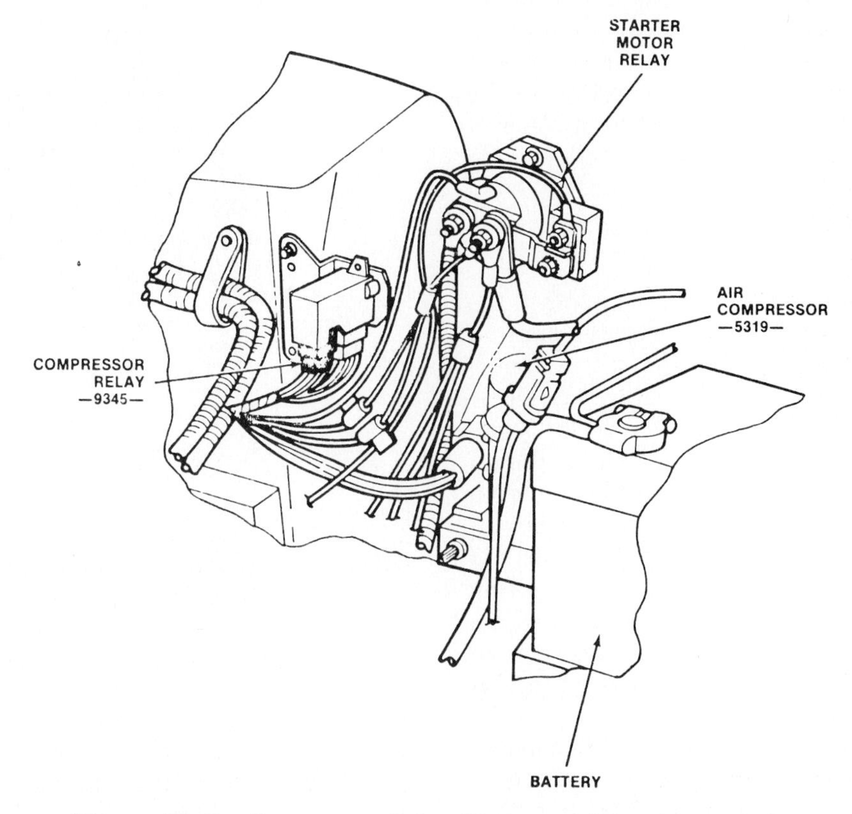

Figure 13-6 Compressor Relay. (Courtesy of Ford Motor Co.)

air suspension system is being serviced, or some other vehicle service is performed, the switch must be in the off position. (Refer to the safety precautions later in this chapter.)

Height Sensors

In the air suspension system there are two front height sensors located between the lower control arms and the cross member. A single rear height sensor is positioned between the suspension arm and the frame. Each height sensor contains a magnet slide that is attached to the upper end of the sensor. This magnet slide moves up and down in the lower sensor housing as changes in vehicle height occur. The lower sensor housing contains two electronic switches that are connected through a wiring harness to the control module. When the vehicle is at trim height, the switches remain closed and the control module receives a trim height signal. If the magnet slide moves upward, the above-trim switch will open. When this signal is received by the module, it will open the appropriate air spring valve and the vent valve. This action will exhaust air from the air spring and correct the above-trim height condition. Downward magnet slide movement closes the above-trim switch and opens the below-trim switch. When the control module receives this signal it will energize the compressor relay, which starts the compressor. The control module will also open the appropriate air spring valve. This action will force air into the air spring and correct the below-trim condition. The height sensors are serviced as a unit. A height sensor is pictured in Figure 13–8.

Warning Lamp

When the control module senses a system defect, the module turns on the air suspension warning lamp in the roof console, which informs the driver that a problem exists. If the air suspension system is working normally, the warning lamp will be on for 1 second when the ignition switch is turned from the off to the run position. After this time the warning lamp should remain off. This lamp does not operate with the ignition switch in the start position. The warning lamp is used during the self-diagnostic procedure, and the spring fill sequence. These procedures are explained later in this chapter. A roof console with the air suspension warning lamp is pictured in Figure 13–9.

The location of all the electronic air suspension system components is illustrated in Figure 13–10.

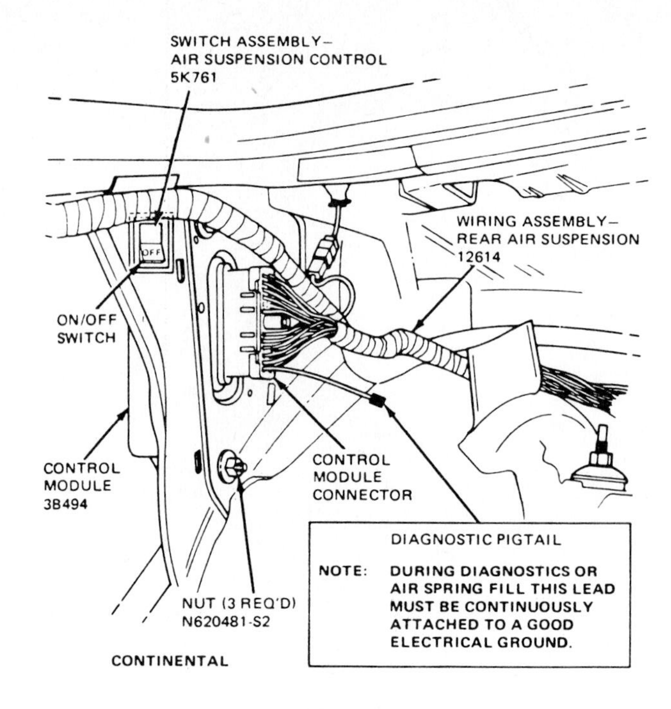

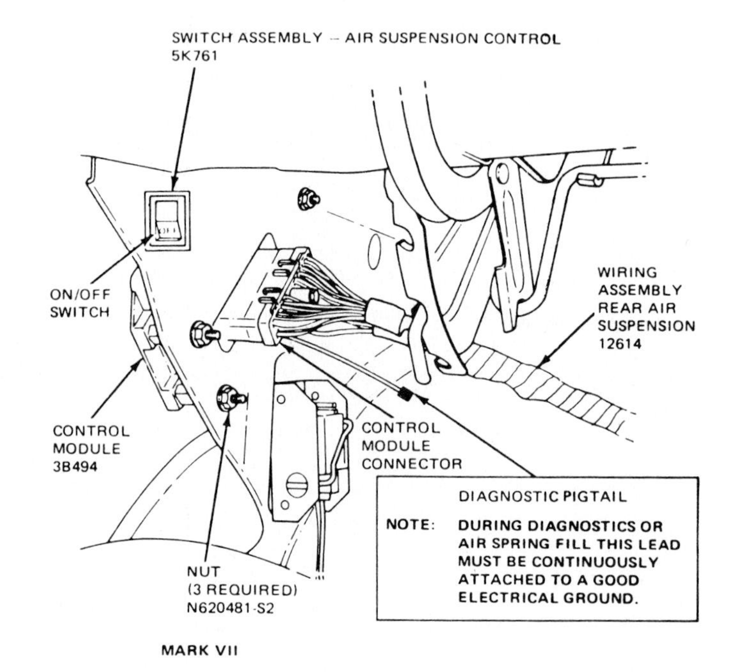

Figure 13–7 Control Module. (Courtesy of Ford Motor Co.)

Electronic Air Suspension System Operation

Ignition Switch in Off Position

The electronic air suspension system is fully operational for 1 hour after the ignition switch is turned from the run to the off position. During this time,

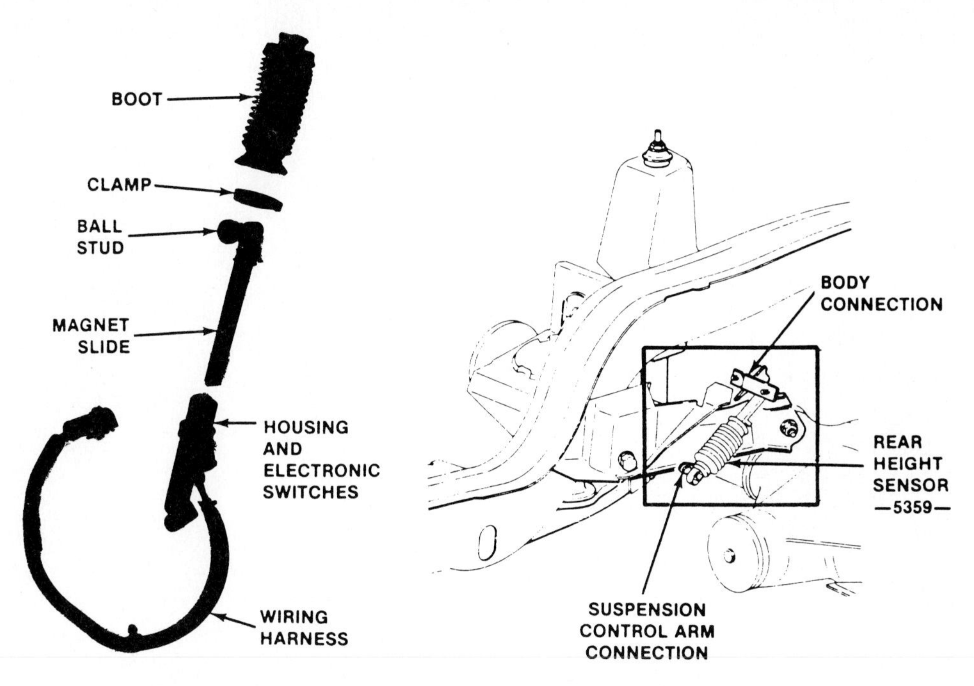

Figure 13-8 Height Sensor. [Courtesy of Ford Motor Co.]

lower vehicle commands will be completed except when a height sensor was providing a high signal when the ignition switch was turned off. After the 1-hour period raise vehicle requests will be completed. The compressor run time is limited to 15 seconds for rear springs and 30 seconds for front springs.

Ignition Switch in Run Position

When the ignition switch has been in the run position for less than 45 seconds, raise vehicle commands will be completed immediately, but lower vehicle commands will be ignored.

If the ignition switch has been in the run position for more than 45 seconds the operation will be as follows:

1. If a door is opened with the brake pedal released, raise vehicle commands will be completed immediately, but lower vehicle requests will only be serviced after the door is closed.
2. If the doors are closed and the brake pedal is released, all commands are serviced by a 45-second averaging method to prevent excessive suspension height corrections on rough road surfaces.
3. If the brake is applied and a door is open, raise vehicle commands are completed immediately, but lower vehicle requests will be ignored.
4. When the doors are closed and the brake pedal is applied, all requests will be ignored by the control module. If a request to raise the rear suspension is in progress under these conditions, this request will be completed. This action prevents correction of front end jounce while braking.

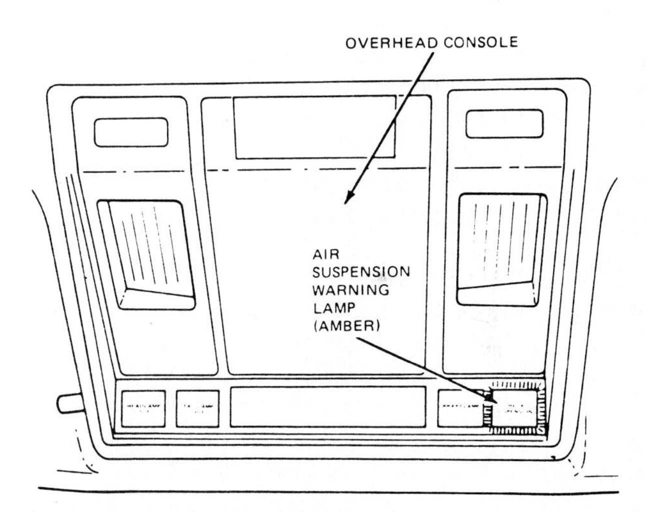

Figure 13-9 Air Suspension Warning Lamp. [Courtesy of Ford Motor Co.]

General Operation

When a height sensor sends a raise vehicle command to the control module and the other input signals are acceptable, the module will ground the compressor relay winding, which starts the compressor. The

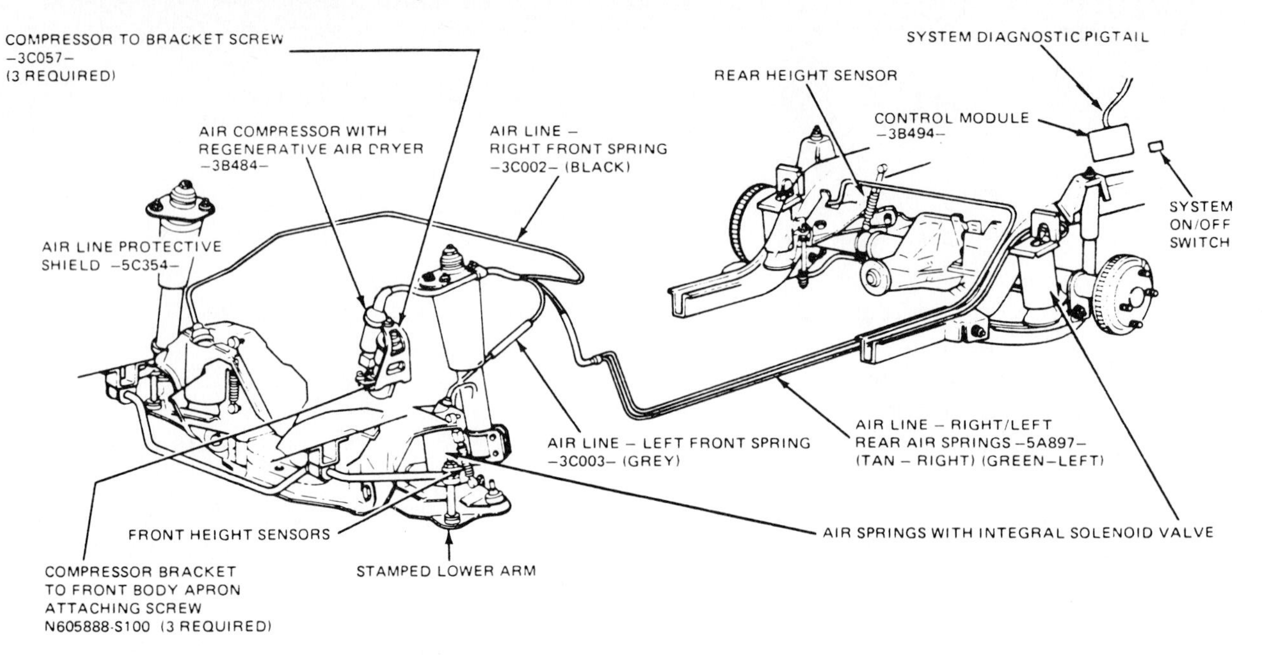

Figure 13-10 Electric Air Suspension System Component Locations. [Courtesy of Ford Motor Co.]

module will also open the appropriate air spring valve to allow air flow into the air spring. The rear air valve solenoids are always operated together, but the front solenoids may be energized independently. When the correct chassis trim height is obtained, the control module will open the circuit from the compressor relay winding to ground and deenergize the air spring valve. This action shuts off the compressor and traps the air in the air spring.

If a lower vehicle request is sent from a height sensor to the control module, and the other input signals are acceptable, the control module will open the air vent valve and appropriate air spring valve. When this occurs, air will be released from the air spring until the correct trim height is obtained. The trim height signal from the height sensor to the module causes the module to close the air vent valve and the air spring valve.

Specific Control Module Operation

The control module will not service lower vehicle requests if any door is open. Commands are completed by the module in this order: rear up, front up, rear down, front down. When the ignition switch is in the run position and a command cannot be completed within 3 minutes, the module will turn on the air suspension warning lamp. This lamp will remain on until the ignition switch is turned off. All the control module memory is erased when the ignition switch is turned off. Therefore, the warning lamp may not indicate a defect immediately if the ignition switch is turned from the off to the run position. When a system defect causes the module to illuminate the warning lamp with the ignition in the run position, other commands may be completed by the module. Commands from the front and rear height sensors will never be completed simultaneously. A complete wiring diagram of the electronic air suspension system is provided in Figure 13-11.

Electronic Air Suspension System Service and Diagnosis

Service Precautions

When vehicles equipped with electronic air suspension systems are being serviced, these service precautions must be observed to prevent personal injury or system damage.

1. The system control switch must be in the off position when system components are being serviced.
2. The system control switch must be turned off prior to hoisting, jacking, or towing the vehicle.
3. When air spring valves are being removed, always rotate the valve to first stage until all the air escapes from the air spring. Never turn the valve to

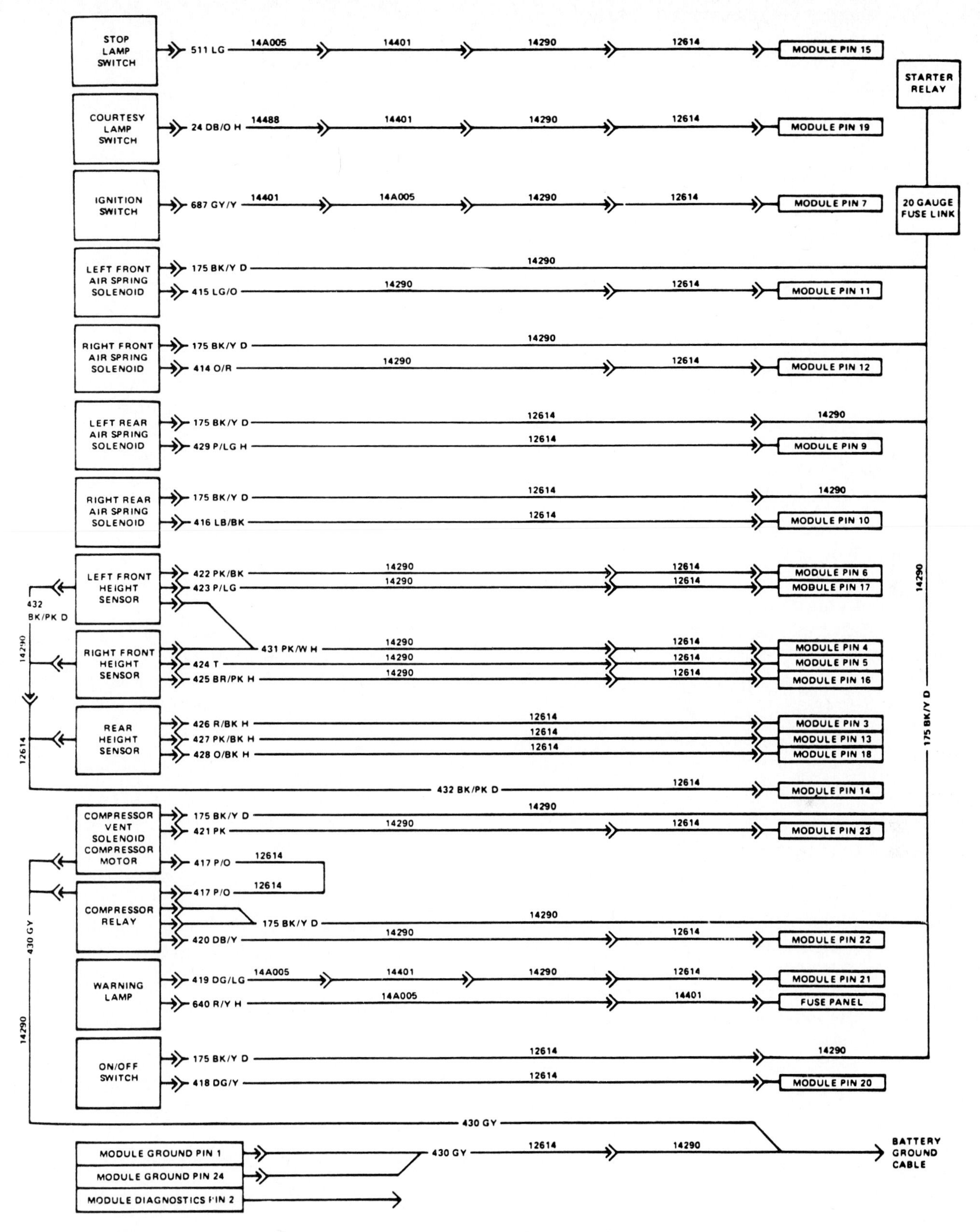

Figure 13–11 Electronic Air Suspension System Wiring Diagram. (Courtesy of Ford Motor Co.)

the second, release, stage until all the air is released from the spring.

4. Do not allow the suspension to compress an air spring until the air spring is inflated. The air spring inflation procedure is provided later in this chapter.

Diagnostic Procedure

When the air suspension warning lamp indicates a system defect, the diagnostic procedure may be entered as follows:

1. Be sure the air suspension system switch is turned on.
2. Turn the ignition switch on for 5 seconds and then turn it off. Leave the driver's door open and the other doors closed.
3. Ground the diagnostic lead located near the control module, and close the driver's door with the window down.
4. Turn the ignition switch on. The warning lamp should blink continuously at 1.8 times per second to indicate that the system is in the diagnostic mode.

There are 10 tests in the diagnostic procedure. The control module switches from one test to the next when the driver's door is opened and closed. The first three tests in the diagnostic procedure are the following:

1. Rear suspension.
2. Right front suspension.
3. Left front suspension.

During these three tests each suspension location should be raised for 30 seconds, lowered for 30 seconds, and raised for 30 seconds. For example, in test 2 this procedure should be followed on the right front suspension. If the expected signal, or an illegal signal, is received during the test procedure, the test will stop and the air suspension warning lamp will be illuminated continuously. If all the signals and commands are normal during the first three tests, the warning lamp will continue to flash at 1.8 times per second.

While tests 4 through 10 are being performed, the air suspension warning lamp will flash the test number. For example, during test 4 the warning lamp will flash 4 times followed by a pause and 4 more flashes. This flash sequence continues while test 4 is completed. The driver's door must be opened and closed to move to the next test. During tests 4 through 10 the technician must listen to, or observe, various components to detect abnormal operation. The warning lamp only indicates the test number during these tests. Actions performed by the control module during tests 4 through 10 are as follows:

Test 4 The compressor is cycled on and off at 0.25 cycles per second. This action is limited to 50 cycles.

Test 5 The vent solenoid is opened and closed at 1 cycle per second.

Test 6 The left front air valve is opened and closed at 1 cycle per second, and the vent solenoid is opened. When this occurs, the left front corner of the vehicle should drop slowly.

Test 7 The right front air valve is opened and closed at 1 cycle per second, and the vent solenoid is opened. This action causes the right front corner of the vehicle to drop slowly.

Test 8 During this test the right rear air valve is opened and closed at 1 cycle per second and the vent valve is opened. This action should cause the right rear corner of the vehicle to drop slowly.

Test 9 The left rear solenoid is opened and closed at 1 cycle per second and the vent valve is opened, which should cause the left rear corner of the vehicle to drop slowly.

Test 10 Return the module from the diagnostic mode to normal operation by disconnecting the diagnostic lead from ground. This mode change will also occur if the ignition switch is turned off, or the brake pedal is depressed.

If defects are found during the test sequence, specific electrical tests may be performed on air valve, or vent valve, windings and connecting wires to locate the problem.

Air Spring Folding Procedure

Before an air spring is installed, it must be properly folded over the piston at the bottom of the membrane, as indicated in Figure 13–12.

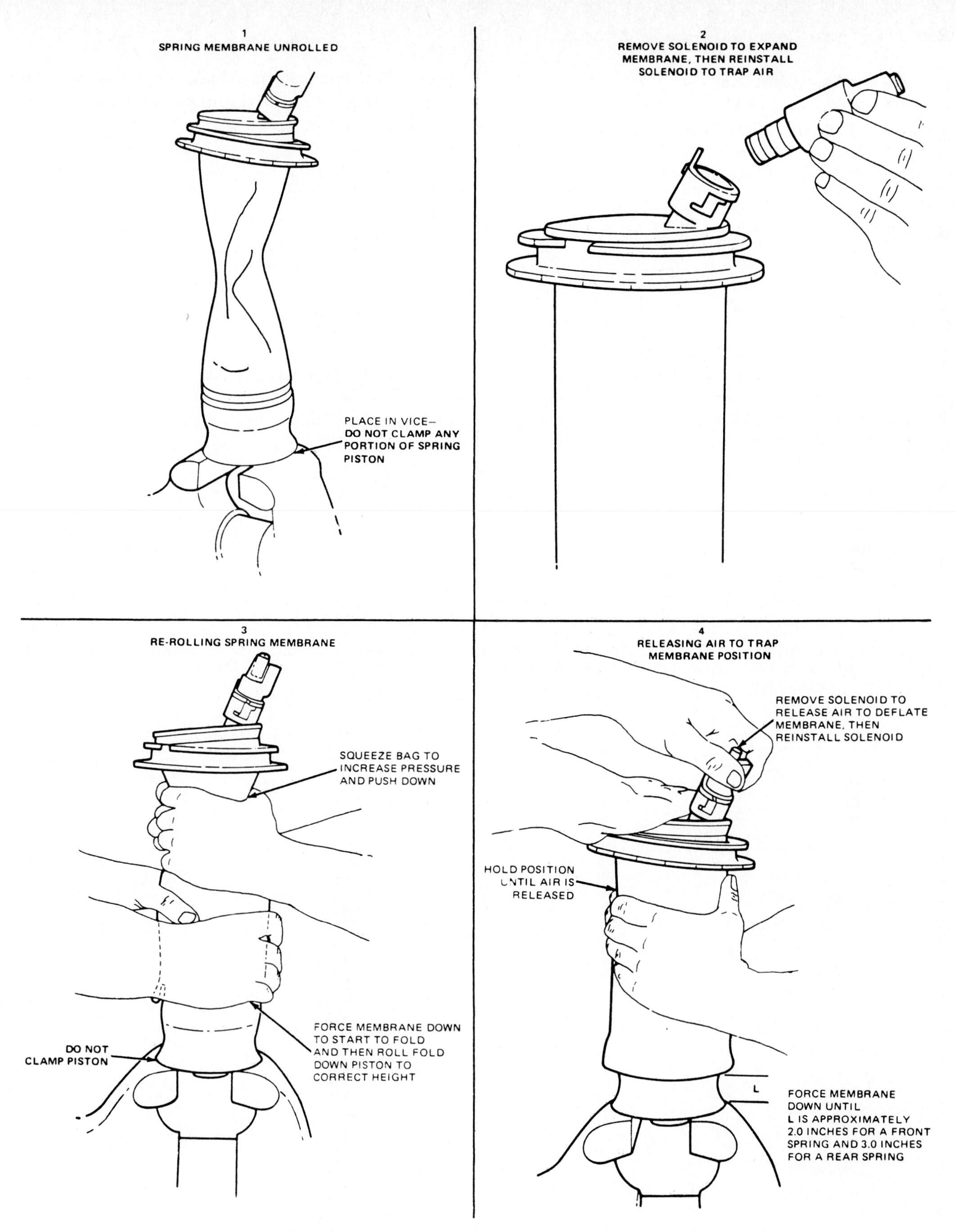
1
SPRING MEMBRANE UNROLLED
PLACE IN VICE—
DO NOT CLAMP ANY
PORTION OF SPRING
PISTON
2
REMOVE SOLENOID TO EXPAND
MEMBRANE, THEN REINSTALL
SOLENOID TO TRAP AIR
3
RE-ROLLING SPRING MEMBRANE
SQUEEZE BAG TO
INCREASE PRESSURE
AND PUSH DOWN
FORCE MEMBRANE DOWN
TO START TO FOLD
AND THEN ROLL FOLD
DOWN PISTON TO
CORRECT HEIGHT
DO NOT
CLAMP PISTON
4
RELEASING AIR TO TRAP
MEMBRANE POSITION
REMOVE SOLENOID TO
RELEASE AIR TO DEFLATE
MEMBRANE, THEN
REINSTALL SOLENOID
HOLD POSITION
UNTIL AIR IS
RELEASED
L
FORCE MEMBRANE
DOWN UNTIL
L IS APPROXIMATELY
2.0 INCHES FOR A FRONT
SPRING AND 3.0 INCHES
FOR A REAR SPRING

Figure 13–12 Air Spring Folding. (Courtesy of Ford Motor Co.)

Air Spring Installation

When an air spring is installed in the front or rear suspension, folds and creases in the membrane must be eliminated, as pictured in Figure 13–13.

Air Spring Inflation

The weight of the vehicle must not be allowed to compress an uninflated air spring. When an air spring is being inflated use this procedure:

1. With the vehicle chassis supported on a hoist, lower the hoist until a slight load is placed on the suspension. Do not lower the hoist until the suspension is heavily loaded.
2. Turn on the air suspension system switch.
3. Turn the ignition switch from off to run for 5 seconds with the driver's door open and the other doors shut. Turn the ignition switch off.
4. Ground the diagnostic lead.
5. Apply the brake pedal and turn the ignition to the run position. The warning lamp will flash every 2 seconds to indicate the fill mode.
6. To fill a rear spring, or springs, close and open the driver's door once. After a 6-second delay the rear spring will be filled for 60 seconds.
7. To fill a front spring, or springs, close and open the driver's door twice. After a 6-second delay the front spring will be filled for 60 seconds.
8. When front and rear springs require filling, fill the rear springs first. Once the rear springs are filled, close and open the driver's door once to begin filling the front springs.
9. The spring fill mode is terminated if the diagnostic lead is disconnected from ground. This action is also obtained if the ignition switch is turned off or the brake pedal is applied.

Ride Height Adjustment

The ride height should be measured on the front and rear suspension at the points indicated in Figure 13–14.

If the rear suspension ride height is not within specifications it may be adjusted by loosening the attaching bolt on the top height sensor bracket. When the bracket is moved one index mark up or down the ride height will be raised or lowered 0.25 in (6.35 mm). The rear ride height adjustment is shown in Figure 13–15.

The front suspension ride height may be adjusted by loosening the lower height sensor attaching bolt. Three adjustment positions are located in the lower

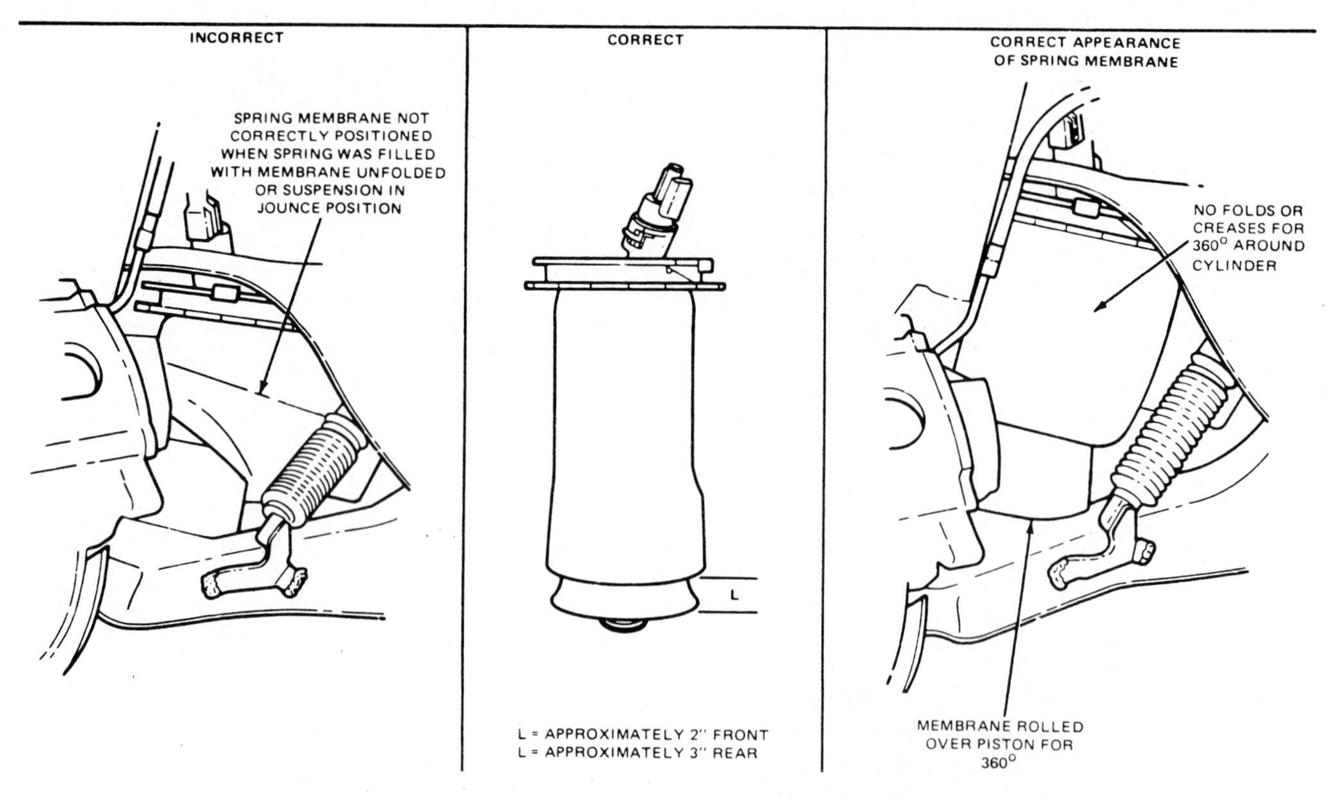

Figure 13–13 Air Spring Installation. (Courtesy of Ford Motor Co.)

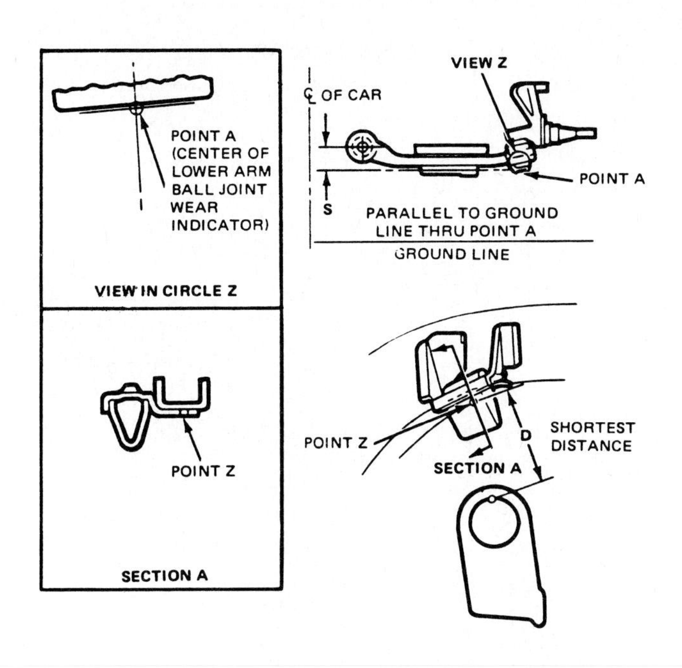

SUSPENSION RIDE HEIGHT

VEHICLE		S	D
MARK VII/ CONTINENTAL	INCHES	0.24	5.06
	MM	6.0	128.6

Figure 13–14 Ride Height Measurement Locations. (Courtesy of Ford Motor Co.)

front height sensor bracket. If the height sensor attaching bolt is moved one position up or down, the front suspension height will be raised or lowered 0.5 in (12.7 mm). The front suspension height adjustment is illustrated in Figure 13–16.

Line Service

Nylon lines on the electronic air suspension system should be released from the air spring valves and the air compressor outlets as shown in Figure 13–17.

The removal procedures for air spring valve quick-connect fittings and the repair of damaged air lines is shown in Figure 13–18.

Test Questions

1. When a new deflated air spring is installed, the complete suspension weight should be applied to the spring prior to spring inflation. T F
2. When a vehicle with electronic air suspension is lifted on a hoist, the:
   (a) ignition switch should be turned on.
   (b) air should be released from the suspension system.
   (c) air suspension system switch should be turned off.
3. Before an air spring valve is removed, the air must be released from the air spring by:
   (a) rotating the air spring valve to the first stage.

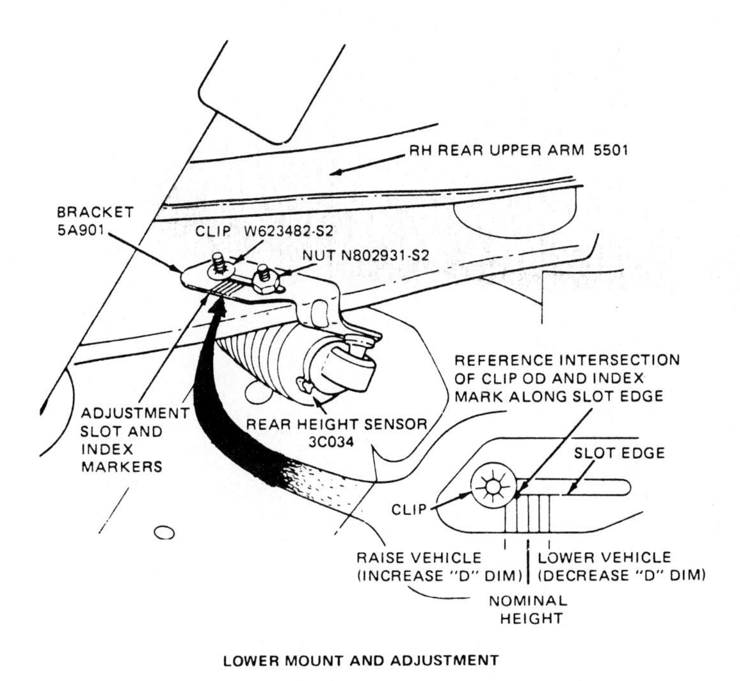

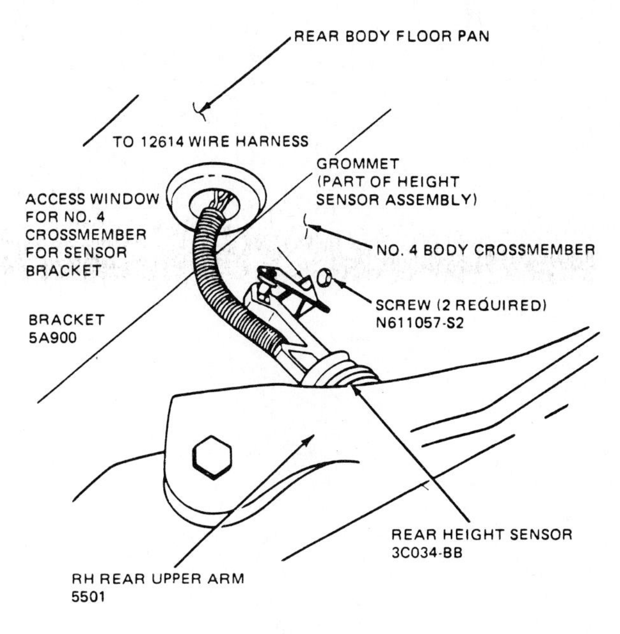

Figure 13–15 Rear Ride Height Adjustment. (Courtesy of Ford Motor Co.)

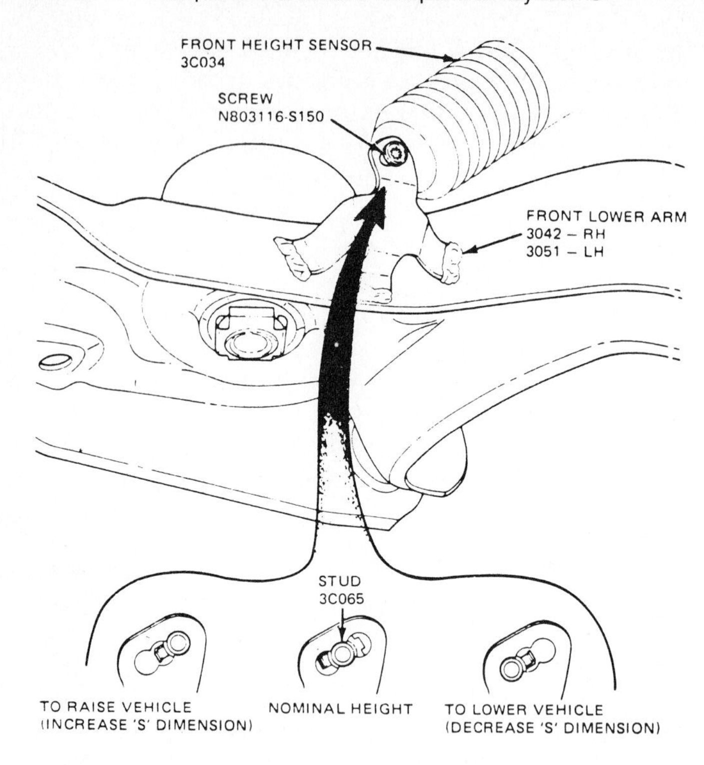

Figure 13-16 Front Ride Height Adjustment. [Courtesy of Ford Motor Co.]

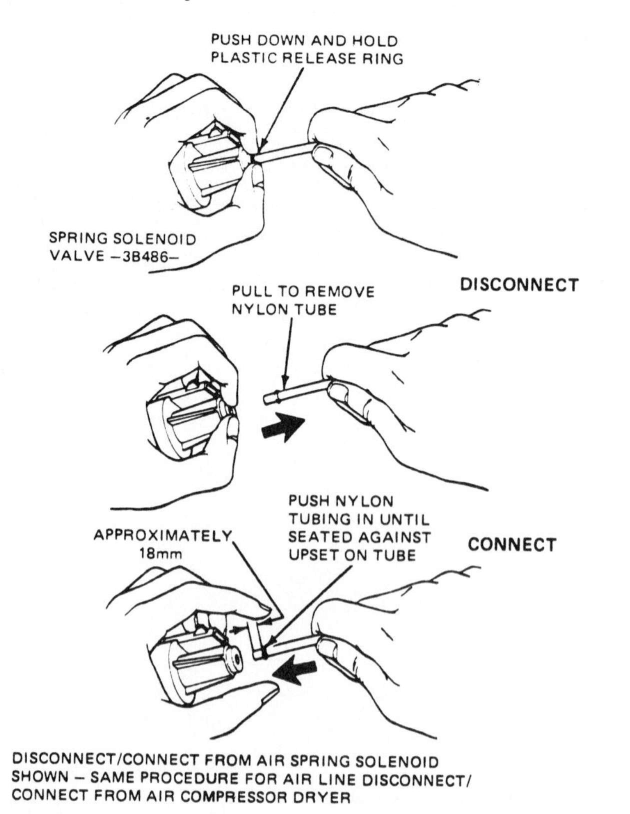

Figure 13-17 Air Line Removal from Air Spring Valves or Compressor Outlets. [Courtesy of Ford Motor Co.]

(b) rotating the air spring valve to the second stage.

(c) removing the nylon line from the air valve quick-connect fitting.

4. If the brakes are applied at a vehicle speed of 50 mph (31 kph), the control module will complete a raise front suspension command. T F
5. The control module will complete a raise suspension command 2 hours after the ignition switch has been turned from the run to the off position. T F
6. When air is exhausted from an air spring, the control module opens the:

   (a) air vent valve.

   (b) air vent valve and air spring valve.

   (c) air spring valve and compressor inlet valve.

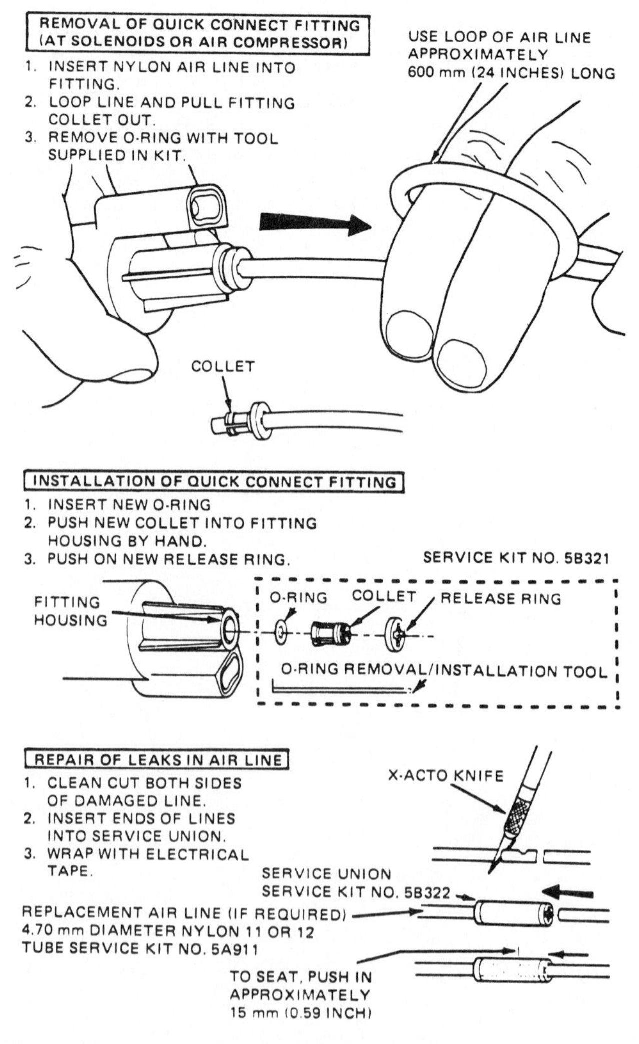

Figure 13-18 Removal of Quick-Connect Fittings and Air Line Repair. [Courtesy of Ford Motor Co.]

14

Multiplex Wiring Systems

Practical Completion Objectives

1. Demonstrate an understanding of the advantages of multiplex wiring systems compared to conventional wiring systems.
2. Demonstrate an understanding of fiber optic and bus data link multiplex wiring systems.
3. Demonstrate a knowledge of the advantages of fiber optic multiplex wiring systems compared to bus data link multiplex systems.

Introduction to Multiplex Systems

Advantages

The trend in recent years toward automobile downsizing has placed a premium on space in the automobile. To meet fuel economy standards, car manufacturers have reduced the exterior dimensions and at the same time maintained interior space at, or close to, the interior size available in the larger cars of past years. In these new downsized cars, doors became thinner and under-dash space was reduced. This design provides less space for wiring harnesses. Simultaneously customer demand for electrical and electronic equipment increased rapidly, which increased the size of many wiring harnesses.

When the number of wires in a harness is increased, and computer-controlled equipment is used, the problem of electromagnetic interference (EMI) becomes critical. The magnetic field around a current carrying wire may build up and collapse across an adjacent wire. When this occurs a voltage is induced in the adjacent wire that may interfere with an electric signal in this adjacent wire. This action is known as EMI, and it may cause malfunctioning of electronic equipment. Many computer-controlled circuits operate on a very low current flow of a few milliamps. The problem of EMI is critical in this type of circuit. Multiplexing of wiring systems is one solution to EMI problems and reduced space in cars. The advantages of multiplex wiring systems are the following:

1. Reduced number of wires.
2. Smaller gauge control wires.
3. Reduced installation space requirements.
4. Lower installation weight.
5. Easily expandable for added functions.
6. EMI problems are reduced or eliminated especially when optical link multiplexing is used.

Automotive electrical loads that are suitable for multiplexing are the following:

1. Headlights.
2. Parking lights.
3. Turn signals.
4. Brake lights.
5. Backup lights.
6. Courtesy lights.
7. Cornering lights.
8. Power seats.
9. Power door locks.
10. Power antenna.
11. Heater/air conditioning blower.
12. Windshield wiper-washer.
13. Power sunroof.
14. Side marker lights.

15. Tail lights.
16. Power windows.
17. Horn.
18. Trunk release.

Multiplex Wiring Systems with Bus Data Links

System Design

In a multiplex wiring system a central transmitter is located in the vehicle, and several receivers are mounted near the electrical loads that they control. The transmitter is a microprocessor that acts as the master controller in the system. A bidirectional bus data link is connected between the master controller and the receivers. Battery power is also supplied to the master controller and the receivers. The location of the multiplex system components and the loads controlled by each receiver are illustrated in Figure 14–1.

The bus data link between the master controller and the receivers requires only two wires, a signal wire and a ground wire. (Different types of data links and data signals are explained in Chapter 2.) A master controller could be capable of controlling 16 receivers. Therefore, the multiplex system could easily be expanded to control additional electrical loads. The switches are connected to the master controller, or transmitter, and each switch has a unique signal.

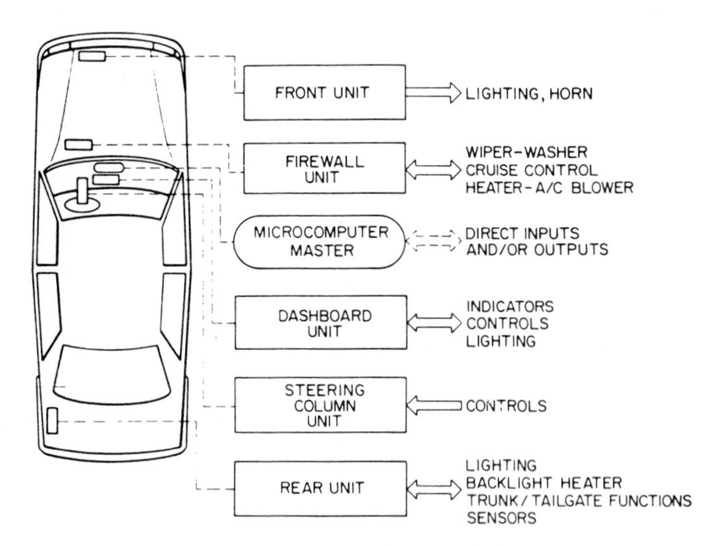

Figure 14–1 Multiplex Wiring System Component Location and Receiver Functions. [Reprinted with Permission of Society of Automotive Engineers © 1984]

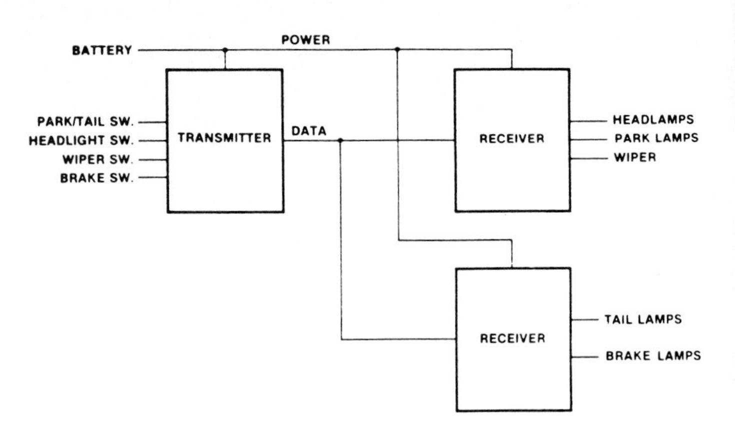

Figure 14–2 Multiplex System Wiring Diagram. [Reprinted with Permission of Society of Automotive Engineers © 1981]

If the master controller receives a switch signal, it transmits a command signal on the bus data link to the appropriate receiver. When this signal is received, the receiver performs the necessary output function. A block diagram of a partial multiplex system is shown in Figure 14–2.

Master Controller

All system intelligence and decision-making ability is concentrated in the master controller. The microprocessor in the master controller provides bus data link control and state sequencing logic.

State sequencing is a software approach that cycles through various tables to determine if the proper conditions exist to advance from one state to the next state. All switch input requests are interpreted by the master controller. These requests may be ignored, passed on as a command to a receiver, or interpreted as a specific command to be issued. For example, a request to raise or lower a power window with the ignition switch off would be ignored, because this action is not permitted. A request to turn on the headlights would be passed on as a command whenever the request is received. The master controller may cancel a command as a result of other requests. For example, if a command has been issued to flash the right signal lights and then a request is received to turn on the stop lights, the master controller will cancel the right rear brake light request until the signal light operation is turned off. All flashers and timers on the vehicle may be replaced by the clock in the master controller. Switches with a low current capacity may be used in the multiplex system because they are only used as signals to the master controller.

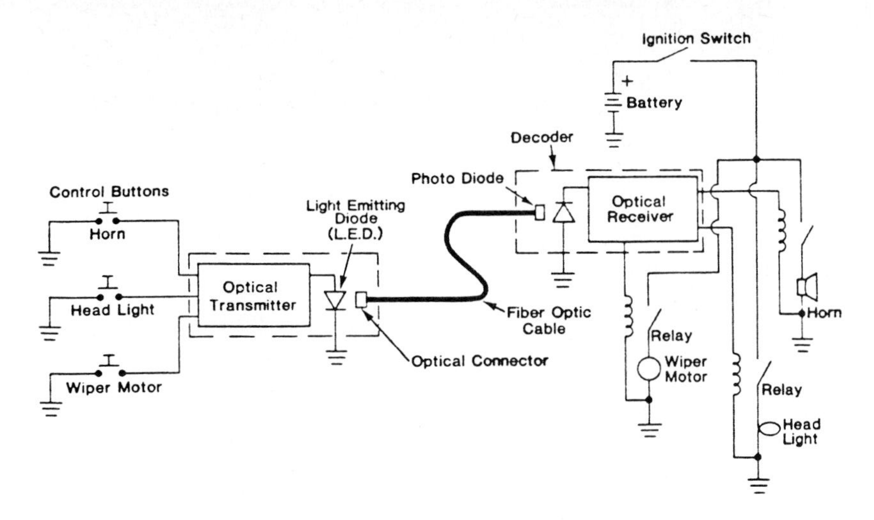

Figure 14-3 Basic Principle of Optical Link Multiplex System. [Reprinted with Permission of Society of Automotive Engineers © 1980]

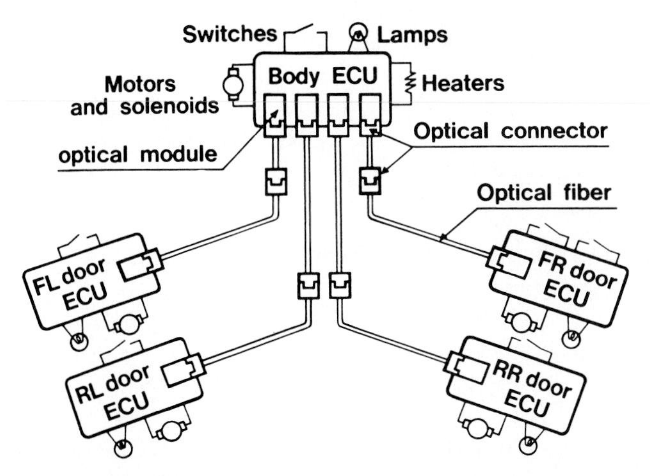

Figure 14-4 Multiplex System with Fiber Optic Cables. [Reprinted with Permission of Society of Automotive Engineers © 1984]

TABLE 14-1 Multiplex System Output Control Functions.

Control functions
1 Power door lock control
2 Power window control
3 Power vent window control
4 Power seat control
5 Illumination control
6 Seat heater control
7 Diagnosis

(Reprinted with Permission of Society of Automotive Engineers © 1984)

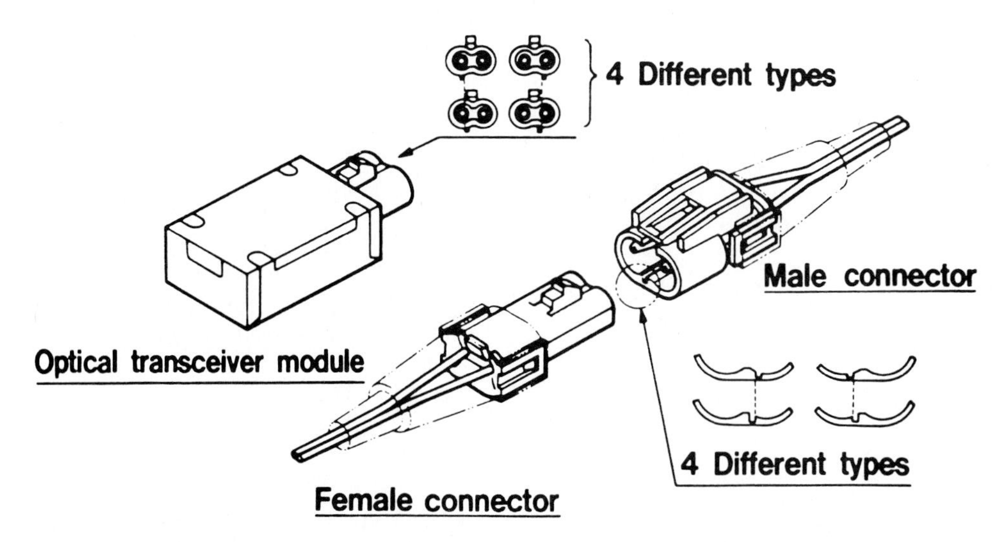

Figure 14-5 Fiber Optic Connectors. [Reprinted with Permission of Society of Automotive Engineers © 1984]

Receivers

The receivers are relatively "dumb" electronic units that perform specific output control functions as commanded by the master controller. Receivers may operate electrical loads directly, or they may control mechanical, or electronic, relays to perform these output functions. The receivers do not have decision-making capabilities. However, the receivers can be used to send signals back to the master controller regarding bulb outages or other system malfunctions. The master controller may be used to illuminate a warning indicator on the instrument panel.

Convenience Features

Many convenience features may be programmed into the microprocessor in the master controller. Examples of such convenience features could be the following:

1. A delayed headlight shut-off could be combined with a delayed backup light turn-off to provide lighting at the front and rear of the vehicle.
2. Headlights may be programmed to come on in the low beam position regardless of their previous state.
3. A combination of wiper high and wiper off buttons touched at the same time could result in the windshield wipers stopping in a nonpark preset position for servicing ease.

Multiplex Wiring Systems with Optical Data Links

System Design

Multiplex wiring systems are not widely used by North American car manufacturers at the present time. However, the major car manufacturers indicate that these systems will be introduced in the near future. Multiplex systems with bus data links may still be affected by EMI. Multiplex systems can be designed with optical data links connected between the master controller and the receivers. The signals through the fiber optic cable cannot be affected by EMI, and this type of cable cannot cause EMI or radio frequency interference (RFI). In a fiber optic system with optical data links, signals are sent from an LED in the master controller, or transmitter, through the fiber optic cable to the receiver. A photo diode in the receiver accepts the signal, and a decoder interprets the signal. This action causes the receiver to perform the requested output control function. The basic principle of the optical link multiplex system is illustrated in Figure 14–3.

Toyota Optical Link Multiplex System

Output Control Functions

A multiplex system with optical data links has been used on the Toyota Century marketed in Japan since 1982. In this system an electronic control unit (ECU) in each door is connected through fiber optic cables to an optical module in a central body ECU, as shown in Figure 14–4.

This multiplex system controls seven functions as indicated in Table 14–1.

The control switches in each door, the switches and sensors in the vehicle, which operate the multiplex system, and the loads controlled by these switches are listed in Table 14–2.

Fiber Optic Cable, Connectors, and Switches

A heat-resistant plastic fiber optic cable is used in the Toyota multiplex system. This cable is surrounded with a polyvinyl chloride tube to prevent impact damage. Two fiber optic cables are connected between each door ECU and the body ECU. Therefore, this multiplex system has bidirectional capabilities, which makes it possible for the body ECU to send output commands to the door ECUs, and the door ECUs can provide signals back to the body ECU for diagnostic purposes. For example, if an abnormal condition exists in a door lock/unlock solenoid, a signal is sent to the diagnostic ECU to report this problem. Special fiber-to-fiber connectors are necessary to prevent loss of the optic signal. The fiber optic connectors are pictured in Figure 14–5, and the bidirectional system capabilities are illustrated in Figure 14–6. Since the switches only require a low current capacity, touch-type switches with conductive rubber contacts are used.

Electrical Circuit

A power supply wire and a power request wire are connected between the body ECU and each door ECU. When a switch in any door is touched, a relay wind-

TABLE 14–2 Multiplex System Switches, Sensors, and Electrical Loads.

	SWITCHES OR SENSORS	*LOADS*
FRONT RIGHT DOOR (DRIVER'S SIDE DOOR)	Door Lock SW Door Key SW Door SW FR Auto P/W SW FR Manual P/W SW FR P/V SW FL P/W SW FL P/V SW RR P/W SW RR P/V SW RL P/W SW RL P/V SW FR S/H SW Rear S/H SW	Door Lock Solenoid P/W Motor P/V Motor FR S/H Indicator Rear S/H Indicator Courtesy Lamp SW Illumination Lamp Ashtray Illumination
FRONT LEFT DOOR	Door Lock SW Door Key SW Door SW P/W SW P/V SW S/H SW	Door Lock Solenoid P/W Motor P/V Motor S/H Indicator Courtesy Lamp SW Illumination Lamp Ashtray Illumination Lamp
REAR RIGHT DOOR	Door SW P/W SW P/V SW S/H SW P/S SW	Door Lock Solenoid P/W Motor P/V Motor S/H Indicator Courtesy Lamp SW Illumination Lamp Ashtray Illumination Lamp
REAR LEFT DOOR	Door SW P/W SW P/V SW S/H SW P/S SW	Door Lock Solenoid P/W Motor P/V Motor S/H Indicator Courtesy Lamp SW Illumination Lamp Ashtray Illumination Lamp
INTERIOR	Ignition SW Light SW Speed Sensor IG Key Sensor P/S Remote Control SW S/H Thermostats Park Position Sensor Neutral Position Sensor	Power Seat Motors Seat Heaters Map Lamps Foot Lamps IG Key Illumination Lamp Door Ajar Warning Lamp

(Reprinted with Permission of Society of Automotive Engineers © 1984)

NOTE: SW — Switch
P/W — Power Window
P/V — Power Vent Window
S/H — Seat Heater
P/S — Power Seat

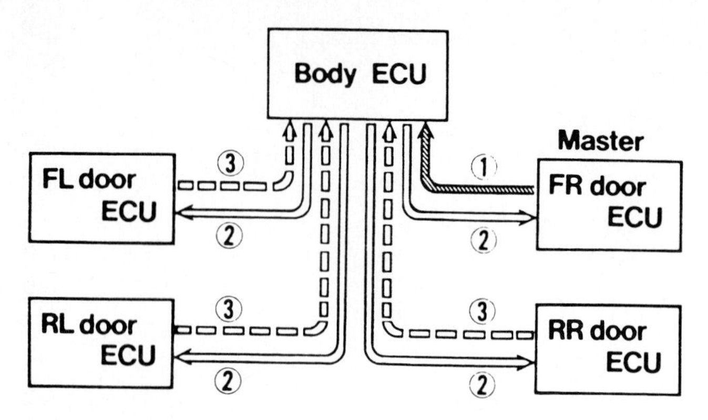

Figure 14–6 Bidirectional Fiber Optic System Capabilities. [Reprinted with Permission of Society of Automotive Engineers © 1984]

ing in the body ECU is energized, which closes the relay points. This action supplies power to the control circuits in the body ECU and the door ECUs. Therefore, power is only supplied to these control circuits when a switch is touched. Once the control circuits are activated, the necessary signal will be sent from the body ECU to the door ECU through the fiber optic cable, and the door ECU will perform the requested output function. The electrical circuit for the fiber optic multiplex system is shown in Figure 14–7.

Advantages of the Toyota Multiplex System

The advantages of the multiplex system with fiber optic cables are the following:

1. The number of wires is reduced. If a conventional electrical system was used in place of the multiplex system, 46 wires would be required in the door on the driver's side. In the multiplex system 8 wires and 2 fiber optic cables enter the driver's door. The total wiring harness weight is reduced by approximately 25 percent.
2. Greater freedom in system design is available. For example, all timing functions can be controlled by the body ECU. In this system the interior lights remain on for a few seconds after all the doors are closed. When the vehicle speed exceeds 12.5 mph (20 kph), all doors will be locked. If the ignition switch is turned off, or left on with the transmission in park, the doors will automatically unlock.
3. Increased flexibility in switch design is possible because the switches only require a low current capacity. With touch-type switches, greater ingenuity may be used in the cosmetic design.
4. There is no possibility of EMI interfering with the signals in the fiber optic cable.

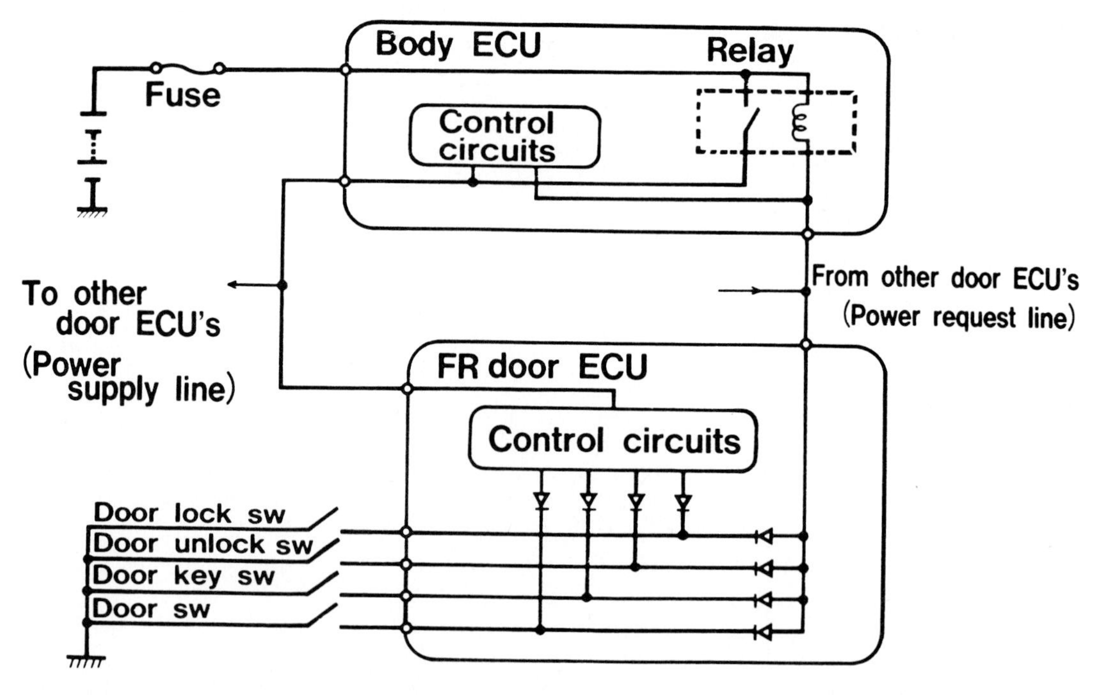

Figure 14–7 Electrical Circuit for Fiber Optic Multiplex System. [Reprinted with Permission of Society of Automotive Engineers © 1984]

Test Questions

1. In a bus data link-type multiplex wiring system, all system intelligence is in the:
   (a) receiver.
   (b) master controller.
   (c) bus data link.
2. Separate flashes and timers for each electric circuit are required with a multiplex wiring system. T F
3. Switches in a multiplex system have high current carrying capacity. T F
4. In a fiber optic type of multiplex system in normal condition, signals are sent through the fiber optic cable from a ____________ in the master controller.
5. Signals through a fiber optic cable may be affected by electromagnetic interference. T F
6. In the Toyota fiber optic type of multiplex system, the power is supplied to the control circuits in the body ECU and door ECUs:
   (a) at all times.
   (b) when the ignition switch is turned on.
   (c) if a system switch is touched.

15

Computer-Controlled Diesel Injection and Emission Systems

Practical Completion Objectives

1. Service fuel filters.
2. Drain and purge water from the fuel system.
3. Bleed air from the fuel system.
4. Test injection nozzles.
5. Install injection pump and drive belt.
6. Check and adjust injection pump timing.

Diesel Engine Principles

Diesel Ignition and Compression Ratio

There are many similarities between the four-stroke cycle diesel engine and the four-stroke cycle gasoline engine.

We will consider the major differences between the two engines. The diesel engine does not require an ignition system to ignite the air-fuel mixture in the cylinders. Instead the diesel engine uses a very high compression ratio, which creates enough heat on the compression stroke to ignite the air–fuel mixture. The average automotive diesel engine would have a compression ratio of 22.5:1 and a cranking compression pressure of 400 psi (2758 kPa). For every psi of the compression pressure, the temperature of the air in the cylinder will increase approximately 2°F. Therefore, a compression pressure of 400 psi (2758 kPa) would create a cylinder temperature of approximately 800°F (426°C). The ignition temperature of diesel fuel is about 750°F (398°C). The diesel engine does not use a throttle to control engine speed. The air intake system is unrestricted, and the speed of the diesel engine is controlled by regulating the amount of fuel that is injected into the cylinders by the injection pump and injectors. Most of the components in a diesel engine are more heavily constructed than the comparable gasoline engine parts because of the higher pressure encountered in the diesel engine. A four-cylinder automotive diesel engine is illustrated in Figure 15–1.

Electronically Controlled Diesel Injection Pump

Electronic Control System

As diesel emission standards became increasingly stringent, especially in California, it has been necessary to introduce diesel injection pumps that are electronically controlled. These pumps are similar to conventional pumps but the injection advance is controlled by an electronic control unit, which controls a solenoid valve in the injection pump. Input signals to the electronic control unit are the following:

1. Coolant temperature sensor.
2. Speed and crankshaft position sensor.
3. Altitude switch.
4. Injection nozzle with inductive transmitter.

On the basis of the input signals that it receives, the electronic control unit controls the pump solenoid valve to provide the precise injection advance required by the engine. The complete injection pump electronic control system is shown in Figure 15–2.

The electronic control unit is mounted in the luggage compartment. A power relay supplies voltage to this control unit when the ignition switch is turned on. If the battery polarity is reversed, a diode in the

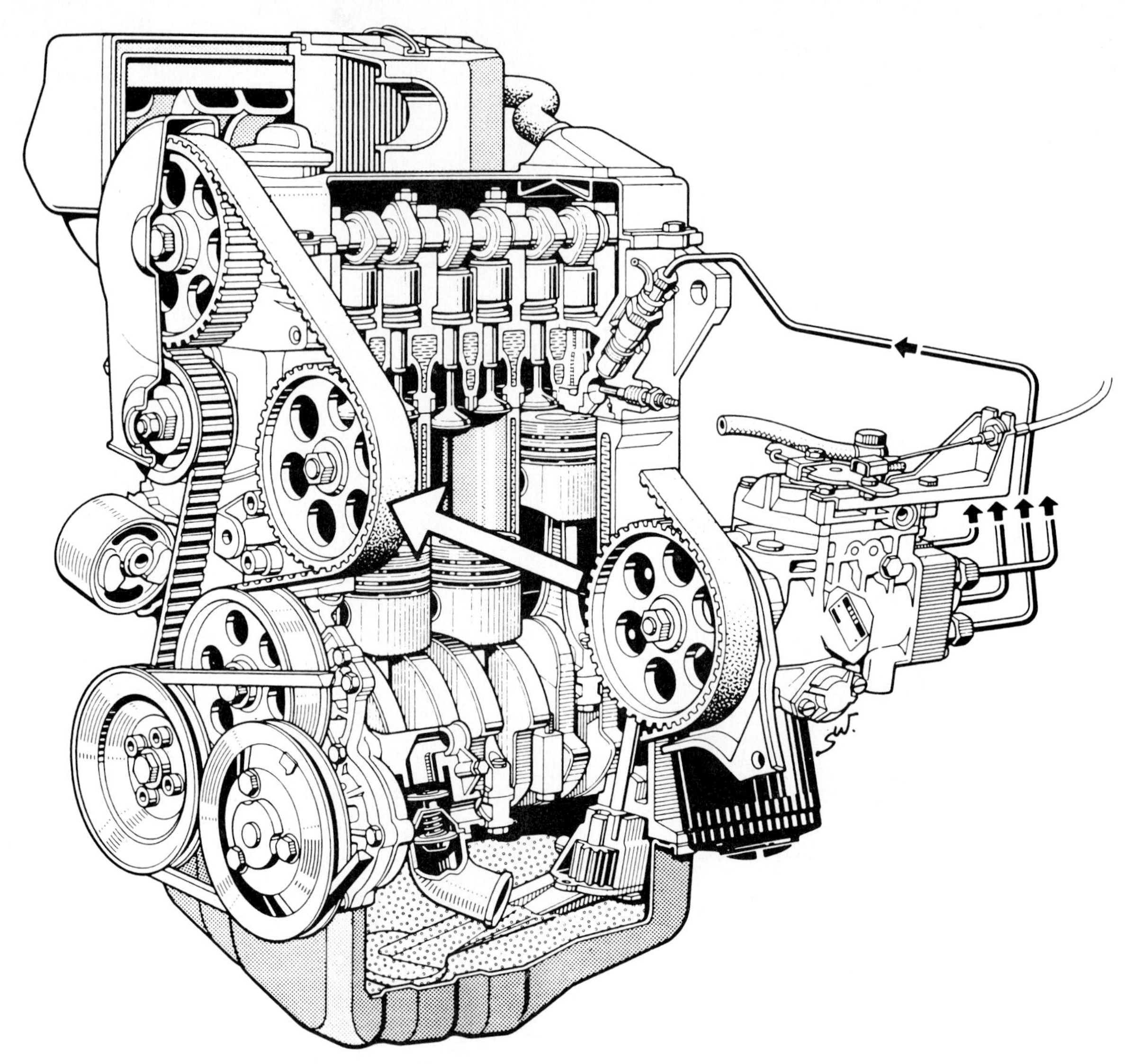

Figure 15–1 Four-Cylinder Automotive Diesel Engine. [Reprinted with Permission of Society of Automotive Engineers © 1977]

power relay prevents the relay from closing and thus protects the electronic control unit. The power relay is illustrated in Figure 15–3.

Fuel System

Operation

An electric lift pump delivers fuel from the fuel tank to the injection pump. A fuel conditioner is connected in the fuel line between the tank and the electronic pump. This conditioner contains a filter that removes contaminants such as dirt and water from the fuel. A transfer pump in the injection pump delivers fuel to the injection pump plunger. High pressure fuel is forced from the pump plunger to the injection nozzles, which inject fuel into the cylinders at the right instant. Excess fuel is returned from the pump and the injectors to the fuel tank. The internal pump components are lubricated and cooled by the diesel fuel. A complete diesel fuel system is pictured in Figure 15–4.

Injection Pump

Transfer Pump

The transfer pump contains four spring-loaded vanes mounted in rotor slots. An eccentric pump cavity surrounds the outer edges of the vanes. Fuel enters the transfer pump from the lift pump, and the transfer pump forces fuel past the pressure-regulating valve to the injection pump plunger. The pressure-regulating valve opens at a specific pressure and returns some of the fuel to the pump inlet, which limits transfer pump pressure. When engine speed and transfer pump speed increase, the transfer pump pressure will gradually increase because of the pressure regulating valve design. A transfer pump with a pressure-regulating valve is shown in Figure 15–5.

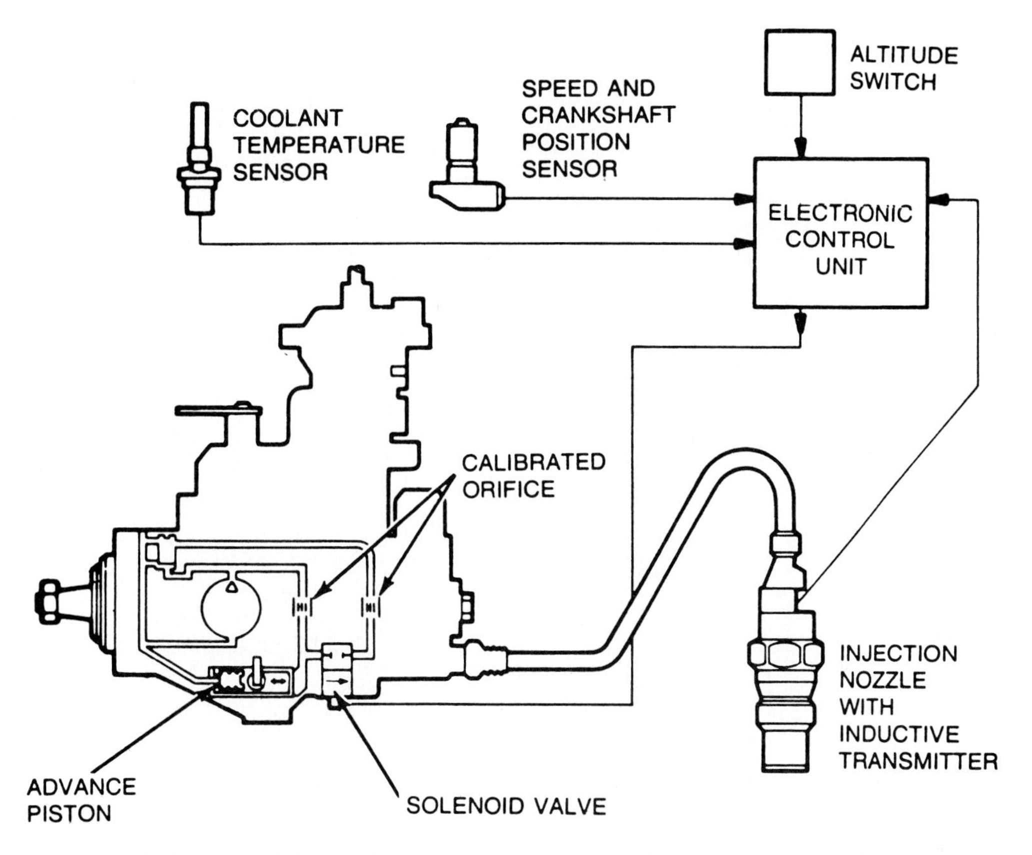

Figure 15–2 Diesel Injection Pump Electronic Control System. (Courtesy of Ford Motor Co.)

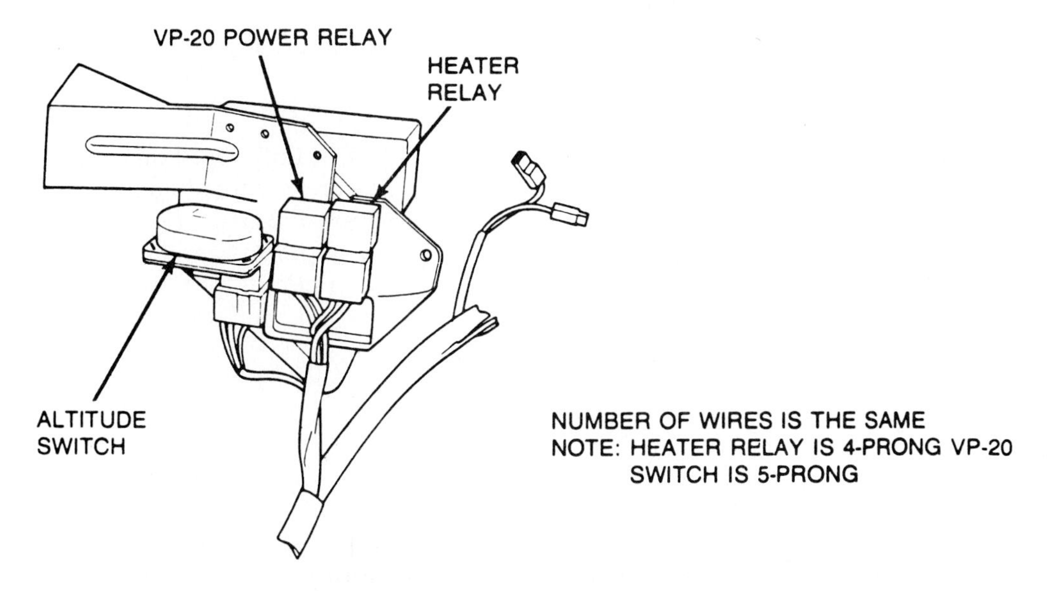

Figure 15–3 Electronic Control Unit Power Relay. (Courtesy of Ford Motor Co.)

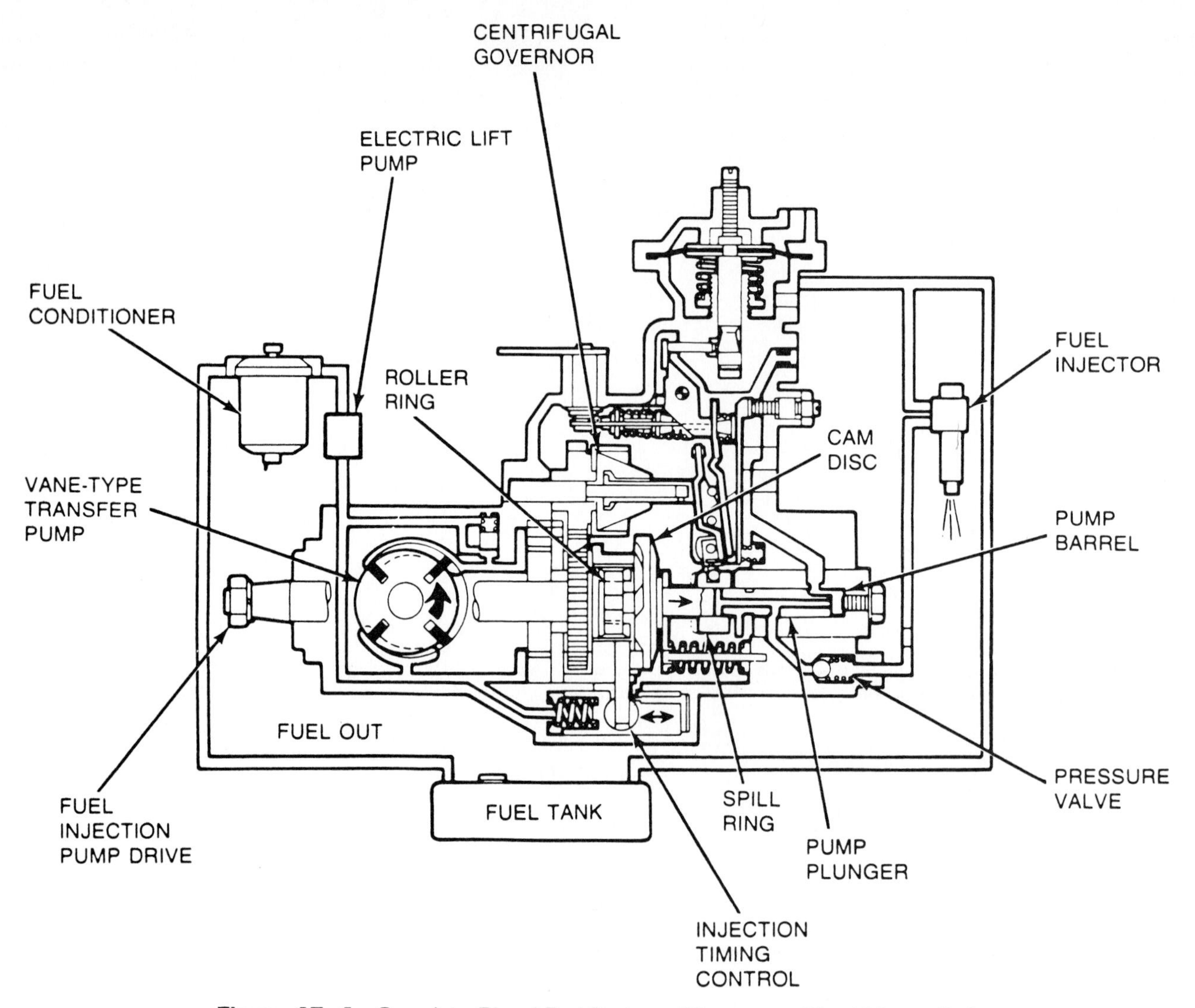

Figure 15-4 Complete Diesel Fuel System. [Courtesy of Ford Motor Co.]

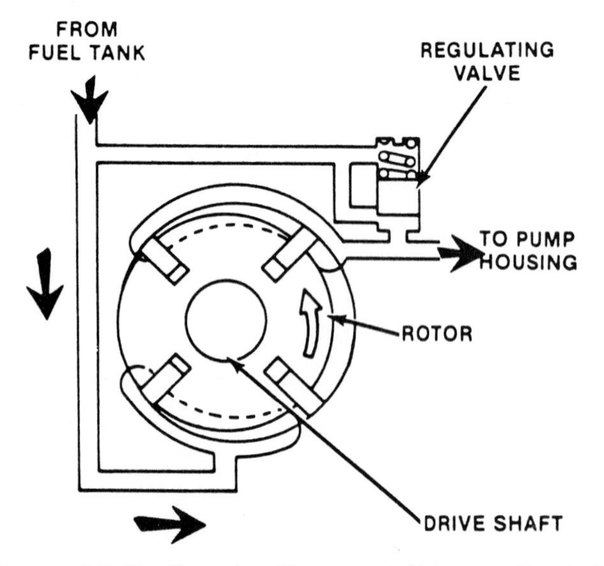

Figure 15-5 Transfer Pump and Pressure-Regulating Valve. [Courtesy of Ford Motor Co.]

High Pressure Pump Plunger

When the ignition switch is turned on, the electric fuel shutoff plunger is lifted, which allows transfer pump pressure to force fuel into the passages in the high pressure plunger. The transfer pump and the high pressure plunger are rotated by the pump drive. A cam disc is mounted on the high pressure plunger, and the cams on this disc contact rollers in a roller ring. When the low points on the cam are in contact with the rollers, the fuel flows from the transfer pump through one of the plunger fill slots into the center of the plunger. The number of rotor fill slots corresponds to the number of engine cylinders. A return spring holds the plunger and cam disc against the rollers. As the disc high points contact the rollers, the plunger is forced ahead, and the plunger seals the inlet port from the transfer pump. Forward plunger movement creates a very high fuel pressure in the center of the plunger, which forces fuel from the outlet port through the pressure valve to the injection nozzles. This extremely high fuel pressure opens the injection nozzles at the right instant, and the nozzle sprays fuel into the combustion chamber. The clearance between the plunger and barrel is in millionths of an inch to prevent fuel leaks between these components. A series of equally spaced outlet ports is located around the plunger, and there is always one outlet port for each engine cylinder. The plunger stroke is shown in Figure 15–6, and the plunger and barrel are illustrated in Figure 15–7.

The amount of fuel injected is controlled by the spill ring position, which determines the effective plunger stroke. When the plunger is moved forward, the vertical fuel passage in the plunger is uncovered at the edge of the spill ring, which allows fuel to be spilled from the center of the plunger. When this occurs, fuel pressure decreases instantly in the plunger and in the injection line, which results in injection nozzle closure. The actual plunger stroke remains constant, but the effective plunger stroke that determines the amount of fuel delivered by the pump is controlled by the spill ring position. If the spill ring

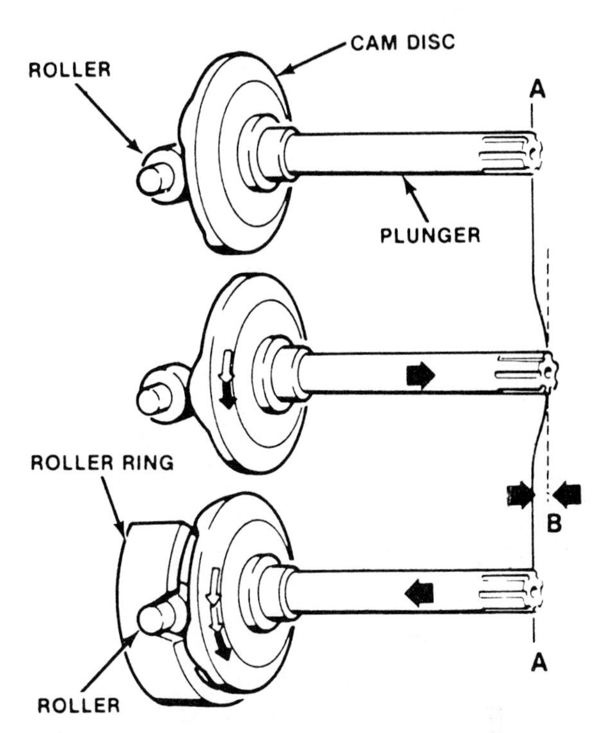

Figure 15–6 Plunger with Cam Disc, Roller, and Roller Ring. [Courtesy of Ford Motor Co.]

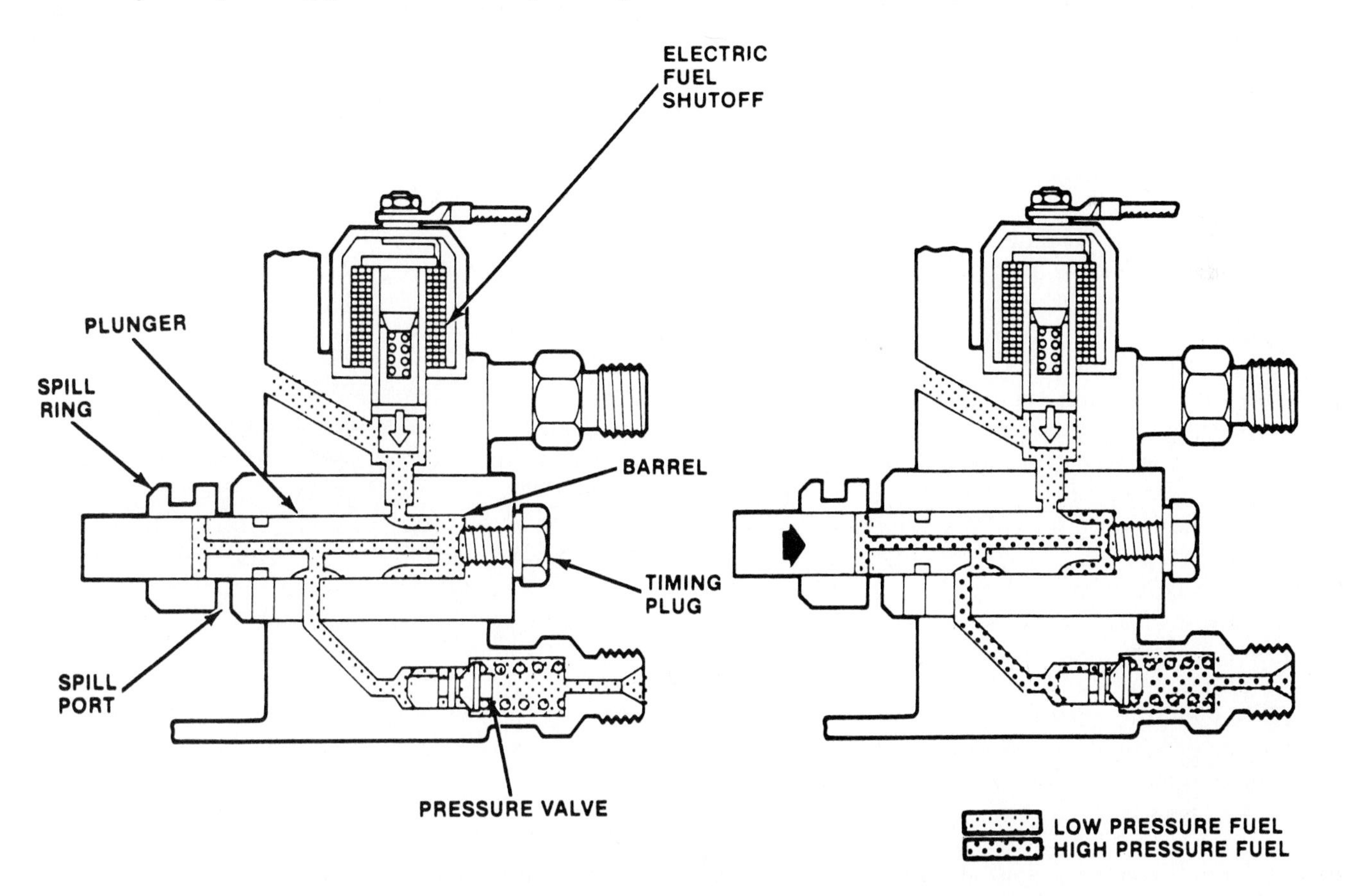

Figure 15–7 Plunger and Barrel Fuel Passages. [Courtesy of Ford Motor Co.]

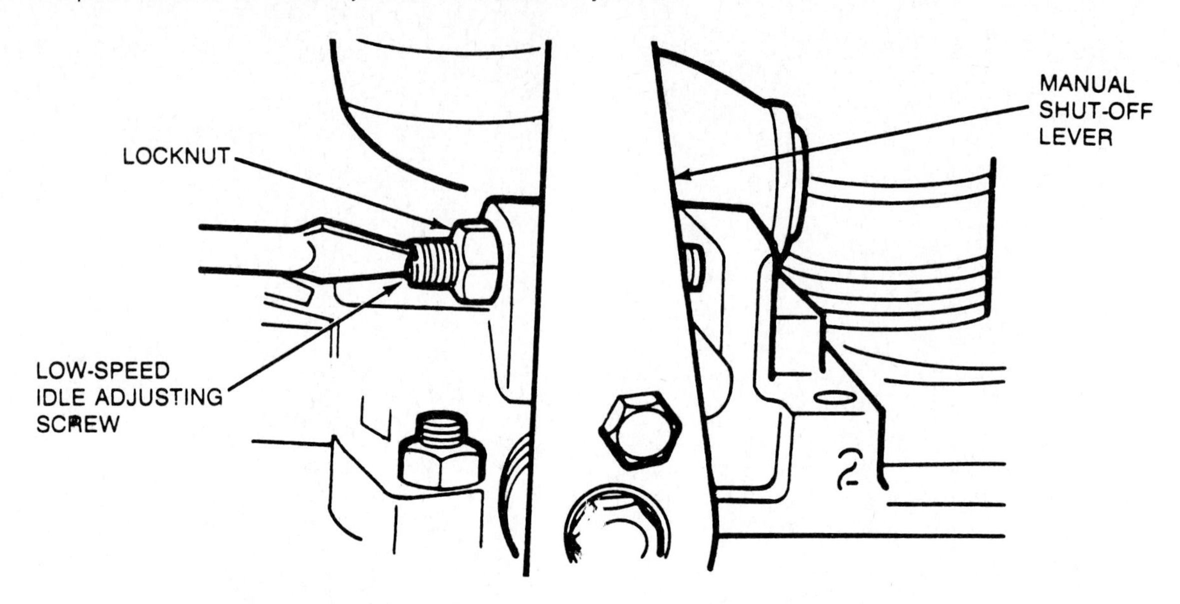

Figure 15-8 Idle Speed Adjustment. (Courtesy of Ford Motor Co.)

is moved toward the cam disc, the effective pump stroke, fuel delivery, and engine speed are reduced. When the spill ring is moved from the cam disc, the effective pump stroke, fuel delivery, and engine speed are increased.

Governor

The position of the spill ring is controlled by the accelerator pedal, and the governor in the pump. A linkage is connected between the accelerator pedal and the governor lever. The governor contains a group of pivoted flyweights, which are rotated by a set of gears. A governor drive gear is mounted on the pump drive shaft, and the driven gear rotates the flyweights and retainer. The accelerator pedal is connected to the governor lever through the linkage and governor spring. When the engine is idling, the governor weights, spring, and linkage position the spill ring to inject the correct amount of fuel to obtain the correct idle speed. A low speed idle adjusting screw on the injection pump is used to adjust the idle speed as indicated in Figure 15–8.

As the accelerator pedal is depressed, the linkage and governor spring will move the spill ring away from the cam disc. This movement increases the effective plunger stroke, which injects more fuel to increase the engine speed. When the engine speed reaches the governed rpm, the outward weight movement overcomes the governor spring tension and forces the lever away from the pump drive. This action moves the spill ring toward the cam disc, which reduces the effective plunger stroke and the amount of fuel injected to limit maximum engine speed. The governor operation is shown in Figure 15–9.

Pressure Valve and Injection Lines

A pressure valve is located in each pump outlet fitting. When the spill port opens at the end of the effective plunger stroke, pressure drops rapidly in the injection line. When this occurs the pressure valve closes, which controls the negative pressure in the injection line. The pressure valve allows enough sudden pressure decrease in the injection line to

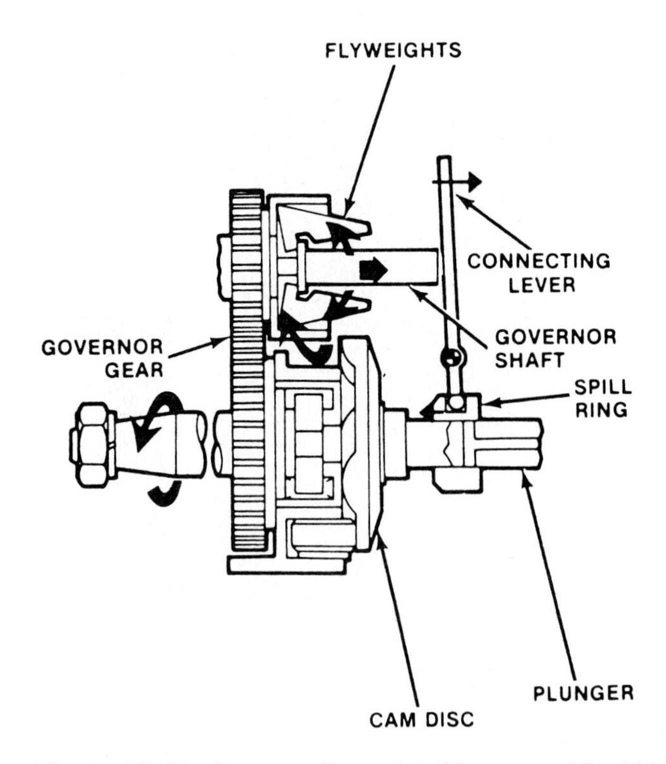

Figure 15-9 Governor Operation. (Courtesy of Ford Motor Co.)

cause rapid injection nozzle closure and prevent fuel dribbling at the injector. Excessive negative pressures in the injection lines can actually tear the fuel apart and create vapor cavities in the fuel. When these vapor cavities are formed, very small particles of metal can be torn from the inside surface of the injection lines. This process is known as "cavitation," and it can severely deteriorate injection lines in a short time. The pressure valve controls negative injection line pressure and prevents cavitation.

Injection lines are manufactured from high pressure steel tubing. These lines are all the same length on each engine. Injection lines of different lengths on the same engine would cause a slight variation in injection timing. Therefore injection lines must never be altered in any way.

Electronic Injection Advance

A timing control piston is mounted in a bore below the roller ring, and a pin is connected from the roller ring to the piston. Transfer pump pressure is applied to the right side of the timing control piston, and a return spring is positioned on the opposite end of the piston. When engine speed and transfer pump pressure increase, the increase in transfer pump pressure forces the timing control piston to the left against the spring tension. This piston movement rotates the roller ring in the opposite direction to plunger and cam disc rotation, which causes the cams to contact the rollers sooner and advance the injection timing. The advance mechanism is illustrated in Figure 15–10.

The injection timing advance solenoid is controlled by the electronic control unit. Some fuel moves from the advance side of the timing control piston through the advance solenoid to the return spring side of the timing control piston. When the input signals inform the electronic control unit that increased injection advance is required, the electronic control unit allows the advance solenoid plunger to move upward. When this occurs, fuel flow past the advance solenoid plunger is reduced, which

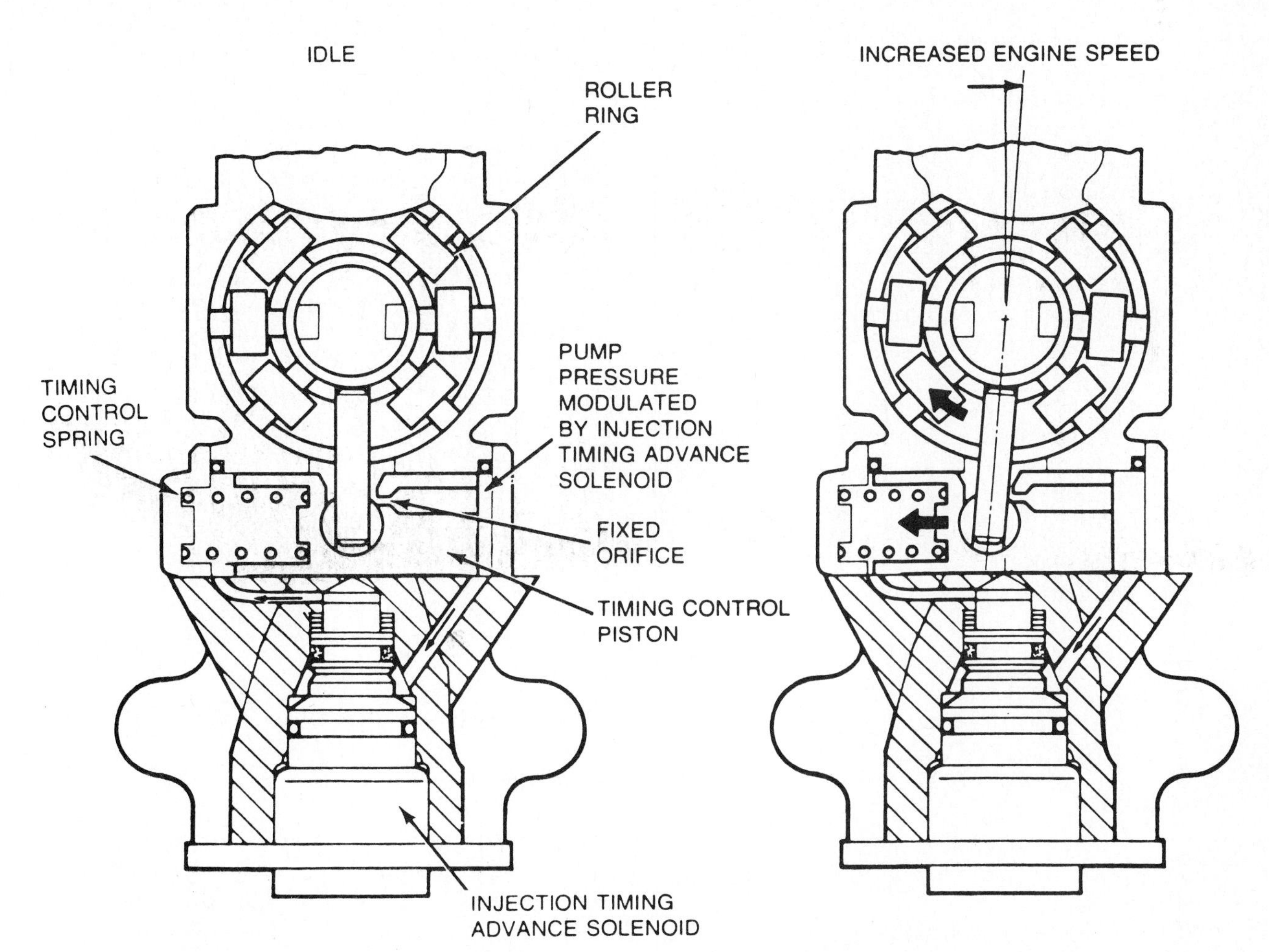

Figure 15–10 Electronically Controlled Injection Advance Mechanism. (Courtesy of Ford Motor Co.)

increases the transfer pump pressure on the timing control piston. This pressure increase moves the timing control piston and roller ring to provide more injection advance.

If a reduction in injection timing is necessary, the electronic control unit moves the timing advance solenoid plunger downward. This action allows more fuel to flow from the advance side of the timing control piston past the advance solenoid plunger to the return spring side of the timing control piston. When this occurs transfer pump pressure applied to the timing control piston is reduced, which allows the roller ring and timing control piston to move to the right. This movement reduces the injection timing advance.

Injection Nozzles

Design and Operation

Many automotive diesel nozzles are threaded into the cylinder head. A sealing washer seals the end of the injector into the cylinder head. The injection line fitting is threaded into the nozzle. A needle valve with a precision-tapered tip is positioned in the spray orifice at the injector tip. This needle valve is seated by a pressure spring. When the injection pump delivers high pressure fuel to the nozzle, the needle valve will be lifted against the spring tension, and fuel will be sprayed from the injector tip into the combustion chamber. The needle valve will close immediately when injection pump pressure decreases. Diesel injection nozzle construction is shown in Figure 15–11.

The clearance between the nozzle valve and the nozzle body is measured in millionths of an inch.

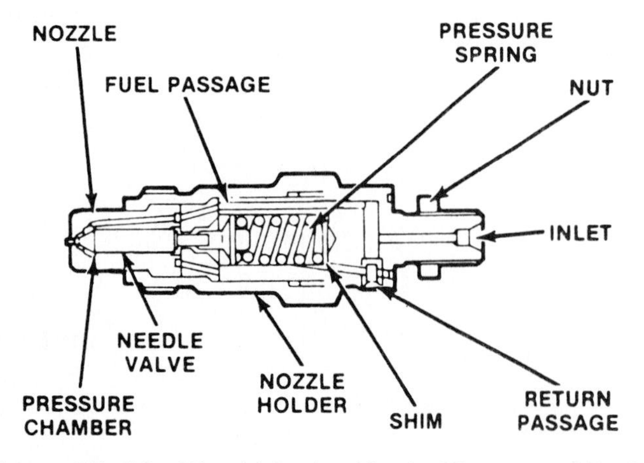

Figure 15–11 Diesel Injection Nozzle. [Courtesy of Ford Motor Co.]

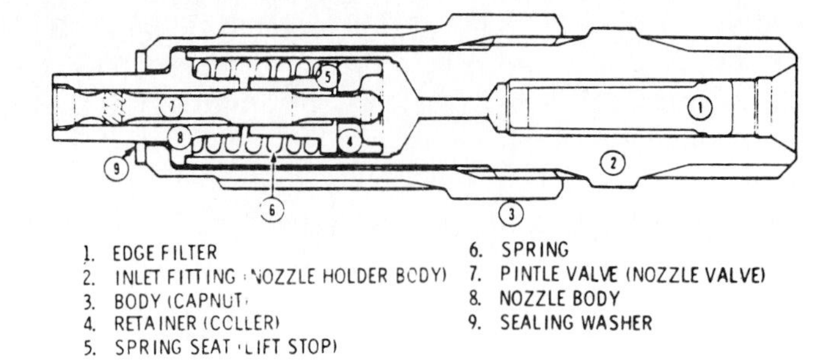

Figure 15–12 Poppet Nozzle Design. [Courtesy of General Motors Corporation]

Therefore, nozzles are precision devices that require careful handling. Some nozzles have several spray orifices in the tip. The actual number and size of the spray orifices vary depending on combustion chamber size.

Some nozzles have a valve with a tapered seat around the circumference of the valve. This type of valve opens outward into the combustion chamber and fuel sprays out around the entire seat circumference. These nozzles are called "poppet nozzles." The internal design of a poppet nozzle is pictured in Figure 15–12.

An internal stop limits the maximum nozzle valve opening. Some nozzles have a return fuel line that returns excess fuel to the fuel tank. In many automotive diesel engines the nozzles inject fuel into a small precombustion chamber. Combustion begins in this chamber with a swirling motion because of the chamber design. The burning air–fuel mixture expands with a strong swirling motion into the main combustion chamber. This action creates increased turbulence in the main combustion chamber to provide more complete burning of the air–fuel mixture. The glow plugs are usually located in the precombustion chamber with the nozzles, as indicated in Figure 15–13.

Nozzles should always be torqued in this cylinder head to manufacturer's specifications. Plastic caps should be placed over disconnected fuel-line connections to keep dirt out of the fuel system.

Glow Plug Circuits

Design and Operation

Current flow through the glow plugs heats the precombustion chamber, which provides easier starting. An electronic control unit operates two relays that supply voltage to the glow plugs. The electronic control unit receives a coolant temperature sensor (CTS)

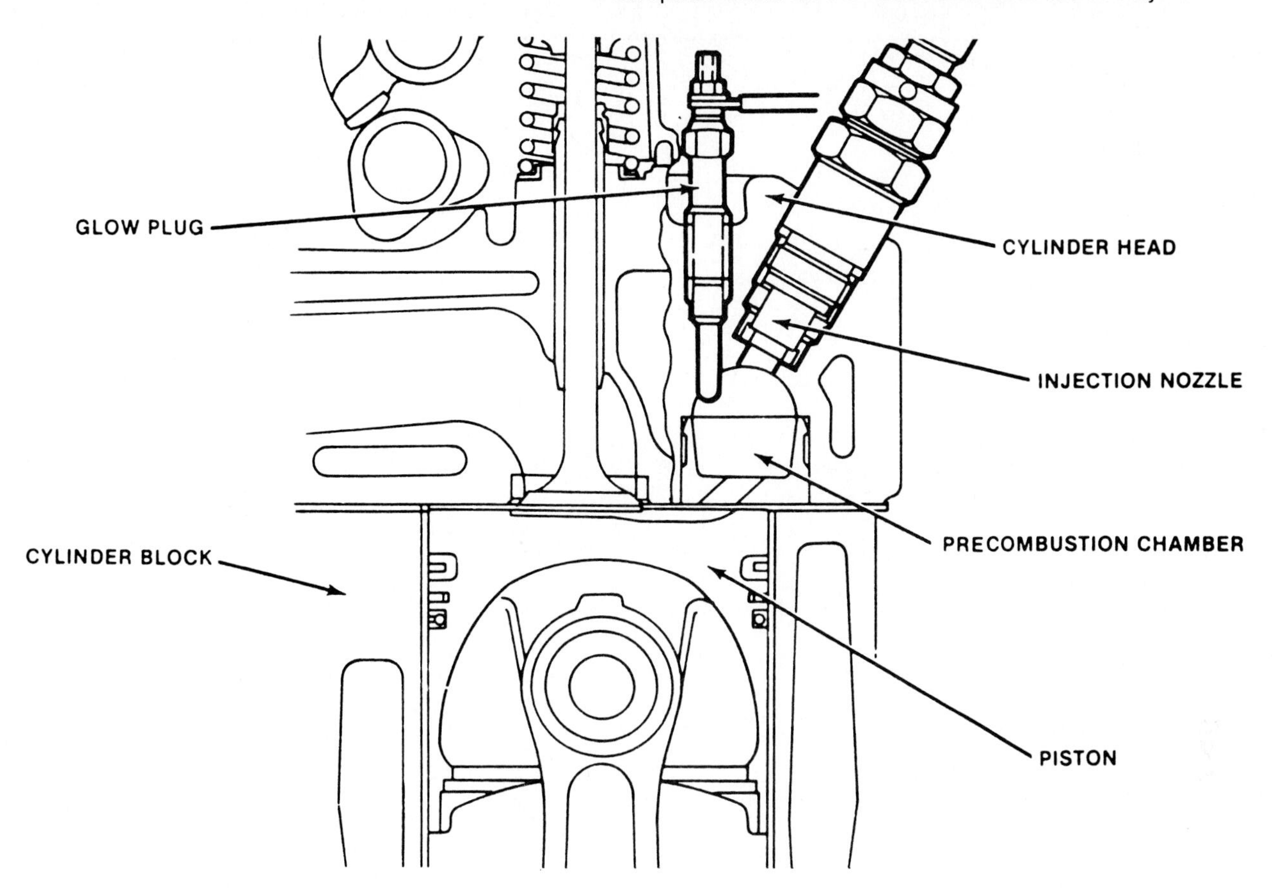

Figure 15–13 Precombustion Chamber with Nozzle and Glow Plug. (Courtesy of Ford Motor Co.)

signal. When the ignition switch is turned on with the coolant temperature below 86°F (30°C), the electronic control unit closes both relays. Under this condition full voltage is supplied to the glow plugs, which heat the precombustion chambers very quickly so the engine may be started immediately. A signal from the alternator informs the electronic control unit when the engine is started. This signal causes the electronic control unit to open relay I and keep relay II closed. A dropping resistor between relay II and the glow plugs reduces the voltage and current flow through the glow plugs. Relay II may remain closed for 30 seconds after the engine has started. The operation of relay II during this time is referred to as an after-glow period, which maintains precombustion chamber temperature to prevent engine stalling. The complete glow plug circuit is shown in Figure 15–14.

The electronic control unit will only keep relay I closed for 3 to 6 seconds after the ignition switch is turned on, whereas the length of the after-glow period will vary depending on engine temperature. If the ignition switch is turned on and the engine is not cranked, the electronic control unit will open both relays in 3 to 6 seconds to prevent battery discharge and glow plug damage. When the coolant temperature is above 86°F (30°C), the electronic control unit will not close relay I. The glow plug circuit shown in Figure 15–14 does not have an instrument panel indicator light. Some glow plug circuits have a wait period before the engine is started. These circuits have an instrument panel indicator lamp that informs the operator when to start the engine. A pulsating voltage is supplied to the glow plugs during the after-glow period in some glow plug circuits. The internal design of a glow plug is illustrated in Figure 15–15.

Computer-Controlled Exhaust-Gas Recirculation System

Electric Circuit and Vacuum Hose Routing

The computer-controlled exhaust-gas recirculation (EGR) system is shown in Figure 15–16.

A diesel engine produces high emissions of oxide of nitrogen (NOx) during idle and off-idle oper-

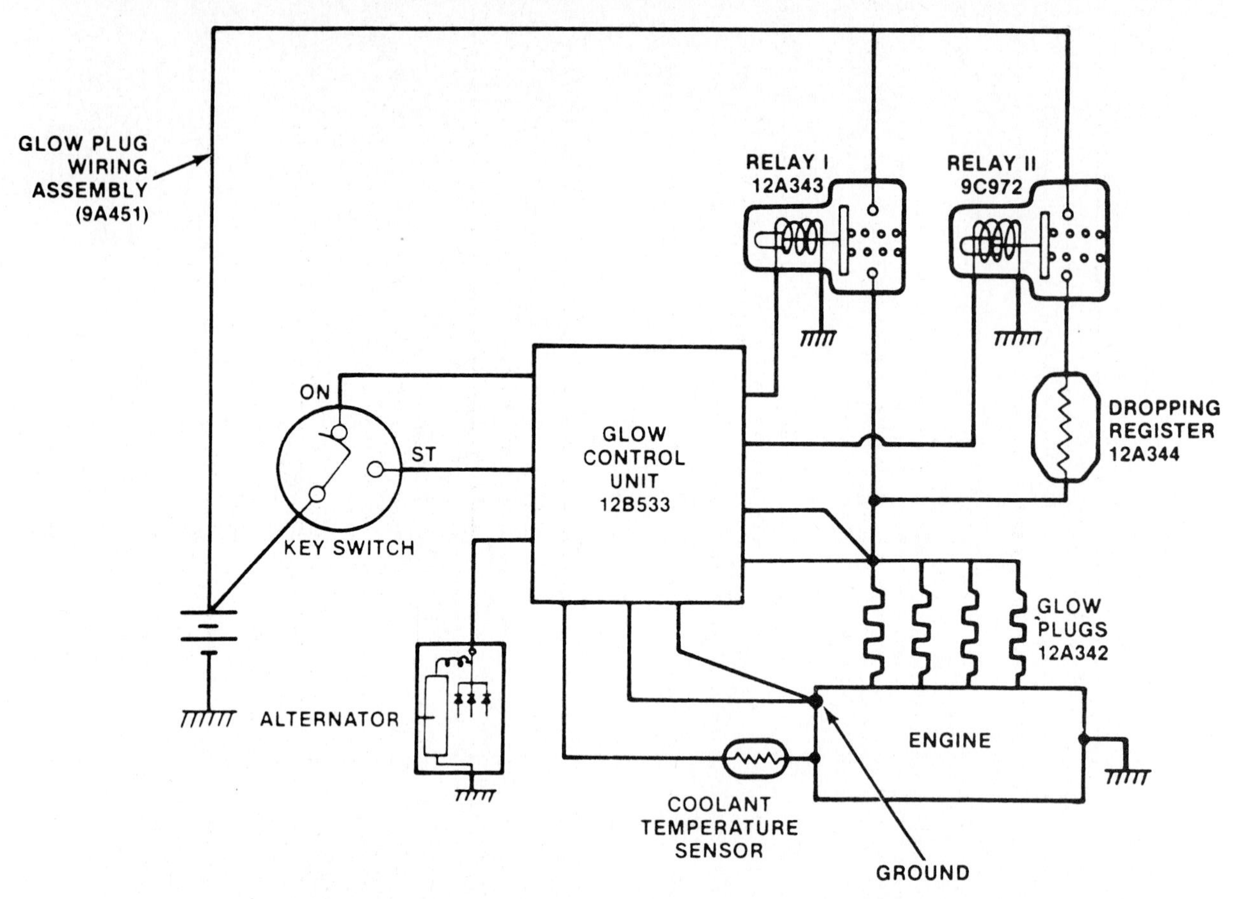

Figure 15-14 Glow Plug Circuit. [Courtesy of Ford Motor Co.]

ation. Therefore, the EGR valve is open during these operating conditions. The purpose of the components in the EGR system may be summarized as follows:

1. Temperature switch—senses coolant temperature. This switch is closed above 172°F (78°C).
2. Vacuum regulator—mounted on the injection pump and operated by the pump linkage to control the EGR flow in relation to fuel delivery.
3. Vacuum reservoir—prevents vacuum pulsations at the vacuum regulator.
4. Vacuum delay valve—delays vacuum supplied to the EGR valve to prevent sudden valve opening and improve driveability.
5. Idle speed switch—informs the injection timing module when the engine is idling. The module uses this signal to control the EGR valve at idle. This switch is mounted on the injection pump.
6. EGR control solenoid—the injection timing module opens and closes this solenoid to supply vacuum to the EGR valve.
7. Vacuum pump—the vacuum pump is mounted beneath the rocker arm cover on the engine, and it provides vacuum to operate the EGR valve.
8. Injection timing module—controls the EGR valve operation. This module also controls injection advance (illustrated previously in Figure 15-2).
9. EGR valve—when this valve is open it recirculates exhaust into the intake manifold.
10. Speed and position sensor—supplies an engine rpm signal to the injection timing module. This sensor is positioned on the lower left corner of the engine block.
11. Altitude switch—informs the injection timing module when the vehicle is operating above 9800 ft (3000 m) elevation.
12. EGR vent valve—if the EGR vacuum supply exceeds 8.6 in hg (29 kPa), this valve vents the vacuum hose from the vacuum pump to the atmosphere.

The computer-controlled EGR system shown in Figure 15-16, and the glow plug system illustrated in Figure 15-14, are used on the Ford 2.4L turbocharged diesel engine.

Diesel Fuel

Types and Ratings

The engine manufacturer's recommendations should be followed when a diesel fuel is selected. Grade 2D diesel fuel is usually recommended for most diesels that operate under medium to heavy loads and uniform speeds in warm climates. If 2D diesel fuel is used in cold weather below 0°F (−18°C) the fuel may thicken and plug the fuel system. This process is referred to as "fuel waxing." Grade ID diesel fuel is recommended in cold climates, and it may also be required in high-speed diesel engines that are often subjected to variations in speed and load. The pour point of diesel fuel is the temperature below which the fuel will no longer flow. The cloud point of the fuel is the temperature at which wax crystals start to form in the fuel. The cloud point is higher than the pour point. Number ID diesel fuel has the lowest cloud and pour points available. The cetane number of the diesel fuel indicates the ability of the fuel to self-ignite in the combustion chamber. The cetane number of the fuel should meet the engine manufacturer's minimum requirements. There is no benefit in using a higher number cetane fuel than called for by the engine manufacturer. When handling or storing

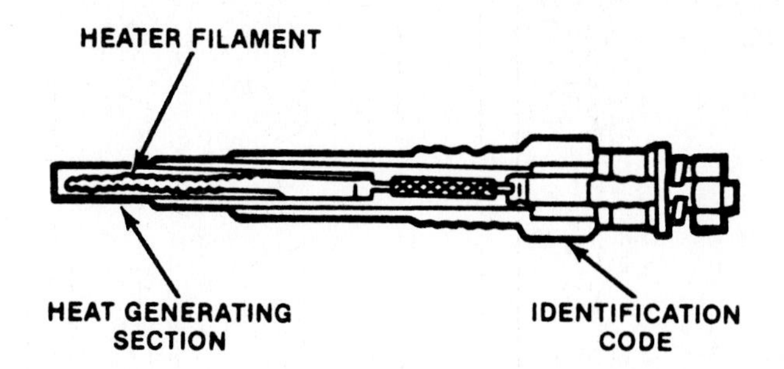

Figure 15–15 Glow Plug Design. (Courtesy of Ford Motor Co.)

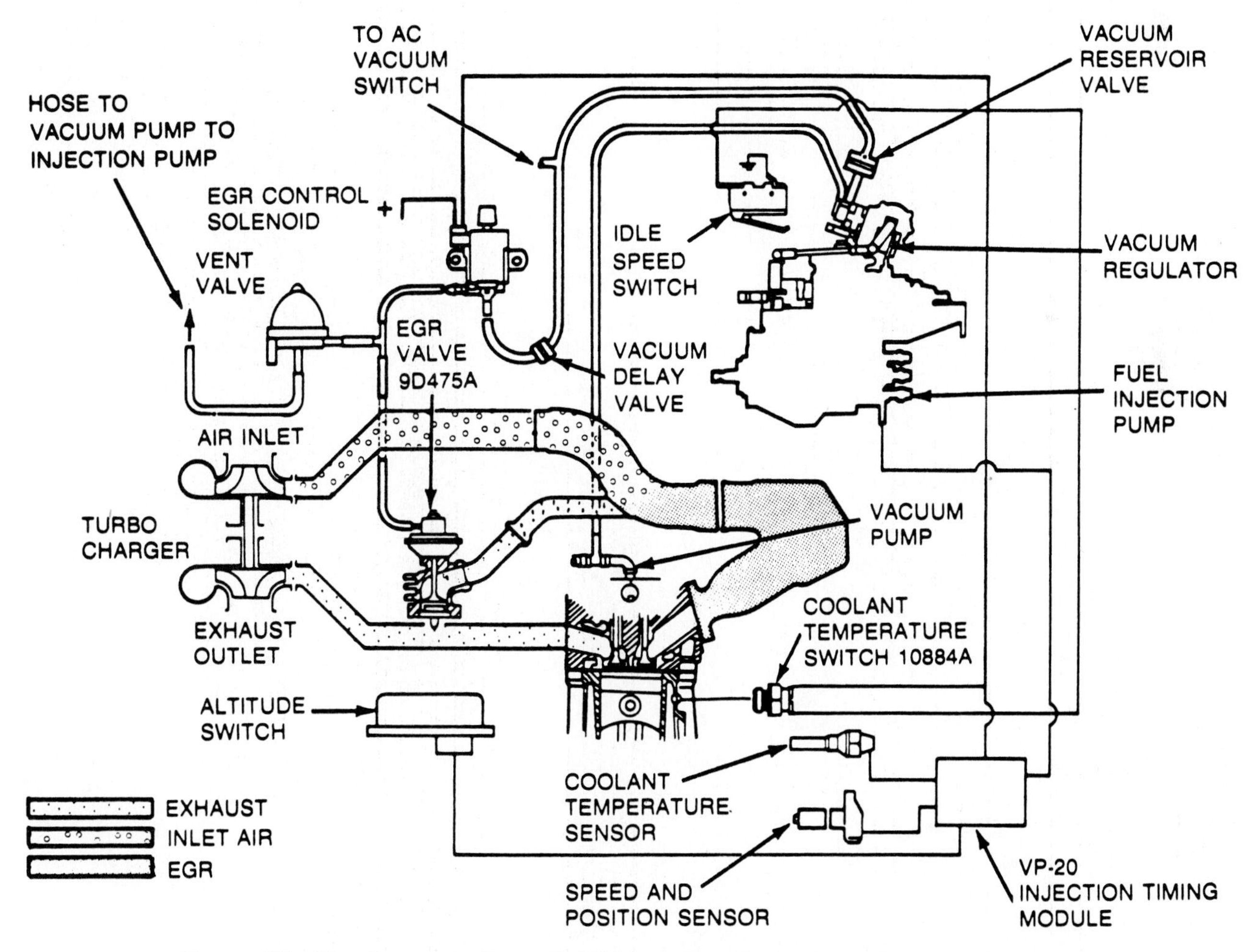

Figure 15–16 Computer-Controlled Exhaust-Gas Recirculation System. (Courtesy of Ford Motor Co.)

diesel fuel it is extremely important that water and other contaminants be kept out of the fuel. Extreme cold weather temperature of −25°F (−31.6°C), or lower, requires special fuels such as arctic or kerosene-treated products, which are often detrimental to the injection pump during extended use.

Diesel Service

Filter Service

Dirt in the fuel can damage nozzles and injection pumps, because these components have precision clearances that are measured in millionths of an inch. Therefore, it is very important to keep dirt out of the fuel system and change the fuel filter at the manufacturer's recommended intervals. The fuel filter replacement procedure is pictured in Figure 15–17.

Some fuel filter assemblies contain a water separator to separate water from the diesel fuel. A water-in-fuel sensor turns on an indicator light on the instrument panel when 50 cc of water is collected in the filter. To purge water from the filter assembly proceed as follows:

1. Turn on the ignition switch to activate the electric lift pump.
2. Place a container below the filter drain.
3. Pull the pull ring to open the drain valve.
4. Release the pull ring to close the drain valve when diesel fuel appears a light amber color.

Air will be purged from the filter assembly when the vent screw is loosened with the ignition switch on and the lift pump running. Close the vent screw when air bubbles no longer appear at the screw. Air purging will be necessary after fuel system repairs, or when the system has run out of fuel. A fuel heater in the filter assembly heats the fuel to prevent fuel waxing when fuel temperature is below 46°F (8°C). Current flow through the fuel heater is controlled by a set of contacts that are closed by a bimetal strip. A fuel filter assembly is shown in Figure 15–18.

Nozzle Testing

The injection nozzles can be tested on a nozzle tester. This tester contains a hand pump and a pressure gauge. A special test fluid is used in the tester. The following nozzle tests should be performed:

1. The spray pattern should be tested by operating the hand pump until the pressure opens the nozzle. The nozzle spray must be directed into a container. Never direct the nozzle spray toward human flesh. The spray pattern varies depending on the type of nozzle. Acceptable and unacceptable

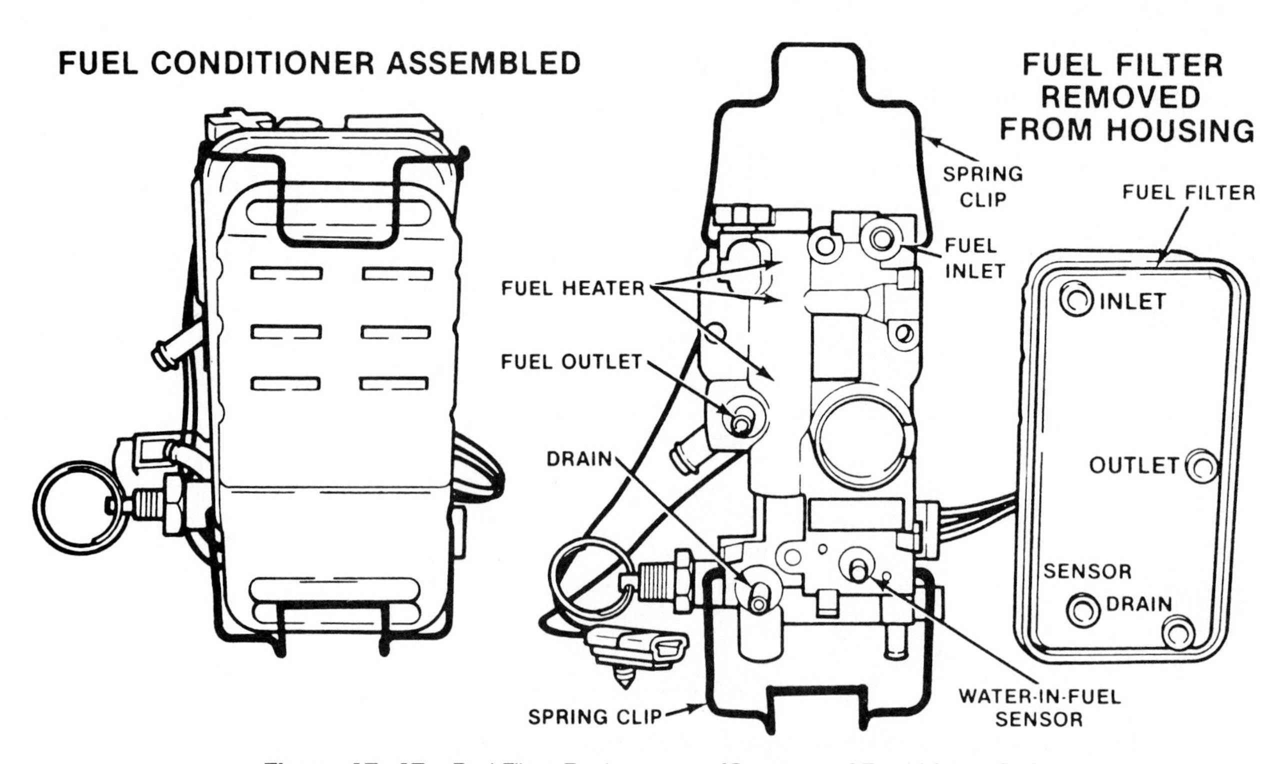

Figure 15–17 Fuel Filter Replacement. (Courtesy of Ford Motor Co.)

spray patterns are illustrated in Figure 15–19.

2. When the hand pump on the tester is operated, the nozzle opening pressure on the tester gauge should equal manufacturer's specifications. On many automotive nozzles, as shown in Figures 15–11 and 15–12, the opening pressure is not adjustable.
3. When the tester hand pump is operated rapidly, the nozzle should chatter as it opens and closes. If the nozzle fails to chatter, the nozzle valve is probably sticking.
4. Nozzle seat leakage should be checked with a tester pressure of 300 psi (2068 kPa) below the nozzle opening pressure applied to the nozzle for 10 seconds. Acceptable and unacceptable seat leakage conditions at the end of the 10-second test are pictured in Figure 15–20.

Nozzle seat leakage causes dribbling after injection, which results in severe engine detonation.

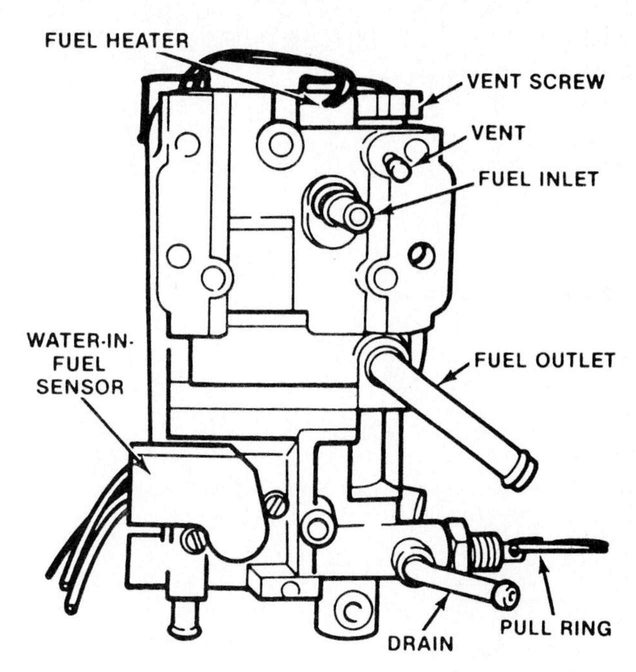

Figure 15–18 Fuel Filter Assembly. (Courtesy of Ford Motor Co.)

Injection Pump Timing

On the Ford 2.4L diesel engine, the injection pump is driven by a cogged belt. This engine also has two silent shafts, which are driven by a separate belt. The correct timing procedure for the silent shaft belt and the injection pump belt is illustrated in Figure 15–21.

The bolt pattern on the injection pump sprocket flange is offset so the sprocket can only be installed in one position, as indicated in Figure 15–22.

The static injection pump timing may be checked with a special adaptor attached to a dial indicator. To check the static injection pump timing, proceed as follows:

1. Position number 1 piston at top dead center (TDC) on the compression stroke.
2. Remove the plug and sealing washer in the injection pump head.

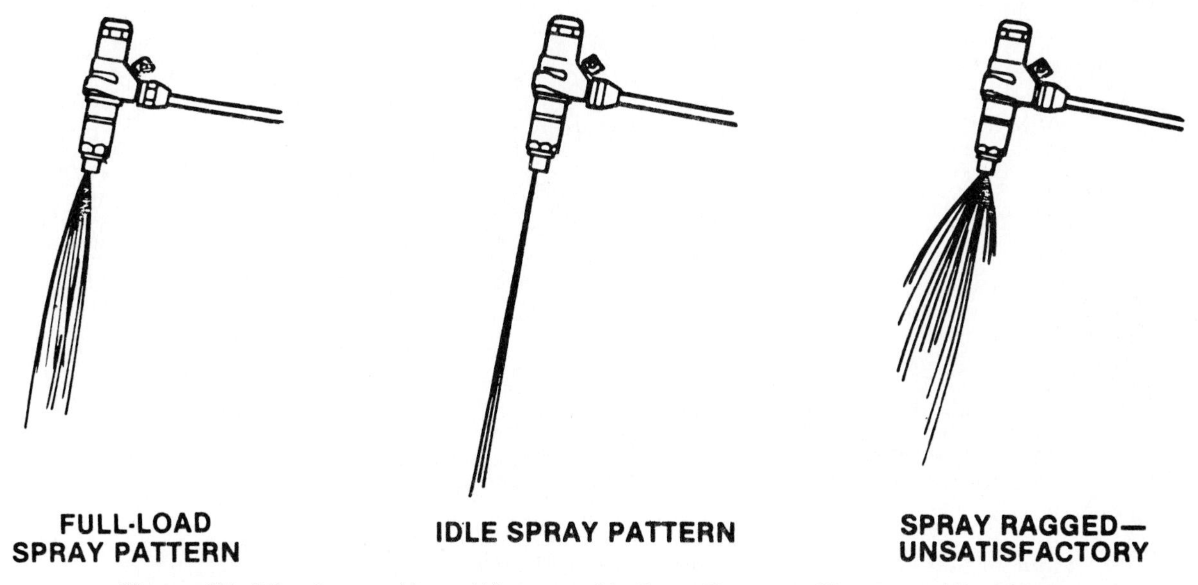

Figure 15–19 Acceptable and Unacceptable Spray Patterns. (Courtesy of Ford Motor Co.)

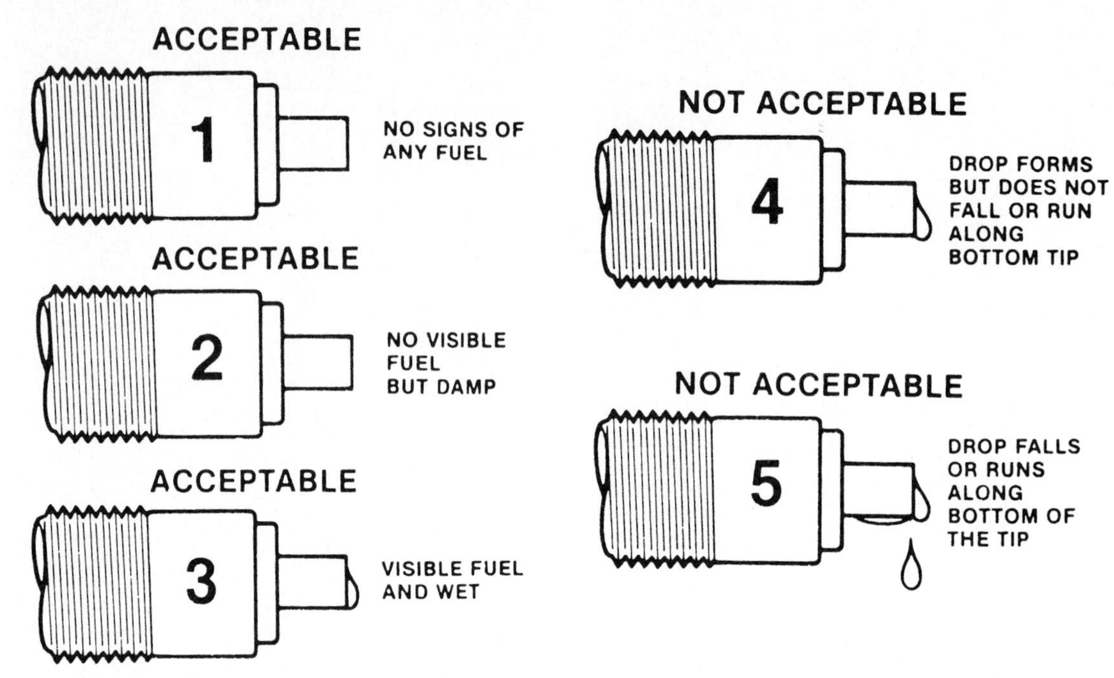

Figure 15-20 Acceptable and Unacceptable Nozzle Seat Leakage Conditions. (Courtesy of Ford Motor Co.)

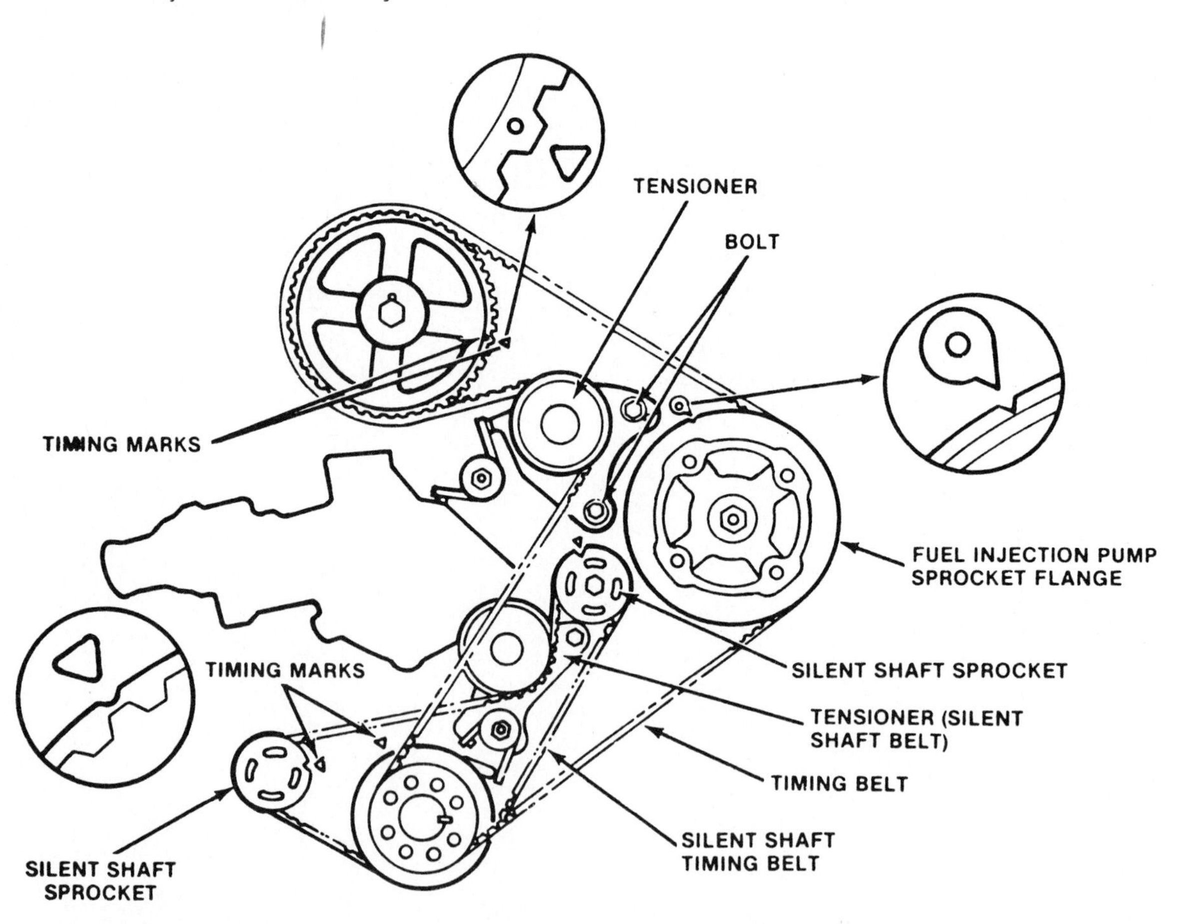

Figure 15-21 Injection Pump Belt and Silent Shaft Belt Timing Procedure. (Courtesy of Ford Motor Co.)

3. Mount the dial indicator and adaptor in the plug opening in the pump head. There must be a minimum of 0.100 in (0.25 mm) preload on the dial indicator.

4. Rotate the crankshaft clockwise until the lowest reading is obtained on the dial indicator, and zero the dial indicator in this position.

5. Continue to rotate the crankshaft clockwise until number 1 piston is at TDC on the compression stroke. The movement on the dial indicator must equal manufacturer's specifications.

6. To adjust the dial indicator reading, loosen the pump mounting bolts and rotate the injection pump clockwise to increase the reading, or counterclockwise to decrease the reading. Never start the engine with the injection pump mounting bolts loose, or severe pump damage and personal injury may result.

7. When the dial indicator reading is correct, remove the dial indicator and install the plug in the pump head. The static timing procedure is shown in Figure 15–23.

Never clean the outside of an injection pump with steam, or hot water, because plunger seizure may occur. A dynamic timing meter may be used to check injection pump timing with the engine running. This meter is connected to a diagnostic connector in the wiring harness, as indicated in Figure 15–24.

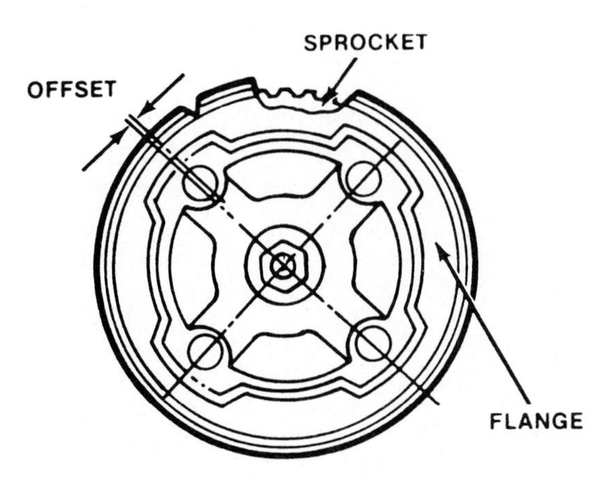

Figure 15–22 Injection Pump Sprocket Mounting Bolts. (Courtesy of Ford Motor Co.)

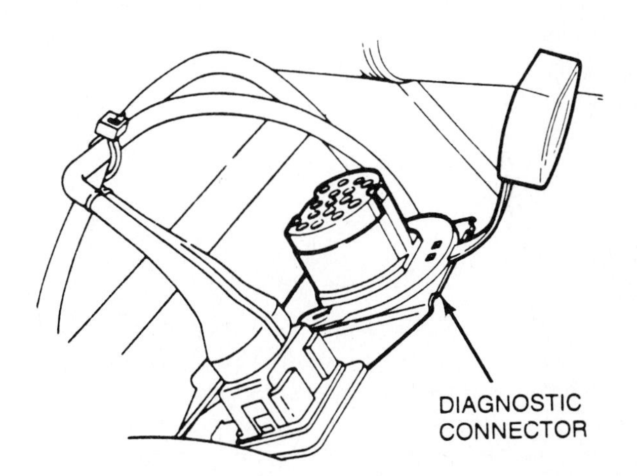

Figure 15–24 Dynamic Timing Meter Connector. (Courtesy of Ford Motor Co.)

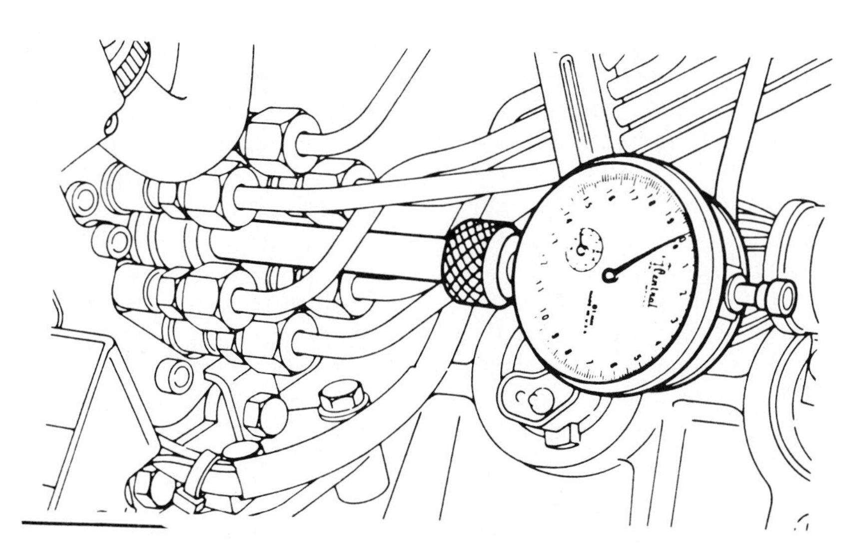

Figure 15–23 Static Injection Pump Timing. (Courtesy of Ford Motor Co.)

Diesel Injection Pump with Computer Control of Fuel Delivery and Injection Advance

Design

The electronic programmable injection control (EPIC) injection pump has internally mounted sensors and actuators and an external microprocessor. Compared to a similar conventional diesel injection pump, the EPIC pump provides lower combustion noise with improved economy and performance. The microprocessor could easily be expanded to control the turbocharger wastegate, intercooler, and exhaust-gas recirculation (EGR). The EPIC injection pump is lighter and more compact than a similar conventional pump. There are only 100 parts in the EPIC injection pump compared to 220 components in a similar conventional pump. The EPIC injection pump is designed primarily for engines in passenger cars and light trucks.

Fuel Control

In many conventional injection pumps, transfer pump pressure supplies fuel past a metering valve to the chamber between the pumping pistons in the rotor. The position of the metering valve controls the amount of fuel injected by regulating fuel delivery to the rotor pumping chamber. Rotation of the metering valve is controlled by the accelerator pedal and the governor. The pumping pistons are forced together by cams on an integral cam ring as the rotor assembly rotates.

The EPIC injection pump has angled ramps on the shoes that fit against the outer ends of the pumping pistons. These ramps match similar ramps on the pump drive shaft. Axial rotor movement varies the pumping piston position and therefore determines the pumping chamber volume, which provides precise fuel delivery control. Transfer pump pressure supplies fuel to the rotor pumping chamber, and this same pressure also delivers fuel through an electro-hydraulic actuator to the front of the rotor. The electronic control unit controls the actuator and the amount of fuel pressure delivered to the front of the rotor. In this way the electronic control unit and the actuator control the rotor axial position to provide precise fuel delivery. A position sensor at the front of the rotor sends a signal to the electronic control unit in relation to rotor axial position. The fuel metering control in a conventional injection pump is compared to the EPIC pump, in Figure 15–25.

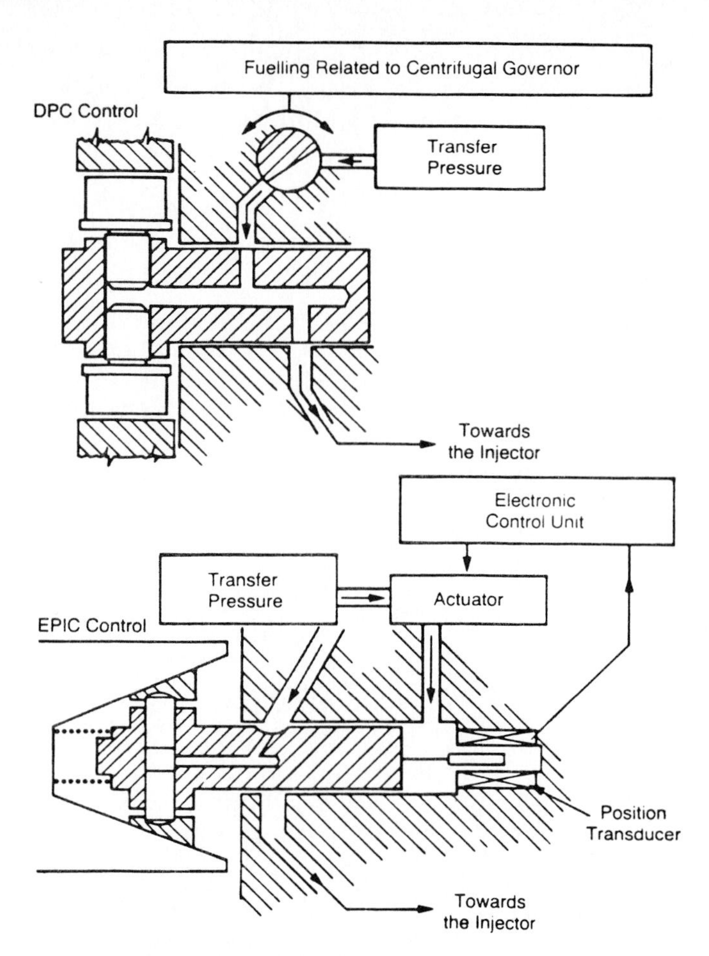

Figure 15–25 Conventional and Electronic Fuel Metering Control. [Reprinted with Permission of Society of Automotive Engineers © 1985]

Injection Timing Control

Transfer pump pressure is also supplied through another electro-hydraulic actuator to the advance piston. When the input signals indicate that more injection advance is required, the electronic control unit operates the actuator to increase the transfer pump pressure supplied to the advance piston. This pressure increase will move the advance piston against the return spring pressure. When this advance piston movement occurs it will rotate the cam ring in opposite direction to rotor rotation. This cam ring rotation causes the pumping piston mechanism to strike the internal cams sooner to advance the injection timing. A position transducer on the return spring side of the advance mechanism sends an input signal to the electronic control unit in relation to advance piston movement. In some injection systems a sensor in the injector informs the electronic control unit when the injector plunger begins to lift. This type of sensor may be used in place of the advance piston position transducer. Both types of sensors are illus-

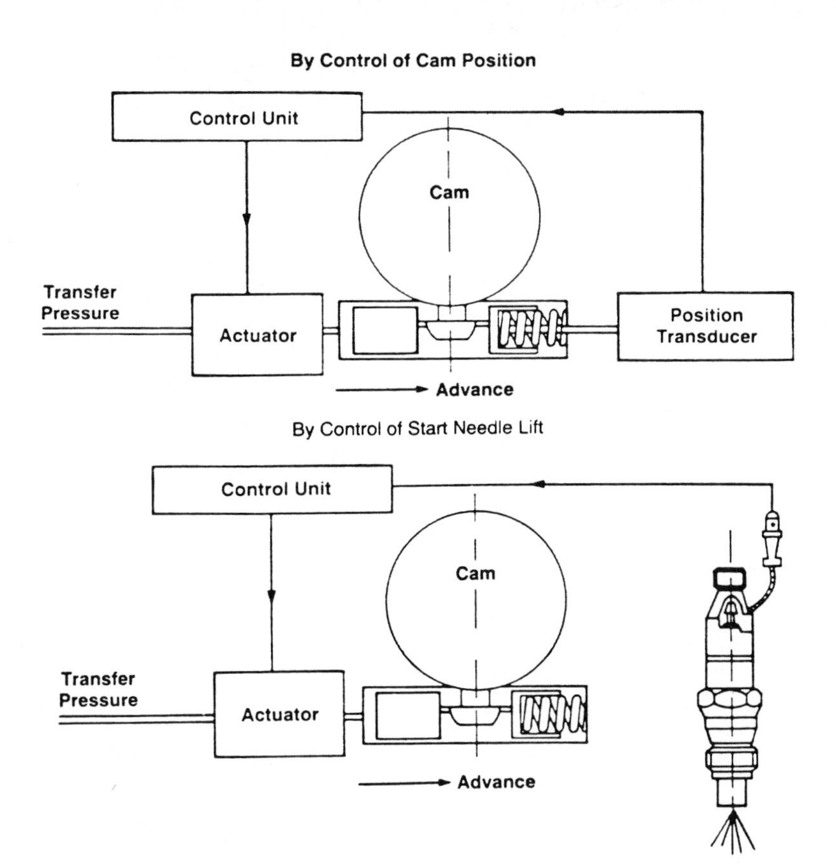

Figure 15–26 Electronic Injection Advance Control. (Reprinted with Permission of Society of Automotive Engineers © 1985)

trated in the injection advance systems in Figure 15–26.

Test Questions

1. The ignition temperature of diesel fuel is approximately:
   (a) 600°F.
   (b) 675°F.
   (c) 750°F.
2. The effective pump stroke is controlled by the:
   (a) spill ring position.
   (b) roller ring position.
   (c) speed of plunger rotation.
3. Excessive negative pressure in diesel injection lines can result in ______________ of the lines.
4. When the ignition switch is turned on and the engine coolant is cold, the electronic control unit in the glow plug circuit will close:
   (a) both relays.
   (b) relay I only.
   (c) relay II only.
5. If the ignition switch is left on and a cold engine is not started, the electronic control unit in a glow plug circuit will:
   (a) keep the glow plugs on until the battery is discharged.
   (b) keep relay II closed.
   (c) shut off both relays in 3 to 6 seconds.
6. A 2D diesel fuel should be used when the temperature is below 0°F (−18°C). T F
7. The injection pump mounting bolts should be loosened with the engine running when the injection timing is being adjusted. T F
8. The outside of the injection pump should be cleaned with a steam cleaner. T F
9. Excessive fuel waxing in cold water may be caused by a defective fuel heater. T F
10. When checking static injection pump timing, number 1 piston should be positioned at:
    (a) TDC on the compression stroke.
    (b) BDC on the intake stroke.
    (c) BDC on the exhaust stroke.
11. Nozzle seat leakage and dribbling can result in engine ______________.

16

Automotive Test Computers

Practical Completion Objectives

1. Connect computer leads to engine.
2. Perform computer diagnosis of ignition system, fuel system, and engine condition.
3. Diagnose defects in ignition system, fuel system, and engine from the computer test results.

Computer Diagnosis of Automotive Systems

Typical Automotive Computer

Many different automotive computer testers are available from the test-equipment industry. A typical automotive computer is shown in Figure 16–1.

The video display will indicate specialized waveforms similar to other oscilloscopes. Meter graphics appear on the display during certain tests to illustrate fast-changing readings or performance ranges. Numbers, or digital information, will also appear on the display when numbers are more precise or easier to use than meters. Menu screens, as shown in Figure 16–1, indicate the possible commands that the operator may use to control the computer.

An audio signal informs the operator if information, or test selections entered, is correct. When the information, or test selections, is out of range for the vehicle being tested, or the test selection is not correct, a different audio signal alerts the operator.

The paper printer types information that appears on the display and also allows the operator to enter information into the computer.

Power Switch Controls

A group of mechanical switches allows the operator to control the main power circuits of the computer.

The function of each power switch is outlined in the following summary.

1. Vacuum valve—regulates vacuum produced by the vacuum pump.
2. Vacuum pump—supplies power to the vacuum pump motor.
3. CO/HC pump—supplies power to the pump in the optional four-gas analyzer.
4. Power—controls all circuits powered by the alternating current (AC) line input, with the exception of the display lights.
5. Display lights—supply power to the merchandising sign, control panel lights, and boom light, without turning on the power switch.
6. Reset—returns the computer to the initial screen regardless of the current test procedure, and erases all test data from computer memory.
7. Volume—controls loudness of audio signal.

Power switches are illustrated in Figure 16-2.

Numeric Keypad

The numeric keypad is used for program control and entering numbers or digital information into the computer. Numbers appear beside each test sequence or entry option on the menu screen. These test sequences or entry options are selected by touching the appropriate number on the numeric keypad.

Numbers can be entered into the computer by touching the desired numbers followed by the ENTER key. If an error is made in numeric entry it can be corrected by touching the CLEAR key before the ENTER is touched.

The word "enter" in small letters on the menu screen indicates that the operator should touch the desired keys on any of the keypads. If the word ENTER appears in capital letters on the menu screen it indicates that the operator should touch the ENTER

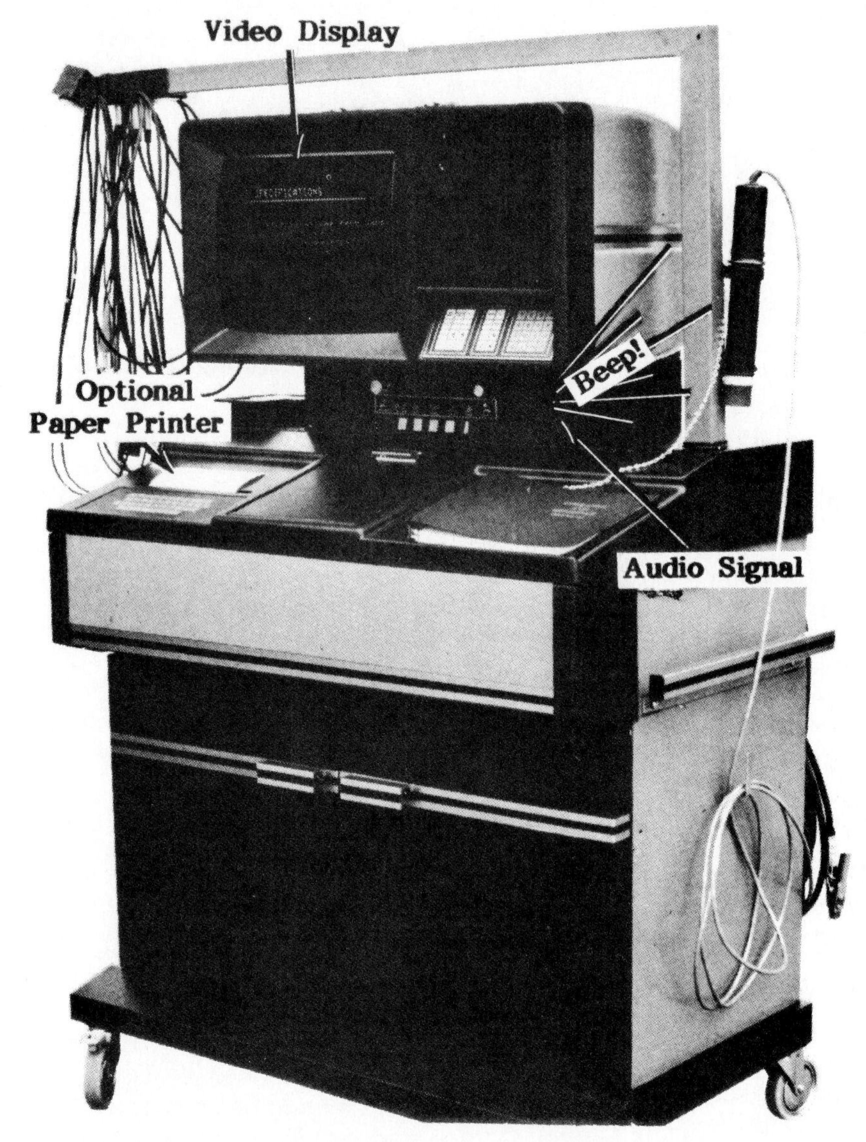

Figure 16–1 Automotive Computerized Equipment (ACE) (Courtesy of Bear Automotive Inc.)

Figure 16–2 Power Switches. (Courtesy of Bear Automotive Inc.)

key on the numeric keypad.

The dot or period may be used as a decimal point when numbers are entered, or as a period when the typewriter is used. During any menu screen display the period may be touched to obtain a printout of date and time.

The CLEAR key erases keyboard input if it is touched before the ENTER key is touched or used by the computer. When menu selections are made, the appropriate number should be touched without touching the ENTER key. In this case the computer uses the ENTER function automatically. The CLEAR key has other uses in some specific tests. The numeric keypad is shown in Figure 16–3.

Program Keypad

The captions on the keys of the program keypad correspond to prompts that appear on the menu screen. These keys allow the operator to lead the computer through the selected test sequence. The keys on the program keypad have the following functions.

1. Abort—exits from the procedure taking place, and the computer returns to the previous logical screen.
2. Repeat—performs the displayed procedure a second time.
3. Print—prints the test results from the screen on the typewriter printout. This key acts as a blank paper feed if a menu screen is displayed.
4. Continue-proceeds to the next test or screen.
5. Kill—stops the engine by disabling the primary ignition circuit. After the engine stops, the primary circuit is reactivated.

Figure 16–3 Numeric Keypad. (Courtesy of Bear Automotive Inc.)

The program keypad is pictured in Figure 16–4.

Waveform Keypad

The waveform keypad is used to control the ignition waveforms on the screen. These waveforms are basically the same as conventional scope patterns. However, there are more options in the waveforms compared to the average oscilloscope, as indicated in the following outline.

1. Primary—displays primary ignition waveforms.
2. Secondary—displays secondary ignition waveforms, which are automatically displayed in the firing order sequence when ignition patterns are requested.
3. Sequential—returns the screen to the previously selected primary or secondary patterns, in firing order sequence.
4. Dual—results in a primary and a secondary waveform for a single cylinder, or cylinders, being displayed at the same time.
5. Raster—results in a vertical display of primary or secondary waveforms in the firing order sequence.
6. Parade—causes a horizontal display of all active cylinders on the same baseline of the screen.
7. Expand—after expand is touched, the arrow keys may be used to expand or contract the length of the waveforms in relation to millisecond timebase lines on the menu screen.
8. Delay—slows the waveform update rate.
9. Freeze—the use of this key changes with the task selected. It is used to stop and hold specific displays.
10. Go—releases freeze action, restores waveform update, and starts automatic power check.

The waveform keypad is displayed in Figure 16–5.

The numeric, program, and waveform keypads are mounted on the front of the ACE. These same keypads are also mounted in a remote control, which allows the operator to control the test sequences from the driver's seat or some other location.

Typewriter Keypad

The typewriter keypad is used by the operator to enter business and personal identification, or information regarding customers and their vehicles includ-

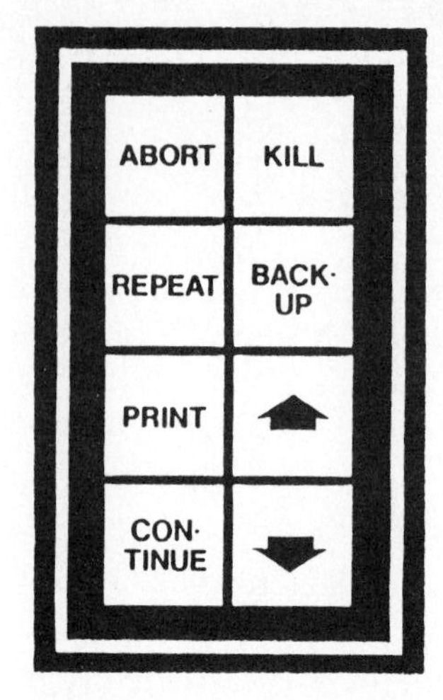

Figure 16–4 Program Keypad. [Courtesy of Bear Automotive Inc.]

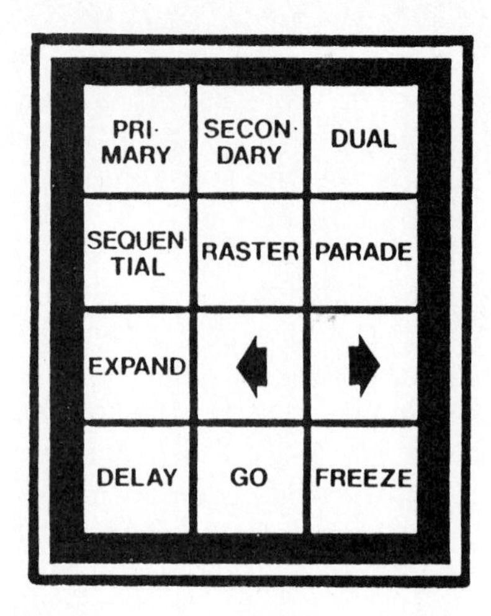

Figure 16–5 Waveform Keypad. [Courtesy of Bear Automotive Inc.]

Figure 16–6 Typewriter Keypad. [Courtesy of Bear Automotive Inc.]

ing date and time of service. This information is used to personalize the permanent records produced by the typewriter-printer. The typewriter-printer will produce as many copies of the typed text from the screen as the operator desires. The typewriter keypad is shown in Figure 16–6.

Computer Test Procedures

Test Selections

The master menu appears first on the screen, which allows the operator to select automatic testing, service tests, typewriter, or utilities by touching the number on the numeric keypad that appears beside the test. Figure 16–7 outlines the test selections on the master, automatic testing, service tests, and utilities menus.

When the operator selects automatic testing on the master menu and dealer ID on the automatic testing menu, the dealer's name, address, and other pertinent information may be entered with the typewriter keypad. The computer will allow a maximum of 34 letters, or numbers, on each line, and a message of 3 lines may be entered after the dealer's name, address and phone number. The operator may enter or change the dealer ID by touching "change" on the dealer ID menu, and the computer guides the operator through the procedure. If the dealer ID has been entered previously, the operator has the option of touching the number beside "print" on the dealer ID menu, and the dealer ID will be typed on the diagnostic printout. Information on the dealer ID screen is stored in the long-term computer memory, and it is saved during power off or reset conditions.

If the operator touches the numeric keypad number that appears beside "customer information" on

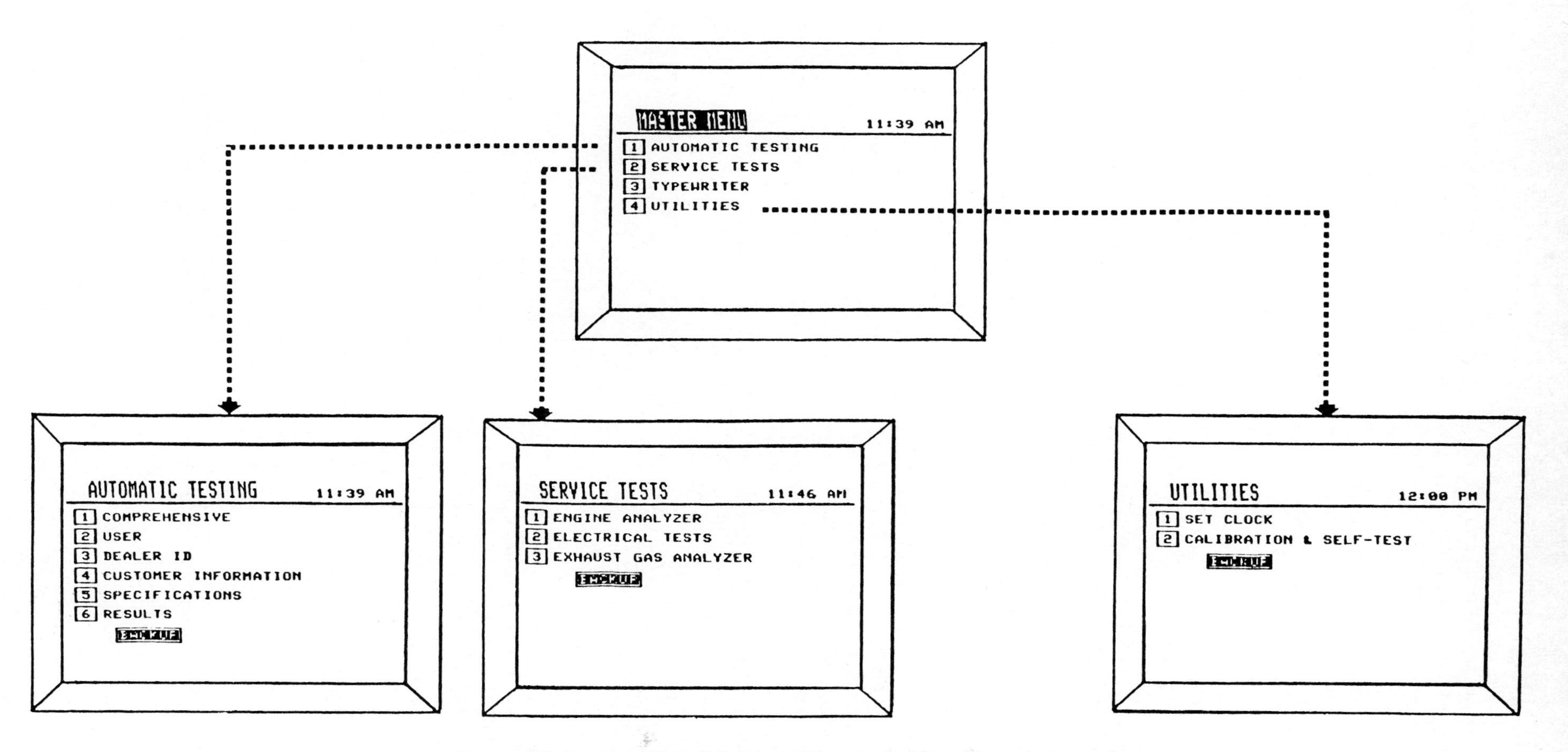

Figure 16-7 Initial Test Selections. (Courtesy of Bear Automotive Inc.)

the automatic testing menu, specific information regarding the customer and his or her vehicle may be entered in the computer with the typewriter keypad. This information is entered by touching number 2, which appears beside "change" on the customer information menu. The computer prompts the operator through the entry or change. A typical customer information screen is shown in Figure 16–8.

When the operator touches number 1 on the numeric keypad while the customer information screen is displayed, the printer will type the customer information on the diagnostic printout. This number appears beside "print" on the customer information screen. The customer information is lost from the computer memory during power off or reset conditions.

The operator may select the specifications screen by touching number 5 on the numeric keypad while the automatic testing menu is displayed. The specifications screen is illustrated in Figure 16–9.

Specifications are entered in the computer by touching number 1 on the numeric keypad while the specifications screen is displayed. This number appears beside "input specifications from wand." The specifications are entered by sliding the wand quickly and lightly across the specification bars on the applicable specifications chart for the car being tested. Prompting by the computer informs the operator to enter the first bar above the steel rod that divides the specification bars followed by the second bar below the steel rod. The specification bars and wand are pictured in Figure 16–10.

The computer screen informs the operator if the specifications have been entered correctly or if the procedure should be repeated. Specifications may be edited by touching number 4 on the numeric keypad when the enter specifications screen is displayed. The first screen of specifications will be displayed with a prompt line on the top line. To keep the specification on which the prompt line appears, touch "enter" on the numeric keypad. If the specification requires changing, the desired numbers must be touched on the numeric keypad followed by "enter." A specifications screen with a prompt bar is illustrated in Figure 16–11.

When "enter" is touched, the prompt bar moves to the next specification line. A number touched by

ENTER SPECIFICATIONS

[1] INPUT SPECIFICATIONS FROM WAND
[2] DISPLAY SPEC TO CRT
[3] DISPLAY SPEC TO PRINTER
[4] EDIT SPECIFICATIONS
[CONTINUE]
[BACKUP]

Figure 16–9 Specifications Screen. (Courtesy of Bear Automotive Inc.)

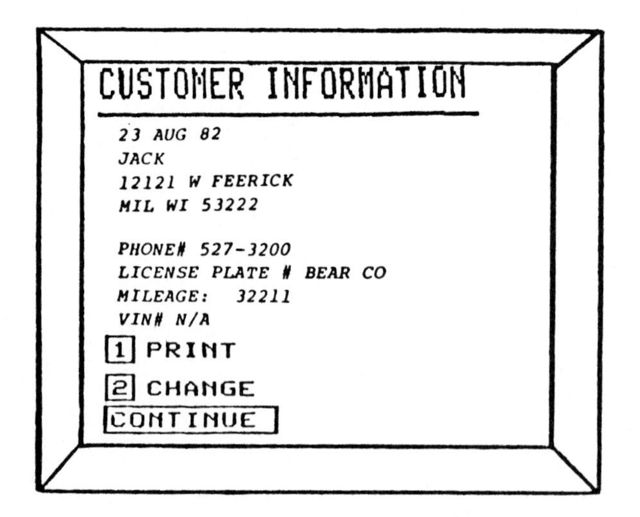

Figure 16–8 Customer Information Screen. (Courtesy of Bear Automotive Inc.)

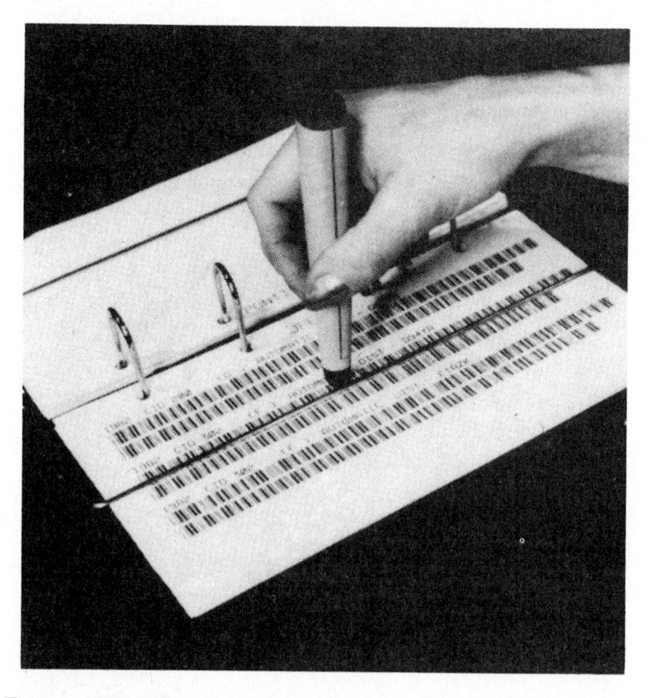

Figure 16–10 Specification Charts and Entry Wand. (Courtesy of Bear Automotive Inc.)

mistake will be erased if "clear" on the numeric keypad is touched. The prompt bar will move back to the previous specification line if "backup" is touched on the program keypad. It is possible for the operator to exit from the specifications editing program if "abort" on the program keypad is touched.

Comprehensive Test Procedure

Test Lead Connections and Initial Procedure

The test leads on the ACE computer should be connected to the electrical system of the vehicle as shown in Figure 16–12.

The specifications for the vehicle being tested should be entered before the diagnosis is started.

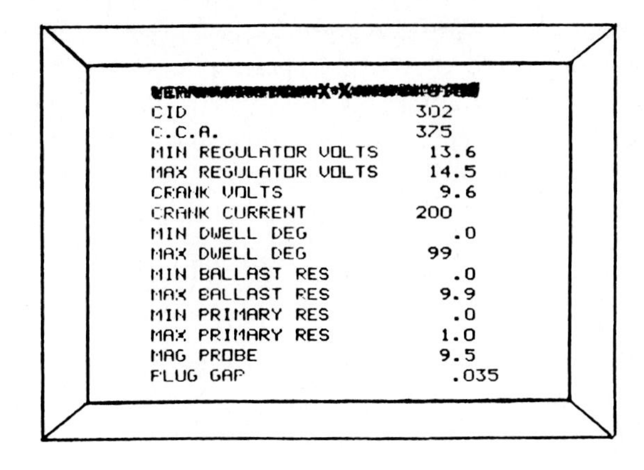

CID	302
C.C.A.	375
MIN REGULATOR VOLTS	13.6
MAX REGULATOR VOLTS	14.5
CRANK VOLTS	9.6
CRANK CURRENT	200
MIN DWELL DEG	.0
MAX DWELL DEG	99
MIN BALLAST RES	.0
MAX BALLAST RES	9.9
MIN PRIMARY RES	.0
MAX PRIMARY RES	1.0
MAG PROBE	9.5
PLUG GAP	.035

Figure 16–11 Specifications Screen with Prompt Line. [Courtesy of Bear Automotive Inc.]

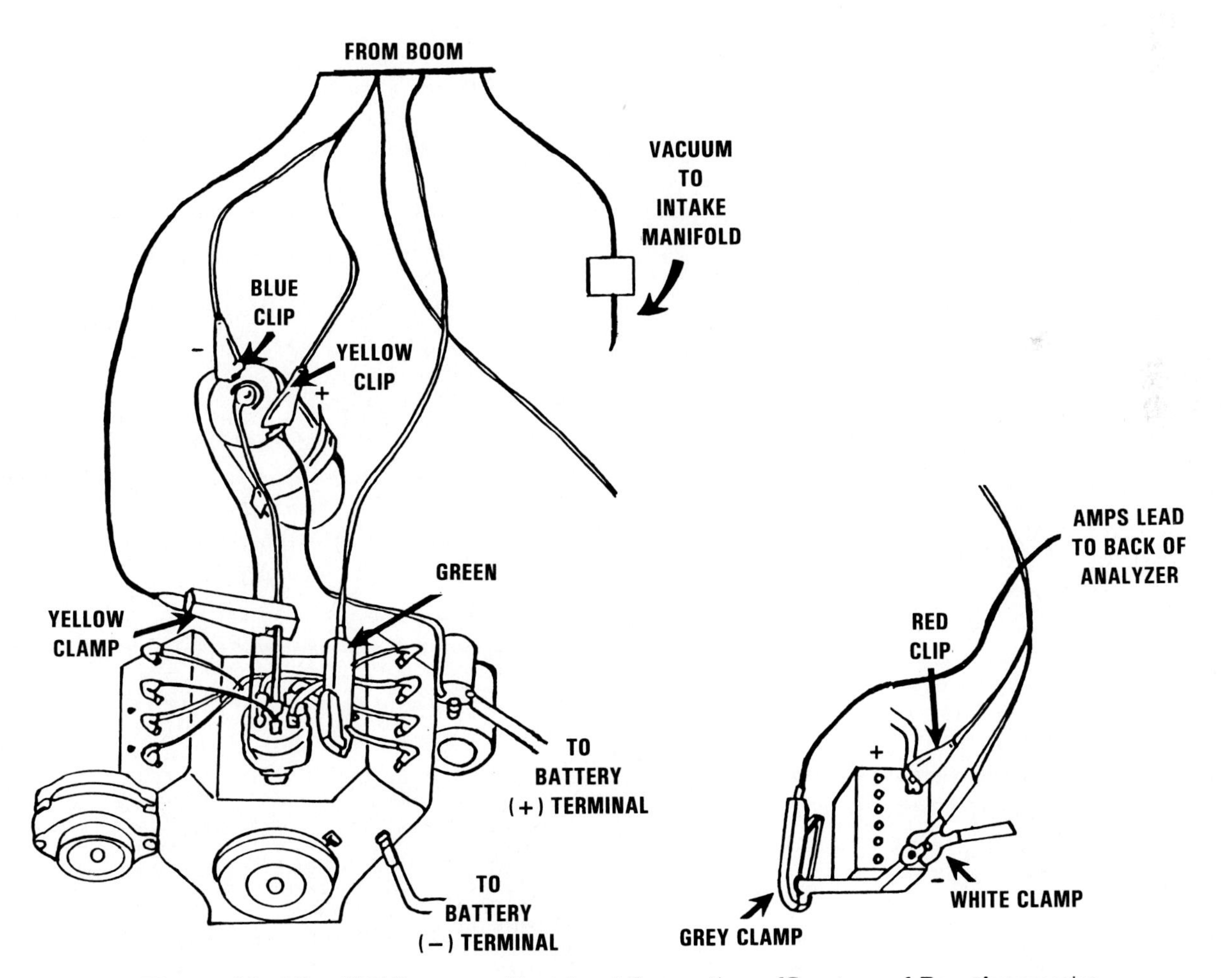

Figure 16–12 ACE Computer Test Lead Connections. [Courtesy of Bear Automotive Inc.]

When the comprehensive tests are selected, as outlined in Figure 16–7, the gas analyzer status will appear on the screen as pictured in Figure 16–13.

When the gas analyzer status appears, the operator has three choices.

1. If the gas analyzer is not ready, it requires 14 minutes warmup time. "Repeat" may be touched at any time to see if the gas analyzer is "OK," or ready.
2. "Continue" may be pressed if the gas analyzer is not ready, but all gas analyzer information must be disregarded, and this information will not be available in the "results" report.
3. The operator may touch "abort" and calibrate the gas analyzer.

The amp readout must be given a zero current indication by disconnecting the amps-probe and then touching "continue" on the program keypad. The prompt on the screen will then inform the operator to reconnect the amp-probe, as indicated in Figure 16–14.

Visual Tests

These tests allow the operator to inspect several important items that must be checked in a complete diagnostic procedure. Each item is marked with a prompt bar on the screen, and the operator touches 1 to indicate good, 2 to indicate bad, or 3 to pass all visual tests. If the battery voltage is too low, the visual test screen will suggest a battery charge. When this occurs the operator may proceed with the test sequence if "continue" is touched, but information regarding the battery test will not be available on the results report. The visual tests are illustrated in Figure 16–15.

Battery Test

When the battery test screen appears, the operator may change the cold cranking amp (CCA) rating of the battery if necessary by touching 1 on the numeric keypad. The battery CCA rating would be entered, but a replacement battery in the vehicle may have a different rating than the original battery. If the bat-

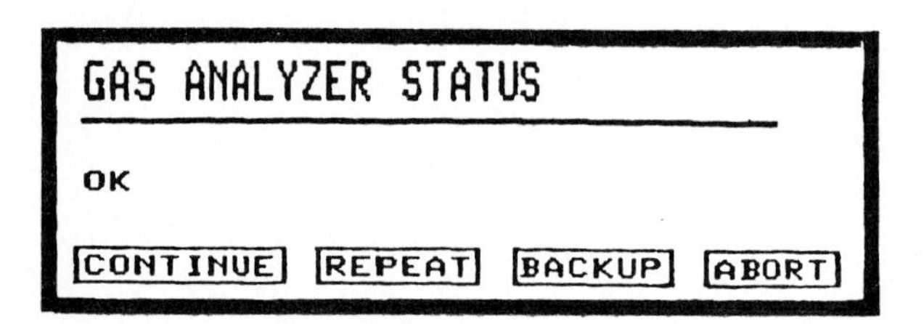

Figure 16–13 Gas Analyzer Status Indicator. (Courtesy of Bear Automotive Inc.)

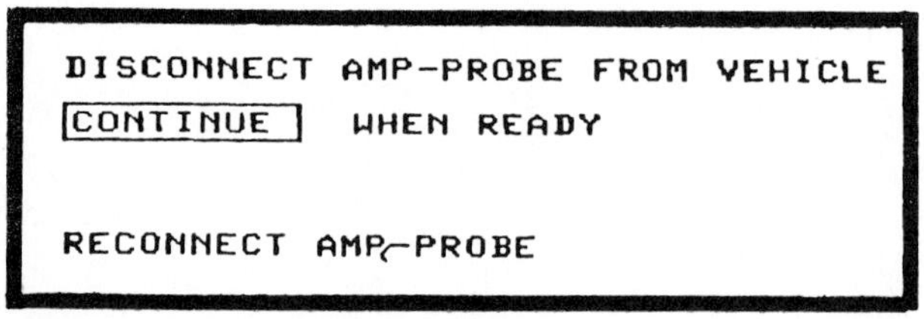

Figure 16–14 Amp Readout Zero Current Indication. (Courtesy of Bear Automotive Inc.)

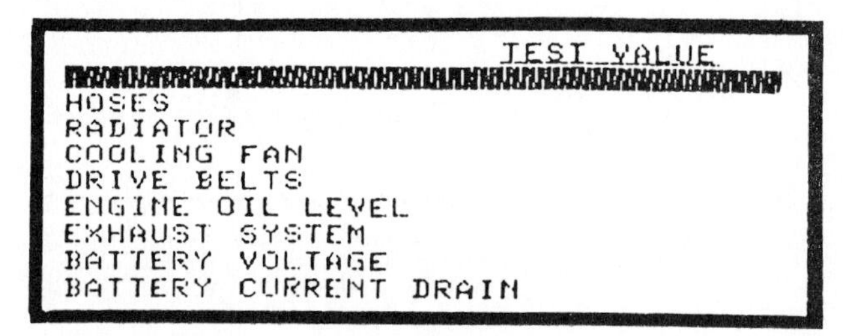

Figure 16–15 Visual Tests. (Courtesy of Bear Automotive Inc.)

tery voltage is too low for test purposes, the voltage will be displayed, and the computer will stop the test procedure. The computer applies a load to the battery for 15 seconds during which time a "battery under load" display appears on the screen. The load test may be repeated if the battery voltage is extremely high because of a surface charge. At the end of the test the results are displayed on the screen as shown in Figure 16–16.

Cranking Test

When "continue" is touched at the end of the battery test, the computer initiates the cranking test. Before this test is started, a probe check appears on the screen. This checks the signals from the current and ignition probes and also displays the trigger mode. Three trigger modes are available in the computer. The standard trigger mode allows the computer to use the primary and secondary probe signals. In the primary 1 trigger mode, the computer is forced to use

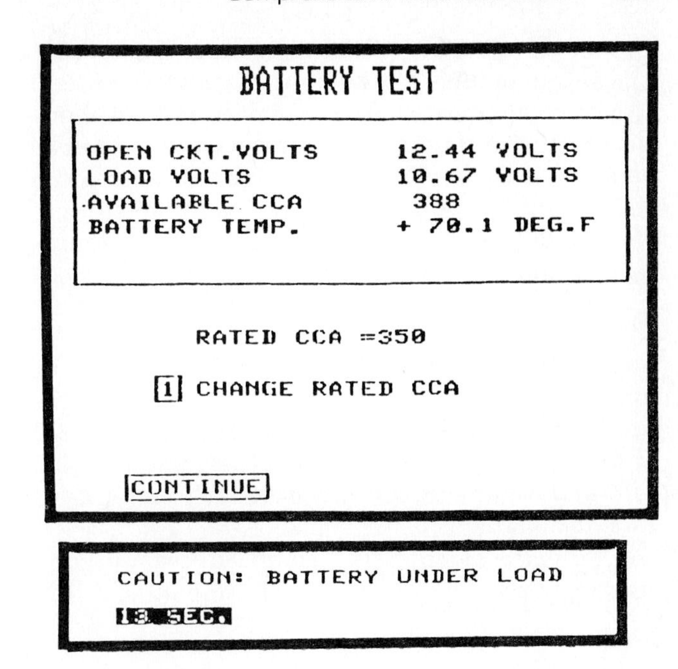

Figure 16–16 Battery Test Display. [Courtesy of Bear Automotive Inc.]

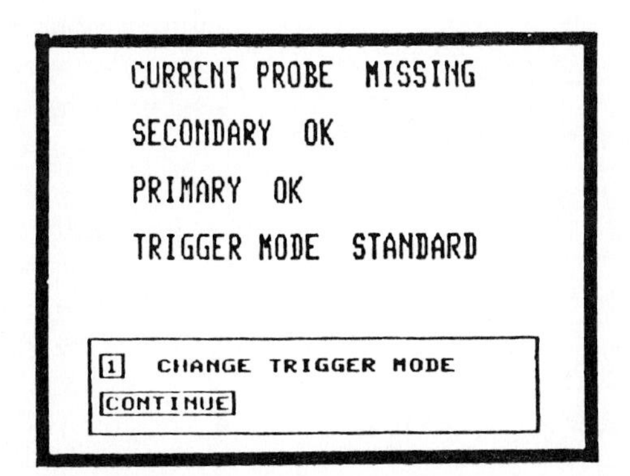

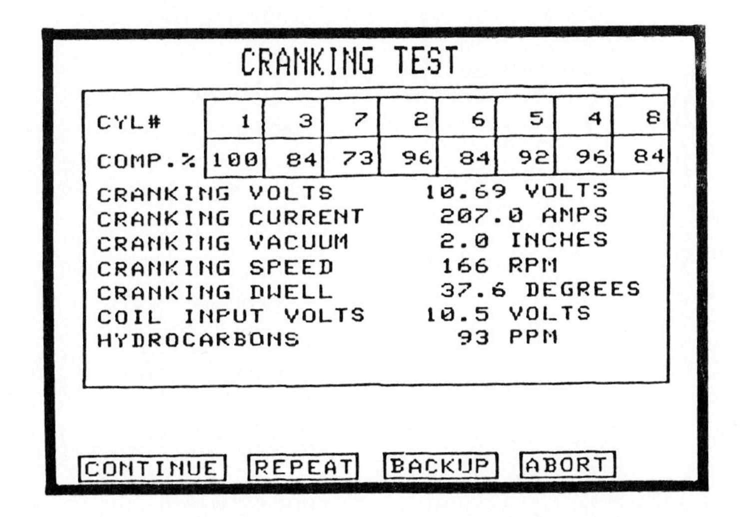

Figure 16–17 Cranking Tests. [Courtesy of Bear Automotive Inc.]

the primary probe signal. The standard trigger mode is used unless uneven triggering suggests the use of primary 1 or primary 2. In the primary 1 trigger mode the computer is forced to use a special primary probe signal. This mode may be necessary on General Motors high-energy ignition (HEI) or Prestolite ignition. The operator may change the trigger mode by touching 1 on the numeric keypad when the probe check appears on the screen. When "continue" is touched with the probe check displayed, the computer will enter the cranking tests. During this test sequence the computer suppresses the coil primary to prevent the engine from starting, and the operator is asked to crank the engine until it starts. The computer will perform the cranking tests displayed in Figure 16–17 while the engine is being cranked.

During the compression (comp. %) test, the computer compares each cylinder with the other cylinders. The best cylinder is given a value of 100, and the computer converts the value of the other cylinders to a percentage of the best cylinder. An "engine won't start" display will appear after the cranking tests if the engine does not start properly. If the engine diesels because of preignition, "engine dieseled" will appear on the screen.

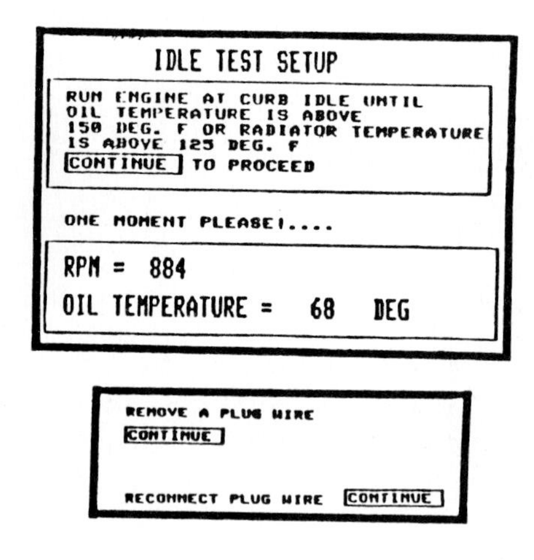

IDLE	TEST VALUE	
COIL INPUT VOLTAGE	14.1	V
COIL OUTPUT VOLTAGE	37.2	KV
COIL PRIMARY RES	2.4	Ω
BALLAST RES	.1	Ω
CURB IDLE SPEED	1372	RPM
IDLE VACUUM	20.6	IN
DWELL	29.5	DEG
HYDROCARBONS	184	PPM
CARBON MONOXIDE	.44	%
OXYGEN	.1	%

CONTINUE REPEAT BACKUP ABORT

Figure 16–18 Idle Test Setup and Report. (Courtesy of Bear Automotive Inc.)

Idle Test Setup and Report

During the idle test setup, a temperature probe is installed through the dipstick tube into the crankcase oil or is strapped to the top radiator hose. The engine must be operated until the oil temperature is above 150°F (65.5°C), or the coolant temperature is above 125°F (51.6°C). On some Chrysler vehicles the computer may prompt the operator to remove a plug wire, reconnect the plug wire, and touch "continue" after each operation. This is necessary because of the longer dwell time on Chrysler electronic ignition. When "continue" is touched with the oil or coolant temperature above the value specified previously, the idle tests illustrated in Figure 16–18 will be performed.

The transmission selector must be in the position stated in the specifications during the idle tests, and this selector must be returned to the neutral position for the remaining comprehensive tests. If "continue" is touched after the idle tests are completed, the computer will enter the next test procedure.

Kilovolt Test

The kilovolt (KV) test supplies individual cylinder dwell degrees and the following KV readings.

1. Average KV—the KV on each cylinder is sampled for 10 firings, and the average KV of these firings is displayed on the screen in the average KV column.
2. Maximum KV—the highest KV reading in the 10 firings on each cylinder is displayed in the maximum KV column.
3. Minimum KV—the lowest KV reading in the 10 firings on each cylinder is shown in the minimum KV column.
4. Snap KV—when the engine is accelerated, the KV of each cylinder is displayed in the snap KV column. The engine should be accelerated with the accelerator pedal. Test sequences may be controlled from the driver's seat with the remote control pendant.

5. Circuit Gap—the voltage required to jump all the air gaps, excluding spark plug gaps, in the secondary circuit to each cylinder is shown in the circuit gap column.

The KV tests are pictured in Figure 16–19.

Fast Idle Test

During the fast idle test, a prompt may appear on the screen for the fast idle screw setting. If this prompt does not appear, the test may be completed at curb idle. A "one moment please" message appears during exhaust gas sampling. When "continue" is touched after the fast idle test procedure is completed, the computer begins the next test. The fast idle test results are illustrated in Figure 16–20.

CYL #	DWELL DEG	AVG KV	MAX KV	MIN KV	SNAP KV	CKT GAP
1	32	9	14	8		
3	32	8	9	8		
7	31	7	8	7		
2	32	10	11	9		
6	32	7	8	7		
5	32	7	9	6		
4	32	8	8	6		
8	31	8	8	7		

ACCELERATE ENGINE RAPIDLY

CYL #	DWELL DEG	AVG KV	MAX KV	MIN KV	SNAP KV	CKT GAP
1	32	9	14	8	17	5
3	32	8	9	8	17	5
7	31	7	8	7	17	5
2	32	10	11	9	22	7
6	32	7	8	7	16	6
5	32	7	9	6	18	5
4	32	8	8	6	23	5
8	31	8	8	7	17	5

CONTINUE REPEAT BACKUP ABORT

Figure 16–19 Kilovolt Tests. (Courtesy of Bear Automotive Inc.)

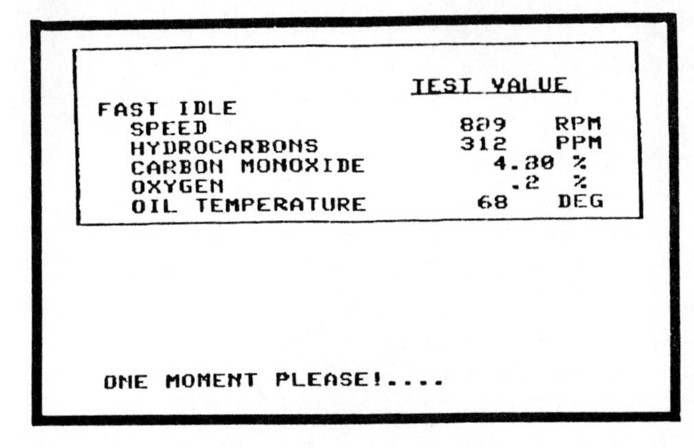

Figure 16–20 Fast Idle Test. (Courtesy of Bear Automotive Inc.)

Power Check

The power check is performed with the engine running at least 900 rpm, and the speed must not exceed 1200 rpm. Initial prompts on the screen will inform the operator to provide the correct rpm and disconnect some other components depending on the engine being tested. Once the engine speed is stabilized, the computer takes a few seconds to perform some calculations before it begins the power check. During the test procedure, the computer suppresses the primary circuit for a brief interval to disable each cylinder in the firing order, and the test results shown in Figure 16–21 will be displayed on the screen.

Some automotive computer systems have an idle speed control motor that is computer controlled. This motor will usually attempt to correct a loss in rpm at low speed. Therefore, the vehicle manufacturer's recommended procedure must be followed to disable this feature during the power check.

High Speed Report and Diode Check

This test is performed at 2000 rpm. The high-speed test results are pictured in Figure 16–22.

When the "caution load" prompt appears, the computer is supplying an electrical load to the charging system for the diode test that follows. The operator touches "continue" to start the diode test.

The operator must compare the alternator waveform on the screen to the satisfactory waveform in Figure 16–23 and enter 1 or 2 on the numeric keypad to indicate a good or bad waveform.

A prompt on the screen will inform the operator if there is no output signal at the red voltmeter clip.

Some defective alternator waveforms are shown in Figure 16–23.

Timing Tests

When "continue" is touched at the end of the diode check, the computer proceeds to the timing tests. If the magnetic timing probe signal is valid, the computer will automatically use this signal, and a magnetic timing probe outline will appear on the screen in place of the timing light picture. The prompt on the screen instructs the operator to increase the engine speed to 2000 rpm. While this speed is maintained, "enter" should be touched on the numeric keypad. If the timing light is used in place of the magnetic probe, the "advance" or "retard" buttons on the light should be touched until the timing mark appears in the zero position. If these buttons are touched momentarily, the timing mark will move in 5° steps, and holding the buttons results in timing mark movements in steps of 3°. When the timing mark reaches the 0° position, touch the "store" button on the timing light. Timing variation indicates

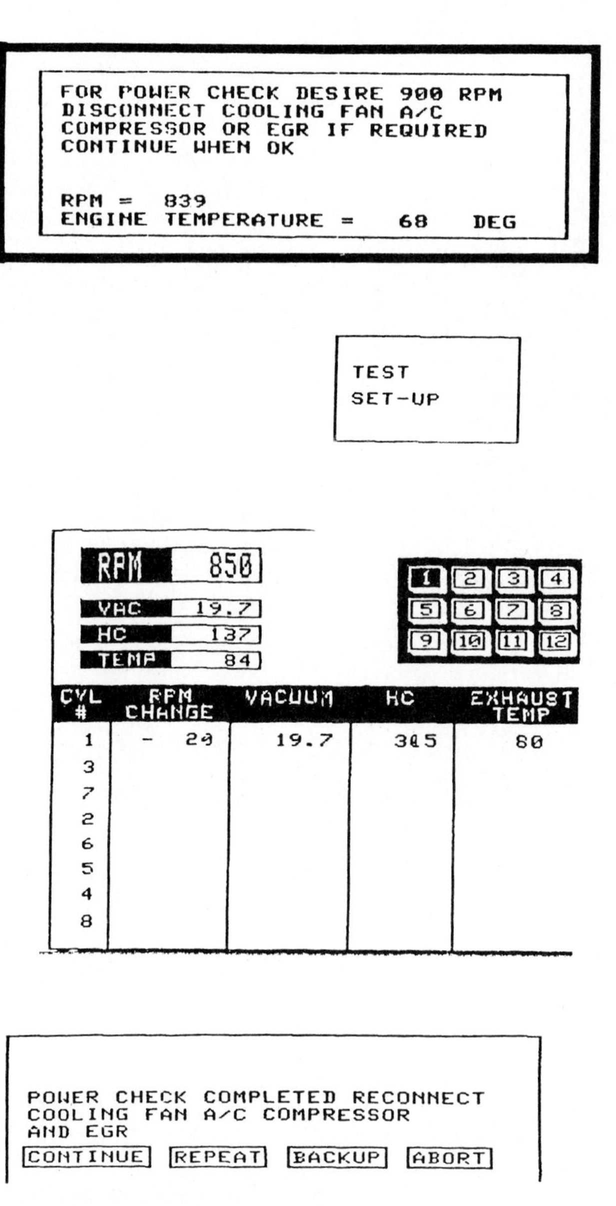

Figure 16–21 Power Check. (Courtesy of Bear Automotive Inc.)

the difference in ignition timing between the earliest and latest firing cylinders. The total advance timing display is shown in Figure 16–24.

A prompt on the screen instructs the operator to repeat the preceding timing advance test with the distributor vacuum advance hose disconnected. This procedure will test the mechanical advance by itself. After the mechanical advance test, another prompt informs the operator to check the basic timing at idle speed. If the timing light is used for this test, the buttons on the light may be used to bring the timing mark to the 0° position. The timing test results will be displayed on the screen as indicated in Figure 16–25.

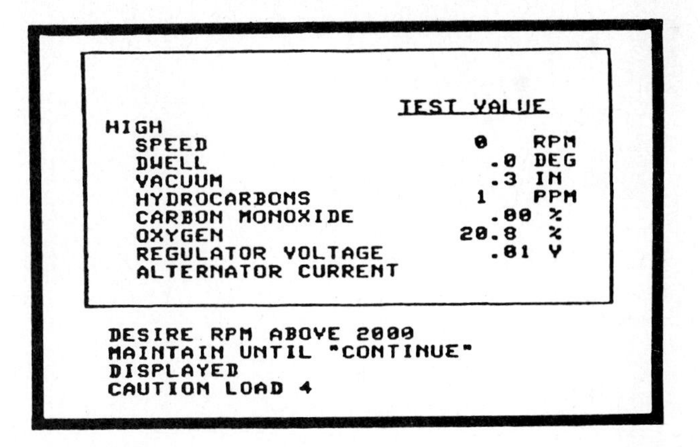

Figure 16–22 High Speed Test. (Courtesy of Bear Automotive Inc.)

Diagnostic Results

When "continue" is touched at the end of timing tests, the computer will return to the automatic testing menu. While the comprehensive tests are being performed, the computer compares all the test results to the specifications. The test results may be obtained if "results" is touched while the automatic testing menu is displayed. The operator has the option of displaying the test results on the screen or printing them on the optional typewriter. If the test results did not meet the specifications, the computer will print out diagnostic messages that help the operator diagnose the cause of the defective readings. A total of 62 diagnostic messages is available.

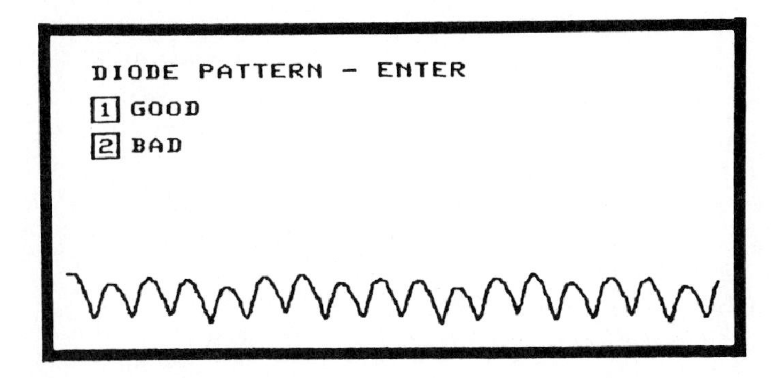

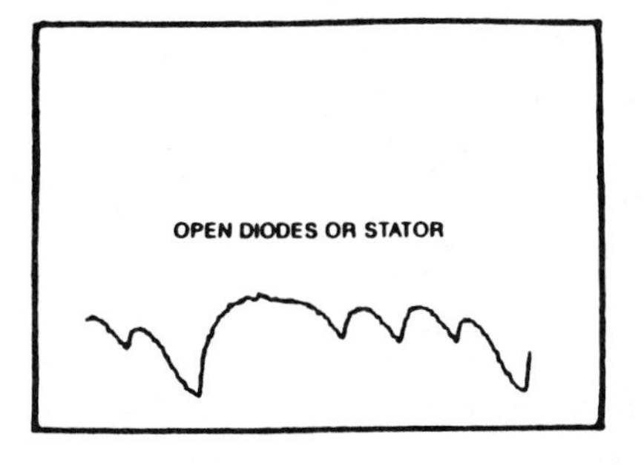

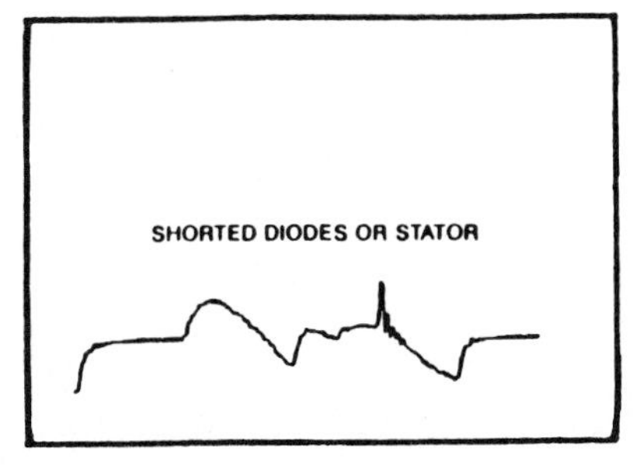

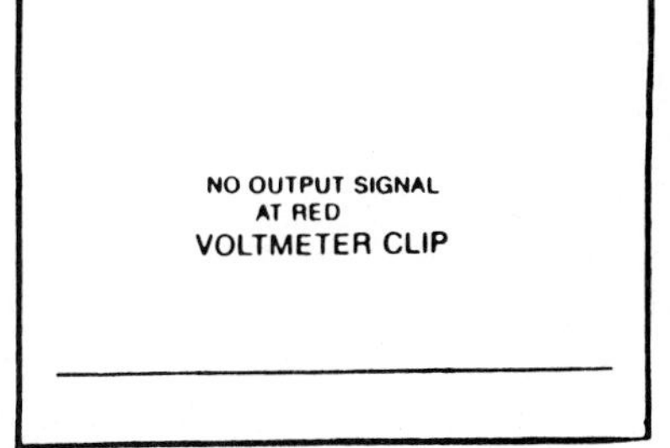

Figure 16–23 Diode Check. (Courtesy of Bear Automotive Inc.)

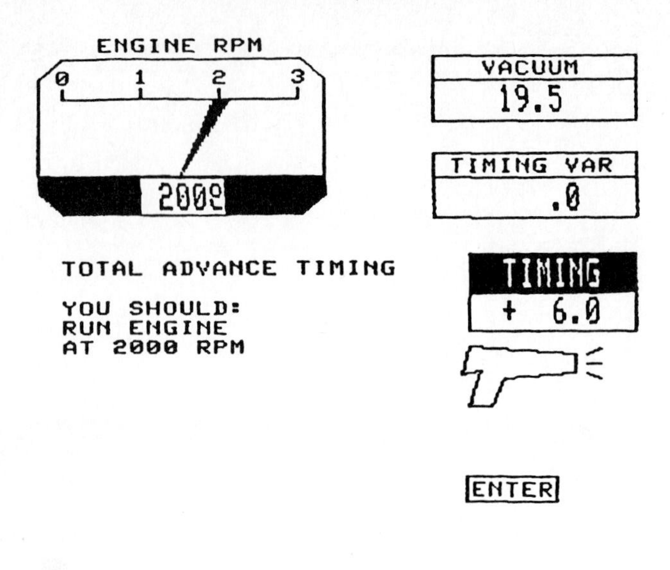

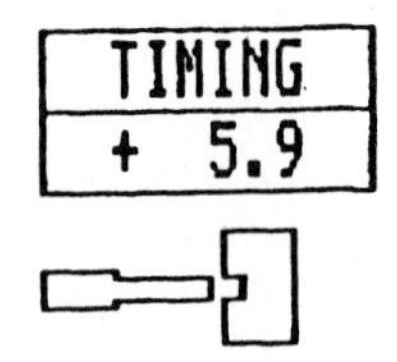

Figure 16-24 Timing Advance Tests. (Courtesy of Bear Automotive Inc.)

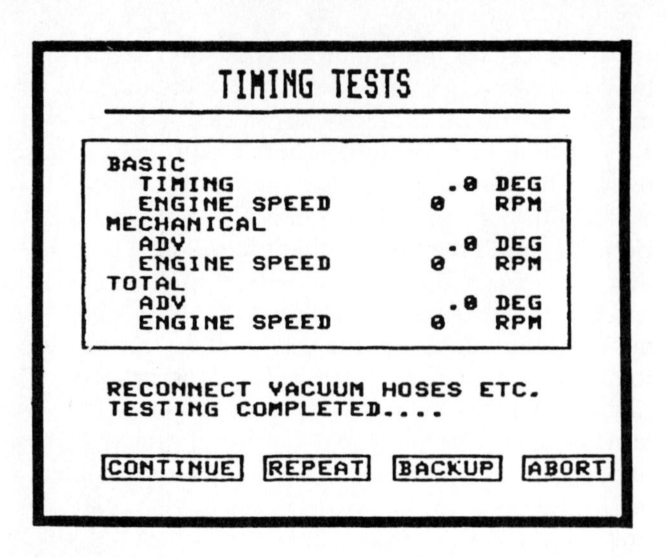

Figure 16-25 Timing Test Results. (Courtesy of Bear Automotive Inc.)

Engine Analyzer Tests

Test Sequence

The engine analyzer tests may be selected by touching number 2 on the numeric keypad when the master menu is displayed on the screen. (Refer to Figure 16-7 for menu screens.) When the service test menu is displayed, the engine analyzer tests are selected if number 1 is touched on the numeric keypad. In the engine analyzer tests, the following tests may be selected by the operator.

1. Ignition patterns
2. Dwell–tach
3. Primary circuit test
4. Power check
5. Secondary KV and dwell
6. Vacuum meter
7. Timing
8. Cranking test
9. Repeat patterns
10. Rough idle
11. Accessory patterns
12. Setup
13. Module test

The ignition patterns are the same as conventional scope patterns. However, a greater variety of patterns is available with the ACE computer compared to some other scopes. The patterns are selected by the waveform keypad. (This keypad is shown in Figure 16-6, and the function of the keys was explained earlier in the chapter.) Scope patterns obtained during the comprehensive test procedure may be obtained after the tests have been completed if the service test and engine analyzer menus are selected, and "repeat patterns" is selected on the engine analyzer menu.

If "setup" is selected on the engine analyzer test menu, the operator may change any of the following items.

1. Number of cylinders
2. Firing order
3. Primary circuit
4. Engine cycle—2 or 4
5. Magnetic timing reference

(Changing the primary circuit input was explained in cranking tests in the comprehensive test procedure.) Many of the tests in the engine analyzer test sequence are similar to the comprehensive tests.

If "continue" is touched with the engine analyzer tests displayed on the screen, the computer will move to the module test screen. The operator touches 1 on the numeric keypad to enter the module and pickup coil procedure. After 1 is touched, the operator must select 1 for general type pickup coils, 2 for Hall Effect pickup assemblies, or 3 for Prestolite pickups.

General pickup coils are pickup coils with a permanent magnet and a winding such as General Motors high-energy ignition (HEI), or early model Chrysler and Ford products. Hall Effect pickups contain a semiconductor material and a permanent magnet. This type of pickup is used in later model Chrysler products and Ford products with an Electronic Engine Control IV (EEC IV) computer system. Some General Motors products with fuel injection use a conventional pickup for starting and a Hall Effect pickup signal while the engine is running. Prestolite pickups were used on some American Motors products and other applications.

After the operator has entered the type of pickup, 1 may be touched for pickup test, or 2 for module test. When the pickup or module test is being performed, the scope primary and secondary leads must be connected to the ignition system. The yellow clip on the computer accessory leads must be connected to one of the distributor pickup leads. The yellow clip is attached to the pickup lead indicated in the following list.

1. Ford—orange pickup wire
2. Chrysler—black pickup wire
3. Hall Effect—grey pickup wire
4. HEI—green pickup wire
5. Prestolite—yellow clip to one pickup wire and black accessory clip to the other pickup wire.

The sharp pin on the yellow accessory lead clip must pierce through the insulation on the pickup lead. On HEI systems the distributor cap must be removed, and a test spark plug should be connected from the center cap terminal to ground during the tests. During the pickup coil test, the engine must be cranked with the ignition switch. Satisfactory pickup coil patterns are shown in Figure 16–26.

When the module test is selected, the ignition switch must be turned on, which causes the computer to pulse the module on and off. Satisfactory module patterns are pictured in Figure 16–27.

The precautions in the following list must be observed while the pickup and module tests are being performed.

1. To avoid blowing the fuse in the accessory leads near the back of the computer, do not allow the accessory clips to touch each other or the vehicle ground.
2. Be certain that the accessory leads are connected only to the pickup leads between the pickup and the module.

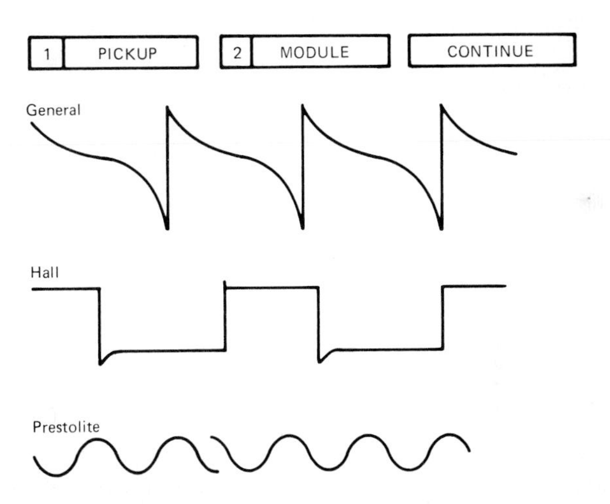

Figure 16–26 Satisfactory Pickup Coil Patterns. (Courtesy of Bear Automotive Inc.)

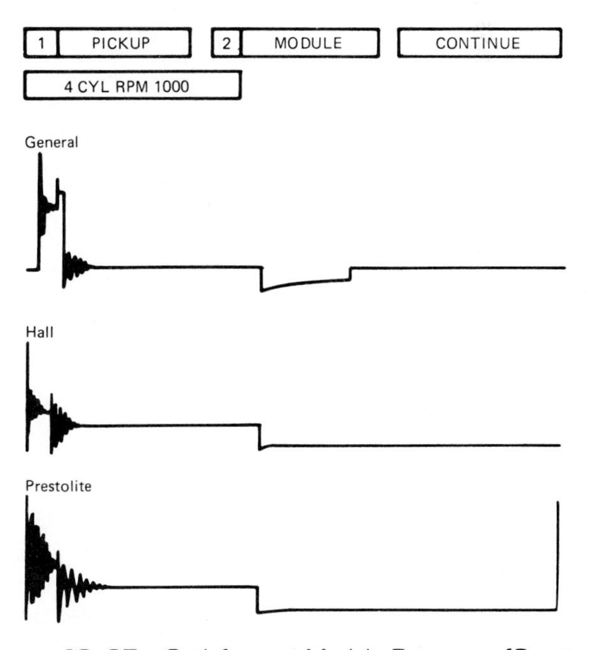

Figure 16–27 Satisfactory Module Patterns. (Courtesy of Bear Automotive Inc.)

3. Disconnect the injectors in the electronic fuel injection systems to avoid over-fueling the engine.

Electrical Tests

Test Procedure

The electrical tests screen is obtained when service tests are selected on the master menu, and electrical tests are chosen from the service tests screen. The following tests are available in the electrical test group.

1. Setup
2. Battery test
3. Charging test
4. Diesel charging test
5. Multimeter
6. Vacuum test
7. Zero amps probe

Many of the tests in the electrical tests sequence are similar to the tests in the comprehensive test procedure. However, if the operator wishes to use the ohmmeter immediately without going through the comprehensive test procedure, he or she may select electrical tests and multimeter. The electrical tests allow immediate access to these test procedures.

Gas Analyzer

Test Procedure

A four-gas analyzer is optional in the ACE computer. This analyzer will read carbon monoxide (CO), carbon dioxide (CO_2), hydrocarbons (HC), and oxygen (O_2). The gas analyzer tests are selected from the service tests screen. If emission test data is chosen on the gas analyzer tests screen, the gas analyzer will indicate the levels of all four gases in the engine exhaust, as shown in Figure 16–28.

If certification test is selected on the gas analyzer screen, the number representing the year of vehicle must be touched when the "select test specification" appears on the screen. The computer will now display the maximum limits for CO and HC emissions for that specific year. The computer will display the CO and HC levels in the engine exhaust under "data," and the maximum limits under "spec." The computer will also indicate if the vehicle passes or fails the emission level test, as illustrated in Figure 16–29.

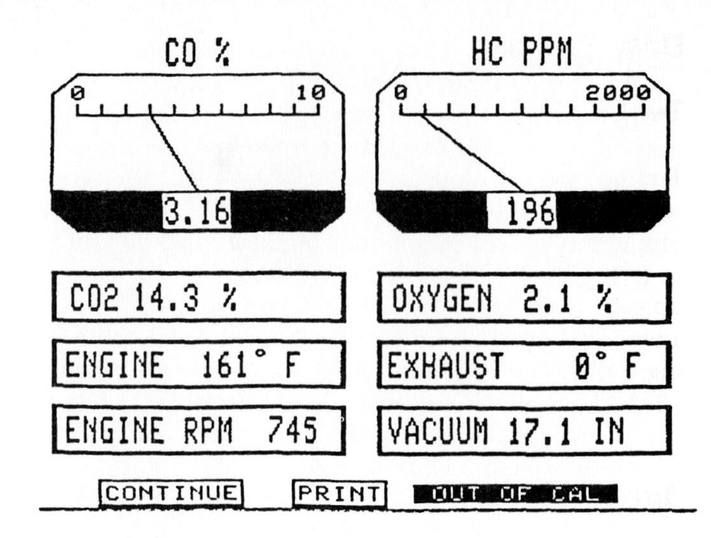

Figure 16–28 Four-Gas Analyzer Tests. [Courtesy of Bear Automotive Inc.]

CERTIFICATION TEST

SELECT TEST SPECIFICATION

1 1900 TO 1964
2 1965 TO 1969
3 1970 TO 1974
4 1975 TO 1979
5 1980 TO 1982
6 SPECIAL

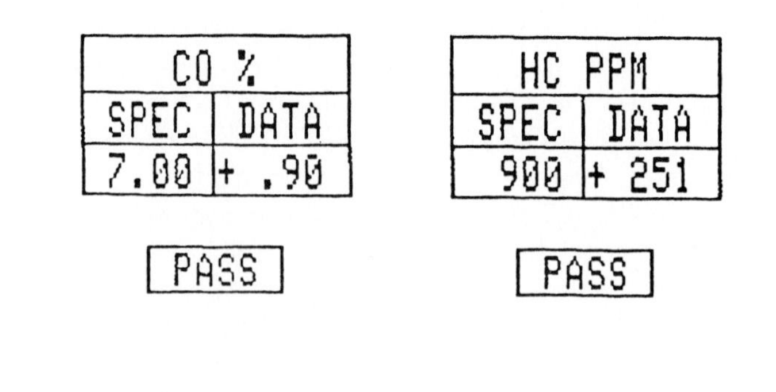

Figure 16–29 Exhaust Emission Certification Test. [Courtesy of Bear Automotive Inc.]

User Tests

Test Purpose

The user test procedure allows the operator to customize the analyzer to meet specific needs. This test procedure allows the operator to create any sequence of tests. For example, the operator may wish to set up a special test sequence that is coordinated with an advertising campaign. The procedure for establishing or reviewing the user tests is outlined in Figure 16–30.

Special Test Options

Automatic Computer System Tests

The ACE computer analyzer has an optional test feature that will diagnose automotive computer systems such as Ford Electronic Engine Control IV (EEC IV), and General Motors Computer Command Control (3C). This option appears on the service tests menu screen. (Refer to Chapters 3 and 4 for conventional diagnosis of these systems.)

Test Questions

1. Before the ACE computer will perform the idle tests in the comprehensive test procedure, the engine oil temperature must be above ______________°F, or the coolant temperature must exceed ______________°F.

2. The pickup coil pattern is the same on an ACE computer regardless of the type of pickup being tested. T F

3. A cylinder with a low compression will provide ______________ rpm change during a power check with an ACE computer.

4. With an ACE computer, the ohmmeter may be used when ______________ is selected on the electrical tests screen.

5. The engine analyzer patterns are selected on an ACE computer by choosing the:
   (a) automatic testing menu.
   (b) service tests menu.
   (c) user tests.

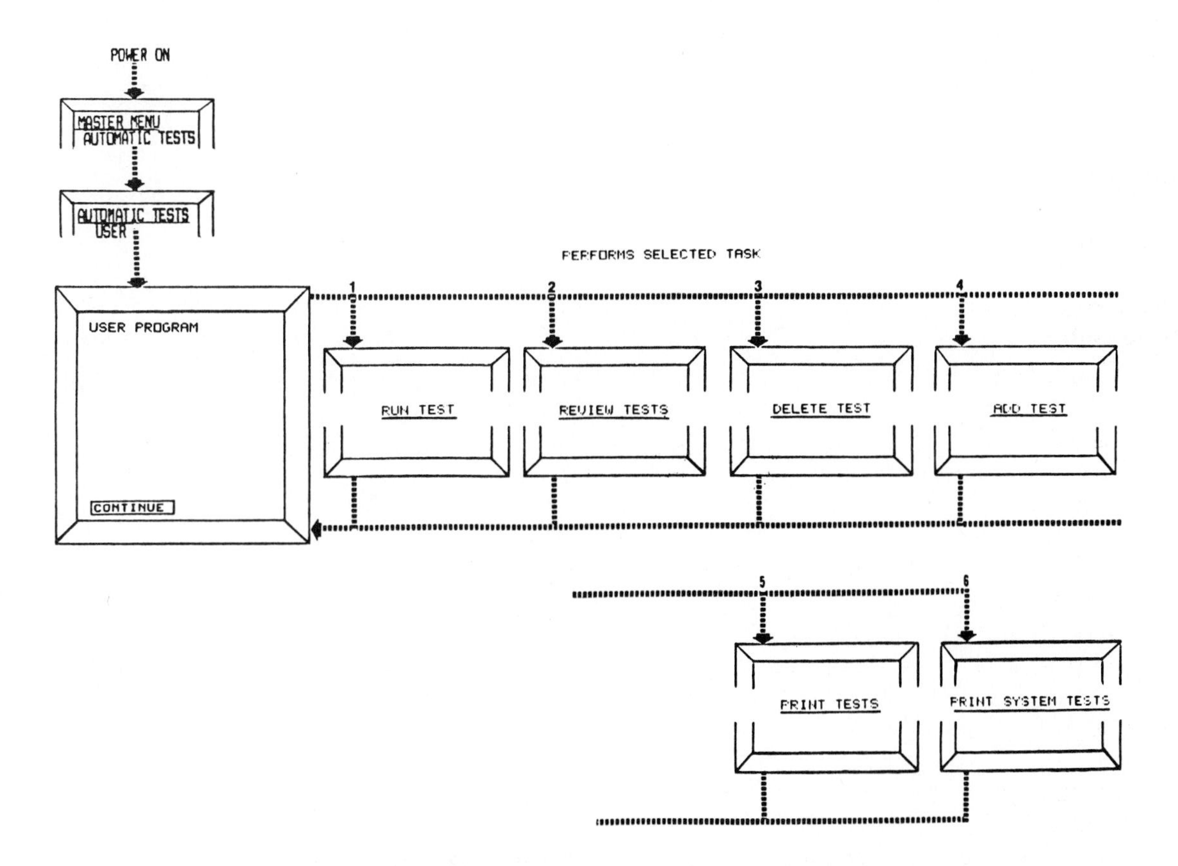

Figure 16–30 User Test Control Chart. (Courtesy of Bear Automotive Inc.)

Index